Lecture Notes in Physics

Founding Editors

Wolf Beiglböck, Heidelberg, Germany

Jürgen Ehlers, Potsdam, Germany

Klaus Hepp, Zürich, Switzerland

Hans-Arwed Weidenmüller, Heidelberg, Germany

Volume 990

Series Editors

Roberta Citro, Salerno, Italy

Peter Hänggi, Augsburg, Germany

Morten Hjorth-Jensen, Oslo, Norway

Maciej Lewenstein, Barcelona, Spain

Angel Rubio, Hamburg, Germany

Wolfgang Schleich, Ulm, Germany

Stefan Theisen, Potsdam, Germany

James D. Wells, Ann Arbor, MI, USA

Gary P. Zank, Huntsville, AL, USA

The series Lecture Notes in Physics (LNP), founded in 1969, reports new developments in physics research and teaching - quickly and informally, but with a high quality and the explicit aim to summarize and communicate current knowledge in an accessible way. Books published in this series are conceived as bridging material between advanced graduate textbooks and the forefront of research and to serve three purposes:

- to be a compact and modern up-to-date source of reference on a well-defined topic;
- to serve as an accessible introduction to the field to postgraduate students and non-specialist researchers from related areas;
- to be a source of advanced teaching material for specialized seminars, courses and schools.

Both monographs and multi-author volumes will be considered for publication. Edited volumes should however consist of a very limited number of contributions only. Proceedings will not be considered for LNP.

Volumes published in LNP are disseminated both in print and in electronic formats, the electronic archive being available at springerlink.com. The series content is indexed, abstracted and referenced by many abstracting and information services, bibliographic networks, subscription agencies, library networks, and consortia.

Proposals should be sent to a member of the Editorial Board, or directly to the responsible editor at Springer:

Dr Lisa Scalone
Springer Nature
Physics
Tiergartenstrasse 17
69121 Heidelberg, Germany
lisa.scalone@springernature.com

More information about this series at https://link.springer.com/bookseries/5304

Gabriel S. Denicol • Dirk H. Rischke

Microscopic Foundations of Relativistic Fluid Dynamics

 Springer

Gabriel S. Denicol
Physics Institute
Fluminense Federal University
Niterói, Brazil

Dirk H. Rischke
Institute for Theoretical Physics
Goethe University Frankfurt
Frankfurt am Main, Germany

ISSN 0075-8450   ISSN 1616-6361   (electronic)
Lecture Notes in Physics
ISBN 978-3-030-82075-6   ISBN 978-3-030-82077-0   (eBook)
https://doi.org/10.1007/978-3-030-82077-0

This Springer imprint is published by the registered company Springer Nature Switzerland AG
The registered company address is: Gewerbestrasse 11, 6330 Cham, Switzerland

Preface

Fluid dynamics is a theory with widespread applications in fundamental research as well as the applied sciences. It relies on the assumption that the details of the microscopic interactions between particles constituting the fluid play no role in the dynamics of the fluid on macroscopic scales. The latter is determined solely by the conservation of charge, energy, and momentum. However, the physics on microscopic scales does affect that on macroscopic scales at several places.

In the absence of dissipation, i.e., for so-called ideal (or perfect) fluids, the solution of the fluid-dynamical equations of motion requires microscopic information in the form of the thermodynamic equation of state of the fluid. The latter determines how much a fluid can be accelerated for a given gradient in density. For dissipative (or non-perfect) fluids, in addition to the equation of state, also other types of microscopic information is required, in the form of the so-called transport coefficients. The most important ones are the bulk and shear viscosity, as well as the charge-diffusion coefficient. These determine how much resistance there is for the fluid to reach thermodynamic equilibrium for given gradients in density or fluid velocity.

The transport coefficients are functions of space-time and, in a microscopic picture of the system as that of many particles scattering with each other, can be derived by integrating out the momentum-space information contained in the Boltzmann equation, which describes the microscopic motion of particles in phase space. Unless one wants to solve the Boltzmann equation directly, which is a task whose complexity grows with the number of constituents of the system, and wants to stick with a fluid-dynamical description, the smaller the system the more important the interactions on microscopic scales become and, consequently, the more transport coefficients need to be known for an accurate description of the dynamics of a given system.

In the non-relativistic context, the theories of ideal and dissipative fluid dynamics are well-established and have been put to test in a multitude of applications, where they have proven their validity. The most traditional theory of dissipative fluid dynamics is the well-known Navier–Stokes theory. The success of this theory in non-relativistic applications led to the belief that fluid dynamics should be constructed from a systematic expansion in gradients in density or fluid velocity. The lowest-order truncation of this expansion gives rise to ideal fluid dynamics, while the first-order truncation leads to Navier–Stokes theory itself. However, higher-order truncations of this picture lead to the Burnett and super-Burnett equations, which were proven to be linearly unstable.

While the relativistic generalization of ideal fluid dynamics is straightforward, a relativistic generalization of Navier–Stokes theory encounters severe difficulties, related to the violation of causality and the occurrence of unstable modes. Following the procedure developed by Grad in the non-relativistic regime, Israel and Stewart were among the first to derive a relativistic theory of dissipative fluid dynamics that is causal and stable. The causality and stability problems of Navier–Stokes theory were essentially solved by introducing a delay in time for the generation of the dissipative currents (i.e., bulk-viscous pressure, charge-diffusion current, and shear-stress tensor) from gradients of the fluid-dynamical variables. In this case, these currents become independent dynamical variables that obey equations of motion which need to be solved in addition to the conservation equations. Besides the transport coefficients related to bulk and shear viscosity, as well as charge diffusion, a multitude of new transport coefficients occur, which describe the coupling of the dissipative currents among each other and with gradients of density and fluid velocity.

Nowadays, the main area of application for Israel–Stewart theory is the field of high-energy nuclear physics. The collective behavior of hot and dense nuclear matter created in collisions of heavy atomic nuclei, so-called heavy-ion collisions, at relativistic energies can be very well described by relativistic fluid dynamics. Due to the smallness of the system and its short lifetime, dissipative effects cannot be neglected, so that Israel–Stewart theory finds a natural application in the description of heavy-ion collisions. Nevertheless, despite its successes, this theory remains an *ad hoc* procedure to derive the equations of motion for the dissipative currents in the sense that it lacks a parameter with which one can power-count and thus judge the validity and, ultimately, systematically improve the fluid-dynamical description. In recent years, this shortcoming has triggered a lot of activity to derive a causal and stable theory of relativistic dissipative fluid dynamics from kinetic theory in a more systematic fashion. The current monograph attempts to summarize these efforts.

There are several monographs dealing with the theory of relativistic fluid dynamics. The earliest is Landau's and Lifshitz's textbook on "Fluid Dynamics" [1], which contains a chapter on relativistic fluids. However, the discussion mostly focuses on ideal fluids, and deals with dissipative fluids only very briefly in the unstable and acausal Navier–Stokes framework. A derivation of fluid dynamics from kinetic theory as the underlying microscopic theory can be found in "Relativistic Kinetic Theory—Principles and Applications" by de Groot, van Leeuwen, and van Weert [2]. This monograph contains a detailed discussion of Chapman–Enskog theory, as well as the mathematical tools to derive dissipative fluid dynamics from kinetic theory applied in this monograph, such as irreducible tensors. However, the derivation of dissipative fluid dynamics is then only performed in the way suggested by Israel and Stewart, which, as explained above, lacks a systematic power-counting scheme and is not systematically improvable. A textbook with similar content as Ref. [2] is the one by Cercignani and Kremer [3].

A discussion of relativistic fluid dynamics with particular emphasis to applications in general relativity is contained in the textbook of Rezzolla and Zanotti [4], but the equations of motion of dissipative fluids are only derived phenomenologically from the second law of thermodynamics. A more recent monograph is the one on "Relativistic Fluid Dynamics In and Out of Equilibrium" by the Romatschke's [5]. The view taken here is that dissipative fluid dynamics can be understood in terms of a gradient expansion around local equilibrium. Although it contains a chapter on kinetic theory, the derivation of dissipative fluid dynamics from the underlying kinetic theory is not presented. In view of the current situation of the literature, we felt that it would be beneficial, both for researchers new to the field of dissipative fluid dynamics as well as a point of reference for experts, to have a monograph which summarizes the last decade's progress in our understanding of relativistic dissipative fluid dynamics as it emerges as a causal and stable theory from the underlying microscopic framework of kinetic theory.

This book is organized as follows: In Chap. 1, we review relativistic fluid dynamics from a phenomenological perspective. We start by deriving the equations of motion of an ideal relativistic fluid. Then, we review how dissipation is phenomenologically introduced in relativistic fluids. The equations of relativistic Navier–Stokes theory are derived via the second law of thermodynamics and, later, extended using the gradient expansion. Next, we discuss the problems of this traditional theory in the relativistic regime, i.e., the acausality of Navier–Stokes theory, and the phenomenological methods to derive a causal theory of relativistic dissipative fluid dynamics. As an example, we review how Israel and Stewart derived a causal fluid-dynamical theory from the second law of thermodynamics. Finally, the end of this chapter contains a discussion about the existence of non-hydrodynamic modes in relativistic fluid dynamics and how they affect the interpretation of this theory.

In Chap. 2, we discuss the causality and stability of relativistic fluid dynamics in the linear regime. We explicitly demonstrate that relativistic Navier–Stokes theory is linearly unstable when perturbed around global equilibrium: a fundamental flaw that prevents this theory from being applied to describe any relativistic fluid in Nature. We further connect this unphysical instability with the acausality displayed by the theory. We then demonstrate that Israel–Stewart theory can be constructed to be linearly stable, as long as the transport coefficients satisfy certain constraints.

In Chap. 3, we expand the analysis performed in the previous chapter by investigating the properties of relativistic fluid dynamics in the nonlinear regime. This can be accomplished by imposing that the fluid expands according to highly symmetrical flow configurations, in such a way that the symmetries of the problem are sufficient to solve for the fluid 4-velocity and the remaining components of the conserved currents can be calculated analytically or, at least, semi-analytically. Such solutions are extremely rare and will be derived in this chapter. These rare solutions of relativistic fluid dynamics are then used to study the asymptotic properties of transient fluid dynamics, to a level that is not possible in the linear regime.

In Chap. 4, we introduce a novel derivation of transient relativistic fluid dynamics from quantum field theory using linear-response theory. In this derivation, we investigate the conditions for the corresponding retarded Green's function under which the linearized equation of motion of a dissipative current can be reduced to a relaxation-type equation, such as the one appearing in Israel–Stewart theory. The power of this new formalism is demonstrated by applying it to two examples: the linearized Boltzmann equation and the linear-response theory with metric perturbations.

In Chap. 5, we continue the discussion of how relativistic fluid dynamics emerges from a microscopic theory, but now from the perspective of relativistic dilute gases. In this case, the derivation of relativistic dissipative fluid dynamics from a microscopic theory, here the relativistic Boltzmann equation, can be carried out in great detail, going beyond the linear regime. This issue will be the main topic for the remainder of this book. This chapter starts this discussion by introducing the two most widespread methods usually employed to derive relativistic fluid dynamics from the Boltzmann equation: the Chapman–Enskog expansion and the method of moments as proposed by Israel and Stewart. These theories will be further improved and studied in the following chapters.

In Chap. 6, a relativistic generalization of the method of moments is formulated. The main goal of this chapter is to describe the moment expansion of the single-particle momentum distribution function and to derive the equations of motion satisfied by the corresponding expansion coefficients, the so-called *irreducible moments*, which turn out to be moments of the deviation of the single-particle distribution function from a chosen reference state. Such a generalization of the method of moments will be essential in understanding how relativistic fluid dynamics actually emerges from the relativistic Boltzmann equation, what its domain of applicability is, and how the equations of motion can be systematically improved. In this chapter, the reference state is assumed to be in local thermodynamical equilibrium. In Chap. 9, we will show how to generalize this to a particular non-equilibrium reference state, exhibiting a spatial anisotropy.

In Chap. 7, we investigate the convergence of the method of moments using the highly symmetric flow configurations described in Chap. 3: the Bjorken- and Gubser-flow scenarios. We shall demonstrate that these simplified, yet non-trivial, scenarios allow us to systematically derive the equations of motion for *all* the moments of the single-particle distribution function. This will enable us to study the convergence of the moment expansion and obtain a grasp of its domain of validity.

In Chap. 8, we show how relativistic fluid dynamics can be derived from kinetic theory, using the method of moments. In this case, we derive the equations of motion in the nonlinear regime, including all the possible nonlinear and higher-order terms. The reduction of the microscopic degrees of freedom of the Boltzmann equation is implemented by identifying its microscopic time scales and considering only the slowest of these to be relevant in the fluid regime. In addition, the equations of motion for the dissipative quantities are truncated according to a systematic power-counting scheme in Knudsen *and* inverse Reynolds numbers. It is mathematically proven that the equations of motion of fluid dynamics can be closed in terms of only 14 dynamical variables, as it happens in Israel–Stewart theory, as long as we only keep terms of first order in either Knudsen or inverse Reynolds number, or the product of both. Furthermore, we show how to systematically improve the applicability of relativistic fluid dynamics by going to higher order in Knudsen number, at the same time keeping the equations of motion hyperbolic and causal. We then investigate the applicability of the fluid-dynamical theory derived from kinetic theory by comparing its solutions with numerical solutions of the relativistic Boltzmann equation.

The final chapter of this book, Chap. 9, is devoted to the discussion of how to generalize the moment expansion around the local-equilibrium state, as performed in Chap. 6, to an expansion around a particular non-equilibrium reference state. This state is supposed to exhibit an anisotropy in a single spatial direction in momentum space, in contrast to the local-equilibrium state, which is isotropic in momentum space in the rest frame of the fluid. This generalization induces several technical complications, as the irreducible tensors used to perform the moment expansion in the local-equilibrium case need to be replaced by new tensors which are irreducible with respect to the subspace orthogonal to *both* the fluid velocity and the 4-vector parametrizing the spatial anisotropy. This also requires novel orthogonality relations for these tensors.

It is needless to say that our discussion of relativistic dissipative fluid dynamics as derived from kinetic theory and its stability and causality properties is far from complete. Several more recent developments are not covered, such as causal and stable first-order theories of relativistic dissipative fluid dynamics, relativistic dissipative magnetohydrodynamics for unpolarized and polarized fluids, and many more. We reserve a more detailed treatment of these novel and very interesting developments for a future edition of this monograph.

Finally, a word of thanks is in order. This book would not have been possible without the unwavering support and countless hours of work of the many colleagues who co-authored the original papers written by us and referenced in the text: Barbara Betz, Ioannis Bouras, Charles Gale, Carsten Greiner, Ulrich Heinz, Sanyong Jeon, Takeshi Kodama, Tomoi Koide, Etele Molnár, Harri Niemi, Jorge Noronha, Shi Pu, Michael Strickland, Zhe Xu, as well as many others with whom we had fruitful and enlightening discussions over the years.

Niterói, Brazil   Gabriel S. Denicol
Frankfurt am Main, Germany   Dirk H. Rischke
April 2021

References

1. Landau, L.D., Lifshitz, E.M.: Fluid Mechanics. Pergamon, New York (1959)
2. De Groot, S.R., Van Leeuwen, W.A., Van Weert, Ch.G.: Relativistic Kinetic Theory—Principles and Applications. North-Holland (1980)

3. Cercignani, C., Kremer, G.M.: The Relativistic Boltzmann Equation: Theory and Applications.
    Birkhauser, Basel (2002)
4. Rezzolla, L., Zanotti, O.: Relativistic Hydrodynamics. Oxford University Press, Oxford (2013)
5. Romatschke, P., Romatschke, U.: Relativistic Fluid Dynamics In and Out of Equilibrium.
    Cambridge University Press, Cambridge (2019) doi:10.1017/9781108651998

Contents

About the Authors

Prof. Gabriel S. Denicol received his Ph.D. in Theoretical Physics (with summa cum laude) from Goethe University Frankfurt am Main, Germany, in 2012. His Ph.D. thesis discussed how relativistic dissipative fluid dynamics emerges from microscopic theory, generalizing the famous and well-known Israel–Stewart theory. He was a Banting postdoctoral fellow at McGill University, Montreal, Canada, from 2012 to 2015, where he focused on the fluid-dynamical description of the hot and dense nuclear matter created in ultrarelativistic heavy-ion collisions. Afterwards, he became research assistant at Brookhaven National Laboratory, where he stayed until 2016. In 2016, he joined the Department of Physics at Federal University Fluminense, Niteroi, Brazil, as a Professor Adjunto. His current research fields are relativistic non-equilibrium phenomena, fluid dynamics, and transport theory. He was awarded the Bolsa de Produtividade fellowship from CNPq, which he maintains since 2016, and the Young Scientist fellowship, from FAPERJ, in 2018.

Prof. Dirk H. Rischke received his Ph.D. in Theoretical Physics (with summa cum laude) from Goethe University Frankfurt am Main, Germany, in 1993. In his Ph.D. thesis, he developed many-body approximations for the equation of state of hot and dense nuclear matter. After graduation, he was awarded a Lynen fellowship from the Alexander von Humboldt foundation, with which he went to Columbia University in the City of New York from 1994 to 1996, where he worked on the hydrodynamical description of ultrarelativistic heavy-ion collisions. Afterwards, he spent one year as a postdoctoral fellow at Duke University, before joining

Brookhaven National Laboratory as a RIKEN-BNL Fellow from 1997 to 2000, where his research focused on color-superconducting quark matter. In 2001, he joined Goethe University as a Full Professor for Theoretical Physics. His current research fields are hadrons in vacuum, hot and dense strongly interacting matter, and the theory of relativistic dissipative fluid dynamics. From 2017 to 2021, he was spokesperson of the DFG-funded Collaborative Research Center—TransRegio 211 *Strong-interaction matter under extreme conditions*. During his career, he has supervised more than 40 Ph.D. students and postdocs, some of whom now hold positions in academia.

Relativistic Fluid Dynamics

1

The physical description of a system consisting of many degrees of freedom is in general quite complicated. However, if one is interested in the large-distance, long-time scale behavior of the system, it becomes possible to devise an effective theory, taking into account only the degrees of freedom that are relevant on these scales. This happens because, on macroscopic time and length scales, we are not able to observe the microscopic degrees of freedom of the underlying theory, but only average quantities resulting from interactions on the microscopic level. Most of the microscopic quantities vary rapidly in space and time, leading to very small changes of the average values, and are not expected to contribute to the macroscopic dynamics. On the other hand, the few variables that do vary slowly, such as conserved quantities, are expected to be relevant for the effective description of the system on the macroscopic length and time scales.

Fluid dynamics is a typical example of such an effective theory. It is a classical field theory that describes the macroscopic dynamics of systems called fluids. A fluid is a continuous system in which every infinitesimal volume element is assumed to be close to thermodynamic equilibrium and to maintain the proximity to equilibrium throughout its evolution. In other words, in the vicinity of each point in space, we define an infinitesimal volume, called a fluid element, in which the matter is taken to be homogeneous, i.e., any spatial gradients can be neglected, and is described by a finite set of thermodynamic variables and currents. This means that each fluid element must be large enough, relative to the microscopic length scales, to guarantee the proximity to thermodynamic equilibrium, and, at the same time, it must be small enough, relative to the macroscopic length scales, to ensure the continuum limit.

At first glance, the simultaneous applicability of the continuous (zero fluid-element volume) and thermodynamic (infinite fluid-element volume) limits might seem contradictory. However, if the microscopic and macroscopic scales of the system are sufficiently far separated, it is always possible to ensure the existence of a

© Springer Nature Switzerland AG 2021

G. S. Denicol and D. H. Rischke, *Microscopic Foundations of Relativistic Fluid Dynamics*,
Lecture Notes in Physics 990, https://doi.org/10.1007/978-3-030-82077-0_1

volume that is small when compared with the macroscopic scales, and large when compared with the microscopic ones. For example, considering the case of water, a fluid element with a volume of about 1 mm^3 is small enough to assure the continuous-limit approximation and large enough to enclose many molecules and to apply the thermodynamic limit. Note that, for small or rapidly changing systems, such a separation of scales may not be so clear, making it difficult to ensure the proximity to local thermal equilibrium. Such types of systems will be discussed in detail later in this book. This discussion is of particular importance when we apply fluid dynamics to describe the hot and dense matter formed in relativistic heavy-ion collisions, where a clear separation between microscopic and macroscopic scales does not exist.

In this chapter, we discuss the basic aspects of relativistic fluid dynamics from a phenomenological perspective. It is organized as follows: In Sect. 1.1, we introduce the basic laws of thermodynamics and derive the thermodynamic relations that are useful for this book. Section 1.2 contains a brief review of relativistic ideal fluid dynamics. We derive the general form of the conserved currents of an ideal fluid and their equations of motion. Then, Sect. 1.3 shows how to introduce dissipation in fluid dynamics. Here, we explain the basic aspects of dissipative fluid dynamics and derive a covariant version of Navier–Stokes theory using the second law of thermodynamics and, also, via the gradient expansion. In Sect. 1.4, we review the problems of Navier–Stokes theory in the relativistic regime, i.e., the acausality and instability of this theory. We also explain how to render Navier–Stokes theory causal and stable, and to derive a consistent theory of fluid dynamics. In Sect. 1.5, we discuss Israel–Stewart theory and show how to derive causal fluid-dynamical equations from the second law of thermodynamics. Finally, Sect. 1.6 contains a discussion about the existence of non-hydrodynamic modes in relativistic fluid dynamics and how it affects the interpretation of this theory.

1.1 Thermodynamics

Thermodynamics is a theory empirically constructed to describe the thermo-dynamical-equilibrium state in macroscopic systems. It attempts to describe such a state in terms of a small set of extensive quantities, such as the total energy, E, volume, V, and (net) number of particles, N, of the system. Thermodynamics is based on four phenomenological laws, obtained over the years by experimental observation [1]:

Zeroth Law: Two systems that are in equilibrium with a third system are in equilibrium with each other.
First Law: Energy is conserved.
Second Law: The change in entropy of a closed thermodynamic system is always positive semi-definite.
Third Law: The difference in entropy between systems connected by a reversible process is zero in the limit of vanishing temperature.

In this book, we shall make use of the first and second laws of thermodynamics and, therefore, it is convenient to discuss them in more detail. The first law of thermodynamics implies that small variations of the state variables, E, V, and N, must be related,

$$\delta E = \delta Q - P\delta V + \mu\delta N , \qquad (1.1)$$

where P and μ are the pressure and chemical potential, respectively. As a conservation law, the first law of thermodynamics postulates that changes in the total energy of the system (δE) must result from mechanical work done by an external force ($-P\delta V$), from particle exchange with an external medium ($\mu\delta N$), or/and from heat exchange (δQ). The heat exchange takes into account the energy variations due to changes of internal degrees of freedom that are not described by the state variables. The heat itself is not a state variable since it can depend on the past evolution of the system and may take several values for the same thermodynamic state.

However, when dealing with time-reversible processes, it becomes possible to assign a state variable related to heat. This variable is the entropy, S, and is defined in terms of the heat exchange as $\delta Q = T\delta S$, with the temperature T being the proportionality constant. Then, when considering variations between equilibrium states that are infinitesimally close to each other, it is possible to write the first law of thermodynamics in terms of differentials of the state variables,

$$dE = TdS - PdV + \mu dN . \qquad (1.2)$$

Using Eq. (1.2), it is possible to identify the intensive quantities T, μ, and P as the following partial derivatives of the energy,

$$T = \left.\frac{\partial E}{\partial S}\right|_{N,V} , \qquad -P = \left.\frac{\partial E}{\partial V}\right|_{S,N} , \qquad \mu = \left.\frac{\partial E}{\partial N}\right|_{S,V} . \qquad (1.3)$$

The first law of thermodynamics can also be written in terms of entropy variations, i.e., $dS = T^{-1}dE + (P/T)\,dV - (\mu/T)\,dN$, in which case the intensive variables can be obtained from partial derivatives of the entropy,

$$\frac{1}{T} = \left.\frac{\partial S}{\partial E}\right|_{N,V} , \qquad \frac{P}{T} = \left.\frac{\partial S}{\partial V}\right|_{E,N} , \qquad \frac{\mu}{T} = -\left.\frac{\partial S}{\partial N}\right|_{E,V} . \qquad (1.4)$$

In the thermodynamical limit, the entropy is an extensive and additive function of the state variables,

$$\lambda S = S\,(\lambda E, \lambda V, \lambda N) . \qquad (1.5)$$

Using this property, it is straightforward to prove that

$$S = \frac{\partial}{\partial\lambda}\,(\lambda S) = \left.\frac{\partial S}{\partial(\lambda E)}\right|_{\lambda N,\lambda V} E + \left.\frac{\partial S}{\partial(\lambda V)}\right|_{\lambda E,\lambda N} V + \left.\frac{\partial S}{\partial(\lambda N)}\right|_{\lambda E,\lambda V} N , \quad (1.6)$$

which holds for any value of λ. Taking $\lambda = 1$ and using Eq. (1.4), we derive the so-called Euler relation,

$$TS = E + PV - \mu N ,\tag{1.7}$$

and using Euler's relation, combined with the first law of thermodynamics, we obtain the Gibbs–Duhem relation,

$$VdP = SdT + Nd\mu .\tag{1.8}$$

Together, Eqs. (1.2), (1.7), and (1.8) allow us to derive the thermodynamic relations satisfied by the energy, entropy, and (net) particle-number densities, $\varepsilon \equiv E/V$, $s \equiv S/V$, and $n \equiv N/V$, respectively. They are

$$\varepsilon + P = Ts + \mu n ,\tag{1.9}$$
$$ds = \beta d\varepsilon + \alpha dn ,\tag{1.10}$$
$$dP = sdT + nd\mu ,\tag{1.11}$$

where we defined the inverse temperature $\beta = 1/T$ and the thermal potential $\alpha = \mu/T$. Then, Eqs. (1.10) and (1.11) can be used to derive the relations between the intensive variables and the densities,

$$\beta = \left.\frac{\partial s}{\partial \varepsilon}\right|_n , \quad \alpha = \left.\frac{\partial s}{\partial n}\right|_\varepsilon , \quad s = \left.\frac{\partial P}{\partial T}\right|_\mu , \quad n = \left.\frac{\partial P}{\partial \mu}\right|_T .\tag{1.12}$$

The second law of thermodynamics dictates that the entropy of an isolated system must either increase or remain constant. This implies that, if a given system is in thermodynamic equilibrium, i.e., if it is in a quasi-stationary state where its extensive and intensive variables no longer change, the entropy of this system must remain constant (as long as the boundary conditions imposed on the system remain fixed). On the other hand, the entropy of a system that is out of equilibrium must always increase. This is a very useful and powerful concept that will be extensively used in this chapter. As we will show later, the second law of thermodynamics can even be used to constrain and, sometimes, derive the equations of motion of a viscous fluid.

These are the basic aspects of thermodynamics that we wanted to address (for a more detailed review, see Ref. [1]). It is worth pointing out that, although the thermodynamic relations specify how the macroscopic variables are related and how they change with time, they are not enough to extract the explicit form of the equation of state, i.e., of the function $s(\varepsilon, n)$. In order to determine the entropy density as a function of the state variables ε, n, a microscopic description of the matter is required, which can only be obtained from a more fundamental approach, such as statistical mechanics.

1.2   Relativistic Ideal Fluid Dynamics

We start our discussion of relativistic fluid dynamics by considering the most simple example: the case of an ideal fluid [2,3]. An ideal fluid is defined by the assumption of *local thermodynamical equilibrium*, i.e., that all fluid elements must be in thermodynamic equilibrium, but not necessarily in the same thermodynamic-equilibrium state (if they are, one speaks of *global* thermodynamical equilibrium). This means that all thermodynamic state variables are functions of the space-time 4-vector $x^\mu \equiv X = (t, \mathbf{x})^T$, e.g., temperature $T(X)$, chemical potential $\mu(X)$. In addition, the fluid is described by a collective velocity field, which is also a function of space-time, $\mathbf{u}(X)$. From now on, the fields T, μ, and $\mathbf{u}$ shall be referred to as primary fluid-dynamical variables.

1.2.1   Conserved Currents in an Ideal Fluid

The state of a fluid is specified by the densities and currents associated with conserved quantities, i.e., energy, momentum, and (net) particle number. For a relativistic fluid, the state variables are the energy-momentum tensor, $T^{\mu\nu}(X)$, and the (net) particle 4-current, $N^\mu(X)$. For an ideal fluid, the general form of these currents can be obtained by performing a Lorentz transformation to the local rest frame of the fluid, in which $\mathbf{u}(X) = 0$. In this frame, the energy-momentum tensor, $T_{RF}^{\mu\nu}$ (the subscript RF indicates the local rest-frame form of this tensor), should have the characteristic form of a system in static equilibrium,

$$T_{RF}^{\mu\nu} = \begin{pmatrix} \varepsilon & 0 & 0 & 0 \\ 0 & P & 0 & 0 \\ 0 & 0 & P & 0 \\ 0 & 0 & 0 & P \end{pmatrix}, \tag{1.13}$$

i.e., in this frame there is no flow of energy ($T_{RF}^{i0} = 0$), and the force per surface element is isotropic and equal to the thermodynamic pressure ($T_{RF}^{ij} = \delta^{ij} P$). In the local rest frame, there should also be no flow of particles and entropy and, consequently, the (net) particle and entropy 4-currents in this frame, N_{RF}^μ and S_{RF}^μ, respectively, take the following simple form

$$N_{RF}^\mu = (n, 0, 0, 0)^T, \tag{1.14}$$

$$S_{RF}^\mu = (s, 0, 0, 0)^T. \tag{1.15}$$

The form of these tensors in a general frame can be derived by applying a Lorentz boost with the fluid velocity $\mathbf{u}$ to N_{RF}^μ, S_{RF}^μ, and $T_{RF}^{\mu\nu}$,

$$N^\mu = \Lambda_\alpha^\mu (\mathbf{u}) N_{RF}^\alpha, \tag{1.16}$$

$$S^\mu = \Lambda_\alpha^\mu (\mathbf{u}) S_{RF}^\alpha, \tag{1.17}$$

$$T^{\mu\nu} = \Lambda_\alpha^\mu (\mathbf{u}) \Lambda_\beta^\nu (\mathbf{u}) T_{RF}^{\alpha\beta}, \tag{1.18}$$

where we remember that the general form of a Lorentz boost is

$$\Lambda^{\mu}_{\nu}\left(\mathbf{u}\right)$$

$$= \begin{pmatrix} \gamma & -u^x & -u^y & -u^z \\ -u^x & 1+(1+\gamma)^{-1}u^x u^x & (1+\gamma)^{-1}u^x u^y & (1+\gamma)^{-1}u^x u^z \\ -u^y & (1+\gamma)^{-1}u^y u^x & 1+(1+\gamma)^{-1}u^y u^y & (1+\gamma)^{-1}u^y u^z \\ -u^z & (1+\gamma)^{-1}u^z u^x & (1+\gamma)^{-1}u^z u^y & 1+(1+\gamma)^{-1}u^z u^z \end{pmatrix},$$

$$(1.19)$$

with $\gamma = \sqrt{1+\mathbf{u}\cdot\mathbf{u}}$ being the Lorentz gamma factor. Then, using a covariant notation, the conserved currents of an ideal fluid can be expressed as

$$N^{\mu}_{\text{ideal}} \equiv N^{\mu}_{(0)} = nu^{\mu} ,$$
$$S^{\mu}_{\text{ideal}} \equiv S^{\mu}_{(0)} = su^{\mu} ,$$
$$T^{\mu\nu}_{\text{ideal}} \equiv T^{\mu\nu}_{(0)} = \varepsilon u^{\mu}u^{\nu} - \Delta^{\mu\nu}p , \qquad (1.20)$$

where u^{μ} is the velocity 4-vector,

$$u^{\mu} = (\gamma, \mathbf{u})^{T} . \qquad (1.21)$$

Note that the velocity 4-vector is constructed to satisfy the normalization condition $u^{\mu}u_{\mu} = 1$, and, therefore, has only three independent components. Also, in Eq. (1.20), we introduced the projection operator onto the 3-space orthogonal to u^{μ},

$$\Delta^{\mu\nu} = g^{\mu\nu} - u^{\mu}u^{\nu} , \qquad (1.22)$$

where $g^{\mu\nu}$ is the space-time metric (in flat space). $\Delta^{\mu\nu}$ satisfies all the properties expected of a projector,

$$u_{\mu}\Delta^{\mu\nu} = u_{\nu}\Delta^{\mu\nu} = 0 , \qquad \Delta^{\mu}_{\lambda}\Delta^{\lambda\nu} = \Delta^{\mu\nu} , \qquad (1.23)$$

and has the following trace:

$$\Delta^{\mu}_{\mu} = 3 . \qquad (1.24)$$

1.2.2   Equations of Motion

The dynamical description of an ideal fluid is obtained using the conservation laws of energy, momentum, and (net) particle number. These conservation laws can be mathematically expressed in terms of the following five continuity equations:

$$\partial_{\mu}N^{\mu}_{(0)} = 0 , \qquad (1.25)$$

$$\partial_{\mu}T^{\mu\nu}_{(0)} = 0 . \qquad (1.26)$$

The partial derivative $\partial_\mu \equiv \partial/\partial x^\mu$ transforms as a covariant vector under Lorentz transformations and, therefore, Eq. (1.26) transforms as a contravariant 4-vector. As a 4-vector, it is convenient to decompose this equation into a part parallel and a part orthogonal to u^μ. The component parallel to the velocity is obtained by contracting the equation of motion with u^μ, $u_\alpha \partial_\beta T_{(0)}^{\alpha\beta}$, while the component orthogonal to the velocity is obtained by contracting it with $\Delta^{\mu\nu}$, $\Delta_\alpha^\mu \partial_\beta T_{(0)}^{\alpha\beta}$. This, together with the conservation law for (net) particle number, leads to the equations of motion of ideal fluid dynamics,

$$u_\alpha \partial_\beta T_{(0)}^{\alpha\beta} = \dot\varepsilon + (\varepsilon + P)\theta = 0 , \tag{1.27}$$

$$\Delta_\alpha^\mu \partial_\beta T_{(0)}^{\alpha\beta} = (\varepsilon + P)\dot u^\mu - \nabla^\mu p = 0 , \tag{1.28}$$

$$\partial_\mu N_{(0)}^\mu = \dot n + n\theta = 0 , \tag{1.29}$$

where we introduced the comoving derivative $u^\mu \partial_\mu A \equiv \dot A$ of any quantity A and the space-like gradient $\Delta_\mu^\lambda \partial_\lambda \equiv \nabla_\mu$. We further defined the expansion rate θ as the 4-divergence of the 4-velocity,

$$\theta \equiv \nabla_\mu u^\mu . \tag{1.30}$$

Note that an ideal fluid is described by four fields, ε, P, n, and u^μ that contain, in total, six independent degrees of freedom. The conservation laws, on the other hand, provide only five equations of motion. To close this system of equations, we must specify the equation of state of the fluid that gives the pressure as a function of the other thermodynamic variables,

$$P = P(\varepsilon, n) . \tag{1.31}$$

The assumption of local thermal equilibrium guarantees the existence of this function and, hence, assures that the equations of ideal fluid dynamics are always closed. In ideal fluid dynamics, the equation of state essentially defines the type of fluid that is being described—it is the only place where (some of) the microscopic properties of the system must be taken into account.

1.2.3   Covariant Thermodynamics and Entropy Production

Using the conserved currents, $N_{(0)}^\mu$, $S_{(0)}^\mu$, and $T_{(0)}^{\mu\nu}$, we can rewrite the equilibrium thermodynamic relations derived in Sect. 1.1, Eqs. (1.9), (1.10), and (1.11), in a covariant form [5–7]. For this purpose, it is convenient to introduce the following 4-vector:

$$\beta^\mu = \frac{u^\mu}{T} . \tag{1.32}$$

Then, following Israel and Stewart [5–7], we postulate a covariant version of the Gibbs–Duhem relation,

$$d\left(p\beta^{\mu}\right) = N_{(0)}^{\mu}d\alpha - T_{(0)}^{\mu\nu}d\beta_{\nu} \, , \tag{1.33}$$

and of Euler's relation,

$$S_{(0)}^{\mu} = P\beta^{\mu} + T_{(0)}^{\mu\nu}\beta_{\nu} - \alpha N_{(0)}^{\mu} \, . \tag{1.34}$$

Equations (1.33) and (1.34) can then be used to derive a covariant form of the first law of thermodynamics (1.10),

$$dS_{(0)}^{\mu} = \beta_{\nu}dT_{(0)}^{\mu\nu} - \alpha dN_{(0)}^{\mu} \, . \tag{1.35}$$

The above covariant thermodynamic relations were constructed in such a way that, when contracted with the fluid 4-velocity, the usual thermodynamic relations, Eqs. (1.33), (1.34), and (1.35), are recovered,

$$u_{\mu}\left(d\left(P\beta^{\mu}\right) - N_{(0)}^{\mu}d\alpha + T_{(0)}^{\mu\nu}d\beta_{\nu}\right) = d\left(P\beta\right) - nd\alpha + \varepsilon d\beta = 0 \, , \tag{1.36}$$

$$u_{\mu}\left(S_{(0)}^{\mu} - P\beta^{\mu} - T_{(0)}^{\mu\nu}\beta_{\nu} + \alpha N_{(0)}^{\mu}\right) = s + \alpha n - \beta\left(\varepsilon + P\right) = 0 \, , \tag{1.37}$$

$$u_{\mu}\left(dS_{(0)}^{\mu} - \beta_{\nu}dT_{(0)}^{\mu\nu} + \alpha dN_{(0)}^{\mu}\right) = ds - \beta d\varepsilon + \alpha dn = 0 \, , \tag{1.38}$$

where we used that $u_{\mu}du^{\mu} = 0$. Note also that the covariant thermodynamic relations do not contain more information than the usual thermodynamic relations. The projection of Eqs. (1.33), (1.34), and (1.35) onto the 3-space orthogonal to u^{μ} just leads to trivial relations,

$$\Delta_{\mu}^{\alpha}\left(d\left(P\beta^{\mu}\right) - N_{(0)}^{\mu}d\alpha_{0} + T_{(0)}^{\mu\nu}d\beta_{\nu}\right) = 0 \quad\Longrightarrow\quad 0 = 0 \, , \tag{1.39}$$

$$\Delta_{\mu}^{\alpha}\left(S_{(0)}^{\mu} - P\beta^{\mu} - T_{(0)}^{\mu\nu}\beta_{\nu} + \alpha N_{(0)}^{\mu}\right) = 0 \quad\Longrightarrow\quad 0 = 0 \, , \tag{1.40}$$

$$\Delta_{\mu}^{\alpha}\left(dS_{(0)}^{\mu} - \beta_{\nu}dT_{(0)}^{\mu\nu} + \alpha dN_{(0)}^{\mu}\right) = 0 \quad\Longrightarrow\quad 0 = 0 \, . \tag{1.41}$$

The first law of thermodynamics, Eq. (1.35), leads to the following equation of motion for the entropy 4-current:

$$\partial_{\mu}S_{(0)}^{\mu} = \beta_{\nu}\partial_{\mu}T_{(0)}^{\mu\nu} - \alpha\partial_{\mu}N_{(0)}^{\mu} \, , \tag{1.42}$$

which, using the conservation of (net) particle number, energy, and momentum, $\partial_{\mu}N_{(0)}^{\mu} = \partial_{\mu}T_{(0)}^{\mu\nu} = 0$, implies the conservation of entropy,

$$\partial_{\mu}S_{(0)}^{\mu} = 0 \, . \tag{1.43}$$

Note that the entropy conservation appeared naturally, as a consequence of (net) particle number, energy-momentum conservation, and the first law of thermodynamics. The equation of motion for the entropy density comes directly from Eq. (1.43),

$$\partial_\mu S^\mu_{(0)} = u^\mu \partial_\mu s + s \partial_\mu u^\mu = \dot{s} + s\theta = 0 \,. \tag{1.44}$$

In this section, we introduced and derived the equations of motion of ideal fluid dynamics. This was done using the conservation laws expected to be valid in a fluid and the assumption of local thermal equilibrium. In the next sections, we show how to introduce dissipation into this scheme.

1.3 Relativistic Dissipative Fluid Dynamics

Relativistic ideal fluid dynamics was derived using the conservation laws, the properties of the Lorentz transformation, and, most importantly, by imposing local thermodynamic equilibrium. While the conservation laws and the properties of the Lorentz transformation are always valid, the assumption of local thermodynamic equilibrium is very restrictive and is never realized in practice. Strictly speaking, a fluid can never maintain exact local thermodynamic equilibrium during the whole course of its dynamical evolution. In this section, we consider a more general theory of fluid dynamics, which attempts to take into account the dissipative processes that must happen in a fluid.

Dissipative effects originate from irreversible thermodynamic processes that occur during the motion of the fluid. In general, each fluid element is not in equilibrium with the rest of the fluid and, in order to approach global equilibrium, it will exchange heat with its surroundings. Furthermore, the fluid elements are in relative motion and can also dissipate energy by friction. All these processes must be included in order to obtain a reasonable description of a relativistic fluid.

The first to propose a covariant formulation of dissipative fluid dynamics were Eckart [4], in 1940, and, later, Landau and Lifshitz [2], in 1959. Both theories, often called first-order theories, are based on a covariant formulation of Navier–Stokes theory. At that time, Navier–Stokes theory had already become a successful theory of fluid dynamics, being able to describe a wide variety of non-relativistic fluids, from weakly coupled gases, such as air, to strongly coupled fluids, such as water. Therefore, a relativistic extension of Navier–Stokes theory was considered to be the most promising way to describe relativistic viscous fluids.

However, the situation was shown to be more subtle since the relativistic version of Navier–Stokes theory is actually intrinsically unstable [8–12]. The source of such instability is well understood and will be discussed in detail in the next chapter. It comes from the inherent acausal behavior of Navier–Stokes theory [13,14], which allows signals to propagate with infinite speed. In non-relativistic theories, this non-intuitive feature does not give rise to an intrinsic problem and can be ignored. On the other hand, in relativistic systems, where causality is a physical property that is naturally preserved, this feature leads to equations of motion that are intrinsically unstable. Nevertheless, first-order theories are an important initial step to illustrate the basic features of relativistic dissipative fluid dynamics and thus shall be reviewed in this section.

Just like for an ideal fluid, the basic equations of motion for viscous fluids are given by the conservation laws of (net) particle number and energy-momentum,

$$\partial_\mu N^\mu = 0 \,, \tag{1.45}$$

$$\partial_\mu T^{\mu\nu} = 0 \,. \tag{1.46}$$

However, in the presence of dissipation, the energy-momentum tensor is no longer diagonal and isotropic in the local rest frame. Also, due to diffusion, we expect (net) particle-number flow to appear in the local rest frame of the fluid element. These effects must be taken into account and are introduced in fluid dynamics by adding dissipative currents, n^μ and $\tau^{\mu\nu}$, to the previously derived ideal currents, $N_{(0)}^\mu$ and $T_{(0)}^{\mu\nu}$,

$$N^\mu = N_{(0)}^\mu + n^\mu = n_0 u^\mu + n^\mu \,, \tag{1.47}$$

$$T^{\mu\nu} = T_{(0)}^{\mu\nu} + \tau^{\mu\nu} = \varepsilon_0 u^\mu u^\nu - \Delta^{\mu\nu} P_0 + \tau^{\mu\nu} \,, \tag{1.48}$$

where we indicated equilibrium quantities with a subscript "0". In order to satisfy angular-momentum conservation, $\tau^{\mu\nu}$ is defined to be a symmetric tensor, $\tau^{\mu\nu} = \tau^{\nu\mu}$. The main problem then becomes to find the dynamical or constitutive equations satisfied by such dissipative currents.

1.3.1 Matching Conditions

The introduction of the dissipative currents renders the equilibrium variables ill-defined, since the fluid can no longer be considered to be in local thermodynamic equilibrium. In a viscous fluid, the thermodynamic variables, α_0, β_0, s_0, P_0, ..., can only be defined in terms of a fictitious equilibrium state (labeled by the subscript "0"), constructed such that the thermodynamic relations are valid as if the fluid were in local thermodynamic equilibrium. The first step to construct such an equilibrium state is to define n_0 and ε_0 as the actual (net) particle density n and the actual energy density ε in the local rest frame of the fluid. This is guaranteed by the so-called *matching conditions*,

$$\varepsilon_0 \equiv \varepsilon \equiv u_\nu u_\mu T^{\mu\nu} \,, \tag{1.49}$$

$$n_0 \equiv n \equiv u_\mu N^\mu \,. \tag{1.50}$$

The matching conditions (1.49) and (1.50) are enforced by applying the following set of constraints to the dissipative currents:

$$u_\nu u_\mu \tau^{\mu\nu} = 0 \,,$$

$$u_\mu n^\mu = 0 \quad \Longleftrightarrow \quad \Delta_\lambda^\mu n^\lambda = n^\mu \,. \tag{1.51}$$

Then, using ε and n we can construct our equilibrium state. The thermodynamic entropy density is determined by the equation of state of the fluid as if in thermodynamic equilibrium,

$$s_0 \equiv s_0(\varepsilon, n) , \tag{1.52}$$

while the remaining thermodynamic variables, e.g., the thermodynamic pressure, temperature, and chemical potential, are defined from the thermodynamic relations derived in Sect. 1.1. The inverse temperature and the ratio of chemical potential over temperature are computed using Eq. (1.12),

$$\beta_0 = \left.\frac{\partial s}{\partial \varepsilon}\right|_n , \qquad \alpha_0 = \left.\frac{\partial s}{\partial n}\right|_\varepsilon , \tag{1.53}$$

and the thermodynamic pressure is extracted via Eq. (1.9),

$$P_0 = -\varepsilon + T_0 s_0 + \mu_0 n . \tag{1.54}$$

Note that, in principle, the thermodynamic pressure can also be expressed as a function of other thermodynamic variables, e.g., $P_0 = P_0(\beta_0, \alpha_0)$.

It is important to emphasize that, while the energy and (net) particle densities are physically well-defined, all other quantities (s_0, P_0, T_0, μ_0, ...) are defined only in terms of a fictitious equilibrium state and do not necessarily retain their usual physical meaning. For example, the second law of thermodynamics does not constrain the production of entropy of the fictitious state: it constrains only the production of the actual non-equilibrium entropy of the fluid—a quantity that can be rather non-trivial to construct, as will be discussed later in this chapter.

1.3.2 Tensor Decomposition of $\tau^{\mu\nu}$

It is convenient to decompose $\tau^{\mu\nu}$ in terms of its irreducible components, i.e., a scalar, a 4-vector, and a traceless and symmetric second-rank tensor. This tensor decomposition must respect the matching (or orthogonality) condition satisfied by $\tau^{\mu\nu}$, Eq. (1.51). For this purpose, we introduce yet another projection operator: the double-symmetric, traceless projection operator orthogonal to u^μ,

$$\Delta^{\mu\nu\alpha\beta} = \frac{1}{2}\left(\Delta^{\mu\alpha}\Delta^{\nu\beta} + \Delta^{\mu\beta}\Delta^{\nu\alpha}\right) - \frac{1}{\Delta^\lambda_\lambda}\Delta^{\mu\nu}\Delta^{\alpha\beta} , \tag{1.55}$$

which satisfies the following properties:

$$\begin{aligned}
\Delta^{(\mu\nu)(\alpha\beta)} &= \Delta^{(\alpha\beta)(\mu\nu)} , \\
\Delta^{\mu\nu}_{\lambda\rho}\Delta^{\lambda\rho}_{\alpha\beta} &= \Delta^{\mu\nu}_{\alpha\beta} , \\
u_\mu\Delta^{\mu\nu\alpha\beta} &= g_{\mu\nu}\Delta^{\mu\nu\alpha\beta} = 0 , \\
\Delta^{\mu\nu}_{\mu\nu} &= 5 .
\end{aligned} \tag{1.56}$$

Then, using $\Delta^{\mu\nu}$ and $\Delta^{\mu\nu}_{\alpha\beta}$, the tensor decomposition of $\tau^{\mu\nu}$ in its irreducible form is implemented as

$$\tau^{\mu\nu} \equiv -\Pi\Delta^{\mu\nu} + 2u^{(\mu}h^{\nu)} + \pi^{\mu\nu}, \qquad (1.57)$$

where the parentheses () denote the symmetrization of all Lorentz indices, $a^{(\mu\nu)} = \left(a^{\mu\nu} + a^{\nu\mu}\right)/2$, and where we defined

$$\Pi \equiv -\frac{1}{3}\Delta_{\alpha\beta}\tau^{\alpha\beta}, \qquad h^{\mu} \equiv \Delta^{\mu}_{\alpha}u_{\beta}\tau^{\alpha\beta}, \qquad \pi^{\mu\nu} \equiv \Delta^{\mu\nu}_{\alpha\beta}\tau^{\alpha\beta}. \qquad (1.58)$$

The scalar term, Π, is the bulk-viscous pressure, the vector term, h^{μ}, is the energy-diffusion 4-current, and the second-rank tensor, $\pi^{\mu\nu}$, is the shear-stress tensor. The properties of the projection operators Δ^{μ}_{ν} and $\Delta^{\mu\nu}_{\alpha\beta}$ given in Eqs. (1.23) and (1.56) imply that h^{μ} and $\pi^{\mu\nu}$ satisfy

$$h^{\mu} = \Delta^{\mu}_{\nu}h^{\nu} \iff u_{\mu}h^{\mu} = 0, \qquad (1.59)$$

$$\pi^{\mu\nu} = \pi^{\langle\mu\nu\rangle} \iff u_{\mu}\pi^{\mu\nu} = 0, \qquad (1.60)$$

$$\pi^{\mu}_{\mu} = 0, \qquad (1.61)$$

where the brackets $\langle\rangle$ denote the following projection of a second-rank tensor, $A^{\langle\mu\nu\rangle} \equiv \Delta^{\mu\nu}_{\alpha\beta}A^{\alpha\beta}$. In summary, the fields Π, h^{μ}, n^{μ}, and $\pi^{\mu\nu}$ are expressed in terms of N^{μ} and $T^{\mu\nu}$ as

$$\Pi = -P_0 - \frac{1}{3}\Delta_{\mu\nu}T^{\mu\nu}, \qquad (1.62)$$

$$h^{\mu} = u_{\alpha}\Delta^{\mu}_{\beta}T^{\alpha\beta}, \qquad (1.63)$$

$$n^{\mu} = \Delta^{\mu}_{\alpha}N^{\alpha}, \qquad (1.64)$$

$$\pi^{\mu\nu} = T^{\langle\mu\nu\rangle}. \qquad (1.65)$$

Note that $T^{\mu\nu}$ is a symmetric second-rank tensor and, thus, N^{μ} and $T^{\mu\nu}$ have 14 independent components (four from N^{μ} and ten from $T^{\mu\nu}$). In order to satisfy their orthogonality to u^{μ}, n^{μ} and h^{μ} can only have three independent components each. The shear-stress tensor is symmetric, traceless, and orthogonal to u^{μ} and, therefore, can have only five independent components. Together with u^{μ}, ε, n, and Π, which have in total six independent components, we find a total of 17 independent components, three more than expected. This happened because, so far, the velocity field itself has not been specified, being introduced just as a general normalized 4-vector. The definition of the velocity field will provide the three missing constraints that will reduce the number of independent components to the correct value.

1.3.3 Definition of the Local Rest Frame and Equations of Motion

The definition of the velocity field is an important step in deriving fluid dynamics. For ideal fluids, the local rest frame was implicitly defined as the frame in which there is no flow of energy and (net) particle number. Due to the presence of energy and particle diffusion in viscous fluids, this definition is no longer possible. From a mathematical point of view, velocity can be defined in numerous ways. From the physical perspective, there are, however, two natural choices. The *Landau frame* [2], in which the velocity is defined by the flow of the total energy,

$$u_\mu T^{\mu\nu} \equiv \varepsilon u^\nu , \tag{1.66}$$

and the *Eckart frame* [3,4], in which the velocity is specified by the flow of (net) particle number,

$$N^\mu \equiv n u^\mu . \tag{1.67}$$

If the system has more than one type of particle (or charge), the Eckart frame must be defined by selecting one of these particle (or charge) types.

Both choices of frame impose different constraints on the dissipative currents introduced in this section. In the Landau frame, the energy diffusion is always set to zero,

$$h^\mu = 0 , \tag{1.68}$$

while in the Eckart frame the particle diffusion is always zero,

$$n^\mu = 0 . \tag{1.69}$$

In other words, in the Landau frame the velocity field is fixed to eliminate any diffusion of energy, while in the Eckart frame it is defined to eliminate any diffusion of particles. In this book, we shall always use the Landau frame, Eq. (1.66), and, therefore, the conserved currents take the following simpler form:

$$N^\mu = n u^\mu + n^\mu , \tag{1.70}$$

$$T^{\mu\nu} = \varepsilon u^\mu u^\nu - \Delta^{\mu\nu} (P_0 + \Pi) + \pi^{\mu\nu} . \tag{1.71}$$

Note that both the Landau and the Eckart choices of frame reduce the number of independent variables to 14, as advertised at the end of the last section.

As for an ideal fluid, we decompose Eq. (1.46) into a part parallel and another one orthogonal to u^μ. As shown in the last section, this is done by projecting and contracting Eq. (1.46) with u_μ and $\Delta^{\mu\nu}$, i.e., by taking $u_\alpha \partial_\beta T^{\alpha\beta}$ and $\Delta^\mu_\alpha \partial_\beta T^{\alpha\beta}$, respectively. Together with Eqs. (1.45), (1.70), and (1.71), this procedure leads to the equations of motion of the fluid,

$$u_\alpha \partial_\beta T^{\alpha\beta} = \dot{\varepsilon} + (\varepsilon + P_0 + \Pi) \theta - \pi^{\alpha\beta} \sigma_{\alpha\beta} = 0 , \tag{1.72}$$

$$\Delta^\mu_\alpha \partial_\beta T^{\alpha\beta} = (\varepsilon + P_0 + \Pi) \dot{u}^\mu - \nabla^\mu (P_0 + \Pi) + \Delta^\mu_\alpha \partial_\beta \pi^{\alpha\beta} = 0 , \tag{1.73}$$

$$\partial_\mu N^\mu = \dot{n} + n\theta + \partial_\mu n^\mu = 0 , \tag{1.74}$$

where we defined the shear tensor

$$\sigma^{\mu\nu} \equiv \partial^{\langle\mu} u^{\nu\rangle} = \frac{1}{2} \left(\nabla^\mu u^\nu + \nabla^\nu u^\mu \right) - \frac{1}{3} \Delta^{\mu\nu} \theta \ . \tag{1.75}$$

Note that the quantities n, ε, P_0, u^μ, Π, n^μ, and $\pi^{\mu\nu}$ introduced in this section were defined from a strict mathematical perspective via the most general tensor decomposition allowed by symmetry. The conservation laws, Eqs. (1.45) and (1.46), the definition of the fictitious equilibrium state, and the definition of the velocity field are also general and valid regardless of the fluid-dynamical regime (even though they make more sense near the fluid-dynamical limit). Thus, by writing down any of the above equations, we have not, by any means, derived fluid dynamics. In order to derive the complete equations of dissipative fluid dynamics, one still has to provide nine additional equations of motion that will close Eqs. (1.72), (1.73), and (1.74). In the end, this corresponds to finding the closed dynamical or constitutive relations satisfied by the dissipative variables Π, n^μ, and $\pi^{\mu\nu}$. Ideal fluid dynamics, discussed in the previous section, corresponds to a trivial example of this procedure, in which the dissipative currents are simply set to zero.

1.3.4   Relativistic Navier–Stokes Theory

In the presence of dissipative currents, the entropy is no longer conserved. Deriving the equation for the entropy 4-current is not trivial for a viscous fluid, since *a priori* we do not know the form of this current. For now, let us take the same steps as in the ideal fluid case and see where we arrive. We start by taking Eq. (1.42),

$$\partial_\mu S^\mu_{(0)} = \beta_0 u_\nu \partial_\mu T^{\mu\nu}_{(0)} - \alpha_0 \partial_\mu N^\mu_{(0)} \ , \tag{1.76}$$

which remains valid in a viscous fluid since, as explained in Sect. 1.3.1, the equilibrium variables and, consequently, the equilibrium currents, were constructed to satisfy thermodynamic relations as if in equilibrium. Now, however, the equilibrium part of the currents are not conserved, $\partial_\mu N^\mu_{(0)} = -\partial_\mu n^\mu \neq 0$ and $u_\nu \partial_\mu T^{\mu\nu}_{(0)} = -\Pi\theta + \pi^{\mu\nu} \neq 0$, cf. Eqs. (1.72) and (1.74), and in a viscous fluid Eq. (1.76) leads to

$$\partial_\mu S^\mu_{(0)} = \alpha_0 \partial_\mu n^\mu + \beta_0 \left(-\Pi\theta + \pi^{\mu\nu} \sigma_{\mu\nu} \right) \ . \tag{1.77}$$

By decomposing the first term on the right-hand side as $\alpha_0 \partial_\mu n^\mu = \partial_\mu \left(\alpha_0 n^\mu \right) - n^\mu \nabla_\mu \alpha_0$, Eq. (1.77) can be written in a more convenient form,

$$\partial_\mu \left(S^\mu_{(0)} - \alpha_0 n^\mu \right) = -n^\mu \nabla_\mu \alpha_0 - \beta_0 \Pi\theta + \beta_0 \pi^{\mu\nu} \sigma_{\mu\nu} \equiv Q \ . \tag{1.78}$$

It is very tempting to identify the term on the left-hand side of Eq. (1.78) as the 4-divergence of the (off-equilibrium) entropy 4-current

$$S^\mu \equiv S^\mu_{(0)} - \alpha_0 n^\mu = s_0 u^\mu - \alpha_0 n^\mu \ , \tag{1.79}$$

and the terms on the right-hand side, Q, as the source terms for entropy production. Note, however, that this is not necessarily the case. Nevertheless, this was the identification proposed by Eckart and by Landau and Lifshitz, and we shall consider it here in order to derive relativistic Navier–Stokes theory.

The relativistic Navier–Stokes theory is then obtained by applying the second law of thermodynamics to each fluid element, i.e., by requiring that the entropy production obtained in Eq. (1.78) must always be positive semi-definite, $Q \geq 0$. The simplest way to satisfy this condition for all possible fluid configurations is to assume that the bulk-viscous pressure, the particle-diffusion 4-current, and the shear-stress tensor are linearly proportional to θ, $\nabla^\mu \alpha_0$, and $\sigma^{\mu\nu}$, respectively,

$$\Pi = -\zeta\theta , \tag{1.80}$$

$$n^\mu = \varkappa \nabla^\mu \alpha_0 , \tag{1.81}$$

$$\pi^{\mu\nu} = 2\eta\sigma^{\mu\nu} . \tag{1.82}$$

The proportionality coefficients ζ, $\varkappa$, and η are the coefficients of bulk viscosity, particle diffusion, and shear viscosity, respectively. Then, substituting Eqs. (1.80), (1.81), and (1.82) into Eq. (1.78), we see that the entropy production becomes a quadratic function of the dissipative currents

$$Q = \frac{\beta_0}{\zeta}\Pi^2 - \frac{1}{\varkappa}n^\mu n_\mu + \frac{\beta_0}{2\eta}\pi_{\mu\nu}\pi^{\mu\nu} . \tag{1.83}$$

Note that $n^\mu n_\mu$ is negative (n^μ is a space-like vector) while $\pi_{\mu\nu}\pi^{\mu\nu}$ is positive and, therefore, as long as ζ, $\varkappa$, $\eta \geq 0$, Q is, in fact, always positive semi-definite.

The equations of fluid dynamics are obtained by substituting Eqs. (1.80), (1.81), and (1.82) into the conservation laws, Eqs. (1.72), (1.73), and (1.74),

$$\dot{\varepsilon} = -(\varepsilon + P_0 - \zeta\theta)\theta + 2\eta\sigma_{\alpha\beta}\sigma^{\alpha\beta} , \tag{1.84}$$

$$(\varepsilon + P_0 - \zeta\theta)\dot{u}^\mu = \nabla^\mu P_0 - \nabla^\mu(\zeta\theta) - 2\Delta^\mu_\alpha \partial_\beta(\eta\sigma^{\alpha\beta}) , \tag{1.85}$$

$$\dot{n} = -n\theta - \partial_\mu(\varkappa\nabla^\mu\alpha_0) . \tag{1.86}$$

The above equations are known as the relativistic Navier–Stokes equations.

This theory was obtained by Landau and Lifshitz. A similar theory was obtained independently by Eckart, using a different definition of the local rest frame. In this formulation, the state of a viscous fluid remains being described by the same variables as in the case of an ideal fluid, i.e., the primary fluid-dynamical variables α_0, β_0, and u^μ. The only difference is the existence of dissipative processes, corresponding to new forms of particle and energy-momentum transfer, which occur due to gradients of the primary fluid-dynamical variables.

As already mentioned, Navier–Stokes theory is acausal and, consequently, unstable. Thus, it is unable to describe any relativistic fluid existing in Nature. The source of the acausality can be understood from the constitutive relations satisfied by the dissipative currents, Eqs. (1.80), (1.81), and (1.82). Such linear relations imply that any inhomogeneity of α_0, β_0, and u^μ will *instantaneously* give rise to a dissipative current. This instantaneous creation of currents from (space-like) gradients of the primary fluid-dynamical variables renders the equations of motion parabolic. In a relativistic theory this leads to instabilities, as will be discussed in Chap. 2.

1.3.5 Gradient Expansion and Navier–Stokes Theory

Navier–Stokes theory can also be derived (and extended) via the so-called gradient expansion [15,16]. In this framework, the bulk-viscous pressure, the particle-diffusion 4-current, and the shear-stress tensor are assumed to be expressable solely in terms of powers of gradients of the primary fluid-dynamical variables α_0, β_0, and u^μ. The dissipative currents can then be schematically written as the series

$$\Pi = \lambda_\Pi^{(1)} \mathcal{O}_1 + \lambda_\Pi^{(2)} \mathcal{O}_2 + \cdots , \tag{1.87}$$

$$n^\mu = \lambda_n^{(1)} \mathcal{O}_1^\mu + \lambda_n^{(2)} \mathcal{O}_2^\mu + \cdots , \tag{1.88}$$

$$\pi^{\mu\nu} = \lambda_\pi^{(1)} \mathcal{O}_1^{\mu\nu} + \lambda_\pi^{(2)} \mathcal{O}_2^{\mu\nu} + \cdots , \tag{1.89}$$

where the quantities $(\mathcal{O}_1, \mathcal{O}_1^\mu, \mathcal{O}_1^{\mu\nu})$ and $(\mathcal{O}_2, \mathcal{O}_2^\mu, \mathcal{O}_2^{\mu\nu})$ correspond to terms of first and second order in gradients of α_0, β_0, and u^μ, respectively, and the dots denote possible higher-order gradient terms.

It is important to remark that when the system exhibits a clear separation between the typical microscopic and macroscopic scales, λ and L, respectively, it may be possible to truncate the expansion on the right-hand sides of Eqs. (1.87), (1.88), and (1.89). The microscopic scale can, for example, be the mean free path for dilute gases or the inverse temperature for conformal fluids, while a macroscopic scale is given by typical time or length scales over which a certain primary fluid-dynamical variable varies, e.g., $L \sim [(\partial a)/a]^{-1}$, with $a = \alpha_0$, β_0, or u^μ. The terms $\mathcal{O}_1$, $\mathcal{O}_1^\mu$, and $\mathcal{O}_1^{\mu\nu}$ are linearly proportional to a gradient of a macroscopic variable, and thus are of order $\sim L^{-1}$. Every additional derivative brings in another inverse power of L and, thus, $\mathcal{O}_n$, $\mathcal{O}_n^\mu$, $\mathcal{O}_n^{\mu\nu} \sim L^{-n}$. The microscopic scale λ is contained in the coefficients, $\lambda_\Pi^{(n)}$, $\lambda_n^{(n)}$, and $\lambda_\pi^{(n)}$. Up to some overall power of λ (which restores the correct scaling dimension), $\lambda_\Pi^{(n)}, \lambda_n^{(n)}, \lambda_\pi^{(n)} \sim \lambda^n$. Therefore, the terms $(\mathcal{O}_1, \mathcal{O}_1^\mu, \mathcal{O}_1^{\mu\nu})$ and $(\mathcal{O}_2, \mathcal{O}_2^\mu, \mathcal{O}_2^{\mu\nu})$, multiplied by their corresponding coefficients in Eqs. (1.87), (1.88), and (1.89) are of order λ/L and $(\lambda/L)^2$, respectively. Subsequent terms would be of higher order in this ratio. This is nothing but a series in powers of the so-called Knudsen number $\mathrm{Kn} \equiv \lambda/L$. If $\mathrm{Kn} \ll 1$ and this series converges, the gradient expansion of the dissipative currents can be truncated at a given order and one obtains a closed macroscopic theory for them. Ideal fluid dynamics corresponds to the zeroth-order truncation of this series, i.e., when no gradient terms are considered. The first-order truncation of the gradient expansion is Navier–Stokes theory, as will be shown below. Higher-order truncations would lead to the relativistic Burnett and super-Burnett equations and so on.

We note that the convergence of the gradient expansion is not well established and is a topic that is still being intensely investigated [17]. We shall discuss the few examples in which this series can be calculated to arbitrarily high order in Chap. 3 and see that it actually diverges in such cases, i.e., it has a vanishing radius of convergence. The divergence of the series considerably modifies how we interpret truncations of the gradient expansion, which no longer have a clear domain of validity in terms of the magnitude of the Knudsen number. Instead, what remains to be done is to determine the optimal truncation of the gradient-expansion series and use that as an effective fluid-dynamical theory. In the following, for the sake of simplicity, we ignore these issues and consider the effective theories that arise as one truncates the gradient expansion.

The first term in the gradient expansion can be obtained by constructing all possible tensors that can be formed from the first-order derivatives of α_0, β_0, and u^μ. These can be easily obtained and are

$$\partial_\mu \alpha_0 , \ \partial_\mu \beta_0 , \text{ and } \partial_\mu u_\nu . \tag{1.90}$$

Next, using these gradients one has to construct tensors that have the same properties as the dissipative currents. There must be a scalar, such as the bulk-viscous pressure, a 4-vector orthogonal to u^μ, such as the particle-diffusion 4-current, and a symmetric, traceless second-rank tensor orthogonal to u^μ, such as the shear-stress tensor. The only possibilities are

$$\text{Scalar} : \theta = \nabla_\mu u^\mu , \tag{1.91}$$

$$\text{Vector} : I^\mu \equiv \nabla^\mu \alpha_0 , \quad J^\mu \equiv \nabla^\mu \beta_0 , \tag{1.92}$$

$$\text{Tensor} : \sigma^{\mu\nu} = \partial^{\langle\mu} u^{\nu\rangle} = \frac{1}{2}\left(\nabla^\mu u^\nu + \nabla^\nu u^\mu\right) - \frac{1}{3}\Delta^{\mu\nu}\theta . \tag{1.93}$$

Then, the most general first-order terms allowed by symmetry are

$$\mathcal{O}_1 = \theta , \tag{1.94}$$

$$\mathcal{O}_1^\mu = I^\mu + \gamma J^\mu , \tag{1.95}$$

$$\mathcal{O}_1^{\mu\nu} = \sigma^{\mu\nu} , \tag{1.96}$$

where the constant γ is introduced in order to restore the correct dimension.

In order to respect the second law of thermodynamics, discussed in the previous section, the transport coefficient γ must always be zero, $\gamma = 0$. This can be seen as follows. One can rewrite the Gibbs–Duhem relation (1.11) with the help of the Euler equation (1.9) in the form

$$d\beta_0 = h_0^{-1} d\alpha_0 - \frac{\beta_0}{\varepsilon + P_0} dP_0 , \tag{1.97}$$

where $h_0 \equiv (\varepsilon + P_0)/n$ is the specific enthalpy. Then, using the hydrodynamic equation (1.73),

$$J^\mu = h_0^{-1} I^\mu - \beta_0 \dot{u}^\mu + \mathcal{O}_2 , \tag{1.98}$$

where the second-order terms involve gradients of dissipative currents or products of dissipative currents with gradients of primary fluid-dynamical variables. Employing Eqs. (1.88) and (1.95), the first term in the entropy production Eq. (1.78) would be $\sim -\lambda_n^{(1)} \left[\left(1 + \gamma h_0^{-1} \right) I^\mu - \gamma \beta_0 \dot{u}^\mu \right\} I_\mu$, which is in general no longer positive definite, unless $\gamma = 0$.

Therefore, the most general relations satisfied by Π, n^μ, and $\pi^{\mu\nu}$, up to first order in Kn, are

$$\Pi = \lambda_\Pi^{(1)} \theta \ ,$$
$$n^\mu = \lambda_n^{(1)} I^\mu \ ,$$
$$\pi^{\mu\nu} = \lambda_\pi^{(1)} \sigma^{\mu\nu} \ ,$$

which corresponds to the relativistic Navier–Stokes theory [2], with the bulk-viscosity coefficient, the diffusion coefficient, and the shear-viscosity coefficient being identified as $\zeta \equiv -\lambda_\Pi^{(1)}$, $\varkappa \equiv \lambda_n^{(1)}$, and $\eta \equiv \lambda_\pi^{(1)}/2$, respectively.

In the framework of the gradient expansion, the relativistic Navier–Stokes theory can be extended by including terms of second order in gradients of α_0, β_0, and u^μ. In order to do so, one has to obtain all possible terms that can contribute to $\mathcal{O}_2$, $\mathcal{O}_2^\mu$, and $\mathcal{O}_2^{\mu\nu}$. These are

Scalar : $\omega_{\mu\nu}\omega^{\mu\nu}, \ \sigma_{\mu\nu}\sigma^{\mu\nu}, \ \theta^2, \ \ I_\mu I^\mu, \ J_\mu J^\mu, \ I_\mu J^\mu, \ \nabla_\mu I^\mu, \ \nabla_\mu J^\mu,$

Vector : $\sigma^{\mu\nu}I_\nu, \ \sigma^{\mu\nu}J_\nu, \ I^\mu\theta, \ J^\mu\theta, \ \omega^{\mu\nu}I_\nu, \ \omega^{\mu\nu}J_\nu, \ \Delta_\alpha^\mu \partial_\nu \sigma^{\alpha\nu}, \ \nabla^\mu\theta,$

Tensor : $\omega_\lambda^{\ \langle\mu}\omega^{\nu\rangle\lambda}, \ \theta\sigma^{\mu\nu}, \ \sigma^{\lambda\langle\mu}\sigma_\lambda^{\ \nu\rangle}, \ \sigma_\lambda^{\ \langle\mu}\omega^{\nu\rangle\lambda}, \ I^{\langle\mu}I^{\nu\rangle}, \ J^{\langle\mu}J^{\nu\rangle},$

$\qquad\qquad I^{\langle\mu}J^{\nu\rangle}, \ \nabla^{\langle\mu}I^{\nu\rangle}, \ \nabla^{\langle\mu}J^{\nu\rangle},$ $\qquad\qquad\qquad\qquad\qquad$ (1.99)

where we introduced the fluid vorticity tensor,

$$\omega^{\mu\nu} \equiv \frac{1}{2}\left(\nabla^\mu u^\nu - \nabla^\nu u^\mu \right) \ . \qquad\qquad (1.100)$$

Then, the most general second-order terms allowed by symmetry are

$$\lambda_\Pi^{(2)}\mathcal{O}_2 = \zeta_1 \omega_{\mu\nu}\omega^{\mu\nu} + \zeta_2 \sigma_{\mu\nu}\sigma^{\mu\nu} + \zeta_3 \theta^2 + \zeta_4 I_\mu I^\mu$$
$$+ \zeta_5 J_\mu J^\mu + \zeta_6 I_\mu J^\mu + \zeta_7 \nabla_\mu I^\mu + \zeta_8 \nabla_\mu J^\mu \ , \qquad (1.101)$$

$$\lambda_n^{(2)}\mathcal{O}_2^\mu = \varkappa_1 \sigma^{\mu\nu}I_\nu + \varkappa_2 \sigma^{\mu\nu}J_\nu + \varkappa_3 I^\mu\theta + \varkappa_4 J^\mu\theta$$
$$+ \varkappa_5 \omega^{\mu\nu}I_\nu + \varkappa_6 \omega^{\mu\nu}J_\nu + \varkappa_7 \Delta_\alpha^\mu \partial_\nu \sigma^{\alpha\nu} + \varkappa_8 \nabla^\mu\theta \ , \qquad (1.102)$$

$$\lambda_\pi^{(2)}\mathcal{O}_2^{\mu\nu} = \eta_1 \omega_\lambda^{\ \langle\mu}\omega^{\nu\rangle\lambda} + \eta_2 \theta\sigma^{\mu\nu} + \eta_3 \sigma^{\lambda\langle\mu}\sigma_\lambda^{\ \nu\rangle} + \eta_4 \sigma_\lambda^{\ \langle\mu}\omega^{\nu\rangle\lambda} + \eta_5 I^{\langle\mu}I^{\nu\rangle}$$
$$+ \eta_6 J^{\langle\mu}J^{\nu\rangle} + \eta_7 I^{\langle\mu}J^{\nu\rangle} + \eta_8 \nabla^{\langle\mu}I^{\nu\rangle} + \eta_9 \nabla^{\langle\mu}J^{\nu\rangle} \ , \qquad (1.103)$$

where we introduced additional transport coefficients, ζ_i, $\varkappa_i$, and η_i. By including the above terms in the expressions for the dissipative currents, Eqs. (1.87), (1.88), and (1.89), we obtain the relativistic Burnett equations [16],

$$\Pi = -\zeta\theta + \zeta_1\omega_{\mu\nu}\omega^{\mu\nu} + \zeta_2\sigma_{\mu\nu}\sigma^{\mu\nu} + \zeta_3\theta^2 + \zeta_4 I_\mu I^\mu$$
$$+ \zeta_5 J_\mu J^\mu + \zeta_6 I_\mu J^\mu + \zeta_7\nabla_\mu I^\mu + \zeta_8\nabla_\mu J^\mu , \tag{1.104}$$

$$n^\mu = \varkappa I^\mu + \varkappa_1\sigma^{\mu\nu}I_\nu + \varkappa_2\sigma^{\mu\nu}J_\nu + \varkappa_3 I^\mu\theta + \varkappa_4 J^\mu\theta$$
$$+ \varkappa_5\omega^{\mu\nu}I_\nu + \varkappa_6\omega^{\mu\nu}J_\nu + \varkappa_7\Delta_\alpha^\mu\partial_\nu\sigma^{\alpha\nu} + \varkappa_8\nabla^\mu\theta , \tag{1.105}$$

$$\pi^{\mu\nu} = 2\eta\sigma^{\mu\nu} + \eta_1\omega_\lambda^{\langle\mu}\omega^{\nu\rangle\lambda} + \eta_2\theta\sigma^{\mu\nu} + \eta_3\sigma^{\lambda\langle\mu}\sigma_\lambda^{\nu\rangle} + \eta_4\sigma_\lambda^{\langle\mu}\omega^{\nu\rangle\lambda}$$
$$+ \eta_5 I^{\langle\mu}I^{\nu\rangle} + \eta_6 J^{\langle\mu}J^{\nu\rangle} + \eta_7 I^{\langle\mu}J^{\nu\rangle} + \eta_8\nabla^{\langle\mu}I^{\nu\rangle} + \eta_9\nabla^{\langle\mu}J^{\nu\rangle} .$$

$$\tag{1.106}$$

In this section, we showed how to extend and derive Navier–Stokes theory via the gradient expansion. Note, however, that these extensions remain acausal and unstable and have no practical purpose. As a matter of fact, the Burnett equations are unstable even in the non-relativistic regime [18].

1.4 Causal Fluid Dynamics

Many theories have been developed to incorporate dissipative effects in fluid dynamics preserving causality: Grad–Israel–Stewart theory [5–7, 19], divergence-type theories [20–22], extended irreversible thermodynamics [23–27], Carter's theory [28], and Öttinger–Grmela theory [29], among others [30, 31]. In Sect. 1.5, we shall briefly review Israel and Stewart's approach.

However, before explaining Israel–Stewart theory, it is useful to discuss a more ad hoc approach to render Navier–Stokes theory causal and stable [32]. For the sake of simplicity, we first illustrate this method using the simple example of a transport equation that shares the same problems as Navier–Stokes theory: the heat-conduction equation.

1.4.1 Diffusion Equation and Acausality in Heat Conduction

The fundamental problem of the diffusion equation comes from the fact that it is a parabolic equation, allowing for signals that can propagate with infinite speed [25–27, 33–35]. In the non-relativistic case, this problem was first addressed by Cattaneo [36, 37] and applied to heat conduction. Cattaneo argued that the problem of acausal propagation in the diffusion equation

$$\partial_t A = D\nabla^2 A \tag{1.107}$$

can be corrected by introducing a term with a second-order time derivative, i.e.,

$$\tau_R \partial_t^2 A + \partial_t A = D\nabla^2 A \;,\tag{1.108}$$

thereby converting a parabolic equation to a hyperbolic one. Above, we introduced the diffusion and the relaxation-time coefficients, D and τ_R, respectively. Equation (1.108) is often referred to as telegraph equation and, for suitable choices of D and τ_R, can lead to causal signal propagation. As a matter of fact, the maximum propagation speed of signals in this theory can be proven to be [38],

$$v_{\max} = \sqrt{D/\tau_R} \;.\tag{1.109}$$

Therefore, as long as $D/\tau_R \le 1$, the telegraph equation is causal. On the other hand, in the limit $\tau_R \to 0$, in which the diffusion equation is recovered, the propagation speed diverges and the theory becomes acausal.

Next, we consider the heat-conduction problem as an example to understand the physical origin of the telegraph equation. The diffusion equation used to describe heat conduction is constructed from two basic features. One is the energy-balance equation,

$$\rho c \partial_t T + \nabla \cdot \mathbf{J} = 0 \;,\tag{1.110}$$

where T is the temperature, $\mathbf{J}$ is the heat flux, ρ is the mass density, and c is the specific heat capacity of the material in question. The other ingredient is Fourier's law,

$$\mathbf{J} = -\varpi \nabla T \;,\tag{1.111}$$

where ϖ is the heat conductivity. Then, Eqs. (1.110) and (1.111) can be combined to eliminate $\mathbf{J}$ from the differential equation (1.110) and describe heat conduction via the following diffusion equation,

$$\partial_t T = \frac{\varpi}{\rho c} \nabla^2 T \;,\tag{1.112}$$

where, for the sake of simplicity, we assumed that ρ and ϖ are spatially constant in the material.

The source of acausality cannot be the energy-balance equation, which is the direct consequence of a conservation law. Then, it must be Fourier's law. Heat conduction should always be induced by an inhomogeneous temperature profile. However, Fourier's law is only an approximation of this process. It does not contain any inertial effects and, consequently, certain perturbations in the temperature distribution are felt instantaneously at every point of the system, i.e., according to Eq. (1.112) every point in the material heats at the same time, no matter how distant from the heat source.

Therefore, even though very successful in non-relativistic applications, Fourier's law cannot be employed in the relativistic limit: in relativistic theories, causality is a physical principle that is naturally preserved by the Lorentz transformations and dictates that the propagation of any current must happen in a nonzero interval of time. This feature can be included in the description of heat conduction via linear response theory, which prescribes a more general expression for $\mathbf{J}$ [39],

$$\mathbf{J}(t) = -\int^{t} ds\, G(t-s)\nabla T(s) , \qquad (1.113)$$

where $G(t)$ is the corresponding retarded Green function. The Green function includes the microscopic time scales that describe the creation of the heat flux from temperature gradients. In non-relativistic systems in which the separation between the microscopic and macroscopic time scales of the system are sufficiently large, the time dependence of the Green function can be approximated by a Dirac delta function, $G(t) \sim \delta(t)$, and we recover Fourier's law. However, when the microscopic and macroscopic time scales are not separated by orders of magnitude, the transient dynamics of the heat flux must be explicitly described and the time dependence of $G(t)$ must be taken into account. In relativistic systems, such microscopic time scales must always be considered in order to preserve causality. When deriving fluid dynamics in the next chapters, we shall address this topic within a more formal framework.

The simplest choice for $G(t)$ is the exponential *ansatz* (this choice can actually be derived in the framework of kinetic theory) [40],

$$G(t) = \frac{\varpi}{\tau_R} e^{-t/\tau_R} , \qquad (1.114)$$

where τ_R is the heat-flux relaxation time. Then, by taking the time derivative of Eq. (1.113) and substituting this *ansatz*, we obtain the following equation of motion for $\mathbf{J}$:

$$\tau_R \partial_t \mathbf{J}(t) + \mathbf{J}(t) = -\varpi \nabla T(t) . \qquad (1.115)$$

This equation is often referred to as Maxwell–Cattaneo equation. In this theory, the heat flux is not created instantaneously from temperature inhomogeneities. For example, when $\nabla T = 0$,

$$\tau_R \partial_t \mathbf{J}(t) + \mathbf{J}(t) = 0 , \qquad (1.116)$$

the heat flux does not vanish instantaneously, as it happened in Fourier's law, but relaxes exponentially to zero on times scales given by the heat-flux relaxation time τ_R.

By using the energy-balance equation (1.110), we can eliminate $\mathbf{J}$ from the divergence of the Maxwell–Cattaneo equation (1.115), and obtain the telegraph equation for the temperature T,

$$\tau_R \partial_t^2 T + \partial_t T = \frac{\varpi}{\rho c} \nabla^2 T . \qquad (1.117)$$

Note that, in this case, the diffusion equation becomes the asymptotic limit of the telegraph equation, attained only for times much longer than the heat-flux relaxation time (as long as ∇T varies slowly in time). As mentioned before, this equation is causal as long as

$$\sqrt{\frac{\varpi}{\rho c \tau_R}} \leqslant 1 . \tag{1.118}$$

1.4.2 Transient Theory of Fluid Dynamics

The same idea as explained in the previous section can be applied to render Navier–Stokes theory causal. In Sect. 1.3.4, Navier–Stokes theory was constructed from the conservation laws of (net) particle number, energy, and momentum, and the constitutive relations satisfied by the dissipative currents, Eqs. (1.80), (1.81), and (1.82). The conservation laws come from general physical principles, valid even outside the fluid-dynamical regime, and cannot be modified. Therefore, in order to improve on Navier–Stokes theory, we must extend Eqs. (1.80), (1.81), and (1.82). We have already argued that such constitutive relations are the source of causality violations in Navier–Stokes theory since, as happens with Fourier's law, they allow for the instantaneous creation of dissipative currents from gradients of α_0, β_0, and u^μ. In this section, we show how to correct this unphysical behavior using the same arguments as previously applied to heat conduction: the inclusion of a time delay in the creation of the dissipative currents from gradients of the primary fluid-dynamical variables.

In the heat-conduction case, inertial effects on the creation of heat flux from temperature inhomogeneities were included by introducing a term with a first-order time derivative of $\mathbf{J}$ in Fourier's law, giving rise to a causal transport equation for heat conduction, Eq. (1.116). Similarly, relaxation effects can be introduced in Navier–Stokes theory by adding a term of first order in the comoving derivative of each dissipative current to the constitutive relation satisfied by this current, i.e., $\dot{\Pi}$, $\Delta^\mu_\alpha \dot{n}^\alpha$, and $\Delta^{\mu\nu}_{\alpha\beta} \dot{\pi}^{\alpha\beta}$. Then, instead of Navier–Stokes theory, we obtain the following transport equations for Π, n^μ, and $\pi^{\mu\nu}$,

$$\tau_\Pi \dot{\Pi} + \Pi = -\zeta \theta + \cdots , \tag{1.119}$$

$$\tau_n \Delta^\mu_\alpha \dot{n}^\alpha + n^\mu = \varkappa \nabla^\mu \alpha_0 + \cdots , \tag{1.120}$$

$$\tau_\pi \Delta^{\mu\nu}_{\alpha\beta} \dot{\pi}^{\alpha\beta} + \pi^{\mu\nu} = 2\eta \, \sigma^{\mu\nu} + \cdots , \tag{1.121}$$

where we introduced the bulk relaxation time, τ_Π, the diffusion relaxation time, τ_n, and the shear relaxation time, τ_π. The dots denote possible nonlinear terms involving the fluid-dynamical quantities and their gradients.

In this formulation, the dissipative fluid-dynamical currents appear as independent dynamical variables, which relax to the values of Navier–Stokes theory on characteristic time scales given by the relaxation times τ_Π, τ_n, and τ_π. Thus, unlike for the gradient expansion, the dissipative currents in this theory do not have to be zero in the absence of gradients. Instead, they decay to zero on the time scales given by the relaxation times. This type of formalism was referred to as transient theory of fluid dynamics by Israel and Stewart [5], since it describes this relaxation (or transient) process of each dissipative current toward its respective (asymptotic) Navier–Stokes value.

One of the features of transient fluid dynamics is that it reduces to Navier–Stokes theory in the limit of vanishing relaxation times. In other words, in Navier–Stokes theory the dissipative currents relax instantaneously to their Navier–Stokes values (also referred to as dissipative forces), which leads to a violation of causality. In many non-relativistic fluids, the relaxation times are very short, and such transient dynamics can be neglected. In this case, the dissipative currents can actually be well approximated by their Navier–Stokes solution. Nevertheless, in the relativistic case, this approximation is not possible since it will render the equations of motion parabolic and unstable.

For fluid dynamics to be causal, it is therefore *necessary* that the relaxation times assume a nonzero value, but this is *not sufficient*. As will be shown in Chap. 2 (see also Refs. [13,14]), causality imposes a *stronger* constraint for transient theories: the ratio of the relaxation times to their respective viscosity coefficients must *exceed certain values*. It will be also shown that, for relativistic fluids, causality implies stability of the fluid-dynamical equations.

Equations (1.119)–(1.121) correspond to the type of relativistic fluid-dynamical theory that we expect to derive from microscopic theory, including all nonlinear and higher-order terms that might appear on the right-hand side. This will be shown explicitly in Chaps. 5 and 6.

1.5 Transient Thermodynamics and Israel–Stewart Theory

It is also possible to derive causal fluid-dynamical equations, with the same structure as Eqs. (1.119)–(1.121), from the second law of thermodynamics. In this section, we review this derivation as first proposed by Israel and Stewart [5,7]. The main idea of their approach is to apply the second law of thermodynamics to a more general expression of the non-equilibrium entropy 4-current. In equilibrium, the entropy 4-current was expressed exactly in terms of the primary fluid-dynamical variables, α_0, β_0, and u^μ. When the fluid deviates from equilibrium, the situation becomes more complicated. Strictly speaking, the entropy 4-current should depend on a very large number of independent dynamical variables (for a dilute gas, these correspond to all the moments of the Boltzmann equation) that are needed in order to characterize the complicated state of a non-equilibrium system. However, it is reasonable to assume that, as the system approaches equilibrium, the number of dynamical variables needed to describe the state of the fluid gradually decreases, until it reaches the variables required by the equilibrium state, α_0, β_0, and u^μ.

In the previous section, we showed that, in order to render the fluid-dynamical equations causal, the dissipative currents must be promoted to independent dynamical variables. Therefore, we expect that a more realistic description of the entropy 4-current can be obtained by considering it to be a function not only of the primary fluid-dynamical variables but also of the dissipative currents,

$$S^\mu = S^\mu(\alpha_0, \beta_0, u^\mu, \Pi, n^\mu, \pi^{\mu\nu}) . \tag{1.122}$$

Mathematically, it is further assumed that the entropy 4-current has the following properties: (i) it is additive; (ii) it is a convex function of the equilibrium variables *and* the dissipative currents; and (iii) the corresponding entropy production is locally positive. We remark that, while here these properties enter as a hypothesis, they can be rigorously derived in the framework of kinetic theory [5–7].

Then, the entropy 4-current can be expanded in terms of powers of the dissipative currents around a (fictitious) equilibrium state [5,7],

$$S^\mu = S^\mu_{(0)} - \alpha_0 n^\mu + Q^\mu + \mathcal{O}_3 , \tag{1.123}$$

where $\mathcal{O}_3$ denotes terms of third order or higher in the dissipative currents and

$$Q^\mu \equiv -\frac{1}{2} u^\mu \left(\delta_0 \Pi^2 - \delta_1 n_\alpha n^\alpha + \delta_2 \pi_{\alpha\beta} \pi^{\alpha\beta} \right) - \gamma_0 \Pi n^\mu - \gamma_1 \pi^\mu_\nu n^\nu \tag{1.124}$$

is of second order, $Q^\mu \sim \mathcal{O}_2$. The expansion coefficients, δ_0, δ_1, δ_2, γ_0, and γ_1, are complicated functions of the temperature and chemical potential of the (fictitious) equilibrium state and can only be obtained by matching this expansion with the underlying microscopic theory. Note that the entropy 4-current used to derive relativistic Navier–Stokes theory is recovered by taking $Q^\mu = 0$. It is important to remember that Q^μ is not orthogonal to the fluid 4-velocity and, consequently,

$$s \equiv u_\mu S^\mu = s_{(0)} + u_\mu Q^\mu \neq s_{(0)} . \tag{1.125}$$

That is, the non-equilibrium entropy density in the local rest frame, s, does not correspond to the entropy density computed using the (fictitious) equilibrium state, $s_0 (n, \varepsilon)$.

The existence of second-order contributions to the entropy 4-current will affect all previous conclusions drawn from the second law of thermodynamics, which can then be understood to be valid only up to *first order* in the dissipative currents (hence the name first-order theory). Next, we re-calculate the entropy production using the more general entropy 4-current introduced in Eq. (1.123),

$$\partial_\mu S^\mu = \beta_0 u_\nu \partial_\mu T^{\mu\nu}_{(0)} - \alpha_0 \partial_\mu N^\mu_{(0)} - \partial_\mu \left(\alpha_0 n^\mu \right) + \partial_\mu Q^\mu , \tag{1.126}$$

where we employed Eq. (1.42). The conservation laws (1.72) and (1.74) lead to the following result:

$$\partial_\mu S^\mu = -\beta_0 \Pi \theta + \beta_0 \pi^{\mu\nu} \sigma_{\mu\nu} - n^\mu \nabla_\mu \alpha_0 + \partial_\mu Q^\mu . \tag{1.127}$$

Using Eq. (1.124), we can derive all terms originating from $\partial_\mu Q^\mu$,

$$
\begin{aligned}
\partial_\mu Q^\mu ={}& -\delta_0 \Pi \dot{\Pi} + \delta_1 n_\mu \dot{n}^\mu - \delta_2 \pi_{\mu\nu} \dot{\pi}^{\mu\nu} \\
& -\frac{1}{2}\left(\Pi^2 \dot{\delta}_0 - n_\mu n^\mu \dot{\delta}_1 + \pi_{\mu\nu}\pi^{\mu\nu}\dot{\delta}_2\right) \\
& -\frac{1}{2}\left(\delta_0 \Pi^2 - \delta_1 n_\mu n^\mu + \delta_2 \pi_{\mu\nu}\pi^{\mu\nu}\right)\theta \\
& -\gamma_0 \Pi \partial_\mu n^\mu - \gamma_0 n^\mu \nabla_\mu \Pi - \Pi n^\mu \nabla_\mu \gamma_0 \\
& -\gamma_1 \pi_{\mu\nu} \nabla^{\langle\mu} n^{\nu\rangle} - \pi_{\mu\nu} n^{\langle\mu} \nabla^{\nu\rangle}\gamma_1 - \gamma_1 n_\nu \partial_\mu \pi^{\mu\nu} .
\end{aligned}
\tag{1.128}
$$

Then, substituting Eq. (1.128) into Eq. (1.127), we obtain the more general entropy-production equation

$$
\begin{aligned}
\partial_\mu S^\mu ={}& \beta_0 \Pi \left(-\theta - \frac{\delta_0}{\beta_0}\dot{\Pi} - \frac{1}{2\beta_0}\Pi\dot{\delta}_0 \right.\\
& \left. -\frac{1}{2\beta_0}\delta_0 \Pi\theta - \frac{\gamma_0}{\beta_0}\partial_\mu n^\mu - \frac{1-r}{\beta_0}n^\mu \nabla_\mu \gamma_0 \right) \\
& + n_\mu \left(-\nabla^\mu \alpha_0 + \delta_1 \Delta^\mu_\alpha \dot{n}^\alpha + \frac{n^\mu}{2}\dot{\delta}_1 + \frac{\delta_1}{2}n^\mu\theta \right.\\
& \left. -\gamma_0 \nabla^\mu \Pi - r\Pi\nabla^\mu \gamma_0 - \gamma_1 \partial_\nu \pi^{\mu\nu} - y\pi^{\mu\nu}\nabla_\nu \gamma_1 \right) \\
& + \beta_0 \pi_{\mu\nu}\left(\sigma^{\mu\nu} - \frac{\delta_2}{\beta_0}\Delta^{\mu\nu}_{\alpha\beta}\dot{\pi}^{\alpha\beta} - \frac{1}{2\beta_0}\pi^{\mu\nu}\dot{\delta}_2 \right.\\
& \left. -\frac{1}{2\beta_0}\delta_2 \pi^{\mu\nu}\theta - \frac{\gamma_1}{\beta_0}\nabla^{\langle\mu} n^{\nu\rangle} - \frac{1-y}{\beta_0}n^{\langle\mu} \nabla^{\nu\rangle}\gamma_1 \right) ,
\end{aligned}
\tag{1.129}
$$

where r, y are arbitrary constants.

As argued before, the only way to explicitly satisfy the second law of thermodynamics is to assure that the entropy production is a positive semi-definite quadratic function of the dissipative currents, i.e.,

$$
\partial_\mu S^\mu \equiv \beta_0 \varpi_\Pi \Pi^2 - \varpi_n n_\mu n^\mu + \beta_0 \varpi_\pi \pi_{\mu\nu}\pi^{\mu\nu} ,
\tag{1.130}
$$

where ϖ_Π, ϖ_n, $\varpi_\pi \geq 0$. This further implies that the dissipative currents must satisfy the following *dynamical* equations

$$
\begin{aligned}
\frac{\delta_0}{\beta_0}\dot{\Pi} + \varpi_\Pi \Pi ={}& -\theta - \frac{1}{2\beta_0}\Pi\dot{\delta}_0 - \frac{1}{2\beta_0}\delta_0 \Pi\theta \\
& -\frac{\gamma_0}{\beta_0}\partial_\mu n^\mu - \frac{1-r}{\beta_0}n^\mu \nabla_\mu \gamma_0 ,
\end{aligned}
$$

$$
\begin{aligned}
\delta_1 \Delta^\mu_\alpha \dot{n}^\alpha + \varpi_n n^\mu ={}& \nabla^\mu \alpha_0 - \frac{1}{2}n^\mu \dot{\delta}_1 - \frac{\delta_1}{2}n^\mu\theta + \gamma_0 \nabla^\mu \Pi \\
& + r\Pi\nabla^\mu \gamma_0 + \gamma_1 \partial_\nu \pi^{\mu\nu} + y\pi^{\mu\nu}\nabla_\nu \gamma_1 ,
\end{aligned}
$$

$$\frac{\delta_2}{\beta_0}\Delta^{\mu\nu}_{\alpha\beta}\dot{\pi}^{\alpha\beta} + \varpi_\pi\pi^{\mu\nu} = \sigma^{\mu\nu} - \frac{1}{2\beta_0}\pi^{\mu\nu}\dot{\delta}_2 - \frac{1}{2\beta_0}\delta_2\pi^{\mu\nu}\theta$$

$$- \frac{\gamma_1}{\beta_0}\nabla^{\langle\mu}n^{\nu\rangle} - \frac{1-y}{\beta_0}n^{\langle\mu}\nabla^{\nu\rangle}\gamma_1 , \qquad (1.131)$$

which are relaxation-type equations, similar to those conjectured in the last section, i.e., Eqs. (1.119)–(1.121). By comparison with those equations, we find ϖ_Π, ϖ_n, and ϖ_π to be related to the viscosity and diffusion coefficients,

$$\varpi_\Pi = \frac{1}{\zeta} , \quad \varpi_n = \frac{1}{\varkappa} , \quad \varpi_\pi = \frac{1}{2\eta} , \qquad (1.132)$$

and identify the relaxation times as

$$\tau_\Pi = \zeta\frac{\delta_0}{\beta_0} , \quad \tau_n = \varkappa\delta_1 , \quad \tau_\pi = 2\eta\frac{\delta_2}{\beta_0} . \qquad (1.133)$$

Since the relaxation times must be positive, the expansion coefficients δ_0, δ_1, and δ_2 must all be larger than zero. Furthermore, we found some of the nonlinear terms that may appear as source terms in the transient equations of motion for the dissipative currents. We shall see in the next chapters, when we derive the equations of fluid dynamics from microscopic theory, that there are still other nonlinear terms that are missing in this type of derivation. A derivation from microscopic theory also allows to uniquely fix the as-of-yet arbitrary constants r, y.

1.6   Non-hydrodynamic Modes and the Origin of the Relaxation Time

One of the main features of Navier–Stokes theory is that all of its modes, $\omega_{NS}(\mathbf{k})$, vanish in the limit of zero wavenumber, i.e.,

$$\lim_{\mathbf{k}\to 0} \omega_{NS}(\mathbf{k}) = 0 .$$

Such modes are called *hydrodynamic modes* [39] and their existence is quite often taken as evidence for fluid-dynamical behavior. Furthermore, modes that do not share this behavior, i.e., modes for which $\lim_{\mathbf{k}\to 0} \omega(\mathbf{k}) \neq 0$, are known as *non-hydrodynamic modes*.

Surprisingly enough, a stable theory of relativistic fluid dynamics cannot be formulated using only hydrodynamic modes. As mentioned above, Navier–Stokes theory (and its extensions via the gradient expansion) is acausal and unstable and, therefore, such theories cannot be directly applied to describe any relativistic fluid existing in Nature. Israel and Stewart formulated a relativistic theory of fluid dynamics that was causal and, consequently, stable [5–7]. The main new ingredient in their theory were the relaxation times, which correspond to the time scales over which the dissipative currents react to gradients of α_0, β_0, and u^μ. The introduction of such relaxation processes transforms the dissipative currents into independent dynamical variables that satisfy partial differential equations instead of constitutive relations.

The dissipative currents are not conserved quantities and, therefore, when they become independent dynamical variables, non-hydrodynamic modes must appear in the theory. In the case of Israel–Stewart theory, the non-hydrodynamic modes describe the relaxation of the dissipative currents toward their respective asymptotic Navier–Stokes solutions and they can be directly related to the relaxation times [41], as we will explicitly show in Chap. 4.

The physical interpretation of the non-hydrodynamic modes, however, is still a subject of debate. Even though the inclusion of non-hydrodynamic modes is essential to restore causality, this has often been interpreted as a regularization procedure to make Navier–Stokes theory stable [30]. From this perspective, only Navier–Stokes theory and its extensions via the gradient expansion are physically relevant and the relaxation times are introduced as a type of regulator to render high frequencies and wavenumbers of fluid dynamics well behaved. This procedure is implemented in such a way that the regularized theory is linearly equivalent to the gradient expansion at low frequencies and wavenumbers. Naturally, different choices of relaxation times describe different transient dynamics while still producing the same asymptotic solution, i.e., the relativistic Burnett theory [42]. Since, in this case, the transient dynamics described by the non-hydrodynamic modes is assumed to involve time scales much shorter than the macroscopic scales of interest, such arbitrariness on the choice of the relaxation time is argued not to be a problem.

In Chap. 4, we show how transient relativistic fluid dynamics can be derived from quantum field theory, in the linear regime, and give a more detailed discussion about the physical origin of the relaxation times. Specifically, we show how causal theories of fluid dynamics can emerge from the underlying microscopic theory without the need for any regularization procedure. In Chap. 5, we shall derive transient relativistic fluid dynamics from kinetic theory. In this case, we derive the theory not only in the linear regime but including all possible nonlinear and higher-order terms.

References

1. Reichl, L.: A Modern Course in Statistical Physics. Wiley-VCH (2004)
2. Landau, L.D., Lifshitz, E.M.: Fluid Mechanics. Pergamon, New York (1959)
3. Weinberg, S.: Gravitation and Cosmology. Wiley, New York (1972)
4. Eckart, C.: Phys. Rev. **58**, 919 (1940)
5. Israel, W.: Ann. Phys. (N.Y.) **100**, 310 (1976)
6. Stewart, J.M.: Proc. Roy. Soc. A **357**, 59 (1977)
7. Israel, W., Stewart, J.M.: Ann. Phys. (N.Y.) **118**, 341 (1979)
8. Hiscock, W., Lindblom, L.: Ann. Phys. (N.Y.) **151**, 466 (1983)
9. Hiscock, W., Lindblom, L.: Phys. Rev. D **31**, 725 (1985)
10. Hiscock, W., Lindblom, L.: Phys. Rev. D **35**, 3723 (1987)
11. Hiscock, W., Lindblom, L.: Phys. Lett. A **131**, 509 (1988)
12. Hiscock, W., Olson, T.: Phys. Lett. A **141**, 125 (1989)
13. Denicol, G.S., Kodama, T., Koide, T., Mota, Ph.: J. Phys. G **35**, 115102 (2008)

14. Pu, S., Koide, T., Rischke, D.H.: Phys. Rev. D **81**, 114039 (2010)
15. Chapman, S., Cowling, T.G.: The Mathematical Theory of Non-Uniform Gases, 3rd edn. Cambridge University Press, New York (1974)
16. Burnett, D.: Proc. Lond. Math. Soc. **39**, 385 (1935); Proc. Lond. Math. Soc. **40**, 382 (1936)
17. Florkowski, W., Heller M.P., Spalinski, M.: Rept. Prog. Phys. **81**(4), 046001 (2018). https://doi.org/10.1088/1361-6633/aaa091[arXiv:1707.02282 [hep-ph]]
18. Bobylev, A.V.: Sov. Phys. Dokl. **27**, 29 (1982)
19. Grad, H.: Commun. Pure Appl. Math. **2**, 331 (1949)
20. Müller, I.: Living Rev. Rel. **2**, 1 (1999)
21. Müller, I.: Z. Phys. **198**, 329 (1967)
22. Liu, I.S., Müller, I., Ruggeri, T.: Ann. Phys. (N.Y.) **169**, 191 (1986)
23. Pavón, D., Jou, D., Casas-Vázquez, J.: J. Phys. A. **13**, L77 (1980)
24. Pavón, D., Jou, D., Casas-Vázquez, J.: Ann. Inst. H. Poincaré A **36**, 79 (1982)
25. Jou, D., Casas-Vázquez, J., Lebon, G.: Rep. Prog. Phys. **51**, 1105 (1988)
26. Jou, D., Casas-Vázquez, J., Lebon, G.: Rep. Prog. Phys. **62**, 1035 (1999)
27. Jou, D., Casas-Vázquez, J., Lebon, G.: Extended Irreversible Thermodynamics, 2nd edn. Springer, Berlin (1996)
28. Carter, B.: Proc. R. Soc. Lond. Ser A, **433**, 45 (1991)
29. Grmela, M., Öttinger, H.C.: Phys. Rev. E **56**, 6620 (1997)
30. Baier, R., Romatschke, P., Son, D.T., Starinets, A.O., Stephanov, M.A.: JHEP **0804**, 100 (2008)
31. Bemfica, F.S., Disconzi M.M., Noronha, J.: Phys. Rev. D **98**(10), 104064 (2018). https://doi.org/10.1103/PhysRevD.98.104064[arXiv:1708.06255 [gr-qc]]
32. Koide, T., Denicol, G., Mota, Ph., Kodama, T.: Phys. Rev. C **75**, 034909 (2007)
33. Joseph, D., Preziosi, L.: Rev. Mod. Phys. **61**, 41 (1989)
34. Joseph, D., Preziosi, L.: Rev. Mod. Phys. **62**, 375 (1990)
35. Maartens, R.: In: Maharaj, S.D. (ed.) Proceedings of the Hanno Rund Conference on Relativity and Thermodynamics (1996). ArXiv:astro-ph/9609119
36. Cattaneo, C.: Atti. Semin. Mat. Fis. Univ. Modena **3**, 3 (1948)
37. Vernotte, P.: Compt. R. Acad. Sci. Paris **246**, 3154 (1958)
38. Morse, P.M., Feshbach, H.: Methods of Theoretical Physics. McGraw-Hill. Science (1953)
39. Forster, D.: Hydrodynamic Fluctuations, Broken Symmetry, and Correlation Functions (Advanced Book Classics). Westview Press (1995)
40. Such a memory function was also used in the hydrodynamic equations to take into account the time delay necessary for micro-turbulences to achieve thermalization in large-scale systems, such as the one found in a supernova explosion. See, T., Kodama, R. Donangelo., Guidry, M.: Int. J. Mod. Phys. C **9**, 745 (1998) for details
41. Denicol, G.S., Noronha, J., Niemi, H., Rischke, D.H.: Phys. Rev. D **83**, 074019 (2011)
42. Denicol, G.S., Noronha, J., Niemi, H., Rischke, D.H.: J. Phys. G **38**, 124177 (2011). [arXiv:1108.6230 [nucl-th]]

Linear Stability and Causality

It was emphasized in the previous chapter that relativistic generalizations of Navier–Stokes theory, derived by Landau and Lifshitz [1] and, independently, by Eckart [2], are ill-defined, containing unphysical instabilities when perturbed around an arbitrary global-equilibrium state [3–5]. Such instabilities are intrinsically related to the acausal nature of Navier–Stokes theory, which allows for perturbations that propagate with an infinite speed. These fundamental problems prohibit the application of Navier–Stokes theory to describe any practical fluid-dynamical problem, be it in the description of neutron-star mergers or in the description of the quark–gluon plasma produced in heavy-ion collisions.

In order to address these fundamental issues, Israel and Stewart constructed stable and causal theories of relativistic fluid dynamics, following the procedure initially developed by Grad [6] for non-relativistic systems. Israel and Stewart performed this task in two distinct ways [7]. The first is a phenomenological derivation, based on the second law of thermodynamics, which was discussed in detail in Chap. 1. The second is a microscopic derivation starting from the relativistic Boltzmann equation, which will be discussed thoroughly in Chap. 5. Similar theories have been widely developed in the past decades [8–16], but all carry the same fundamental aspect: in contrast to Navier–Stokes theory, such causal theories of fluid dynamics include in their description the transient dynamics of the non-conserved dissipative currents. For this reason, they were initially named by Israel and Stewart as transient fluid dynamics (nowadays, they are often referred to as *second-order theories*).

However, it is important to remark that the theory formulated by Israel and Stewart is not guaranteed to be causal and stable. As was first shown by Hiscock, Lindblom and, later, by Olson, such transient theories of fluid dynamics are only causal and stable if their transport coefficients satisfy certain conditions [17–19]. Such conclusions were obtained by analyzing the properties of the theory in the linear regime and by imposing that the perturbations around a global-equilibrium state are stable and propagate subluminally. Such stability analyses were more recently revisited in Ref. [4], including only the effects of bulk viscosity, and in Ref. [5], which included the effects of both shear and bulk viscosity. In both these papers, constraints for the shear and bulk relaxation times were explicitly derived. Nowadays, the causality of relativistic fluid-dynamical theories has been investigated even in the nonlinear regime [20, 21] (including the effects of shear and bulk viscosity), where more general inequalities required to ensure causality were derived. In the latter case, the inequalities constrain not only the transport coefficients but also the values of the dissipative currents (in the linear regime, the inequalities derived in Ref. [21] reduce to those derived in Refs. [4, 5]). Such constraints are relevant for, e.g., fluid-dynamical applications to heavy-ion collisions, since the transport coefficients of QCD matter are not precisely known (often, they are completely unknown), and such fundamental constraints on transport coefficients (and the values of the dissipative currents) can be extremely useful.

In this chapter, we perform a linear stability analysis of relativistic Navier–Stokes theory and of Israel–Stewart theory around a general global-equilibrium state. We then demonstrate explicitly the results described above. First, we prove that the general global-equilibrium state in relativistic Navier–Stokes theory is unstable, due to the appearance of non-hydrodynamic modes that grow exponentially on microscopic time scales. We then demonstrate that the same does not necessarily occur in Israel–Stewart theory, as long as the transport coefficients of the linearized theory satisfy certain constraints.

This chapter is organized as follows. In Sect. 2.1, we first write the fluid-dynamical equations by approximating them to linear order in perturbations around a global-equilibrium background. In Sect. 2.2, we transform these equations into Fourier space, with the purpose of determining the dispersion relations satisfied by the perturbations. Section 2.3 discusses these dispersion relations for ideal fluid dynamics, where one just has stable sound modes. In Sect. 2.4, we investigate the causality and stability of the linearized fluid-dynamical equations in the Navier–Stokes limit. Although these equations appear to be stable in a static background, they will become unstable when the perturbations are performed on a moving background. Section 2.5 then discusses stability and causality of transient theories of relativistic fluid dynamics. Both in a static as well as in a moving background these theories are causal and stable, provided that certain asymptotic causality conditions are fulfilled. A summary of our results concludes this chapter in Sect. 2.6.

2.1   Fluid-Dynamical Equations Linearized Around Global Equilibrium

In this section, we linearize the fluid-dynamical equations described in the previous chapter around a global-equilibrium state. In their linearized form, the equations simplify considerably and *some* of their properties can be studied systematically. In particular, our goal is to discuss the stability of relativistic fluids and to verify under which circumstances acausal modes appear in the *linearized* theory. Here, we consider small fluid-dynamical perturbations around a global-equilibrium state, with inverse temperature $\beta_0 \equiv 1/T_0$, thermal potential $\alpha_0 = \beta_0 \mu_0$, where μ_0 is the chemical potential, and a velocity u_0^μ, satisfying the normalization condition, $u_{0\mu} u_0^\mu = 1$. The pressure is obtained using an arbitrary equation of state, $P_0 \equiv P(\beta_0, \alpha_0)$. Naturally, in this state the dissipative currents appearing in the net-charge 4-current, N^μ, and the energy-momentum tensor, $T^{\mu\nu}$, all vanish: $\Pi_0 = \pi_0^{\mu\nu} = n_0^\mu = q_0^\mu = 0$.

In Chap. 1, we derived the equations of motion of a relativistic fluid and discussed two possible definitions of the fluid 4-velocity: one proposed by Landau [1] and another by Eckart [2]. We note that the prescription chosen for the velocity field can affect some aspects of the linear stability analysis [19]. Here, we perform our analysis using Landau's prescription for the velocity field, i.e., $T^{\mu\nu} u_\nu \equiv \varepsilon u^\mu$. In this case, the continuity equations related to energy, momentum, and net-charge conservation are given by

$$D\varepsilon + (\varepsilon + P + \Pi)\theta - \pi^{\alpha\beta}\sigma_{\alpha\beta} = 0 , \qquad (2.1)$$

$$(\varepsilon + P + \Pi) Du^\mu - \nabla^\mu (P + \Pi) - \pi^{\mu\beta} Du_\beta + \Delta_\alpha^\mu \nabla_\beta \pi^{\alpha\beta} = 0 , \qquad (2.2)$$

$$Dn + n\theta - n^\mu Du_\mu + \nabla_\mu n^\mu = 0 . \qquad (2.3)$$

We use the same notation as in the previous chapter, that is, ε is the energy density, n is the net-charge density, $D \equiv u^\mu \partial_\mu$ is the comoving derivative, $\theta \equiv \partial_\mu u^\mu$ is the expansion scalar, $\sigma^{\mu\nu} \equiv \partial^{\langle \mu} u^{\nu \rangle}$ is the shear tensor, $\Delta_\nu^\mu = g_\nu^\mu - u^\mu u_\nu$ is the projection operator onto the 3-space orthogonal to u^μ, and $\nabla^\mu \equiv \partial^{\langle \mu \rangle}$, $A^{\langle \mu \rangle} \equiv \Delta_\nu^\mu A^\nu$, $A^{\langle \mu\nu \rangle} \equiv \Delta_{\alpha\beta}^{\mu\nu} A^{\alpha\beta}$, with $\Delta^{\mu\nu\alpha\beta} = \frac{1}{2} \left(\Delta^{\mu\alpha} \Delta^{\nu\beta} + \Delta^{\mu\beta} \Delta^{\nu\alpha} \right) - \frac{1}{3} \Delta^{\mu\nu} \Delta^{\alpha\beta}$.

As discussed in Sect. 1.4.2, in transient theories of fluid dynamics the dissipative currents are determined from evolution equations, and not by constitutive relations. These equations of motion were already derived in the previous chapter and are given by

$$\tau_\Pi D\Pi + \Pi = -\zeta\theta + \cdots , \qquad (2.4)$$

$$\tau_n \Delta_\alpha^\mu Dn^\alpha + n^\mu = \varkappa \nabla^\mu \alpha_0 + \cdots , \qquad (2.5)$$

$$\tau_\pi \Delta_{\alpha\beta}^{\mu\nu} D\pi^{\alpha\beta} + \pi^{\mu\nu} = 2\eta\sigma^{\mu\nu} + \cdots . \qquad (2.6)$$

Above, the dots indicate possible second-order terms [16, 22, 23] which will be neglected in the following linear stability analysis. Note that most second-order terms are nonlinear and do not contribute in the linearized regime (at least when considering perturbations around an equilibrium state).

We consider perturbations of all fluid-dynamical fields around the global-equilibrium state described above,

$$\varepsilon = \varepsilon_0 + \delta\varepsilon \,, \tag{2.7}$$

$$n = n_0 + \delta n \,, \tag{2.8}$$

$$u^\mu = u_0^\mu + \delta u^\mu \,, \tag{2.9}$$

$$\Pi = \delta\Pi \,, \tag{2.10}$$

$$n^\mu = \delta n^\mu \,, \tag{2.11}$$

$$\pi^{\mu\nu} = \delta\pi^{\mu\nu} \,, \tag{2.12}$$

where $n_0 \equiv n(\beta_0, \alpha_0)$ and $\varepsilon_0 \equiv \varepsilon(\beta_0, \alpha_0)$ are constants. The linearization procedure consists of substituting Eqs. (2.7)–(2.12) into the exact fluid-dynamical equations (2.1)–(2.6) and only retaining the terms that are linear, i.e., of first order, in the deviations from the equilibrium state.

Since the fluid 4-velocity is normalized, i.e., $u_\mu u^\mu = 1$, it is straightforward to demonstrate that the perturbations of the fluid velocity satisfy

$$\delta u_\mu u_0^\mu = \mathcal{O}_2 \approx 0 \,, \tag{2.13}$$

where $\mathcal{O}_2$ denotes terms that are of second or higher order in perturbations of the fluid-dynamical fields. That is, up to first order in perturbations, the fluctuations of the fluid 4-velocity are orthogonal to the background 4-velocity. Similarly, it is possible to obtain orthogonality relations satisfied by the dissipative currents, δn^μ and $\delta\pi^{\mu\nu}$. As shown in the previous chapter, the dissipative currents are constructed to be orthogonal to the 4-velocity field, $u_\mu n^\mu = u_\mu \pi^{\mu\nu} = 0$. Since n^μ and $\pi^{\mu\nu}$ are at least of first order (in this power-counting scheme), these relations lead to

$$u_\mu^0 \delta\pi^{\mu\nu} = \mathcal{O}_2^\nu \approx 0 \,, \tag{2.14}$$

$$u_\mu^0 \delta n^\mu = \mathcal{O}_2 \approx 0 \,. \tag{2.15}$$

Thus, the perturbations of the dissipative currents are, to first order, also orthogonal to the background velocity field. Due to these orthogonality relations, it is convenient to introduce a projection operator onto the 3-space orthogonal to the background velocity,

$$\Delta_0^{\mu\nu} \equiv g^{\mu\nu} - u_0^\mu u_0^\nu \,, \tag{2.16}$$

and, similarly, a rank-4 symmetric, traceless projection operator

$$\Delta_0^{\alpha\beta\mu\nu} \equiv \frac{1}{2} \left(\Delta_0^{\mu\alpha} \Delta_0^{\nu\beta} + \Delta_0^{\mu\beta} \Delta_0^{\nu\alpha} \right) - \frac{1}{3} \Delta_0^{\mu\nu} \Delta_0^{\alpha\beta} \,. \tag{2.17}$$

Finally, we define a comoving derivative relative to the background velocity field, $D_0 \equiv u_0^\mu \partial_\mu$.

Using this notation, the equations of motion for the energy density, local velocity field, and net-charge density, linearized around the global-equilibrium state, become

$$D_0 \delta\varepsilon + (\varepsilon_0 + P_0)\,\partial_\mu \delta u^\mu = \mathcal{O}_2 \approx 0 \,, \tag{2.18}$$

$$(\varepsilon_0 + P_0)\,D_0 \delta u^\mu - \Delta_0^{\mu\nu}\partial_\nu\,(\delta P + \delta\Pi) + \partial_\nu \delta\pi^{\mu\nu} = \mathcal{O}_2^\mu \approx 0 \,, \tag{2.19}$$

$$D_0 \delta n + n_0 \partial_\mu \delta u^\mu + \partial_\mu \delta n^\mu = \mathcal{O}_2 \approx 0 \,, \tag{2.20}$$

while the linearized equations for the dissipative currents are

$$\tau_\Pi\,D_0 \delta\Pi + \delta\Pi + \zeta \partial_\mu \delta u^\mu = \mathcal{O}_2 \approx 0 \,, \tag{2.21}$$

$$\tau_n\,D_0 \delta n^\mu + \delta n^\mu - \varkappa \Delta_0^{\mu\nu}\partial_\nu \delta\alpha = \mathcal{O}_2^\mu \approx 0 \,, \tag{2.22}$$

$$\tau_\pi\,D_0 \delta\pi^{\mu\nu} + \delta\pi^{\mu\nu} - 2\eta \Delta_0^{\mu\nu\alpha\beta}\partial_\alpha \delta u_\beta = \mathcal{O}_2^{\mu\nu} \approx 0 \,, \tag{2.23}$$

where ζ, $\varkappa$, η, τ_Π, τ_n, and τ_π are the transport coefficients as functions of the *background* temperature and thermal potential. We note that, in the linear regime, pressure perturbations can be expressed as

$$\delta P = \frac{\partial P_0}{\partial\alpha_0}\bigg|_{\beta_0}\delta\alpha + \frac{\partial P_0}{\partial\beta_0}\bigg|_{\alpha_0}\delta\beta + \mathcal{O}_2 = \frac{n_0}{\beta_0}\delta\alpha - \frac{h_0 n_0}{\beta_0}\delta\beta + \mathcal{O}_2 \,, \tag{2.24}$$

where we have used Eq. (1.97). Similar relations apply to energy-density and net-charge density fluctuations.

2.2 Linearized Fluid-Dynamical Equations in Fourier Space

In order to investigate the (propagating and exponentially increasing or decreasing) modes of the fluid-dynamical equations, it is convenient to Fourier-transform the space-time dependence of the linearized equations. We define the Fourier transformation using the following convention:

$$\tilde{A}(k^\mu) = \int d^4x \exp\left(-ix_\mu k^\mu\right) A(x^\mu) \,, \tag{2.25}$$

$$A(x^\mu) = \int \frac{d^4k}{(2\pi)^4} \exp\left(ix_\mu k^\mu\right) \tilde{A}(k^\mu) \,. \tag{2.26}$$

Here, $k^\mu = (\omega, \mathbf{k})^T$ is a 4-vector, with ω being the frequency and $\mathbf{k}$ the wavenumber 3-vector of the fluctuation, and $x^\mu = (t, \mathbf{x})^T$ is the usual space-time coordinate 4-vector.

Let us define the covariant variables

$$\Omega \equiv u_0^\mu k_\mu \,, \quad \kappa^\mu \equiv \Delta_0^{\mu\nu}k_\nu \,, \tag{2.27}$$

where Ω is the frequency of oscillations in the local rest frame of the background fluid while κ^μ is the wavenumber 4-vector of oscillations in that frame. The linearized conservation laws (2.18)–(2.20) can then be written in Fourier space in the simple form

$$\Omega \delta\tilde{\varepsilon} + (\varepsilon_0 + P_0)\,\kappa_\mu \delta\tilde{u}^\mu = 0 \,, \tag{2.28}$$

$$(\varepsilon_0 + P_0)\,\Omega \delta\tilde{u}^\mu - \kappa^\mu \left(\delta\tilde{P} + \delta\tilde{\Pi} \right) + \kappa_\nu \delta\tilde{\pi}^{\mu\nu} = 0 \,, \tag{2.29}$$

$$\Omega \delta\tilde{n} + n_0 \kappa_\mu \delta\tilde{u}^\mu + \kappa_\mu \delta\tilde{n}^\mu = 0 \,, \tag{2.30}$$

and the linearized equations of motion (2.21)–(2.23) for the dissipative currents become

$$(1 + i\tau_\Pi \Omega)\,\delta\tilde{\Pi} = -i\zeta \kappa_\mu \delta\tilde{u}^\mu \,, \tag{2.31}$$

$$(1 + i\tau_n \Omega)\,\delta\tilde{n}^\mu = i\varkappa \kappa^\mu \delta\tilde{\alpha} \,, \tag{2.32}$$

$$(1 + i\tau_\pi \Omega)\,\delta\tilde{\pi}^{\mu\nu} = i\eta \left(\kappa^\mu \delta\tilde{u}^\nu + \kappa^\nu \delta\tilde{u}^\mu - \frac{2}{3}\Delta_0^{\mu\nu} \kappa_\lambda \delta\tilde{u}^\lambda \right) \,. \tag{2.33}$$

We further define the scalar κ as the modulus of κ^μ, $\kappa_\mu \kappa^\mu \equiv -\kappa^2$. Note that the usual dispersion relation of perturbations in a fluid, $\omega(\mathbf{k})$ (with $\omega \equiv k^0$) is not a Lorentz scalar and will change depending on the magnitude of the background velocity u_0^μ.

2.2.1 Tensor Decomposition in Fourier Space

It is also convenient to decompose fluctuations of tensors into components parallel and orthogonal to κ^μ. For this purpose, we introduce another projection operator, now onto the 3-space orthogonal to $\hat{\kappa}^\mu$,

$$\Delta_\kappa^{\mu\nu} = g^{\mu\nu} + \hat{\kappa}^\mu \hat{\kappa}^\nu \,, \tag{2.34}$$

where $\hat{\kappa}^\mu \equiv \kappa^\mu/\kappa$ is a space-like 4-vector normalized to $\hat{\kappa}_\mu \hat{\kappa}^\mu = -1$. Within this scheme, an arbitrary 4-vector A^μ is decomposed as

$$A^\mu = A_\parallel \hat{\kappa}^\mu + A_\perp^\mu \,, \tag{2.35}$$

with components $A_\parallel \equiv -\hat{\kappa}_\mu A^\mu$ and $A_\perp^\mu \equiv \Delta_\kappa^{\mu\lambda} A_\lambda$. We shall refer to $A_\parallel$ as the longitudinal component of the corresponding 4-vector and $A_\perp^\mu$ as its transverse components. Similarly, an arbitrary traceless, symmetric second-rank tensor, $A^{\mu\nu}$, is decomposed as

$$A^{\mu\nu} = A_\parallel \hat{\kappa}^\mu \hat{\kappa}^\nu + \frac{A_\parallel}{3}\Delta_\kappa^{\mu\nu} + A_\perp^\mu \hat{\kappa}^\nu + A_\perp^\nu \hat{\kappa}^\mu + A_\perp^{\mu\nu} \,, \tag{2.36}$$

with $A_{\parallel} \equiv \hat{\kappa}^{\mu} \hat{\kappa}^{\nu} A_{\mu\nu}$, $A_{\perp}^{\lambda} \equiv -\hat{\kappa}^{\nu} \Delta_{\kappa}^{\mu\lambda} A_{\mu\nu}$, and $A_{\perp}^{\mu\nu} \equiv \Delta_{\kappa}^{\mu\nu\alpha\beta} A_{\alpha\beta}$, where we defined the rank-4 symmetric, traceless projection operator $\Delta_{\kappa}^{\mu\nu\alpha\beta} \equiv \frac{1}{2}(\Delta_{\kappa}^{\mu\alpha} \Delta_{\kappa}^{\nu\beta} + \Delta_{\kappa}^{\mu\beta} \Delta_{\kappa}^{\nu\alpha}) - \frac{1}{3} \Delta_{\kappa}^{\mu\nu} \Delta_{\kappa}^{\alpha\beta}$. In this case, $A_{\parallel}$ is the longitudinal component of the tensor, $A_{\perp}^{\mu}$ correspond to its partially transverse components, and $A_{\perp}^{\mu\nu}$ are its fully transverse (and traceless) components.

2.2.2 Longitudinal and Transverse Components

We now project Eqs. (2.28)–(2.33) onto their components longitudinal and transverse to $\hat{\kappa}^{\mu}$. This is useful because the longitudinal and transverse projections of the equations decouple from each other, and can be solved independently to obtain the dispersion relations satisfied by the perturbations. We first consider the equations for the longitudinal components of the fluctuations.

First, we note that Eqs. (2.28), (2.30), and (2.31) are already written in terms of longitudinal perturbations since they describe perturbations of scalar quantities, i.e., energy density, net-charge density, and bulk-viscous pressure. The longitudinal component of Eqs. (2.29) and (2.32) are obtained by projecting each equation onto $\hat{\kappa}_{\mu}$, while the longitudinal component of Eq. (2.33) is obtained by projecting it onto $\hat{\kappa}_{\mu}\hat{\kappa}_{\nu}$. The result is a set of coupled equations for the perturbations $\delta\tilde{\varepsilon}$, $\delta\tilde{P}$, δn, $\delta\tilde{\alpha}$, $\delta\tilde{u}_{\parallel}$, $\delta\tilde{\Pi}$, $\delta\tilde{n}_{\parallel}$, and $\delta\tilde{\pi}_{\parallel}$,

$$\Omega\delta\tilde{\varepsilon} - \kappa\left(\varepsilon_0 + P_0\right)\delta\tilde{u}_{\parallel} = 0, \tag{2.37}$$

$$\left(\varepsilon_0 + P_0\right)\Omega\delta\tilde{u}_{\parallel} - \kappa\left(\delta\tilde{P} + \delta\tilde{\Pi} + \delta\tilde{\pi}_{\parallel}\right) = 0, \tag{2.38}$$

$$\Omega\delta\tilde{n} - \kappa\left(n_0\delta\tilde{u}_{\parallel} + \delta\tilde{n}_{\parallel}\right) = 0, \tag{2.39}$$

$$\left(1 + i\tau_{\Pi}\Omega\right)\delta\tilde{\Pi} - i\zeta\kappa\delta\tilde{u}_{\parallel} = 0, \tag{2.40}$$

$$\left(1 + i\tau_n\Omega\right)\delta\tilde{n}_{\parallel} - i\varkappa\kappa\delta\tilde{\alpha} = 0, \tag{2.41}$$

$$\left(1 + i\tau_{\pi}\Omega\right)\delta\tilde{\pi}_{\parallel} - \frac{4}{3}i\eta\kappa\delta\tilde{u}_{\parallel} = 0. \tag{2.42}$$

The fluctuations of energy density, net-charge density, and pressure can be converted into fluctuations of inverse temperature and thermal potential, $\delta\tilde{\beta}$ and $\delta\tilde{\alpha}$, respectively. Solving these equations leads to six different modes of the theory, two of them related to the propagation of sound waves, one related to the diffusion of net-charge density fluctuations, and the remaining three being new non-propagating modes due to fluctuations of the dissipative currents that are not allowed in Navier–Stokes theory. The latter modes are *non-hydrodynamic*, i.e., they are modes that do not vanish when the wavenumber is taken to zero, and describe the relaxation of the dissipative currents back to global equilibrium. Such modes were thought to never exist in fluid-dynamical theories, but we shall see that they are required in the relativistic regime in order to restore causality—and that they even appear in Navier–Stokes theory in the relativistic regime.

The equation of motion for the transverse components of the velocity field is obtained by projecting Eq. (2.29) with $\Delta^\lambda_{\kappa\mu}$ while the equation of motion for the partially transverse components of the shear-stress tensor is obtained by projecting Eq. (2.33) with $\hat{\kappa}_\mu \Delta^\lambda_{\kappa\nu}$. The resulting equations are coupled,

$$(\varepsilon_0 + P_0)\, \Omega \delta\tilde{u}^\lambda_\perp - \kappa \delta\tilde{\pi}^\lambda_\perp = 0 \,, \tag{2.43}$$
$$(1 + i\tau_\pi \Omega)\, \delta\tilde{\pi}^\lambda_\perp - i\eta\kappa \delta\tilde{u}^\lambda_\perp = 0 \,. \tag{2.44}$$

Such transverse modes are usually referred to as shear modes, since they do not display any contributions from bulk-viscous pressure or diffusion 4-current. They describe the diffusion of the velocity field and, as long as the relaxation times remain finite, they also contain a non-hydrodynamic mode. Since $\delta\tilde{u}^\mu_\perp$ and $\delta\tilde{\pi}^\mu_\perp$ carry a total of four independent degrees of freedom, each mode obtained from the above equations will have a twofold degeneracy.

Finally, we have the two equations of motion for the transverse fluctuations of the diffusion 4-current and the fully transverse fluctuations of the shear-stress tensor, also containing two modes each,

$$(1 + i\tau_n \Omega)\, \delta\tilde{n}^\lambda_\perp = 0 \,, \tag{2.45}$$
$$(1 + i\tau_\pi \Omega)\, \delta\tilde{\pi}^{\alpha\beta}_\perp = \frac{2}{9} i\eta\kappa \delta\tilde{u}_\parallel \left(\Delta^{\alpha\beta}_\kappa - 3u^\alpha_0 u^\beta_0 \right) \,. \tag{2.46}$$

The first equation is obtained by projecting Eq. (2.32) with $\Delta^\lambda_{\kappa\mu}$ while the second equation is obtained by projecting Eq. (2.33) with $\Delta^{\alpha\beta}_{\kappa\mu\nu}$. The first equation gives rise to a twofold degenerate mode,

$$\Omega = \frac{i}{\tau_n} \,, \tag{2.47}$$

which, in the rest frame of the global-equilibrium background, $u^\mu_0 = (1, 0, 0, 0)^T$, has no wavenumber dependence, i.e., it is non-propagating. Such a mode is obviously *non-hydrodynamic*, i.e., it has a frequency that does not vanish when the wavenumber is taken to zero in the local rest frame. It is purely imaginary, with a positive imaginary part, and thus related to the exponential damping of the dissipative currents toward the equilibrium state. Such modes are rather simple and will not be discussed further.

Equation (2.46) also gives rise to a twofold degenerate mode. With the help of Eq. (2.42), we can write this equation as

$$\delta\tilde{\pi}^{\alpha\beta}_\perp = \frac{1}{6} \delta\tilde{\pi}_\parallel \left(\Delta^{\alpha\beta}_\kappa - 3u^\alpha_0 u^\beta_0 \right) \,, \tag{2.48}$$

which shows that the fully transverse fluctuations of $\pi^{\alpha\beta}$ simply follow the time dependence of the longitudinal one. In the next sections, we discuss the solutions for the longitudinal modes and the remaining transverse modes for ideal fluids, viscous fluids in the Navier–Stokes limit, and for the transient fluid-dynamical equations.

2.3   Ideal Fluid Dynamics

The ideal-fluid limit is obtained by setting the viscosity and relaxation-time coefficients to zero. In this case, one obtains the following set of equations for the longitudinal perturbations

$$\Omega \delta\tilde{\varepsilon} - (\varepsilon_0 + P_0)\,\kappa \delta\tilde{u}_\| = 0\,, \tag{2.49}$$

$$(\varepsilon_0 + P_0)\,\Omega \delta\tilde{u}_\| - \kappa \delta\tilde{P} = 0\,, \tag{2.50}$$

$$\Omega \delta\tilde{n} - n_0\kappa \delta\tilde{u}_\| = 0\,, \tag{2.51}$$

and one equation for the transverse velocity perturbation

$$(\varepsilon_0 + P_0)\,\Omega \delta\tilde{u}_\perp^\mu = 0\,. \tag{2.52}$$

The set of equations for the longitudinal perturbations can be cast into the following matrix form:

$$\begin{pmatrix} \Omega \left.\dfrac{\partial\varepsilon_0}{\partial\beta_0}\right|_{\alpha_0} & \Omega \left.\dfrac{\partial\varepsilon_0}{\partial\alpha_0}\right|_{\beta_0} & -(\varepsilon_0 + P_0)\,\kappa \\[2mm] \kappa \dfrac{h_0 n_0}{\beta_0} & -\kappa \dfrac{n_0}{\beta_0} & (\varepsilon_0 + P_0)\,\Omega \\[2mm] \Omega \left.\dfrac{\partial n_0}{\partial\beta_0}\right|_{\alpha_0} & \Omega \left.\dfrac{\partial n_0}{\partial\alpha_0}\right|_{\beta_0} & -n_0\kappa \end{pmatrix} \begin{pmatrix} \delta\tilde{\beta} \\[2mm] \delta\tilde{\alpha} \\[2mm] \delta\tilde{u}_\| \end{pmatrix} = 0\,, \tag{2.53}$$

where the energy-density, net-charge density, and pressure perturbations were decomposed in terms of temperature and thermal-potential perturbations. Note that, for the pressure perturbation, we have used Eq. (2.24).

The solution for the transverse mode is simply $\Omega = 0$. The remaining modes from the longitudinal fluctuations are obtained by finding the roots of the determinant of the matrix in Eq. (2.53), leading to the following equation:

$$\Omega\left(\Omega^2 - c_s^2\kappa^2\right) = 0\,, \tag{2.54}$$

where we identify the sound velocity as

$$c_s^2 = \left.\frac{\partial P_0}{\partial\varepsilon_0}\right|_{s_0/n_0} = \frac{n_0}{\beta_0}\,\frac{h_0^{-1}\left.\dfrac{\partial\varepsilon_0}{\partial\beta_0}\right|_{\alpha_0} - \left.\dfrac{\partial n_0}{\partial\beta_0}\right|_{\alpha_0} + \left.\dfrac{\partial\varepsilon_0}{\partial\alpha_0}\right|_{\beta_0} - h_0 \left.\dfrac{\partial n_0}{\partial\alpha_0}\right|_{\beta_0}}{\left.\dfrac{\partial\varepsilon_0}{\partial\beta_0}\right|_{\alpha_0}\left.\dfrac{\partial n_0}{\partial\alpha_0}\right|_{\beta_0} - \left.\dfrac{\partial\varepsilon_0}{\partial\alpha_0}\right|_{\beta_0}\left.\dfrac{\partial n_0}{\partial\beta_0}\right|_{\alpha_0}}\,. \tag{2.55}$$

In the rest frame of the global-equilibrium state, i.e., $u_0^\mu = (1, 0, 0, 0)^T$, the scalars Ω and κ reduce to the usual frequency and wavenumber of the perturbations, respectively, $\Omega = \omega$ and $\kappa = k$. In this case, one recovers the well-known expressions for the modes of an ideal fluid,

$$\omega(k) = 0\,, \quad \omega(k) = \pm c_s k\,. \tag{2.56}$$

The static mode appears because we included net-charge fluctuations in the analysis. If one considers a system with zero net charge, the static mode disappears, with the sound modes remaining in the same form as obtained above, but with a different speed of sound.

For a nonzero background velocity (without loss of generality, we can always set the velocity to point into the x-direction), $u_0^\mu = (\gamma, \gamma V, 0, 0)^T$, we have $\Omega = \gamma\omega - \gamma V k^x$ and $\kappa^2 = \Omega^2 - k_\mu k^\mu = \gamma^2 (\omega V - k^x)^2 + k_T^2$. Naturally, a nonzero background velocity introduces an anisotropy in the system, with the modes no longer depending on the modulus of the wavenumber, but displaying a separate dependence on k^x and $k_T = \sqrt{k_y^2 + k_z^2}$. For the sake of simplicity, we consider waves traveling only in the x-direction, i.e., with $k_T = 0$, in which case the frequency will depend solely on $k \equiv k^x$. Then the dispersion relation is obtained by solving the equation

$$(\omega - Vk)\left[\left(1 - c_s^2 V^2\right)\omega^2 - 2\left(1 - c_s^2\right)Vk\omega + \left(V^2 - c_s^2\right)k^2\right] = 0 . \qquad (2.57)$$

This equation has three solutions,

$$\omega(k) = Vk , \qquad (2.58)$$

$$\omega_\pm(k) = \frac{V \pm c_s}{1 \pm V c_s}k . \qquad (2.59)$$

We see that the frequencies of oscillations are both linear in k. As expected, the mode that previously vanished now starts to move with the background velocity. Meanwhile, each sound mode has now a different propagation speed, expressed by the relativistic velocity-addition rule (one mode is moving in the direction of the unperturbed fluid and the other in the opposite direction),

$$v_s^\pm = \frac{V \pm c_s}{1 \pm V c_s} .$$

Naturally, the waves propagating against the flow of the background will propagate slower than those moving in the same direction as the background fluid.

2.4   Relativistic Navier–Stokes Theory

The modes related to relativistic Navier–Stokes theory are obtained from the expressions derived in Sect. 2.2 by taking the limit of vanishing relaxation times, $\tau_\pi = \tau_n = \tau_\Pi = 0$. In this limit, Eqs. (2.40), (2.41), (2.42), and (2.44) dictate that the fluctuations of the dissipative currents are expressed in terms of fluctuations of velocity and thermal potential in the following way:

$$\delta \tilde{\Pi} = i \zeta \kappa \delta \tilde{u}_\| \ , \tag{2.60}$$

$$\delta \tilde{n}_\| = i \varkappa \kappa \delta \tilde{\alpha} \ , \tag{2.61}$$

$$\delta \tilde{\pi}_\| = \frac{4}{3} i \eta \kappa \delta \tilde{u}_\| \ , \tag{2.62}$$

$$\delta \tilde{\pi}_\perp^\lambda = i \eta \kappa \delta \tilde{u}_\perp^\lambda \ . \tag{2.63}$$

The equations of motion for the longitudinal perturbations then reduce to

$$\Omega \delta \tilde{\varepsilon} - \kappa \left(\varepsilon_0 + P_0 \right) \delta \tilde{u}_\| = 0 \ , \tag{2.64}$$

$$\left(\varepsilon_0 + P_0 \right) \Omega \delta \tilde{u}_\| - \kappa \delta \tilde{P} - i \kappa^2 \left(\zeta + \frac{4}{3} \eta \right) \delta \tilde{u}_\| = 0 \ , \tag{2.65}$$

$$\Omega \delta \tilde{n} - \kappa n_0 \delta \tilde{u}_\| - i \kappa^2 \varkappa \delta \tilde{\alpha} = 0 \ , \tag{2.66}$$

while the transverse modes satisfy the simple relation

$$\left(\Omega - i \kappa^2 \frac{\eta}{\varepsilon_0 + P_0} \right) \delta \tilde{u}_\perp^\mu = 0 \ . \tag{2.67}$$

There are no transverse perturbations of the diffusion current, while on account of Eq. (2.48) the fully transverse perturbations of the shear-stress tensor follow that of the longitudinal one, Eq. (2.62).

2.4.1   Transverse Modes

Let us first discuss the transverse (i.e., shear) modes since their dispersion relations are more simple and easier to obtain. From Eq. (2.67), we can immediately infer that Ω and κ obey a diffusion-type dispersion relation

$$\Omega = i \tau_\eta \kappa^2 \ , \tag{2.68}$$

where we defined

$$\tau_\eta \equiv \frac{\eta}{\varepsilon_0 + P_0} \ . \tag{2.69}$$

We note that τ_η is a time scale that naturally appears in the shear modes of Navier–Stokes theory and encompasses all effects of dissipation on the transverse perturbations. We shall soon demonstrate that this time scale is also related to the exponential growth of unstable non-hydrodynamic modes that will appear when perturbing the theory in a moving background. These unstable modes appear only due to the parabolic nature of the relativistic Navier–Stokes theory, which ends up generating new modes in boosted frames.

When $u_0^\mu = (1, 0, 0, 0)^T$, one simply obtains a non-propagating, exponentially damped mode with twofold degeneracy,

$$\omega(k) = i \tau_\eta k^2 \ . \tag{2.70}$$

This relation is the same as the one obtained by solving the diffusion equation, with τ_η playing the role of the diffusion coefficient. That is, the shear modes simply describe a diffusion process of the velocity field.

Now we consider propagation of shear perturbations in a moving background. As done for the case of an ideal fluid, we choose our coordinate system in such a way that the velocity of the background is in the x-direction, i.e., $u_0^\mu = (\gamma, \gamma V, 0, 0)^T$. We also only consider the case of perturbations that travel in the x-direction, i.e., $k^\mu = (\omega, k, 0, 0)^T$. The dispersion relation (2.68) then becomes

$$\omega^2 V^2 + \left(\frac{i}{\gamma \tau_\eta} - 2Vk\right)\omega - \frac{i}{\gamma \tau_\eta}Vk + k^2 = 0 . \tag{2.71}$$

A nonzero background velocity has the effect of mixing the contributions of frequency, ω, and wavenumber, k, which are contained in the Lorentz scalars Ω and κ. Therefore, for a moving background the quadratic term in κ in the dispersion relation will not only carry contributions that are quadratic in k, as in the $V = 0$ case, but will also carry contributions that are quadratic in ω. This leads to the appearance of the quadratic term $\omega^2 V^2$ in the dispersion relation, producing an additional transverse mode in the theory. The two solutions are

$$\omega(k) = \frac{1 + 2i\gamma V\tau_\eta k \pm \sqrt{1 + 4iV\tau_\eta k/\gamma}}{2i\gamma V^2\tau_\eta} . \tag{2.72}$$

The analytic solution in the limit of small wavenumbers, $k \to 0$, is

$$\omega(k \to 0) = \begin{cases} -\dfrac{i}{\gamma V^2\tau_\eta} , \\ Vk + i\tau_\eta k^2/\gamma^3 . \end{cases} \tag{2.73}$$

The second mode corresponds to the one of Eq. (2.70), since it tends to $ik^2\tau_\eta$ when the background velocity goes to zero. On the other hand, the first mode is intrinsically new and is non-hydrodynamic, non-propagating, and, because of a negative imaginary part, unstable. This mode describes perturbations that grow exponentially on time scales of the order of $\gamma V^2\tau_\eta$. This *additional* mode does not appear when the velocity of the unperturbed fluid is zero and, for this reason, this problematic feature of Navier–Stokes theory remains largely unperceived in the literature. Nevertheless, the emergence of unstable non-hydrodynamic modes in relativistic Navier–Stokes theory is a fundamental issue that must be fixed, in order to obtain a relativistic theory of fluid dynamics. We remark that this is a problem of the relativistic version of Navier–Stokes theory—the non-relativistic version of the theory is stable, even when perturbed around a *moving* global-equilibrium state.

2.4.2 Longitudinal Modes

Next, we consider the longitudinal modes. For the sake of simplicity, we restrict our analysis to the limit of vanishing net-charge fluctuations. In this case, Eqs. (2.64) and (2.65) simplify to

$$
\begin{pmatrix} \Omega & -\kappa \\ -c_s^2\kappa & \Omega - i\tau_{\text{eff}}\kappa^2 \end{pmatrix} \begin{pmatrix} \dfrac{\delta\tilde{\varepsilon}}{\varepsilon_0 + P_0} \\ \delta\tilde{u}_\parallel \end{pmatrix} = 0 \,,
\tag{2.74}
$$

where we used the definition of sound velocity at zero chemical potential, $c_s^2 = dP_0/d\varepsilon_0$, and defined the effective time scale

$$
\tau_{\text{eff}} \equiv \frac{\zeta + \frac{4}{3}\eta}{\varepsilon_0 + P_0} \,,
\tag{2.75}
$$

which plays a very similar role as the variable τ_η that appeared in the previous subsection when discussing the shear modes. As before, the dispersion relations satisfied by the perturbations are found by setting the determinant of the matrix in Eq. (2.74) to zero. The equation satisfied by Ω and κ then is

$$
\Omega^2 - i\tau_{\text{eff}}\Omega\kappa^2 - c_s^2\kappa^2 = 0 \,.
\tag{2.76}
$$

In the case where the velocity of the unperturbed system is zero, i.e., $u_0^\mu = (1,0,0,0)^T$, one simply has $\Omega = \omega$ and $\kappa = k$ and Eq. (2.76) simplifies to

$$
\omega^2 - i\tau_{\text{eff}}\omega k^2 - c_s^2 k^2 = 0 \quad \Longrightarrow \quad \omega(k) = i\frac{\tau_{\text{eff}}}{2}k^2 \pm k\sqrt{c_s^2 - \frac{\tau_{\text{eff}}^2 k^2}{4}} \,.
\tag{2.77}
$$

These are the sound modes of the theory and they reduce to the solution found for ideal fluids when $\tau_{\text{eff}} = 0$. Defining the critical wavenumber

$$
k_c^{\text{NS}} = 2c_s/\tau_{\text{eff}} \,,
\tag{2.78}
$$

we see that for $k < k_c^{\text{NS}}$ the sound modes have a part propagating with the reduced speed of sound $c_{s,\text{eff}} \equiv \sqrt{c_s^2 - \tau_{\text{eff}}^2 k^2/4} \leq c_s$. They also have a non-propagating part which describes how such modes are damped by viscosity. On the other hand, for wavenumbers larger than $k \geq k_c^{\text{NS}}$, the frequencies become purely imaginary and there are no propagating modes. In Fig. 2.1, we show the real (left panel) and imaginary (right panel) parts of $\omega(k)$ in units of temperature, for $\tau_{\text{eff}} = 2/T$.

The solution at small wavenumbers, $k \to 0$, can be written in the simple form

$$
\omega(k) = \pm c_s k + i\frac{\tau_{\text{eff}}}{2}k^2 + \mathcal{O}(k^3) \,,
\tag{2.79}
$$

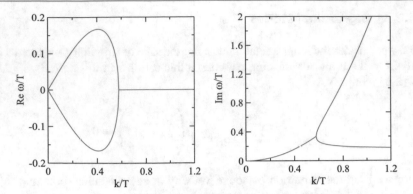

Fig. 2.1 The real parts (left panel) and the imaginary parts (right panel) of the dispersion relations for the longitudinal modes of Navier–Stokes theory for a static background. We set $\tau_{\text{eff}} = 2\beta_0$

from which we conclude that, to leading order, the sound modes propagate with the usual speed of sound c_s, damped by the effective specific viscosity $\tau_{\text{eff}}/2$. On the other hand, at large wavenumbers, $k \to \infty$,

$$\omega(k) = i \frac{\tau_{\text{eff}}}{2} k^2 (1 \pm 1) + \mathcal{O}(1) . \tag{2.80}$$

That is, for large values of wavenumbers, the mode is either zero or appears to be purely diffusive, $\omega(k) \sim i\tau_{\text{eff}} k^2$.

For the same moving background as considered for the shear modes, $u_0^\mu = \gamma (1, V, 0, 0)^T$ and $k^\mu = (\omega, k, 0, 0)^T$, the dispersion relation (2.76) assumes the form

$$(\omega - kV)^2 - i\tau_{\text{eff}}\gamma (\omega - kV) (\omega V - k)^2 - c_s^2 (\omega V - k)^2 = 0 . \tag{2.81}$$

In this case, we observe that the term $\Omega \kappa^2$ in Eq. (2.76) will make the equation *cubic* in ω and, as happened with the shear modes, a new solution will emerge. It is easy to see that such a new solution will be an unstable non-hydrodynamic mode. For this purpose, one can just look for the solutions of the equation at $k = 0$,

$$\omega (k = 0) = \begin{cases} -i \dfrac{1 - c_s^2 V^2}{\tau_{\text{eff}} \gamma V^2} , \\ \pm 0 . \end{cases} \tag{2.82}$$

The last two solutions are simply zero and can be identified with the usual sound modes that already appeared when the unperturbed global-equilibrium state was at rest. The additional solution does not vanish at $k = 0$, i.e., it is a non-hydrodynamic mode, and has a negative imaginary part, i.e., it is an unstable exponentially growing mode. Once more, we see that relativistic Navier–Stokes theory is unstable when perturbed around global equilibrium: a fundamental problem that simply cannot be ignored. In the next subsection, we briefly discuss the origins behind this issue.

2.4.3 Causality and Stability of Navier–Stokes Theory

In the preceding subsections, it was shown that relativistic Navier–Stokes theory is unstable, since perturbations on a moving fluid in global equilibrium grow exponentially, within microscopic time scales of the order of $\tau_\eta \sim \tau_{\text{eff}} \sim \eta/(\varepsilon_0 + P_0)$. This analysis was performed fixing the direction of the background velocity to be in the x-direction and only considering perturbations traveling in the same direction, i.e.,

$$u_0^\mu = \gamma \, (1, V, 0, 0)^T \ , \tag{2.83}$$

$$k^\mu = (\omega, k, 0, 0)^T \ .$$

In this scenario, the covariant frequency, Ω, and wavenumber, κ, of the oscillations satisfy

$$\Omega = \gamma \, (\omega - Vk) \ , \tag{2.84}$$

$$\kappa^2 = \gamma^2 \, (\omega V - k)^2 \ , \tag{2.85}$$

which is equivalent to a 1+1-dimensional Lorentz boost with velocity V of a 4-vector made of ω and k. Therefore, for all practical purposes, the dispersion relation for modes obtained for perturbations of a moving fluid can be obtained by applying a Lorentz transformation to the same dispersion relation obtained for perturbations on a fluid at rest. This can help us understand the connection between the acausal nature of the theory and its perturbative instability when considering moving background fluids. We shall make this argument also noting a connection between the modes of Navier–Stokes theory, at asymptotically large values of wavenumber, and the dispersion relation of the diffusion equation.

We already demonstrated in the previous subsections that, for perturbations of a static fluid, both the transverse and longitudinal modes of Navier–Stokes theory behave in the same way when $k \to \infty$, $\omega(k) \sim i\tau_{\text{eff}}k^2/2 \sim i\tau_\eta k^2$, being purely non-propagating modes that are quadratic in k. This behavior is identical to the one found in the diffusion equation (1.107), in which case there is a single non-propagating mode with dispersion relation $\omega(k) = i Dk^2$, with D being the diffusion coefficient. It is well known that the diffusion equation is acausal and, consequently, we conjecture that a k^2-dependence of any non-propagating mode can also be considered a sign of acausality.

Now we show that the parabolic nature of diffusion-like dispersion relations (given by their quadratic dependence on k) will lead to additional modes once a Lorentz transformation is performed. Let us consider the Lorentz boost of a 4-vector with time component ω and spatial component k in boost direction,

$$\begin{pmatrix} \omega \\ k \end{pmatrix} = \begin{pmatrix} \gamma & -\gamma V \\ -\gamma V & \gamma \end{pmatrix} \begin{pmatrix} \omega' \\ k' \end{pmatrix} = \begin{pmatrix} \gamma\omega' - \gamma Vk' \\ -\gamma V\omega' + \gamma k' \end{pmatrix} \ . \tag{2.86}$$

Then, we substitute the boosted variables into the diffusion dispersion relation, which changes in the following way:

$$\omega = iDk^2 \quad \longrightarrow \quad \omega' - Vk' = i\gamma D \left(k' - V\omega' \right)^2 \ . \tag{2.87}$$

It is straightforward to see that, in the limit of vanishing k', there is always a non-hydrodynamic mode given by

$$\omega'(k' = 0) = -\frac{i}{\gamma D V^2}\,, \tag{2.88}$$

which has a negative imaginary part, i.e., it is exponentially growing and thus unstable. Note that all unstable modes obtained so far in Navier–Stokes theory have the structure above, i.e., at $k = 0$ they always look like boosted modes of the diffusion equation. In this sense, it is not too surprising that a theory with parabolic, diffusion-type modes in a non-moving background will end up featuring unstable modes in a moving background. Naturally, non-relativistic theories will never display this property since the modes transform following Galileo's transformation, which never mixes frequency and wavenumber.

It is important to note how the structure of Lorentz boosts ends up connecting the *infinite-wavenumber behavior* of a mode in a static background (usually considered to be irrelevant for macroscopic dynamics) with the *vanishing wavenumber* behavior of the same mode in a moving background. This is partially why the acausality of a mode, which is an asymptotic feature, can affect its stability, which is a feature that affects the small-wavenumber behavior of the fluctuation. This is certainly an interesting aspect of relativistic theories.

Finally, we shall see in the next section that a transient theory of fluid dynamics does not display any non-propagating modes that match those of the diffusion equation. Also, we shall see that intrinsically unstable modes do not exist —the modes can be rendered stable as long as the transport coefficients satisfy a set of conditions that will be derived below.

2.5   Transient Theory of Fluid Dynamics

We now analyze the modes of the complete theory, which includes the effect of the relaxation times. For the sake of simplicity, we shall perform this analysis neglecting contributions from net-charge fluctuations and in the conformal limit (vanishing bulk-viscous pressure, $\delta\tilde{\Pi} = 0$, and vanishing bulk viscosity, $\zeta = 0$). In this case, Eqs. (2.37), (2.38), and (2.42) for the longitudinal fluctuations simplify to

$$\Omega\delta\tilde{\varepsilon} - \kappa\,(\varepsilon_0 + P_0)\,\delta\tilde{u}_\parallel = 0\,, \tag{2.89}$$

$$(\varepsilon_0 + P_0)\,\Omega\,\delta\tilde{u}_\parallel - \kappa\,\left(c_s^2\delta\tilde{\varepsilon} + \delta\tilde{\pi}_\parallel\right) = 0\,, \tag{2.90}$$

$$(1 + i\tau_\pi\Omega)\,\delta\tilde{\pi}_\parallel - \frac{4}{3}i\eta\kappa\delta\tilde{u}_\parallel = 0\,. \tag{2.91}$$

The remaining equations for the shear modes do not change under these assumptions and are still given by Eqs. (2.43), (2.44), and (2.46). As already discussed above, the solution of Eq. (2.46) is already determined by Eq. (2.91) and does not need to be addressed further.

2.5.1 Transverse Modes in the Rest Frame

The equations of motion (2.43), (2.44) for the transverse (i.e., shear) modes can be cast into the following matrix form:

$$\begin{pmatrix} \Omega & -\kappa \\ -i\tau_\eta\kappa & 1 + i\tau_\pi\Omega \end{pmatrix} \begin{pmatrix} \delta\tilde{u}^\mu_\perp \\ \frac{\delta\tilde{\pi}^\mu_\perp}{\varepsilon_0 + P_0} \end{pmatrix} = 0 , \tag{2.92}$$

leading to the following dispersion relation:

$$\Omega\left(1 + i\tau_\pi\Omega\right) - i\tau_\eta\kappa^2 = 0 . \tag{2.93}$$

Naturally, this equation will have two solutions in both the moving background and the one at rest, since its highest power in Ω is the same as its highest power in κ, i.e., κ^2.

When the background fluid is at rest, i.e., when we set $u_0^\mu = (1, 0, 0, 0)^T$, the dispersion relation becomes

$$\omega\left(1 + i\tau_\pi\omega\right) - i\tau_\eta k^2 = 0 , \tag{2.94}$$

with two distinct solutions

$$\omega(k) = \frac{i}{2\tau_\pi} \pm k\sqrt{\frac{\tau_\eta}{\tau_\pi}}\sqrt{1 - \left(\frac{\kappa_c^{\text{shear}}}{k}\right)^2} , \tag{2.95}$$

where similar to Navier–Stokes theory we defined a critical wavenumber

$$\kappa_c^{\text{shear}} \equiv \frac{1}{\sqrt{4\tau_\eta\tau_\pi}} . \tag{2.96}$$

We see that, due to the nonzero relaxation time τ_π, the transverse shear modes become propagating for $k > \kappa_c^{\text{shear}}$, with propagation speed $\sqrt{\tau_\eta/\tau_\pi} \times \sqrt{1 - \left(\kappa_c^{\text{shear}}/k\right)^2}$. They also have a non-vanishing imaginary part, describing damping on a time scale $\sim 2\tau_\pi$.

At small wavenumbers, $k \to 0$, the shear modes become

$$\omega\,(k \to 0) = \begin{cases} i\tau_\eta k^2 + \mathcal{O}(k^4) , \\ \dfrac{i}{\tau_\pi} + \mathcal{O}(k^2) . \end{cases} \tag{2.97}$$

We see that, for transient theories of fluid dynamics, we already have a non-hydrodynamic mode even for perturbations on a background that is at rest. But such a mode is stable and decays exponentially to zero within a time scale τ_π—the new

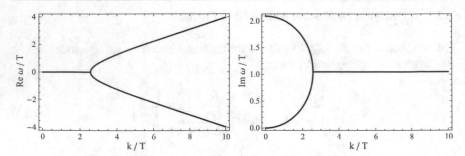

Fig. 2.2 The real parts (left panel) and the imaginary parts (right panel) of the dispersion relations for the transverse modes in a static background. We fixed the relaxation time as $\tau_\pi = 6\tau_\eta$ and $\eta/s_0 = 1/(4\pi)$. Figure taken from Ref. [5]

transport coefficient introduced when constructing transient theories of fluid dynamics. The other mode is hydrodynamical and, at small wavenumbers, is identical to the one obtained from relativistic Navier–Stokes theory. Therefore, at small wavenumbers, the transient theory is very similar to Navier–Stokes theory, the only difference being the appearance of a damped (and therefore stable) non-hydrodynamical mode $\sim i/\tau_\pi$.

At large values of the wavenumber, $k \to \infty$, the solution becomes

$$\omega(k \to \infty) = \pm\sqrt{\frac{\tau_\eta}{\tau_\pi}}k + \mathcal{O}(1) . \tag{2.98}$$

We see that this mode is linear in the wavenumber and, contrary to Navier–Stokes theory, the order of the dispersion relation in powers of frequency will not be modified by a Lorentz boost. As discussed, this asymptotic behavior of the mode may provide a causal and stable mode.

In Fig. 2.2, we show the real (left panel) and imaginary (right panel) parts of the transverse modes in units of temperature. In this plot, we parametrize the relaxation time as $\tau_\pi = 6\tau_\eta$ [7] and fix the shear viscosity to be $\eta/s_0 = 1/(4\pi)$, a value usually associated with strongly coupled conformal fluids [24]. In Fig. 2.3, we show the same result for $\tau_\pi = \tau_\eta$, i.e., for a smaller value of the relaxation time. We see that the result is qualitatively the same.

The dispersion relations for the shear modes resulting from Eq. (2.95) change their behavior from non-propagating to propagating at the critical wavenumber (2.96), as shown in Figs. 2.2 and 2.3. In this sense, it may be instructive to address the issue of causality by looking at the group velocity of such propagating modes. For wavenumbers larger than κ_c^{shear}, the (modulus of the) group velocity of the propagating shear mode is

$$v_g(k) = \frac{\partial \text{Re}\,\omega(k)}{\partial k} = v_{g,\text{shear}}^{\text{as}} \frac{k/\kappa_c^{\text{shear}}}{\sqrt{(k/\kappa_c^{\text{shear}})^2 - 1}} , \tag{2.99}$$

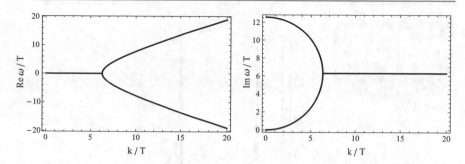

Fig. 2.3 The real parts (left panel) and the imaginary parts (right panel) of the dispersion relations for the transverse modes in a static background. We fixed the relaxation time as $\tau_\pi = \tau_\eta$ and $\eta/s_0 = 1/(4\pi)$. Figure taken from Ref. [5]

where we defined the asymptotic group velocity of this mode,

$$v_{g,\text{shear}}^{\text{as}} \equiv \sqrt{\frac{\tau_\eta}{\tau_\pi}}, \qquad (2.100)$$

as the asymptotic value of v_g in the large-wavenumber limit. The group velocity as a function of the wavenumber is shown in Fig. 2.4.

We can see from Eq. (2.99) (and also from Fig. 2.4) that the group velocity diverges near the critical wavenumber κ_c and approaches its asymptotic value ($k \to \infty$) from above. In Sect. 2.5.4, we shall show that this apparent violation of causality of the group velocity does not cause the theory as a whole to become acausal. The necessary condition required in order to insure that the theory is causal is actually that the asymptotic group velocity (2.100) does not exceed the speed of light,

$$\lim_{k \to \infty} \frac{\partial \text{Re}\,\omega(k)}{\partial k} = v_{g,\text{shear}}^{\text{as}} \leq 1 . \qquad (2.101)$$

This condition, which we shall refer to as *asymptotic causality condition*, is not automatically satisfied by the linearized Israel–Stewart theory. It actually must be imposed additionally, leading to the following constraint that must be satisfied by the shear relaxation time (if the effects of bulk-viscous pressure and net-charge diffusion were included, additional constraints for the relaxation times related to these quantities would also have been obtained [4, 19]),

$$\frac{\tau_\eta}{\tau_\pi} \leq 1 \quad \Longleftrightarrow \quad \tau_\pi \geq \frac{\eta}{\varepsilon_0 + P_0} . \qquad (2.102)$$

That is, the relaxation time must be larger than the time scale τ_η.

When analyzing the longitudinal (sound) modes, we shall find that the causality condition (2.102) is actually not sufficient to guarantee the stability and causality of the sound modes. Since all modes must be causal and stable, the validity of Eq. (2.102) does not guarantee the causality of the linearized theory as a whole.

Fig. 2.4 The group velocity for the transverse mode for $\tau_\pi = 6\tau_\eta$, $c_s^2 = 1/3$, and $\eta/s_0 = 1/(4\pi)$ (full line), $1/4$ (dashed line), and 1 (dotted line). Figure taken from Ref. [5]

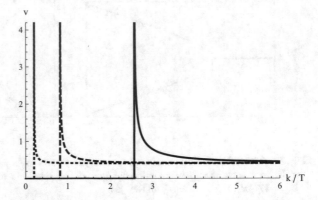

2.5.2 Longitudinal Modes in the Rest Frame

The equations of motion (2.89)–(2.91) for the longitudinal modes can be cast into the following matrix form:

$$
\begin{pmatrix}
\Omega & -\kappa & 0 \\
-c_s^2\kappa & \Omega & -\kappa \\
0 & -i\tau_{\text{eff}}\kappa & 1+i\tau_\pi\Omega
\end{pmatrix}
\begin{pmatrix}
\dfrac{\delta\tilde{\varepsilon}}{\varepsilon_0 + P_0} \\
\dfrac{\delta\tilde{u}_\parallel}{\varepsilon_0 + P_0} \\
\dfrac{\delta\tilde{\pi}_\parallel}{\varepsilon_0 + P_0}
\end{pmatrix}
= 0 ,
\qquad (2.103)
$$

with the dispersion relations being given by

$$
\left(\Omega^2 - c_s^2\kappa^2\right)\left(1 + i\tau_\pi\Omega\right) - i\tau_{\text{eff}}\Omega\kappa^2 = 0 .
\qquad (2.104)
$$

Naturally, this equation will have three solutions in both the moving background and the static one, since its highest power in Ω (Ω^3) is the same as its highest power in both κ and Ω ($\Omega\kappa^2$).

When the unperturbed fluid is at rest, the dispersion relation becomes

$$
\left(\omega^2 - c_s^2 k^2\right)\left(1 + i\tau_\pi\omega\right) - i\tau_{\text{eff}}\omega k^2 = 0 .
\qquad (2.105)
$$

The longitudinal modes are the solutions of a cubic equation, which cannot be expressed in a simple analytical form. In the following, we initially restrict our discussion to the asymptotic form of these modes. The analytic solutions in the limit of small wavenumber k are

$$
\omega(k) =
\begin{cases}
\dfrac{i}{\tau_\pi} , \\[2mm]
\pm c_s\, k + i\, \dfrac{\tau_{\text{eff}}}{2}\, k^2 ,
\end{cases}
\qquad (2.106)
$$

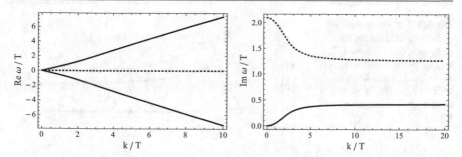

Fig. 2.5 The real parts (left panel) and the imaginary parts (right panel) of the dispersion relations for the sound modes (full lines) and the non-propagating mode (dashed line). The parameters are $\eta/s_0 = 1/(4\pi)$, $\tau_\pi = 6\tau_\eta$, $c_s^2 = 1/3$. Figure taken from Ref. [5]

while for large wavenumber one obtains

$$
\omega(k) =
\begin{cases}
\dfrac{i}{\tau_\pi} \left[1 + \dfrac{\tau_{\text{eff}}}{\tau_\pi c_s^2} \right]^{-1}, \\[3ex]
\pm c_s\, k \sqrt{1 + \dfrac{\tau_{\text{eff}}}{\tau_\pi c_s^2}} + \dfrac{i}{2\tau_\pi} \left[1 + \dfrac{\tau_\pi c_s^2}{\tau_{\text{eff}}} \right]^{-1}.
\end{cases}
\tag{2.107}
$$

This corresponds to one non-propagating mode and two propagating (sound) modes. As observed for the transverse modes, at small wavenumber the longitudinal propagating modes are exactly the same as the ones obtained from Navier–Stokes theory. Also, we note that all imaginary parts are positive and therefore all longitudinal modes are stable (as it was the case for the transverse modes).

The non-propagating mode does not have a real part, Re $\omega = 0$, and, hence, we cannot discuss the causality of this mode using the group velocity. This is similar to what happened in Navier–Stokes theory, where all modes became purely imaginary above a critical wavenumber. For this reason, we analyze this mode by comparing it to the mode of the diffusion equation (1.107)—the same procedure adopted with Navier–Stokes theory. As already stated, the diffusion equation is known to be acausal and has a single non-propagating mode with dispersion relation $\omega = i D k^2$. We have already demonstrated that such a k^2-dependence in any non-propagating mode (even if in the limit of infinite wavenumber) will lead to unstable non-hydrodynamic modes, once boosted to a moving frame. However, the non-propagating mode (2.107) is independent of k in the asymptotic limit (cf. Fig. 2.5) and should not lead to acausal signal propagation. Furthermore, this lack of dependence on the wavenumber leads to a dispersion relation that is well-behaved under Lorentz transformations, i.e., no additional mode will appear when k^μ is boosted.

The dispersion relations resulting from Eq. (2.105) are shown in Fig. 2.5, and the corresponding group velocity in Fig. 2.6. The group velocity has a maximum for a finite value of k/T and approaches its asymptotic value ($k \to \infty$) from above. But it does not diverge for any finite value of k, as happened with the group velocity for the shear modes. We see that for small values of the ratio τ_π/τ_η the group velocity

Fig. 2.6 The group velocity of longitudinal modes for $\eta/s_0 = 1/(4\pi)$, $c_s^2 = 1/3$, and $\tau_\pi = 6\tau_\eta$ (full line), $\tau_\pi = 2\tau_\eta$ (dashed line), as well as $\tau_\pi = 3\tau_\eta/2$ (dotted line). Figure taken from Ref. [5]

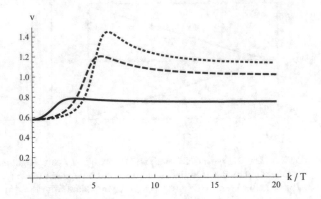

can become superluminal in some wavenumber domains. Nevertheless, as will be discussed in Sect. 2.5.4, this is not necessarily a problem since only the asymptotic value of the group velocity determines whether the mode as a whole violates causality or not. However, if $\tau_\pi/\tau_{\text{eff}}$ is sufficiently small, even the asymptotic group velocity can become larger than the velocity of light. As a matter of fact, the asymptotic value of the group velocity for the longitudinal modes is

$$v_{g,\text{sound}}^{\text{as}} = \lim_{k\to\infty} \frac{\partial \text{Re}\,\omega(k)}{\partial k} = c_s \sqrt{1 + \frac{\tau_{\text{eff}}}{\tau_\pi c_s^2}} \,. \tag{2.108}$$

Consequently, for the asymptotic group velocity of sound waves to be less than the speed of light, τ_π and τ_{eff} should satisfy the following asymptotic causality condition [5]:

$$\frac{\tau_{\text{eff}}}{\tau_\pi} \leq 1 - c_s^2 \,. \tag{2.109}$$

This is similar to the causality condition for the group velocity in the case of bulk viscosity, found originally in Ref. [4]. For conformal fluids, where $c_s^2 = 1/3$, the condition (2.109) simplifies to

$$\frac{\tau_{\text{eff}}}{\tau_\pi} \leq \frac{2}{3} \,. \tag{2.110}$$

Since for conformal fluids $\zeta = 0$, we also have $\tau_{\text{eff}} = 4\tau_\eta/3$, and thus

$$\frac{\tau_\eta}{\tau_\pi} \leq \frac{1}{2} \,, \tag{2.111}$$

which is more restrictive than the asymptotic causality condition (2.102) for the shear modes. We note that, for the values of η and τ_π deduced from the AdS/CFT correspondence [24], $\eta = s_0/(4\pi)$, $\tau_\pi = \beta_0(2 - \ln 2)/(2\pi)$, the condition (2.109) is always satisfied,

$$\frac{\tau_\eta}{\tau_\pi} = \frac{\eta}{\varepsilon_0 + P_0} \frac{2\pi}{\beta_0(2 - \ln 2)} = \frac{\eta}{s_0} \frac{2\pi}{2 - \ln 2} = \frac{1}{2(2 - \ln 2)} \simeq \frac{1}{2.6} < \frac{1}{2} \,.$$

2.5.3   Stability for a Moving Background

So far, we have studied the dispersion relations of transient fluid dynamics for perturbations of a fluid at rest. In this scheme, it was demonstrated that such theories are stable, and also causal, if the relaxation times satisfy a certain asymptotic causality condition. Now we must check if this situation changes when we consider perturbations on a moving (or Lorentz boosted) background—in Navier–Stokes theory, this completely changed the behavior with the appearance of unstable non-hydrodynamic modes.

We consider the same moving background used when analyzing the longitudinal and transverse modes of Navier–Stokes theory, which we parametrized as $u_0^\mu = \gamma\,(1, V, 0, 0)^T$ and $k^\mu = (\omega, k, 0, 0)^T$. This corresponds to fluctuations traveling parallel to the background velocity. The dispersion relation for the transverse modes, Eq. (2.93), transforms to

$$(\omega - Vk)\left[1 + i\gamma\tau_\pi\,(\omega - Vk)\right] - i\tau_\eta\gamma\,(\omega V - k)^2 = 0 \,, \qquad (2.112)$$

where we used Eqs. (2.84) and (2.85). This is a quadratic equation and, consequently, it has two solutions, each with a twofold degeneracy. These are

$$\omega_\pm(k) = \frac{1}{2(\tau_\pi - V^2\tau_\eta)\gamma}$$
$$\times \left[i - 2\gamma(\tau_\eta - \tau_\pi)kV \pm \sqrt{-1 - 4i\gamma^{-1}\tau_\eta Vk + 4\gamma^{-2}\tau_\eta\tau_\pi k^2}\right].$$
$$(2.113)$$

At $k = 0$, it is straightforward to see that the two solutions simplify to

$$\omega\,(k = 0) = \begin{cases} 0 \,, \\ \dfrac{i}{\gamma\,(\tau_\pi - V^2\tau_\eta)} \,, \end{cases} \qquad (2.114)$$

which are very similar to the solution (2.97) obtained in a non-moving background. They are also quite similar to the results obtained from Navier–Stokes theory in a moving background, Eq. (2.73). The main and fundamental difference is that the non-propagating mode that appears in Navier–Stokes theory, when perturbed around a moving fluid, always has a negative imaginary part. As can be seen from the expression above, it is the introduction of a shear relaxation time that makes it possible to change the sign of the imaginary part of such a non-propagating mode. As a matter of fact, we observe that, as long as $\tau_\pi \geq \tau_\eta$, this mode is always stable: a condition that coincides with the asymptotic causality condition (2.102). So we see once more, in a more practical setting, the connection between causality and stability.

Similarly, the dispersion relation for the longitudinal modes, Eq. (2.104), transforms to

$$0 = \left[(\omega - Vk)^2 - c_s^2 (\omega V - k)^2\right]\left[1 + i\gamma\tau_\pi (\omega - Vk)\right]$$
$$- i\gamma\tau_{\text{eff}} (\omega - Vk)(\omega V - k)^2 . \tag{2.115}$$

This equation is cubic and has three solutions, which are too long to write down in
this context. Once more, we restrict ourselves to study the solution at $k = 0$, where
the instability of the non-propagating modes was already obvious in Navier–Stokes
theory (for a moving background). In this case,

$$\omega (k = 0) = \begin{cases} \pm 0 , \\ \dfrac{i\left(1 - c_s^2 V^2\right)}{\gamma\left[\tau_\pi\left(1 - c_s^2 V^2\right)\tau_\pi - \tau_{\text{eff}} V^2\right]} . \end{cases} \tag{2.116}$$

This is very similar to what we found for the shear modes. We have one non-vanishing
solution, corresponding to the mode that was unstable in Navier–Stokes theory. Here,
we see once more that the shear relaxation time can make this mode stable. For
any value of $0 \le V < 1$, if $\tau_{\text{eff}} \le \tau_\pi (1 - c_s^2)$, the imaginary part of this mode will
always be positive and, thus, it will be linearly stable. This condition corresponds
to the asymptotic causality condition (2.109) derived for the longitudinal modes for
a background fluid at rest, confirming that, once causality is assured, the boosted
fluctuations will be stable.

In Fig. 2.7, the dependence of the group velocity on the wavenumber is shown
for various values of the boost velocity V. The left panel shows the behavior of one
of the shear modes and the right panel one of the sound modes. The parameter set
used here is $\eta/s_0 = 1/(4\pi)$, $\tau_\pi = 6\tau_\eta$, $c_s^2 = 1/3$, which satisfies the asymptotic
causality condition. We observe that the divergence of the group velocity of the
shear mode in the rest frame is tampered by the movement of the unperturbed fluid,
resulting in a peak of finite height. However, the group velocity may still exceed the

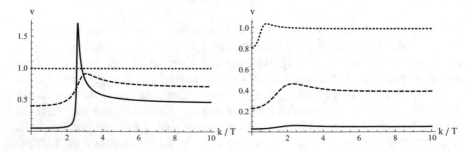

Fig. 2.7 The group velocity calculated for one of the shear modes (left panel) and one of the sound
modes (right panel). We set $\eta/s_0 = 1/(4\pi)$, $\tau_\pi = 6\tau_\eta$, $c_s^2 = 1/3$. The solid line is for a boost
velocity $V = 0.05$, the dashed line for $V = 0.4$, and the dotted line for $V = 0.99$, respectively.
Figure taken from Ref. [5]

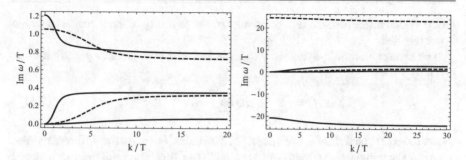

Fig. 2.8 The imaginary parts of the dispersion relations for a boost in x-direction with velocity $V = 0.9$. The left panel shows the results for the parameter set $\eta/s_0 = 1/(4\pi)$, $\tau_\pi = 6\tau_\eta$, $c_s^2 = \frac{1}{3}$, which fulfills the asymptotic causality condition, while the right panel is for $\eta/s_0 = 1/(4\pi)$, $\tau_\pi = \tau_\eta$, $c_s^2 = \frac{1}{3}$, which violates this condition. The dashed lines are for the shear modes, while the solid lines are for the sound modes. Figure taken from Ref. [5]

speed of light in a certain range of wavenumbers. Nevertheless, we note that as we increase the velocity of the unperturbed fluid, the peak height gradually diminishes, until the group velocity remains below the speed of light for all wavenumbers. However, the same does not occur with the longitudinal group velocity, which we found to be superluminal in some ranges of wavenumber even when the velocity of the unperturbed fluid becomes close to the speed of light.

Although the group velocity of the shear or the sound mode may exceed the speed of light, the theory is still stable as long as the asymptotic causality condition is fulfilled. This is demonstrated in the left panel of Fig. 2.8, where the imaginary parts of the modes are shown for the parameter set $\eta/s_0 = 1/(4\pi)$, $\tau_\pi = 6\tau_\eta$, $c_s^2 = 1/3$. We observe that all imaginary parts are positive, indicating the stability of the theory.

In contrast to the case where the background fluid is at rest, where the theory was found to be stable even for parameters which violate the asymptotic causality condition (2.109), this is no longer the case for perturbations performed on a moving fluid. In the right panel of Fig. 2.8, the imaginary parts of the modes are calculated with the parameter set $\eta/s_0 = 1/(4\pi)$, $c_s^2 = 1/3$, and $\tau_\pi = \tau_\eta$—the latter violating the causality and stability conditions derived in this section. Now one observes the appearance of negative imaginary parts, indicating that the theory becomes unstable.

2.5.4 Causality of Wave Propagation

In the preceding discussion, we have seen that transient theories of fluid dynamics can be designed to be stable as long as the transverse and longitudinal modes fulfill asymptotic causality conditions. In this section, we shall show that the *causality* of the theory as a whole is guaranteed if the asymptotic stability condition is fulfilled. The group velocity may become superluminal, or even diverge, as long as this apparent violation of causality is restricted to a finite range of wavenumbers. The argument leading to this conclusion is analogous to that of Sommerfeld and Brillouin in classical electrodynamics [25, 26]. For instance, in the case of anomalous dispersion the group velocity may become superluminal, but the causality of the theory as a whole is not affected.

The change in a fluid-dynamical variable induced by a general perturbation is given by

$$\delta X(x, t) = \sum_j \int d\omega \, \widetilde{\delta X}_j(\omega) \, e^{i\omega t - i k_j(\omega) x} \,, \tag{2.117}$$

where $\delta X(x, t)$ stands for $\delta \varepsilon$, δu^μ, or $\delta \pi^{\mu\nu}$. The index j denotes the different modes, e.g., the shear modes, the sound modes, etc. The function $k_j(\omega)$ is the inverted dispersion relation $\omega_j(k)$ of the respective mode. The Fourier components are given by

$$\sum_j \widetilde{\delta X}_j(\omega) = \frac{1}{2\pi} \int\limits_{-\infty}^{\infty} dt \, \delta X(0, t) \, e^{-i\omega t} \,. \tag{2.118}$$

We assume that the incident wave has a well-defined front that reaches $x = 0$ not before $t = 0$. Thus $\delta X(0, t) = 0$ for $t < 0$. This condition on $\delta X(0, t)$ ensures that $\sum_j \widetilde{\delta X}_j(\omega)$ is analytic in the lower half of the complex ω-plane [25]. On the other hand, in Sects. 2.5.1 and 2.5.3 we have found that the group velocity of the shear modes diverges for certain values of k. These divergences correspond to singularities in the complex ω-plane. However, if the asymptotic causality condition is fulfilled, the imaginary part of the dispersion relation is always positive, i.e., the singularities only appear in the upper half of the complex ω-plane. In this case, the system is also stable. On the other hand, if the asymptotic causality condition is violated, the singularities may appear also in the lower half-plane, i.e., for negative imaginary part of the dispersion relation, and the system is unstable.

We shall now demonstrate that the divergences in the group velocity do not violate causality as long as the asymptotic causality condition is satisfied, i.e., as long as the asymptotic group velocity remains subluminal. To this end, we compute Eq. (2.117) by contour integration in the complex ω-plane. To close the contour, we have to know the asymptotic behavior of the dispersion relations. In our calculation, we found that the real part of the dispersion relation at large k is proportional to k (see Eqs. (2.98) and (2.107)), with a coefficient which is the large-k limit of the group velocity, i.e., v_{gj}^{as},

$$\lim_{k \to \infty} \text{Re} \, \omega_j(k) = v_{gj}^{as} k \,. \tag{2.119}$$

Then, in the large-k limit, the exponential becomes

$$\exp[i\omega t - i k_j(\omega) x] \to \exp\left[-i \frac{\omega}{v_{gj}^{as}} \left(x - v_{gj}^{as} t\right)\right] \,. \tag{2.120}$$

In the case $x > v_{gj}^{as} t$, we have to close the integral contour in the lower half-plane. If the asymptotic causality condition is fulfilled, there are no singularities in the lower half-plane, and Eq. (2.117) vanishes. On the other hand, the contour should be closed in the upper half-plane if $x \leq v_{gj}^{as} t$. Then, because of the singularities, Eq. (2.117) may have a nonzero value. However, as long as we choose a parameter set for which the asymptotic group velocity v_{gj}^{as} is smaller than the speed of light, i.e., for which the asymptotic causality condition is fulfilled, the signal propagation does not violate causality, since the locations x where the disturbance has traveled lie within the cone given by v_{gj}^{as} which, in turn, lies within the lightcone, q.e.d.

To conclude this section, we have shown that the asymptotic causality condition not only implies stability in a general (Lorentz-boosted) frame but also causality of the theory as a whole.

2.6   Summary

In this chapter, we discussed the properties of linearized fluid-dynamical equations around global equilibrium. The main goal was to determine whether hydrodynamic fluctuations around global equilibrium are stable and causal—two fundamental properties that any physical description should satisfy. We investigated these issues for ideal fluids and the two formulations of relativistic dissipative fluid dynamics considered so far in this book: Navier–Stokes theory and transient fluid dynamics (usually represented in the form of Israel–Stewart theory).

First, we studied the linear properties of ideal fluids and recovered all its basic and well-known features. In this case, transverse modes do not appear and the longitudinal modes are purely propagating, describing the propagation of sound waves. The velocity of sound of relativistic fluids was then obtained, and given by the (square root of the) derivative of the pressure with respect to the energy density, at fixed entropy per particle (or per net-charge). All modes in ideal fluid dynamics are stable and the theory is causal, as long as the velocity of sound $c_s \leq 1$—a property that is satisfied by all known microscopic theories.

Next, we discussed the linear properties of relativistic Navier–Stokes theory. When investigating perturbations performed on a fluid that is initially at rest, the linear properties of relativistic Navier–Stokes theory essentially reduce to those of its non-relativistic version. In this case, transverse modes do appear and describe the diffusion of the velocity field due to viscosity. The longitudinal modes still display the propagation of sound waves, which now exhibit diffusion-type damping due to the shear and bulk viscosity of the fluid. Nevertheless, novel striking features of a relativistic version of Navier–Stokes theory appear when considering perturbations on a moving fluid. In this case, we demonstrated that unstable non-hydrodynamic modes, which simply do not exist in the non-relativistic Navier–Stokes theory, appear in both the transverse and longitudinal degrees of freedom. The existence of such unstable modes renders the general global-equilibrium solution of relativistic Navier–Stokes theory linearly unstable, posing a fundamental problem to the application of this theory to describe any relativistic fluid existing in Nature. It is this fundamental problem that the transient theories of fluid dynamics, constructed in the previous chapter, aim to correct.

We then investigated the transverse and longitudinal modes of (second-order) transient fluid dynamics and demonstrated that such theories can be constructed to be causal and stable, at least when perturbed around global equilibrium. We argued (and later proceeded to demonstrate) that as long as the asymptotic value of the group velocity remains subluminal, $v_g^{as} \leq 1$, the theory is causal. We have shown that the condition $v_g^{as} \leq 1$ is equivalent to the requirement that the relaxation time scale τ_π must not be smaller than the ratio $\tau_{eff}/(1 - c_s^2)$, where $\tau_{eff} \sim \eta/(\varepsilon_0 + P_0) \sim \beta_0 \eta/s_0$ Thus, second order transient theories of fluid dynamics are not *per se* stable and causal; they may become unstable and acausal if this condition is violated. This is an important conclusion for practitioners of fluid dynamics, who frequently consider τ_π and the shear viscosity-to-entropy density ratio η/s_0 to be independent from each other. We have demonstrated that this is not the case if one wants the theory to remain causal. These findings also illuminate, from a different perspective, why Navier–Stokes theory violates causality, because there $\tau_\pi \rightarrow 0$, while η remains nonzero. Therefore, transient theories of fluid dynamics, such as Israel–Stewart theory, can actually be successfully applied to describe the dynamics of relativistic fluids.

We finally recounted a time-honored argument from electrodynamics, proving that causality of a theory is guaranteed if the large-wavenumber limit of the group velocity remains subluminal. Thus, a divergence of the group velocity at some finite wavenumber, which actually occurs for some modes of transient fluid dynamics, does not necessarily imply that the theory becomes acausal.

References

1. Landau, L.D., Lifshitz, E.M.: Fluid Mechanics. Pergamon, New York (1959)
2. Eckart, C.: Phys. Rev. **58**, 919 (1940)
3. Hiscock, W., Lindblom, L.: Phys. Rev. D **31**, 725 (1985)
4. Denicol, G.S., Kodama, T., Koide, T., Mota, P.: J. Phys. G **35**, 115102 (2008)
5. Pu, S., Koide, T., Rischke, D.H.: Phys. Rev. D **81**, 114039 (2010)
6. Grad, H.: Commun. Pure Appl. Math. **2**, 331 (1949)
7. Israel, W., Stewart, J.M.: Ann. Phys. (N.Y.) **118**, 341 (1979)
8. Müller, I.: Living Rev. Rel. **2**, 1 (1999)
9. Müller, I.: Z. Phys. **198**, 329 (1967)
10. Liu, I.S., Müller, I., Ruggeri, T.: Ann. Phys. (N.Y.) **169**, 191 (1986)
11. Carter, B.: Proc. R. Soc. Lond., Ser A, **433**, 45 (1991)
12. Grmela, M., Öttinger, H.C.: Phys. Rev. E **56**, 6620 (1997)
13. Baier, R., Romatschke, P., Son, D.T., Starinets, A.O., Stephanov, M.A.: JHEP **0804**, 100 (2008)
14. Koide, T., Denicol, G., Mota, Ph., Kodama, T.: Phys. Rev. C **75**, 034909 (2007)
15. Bemfica, F.S., Disconzi, M.M., Noronha, J.: Phys. Rev. D **98**(10), 104064 (2018). arXiv:1708.06255 [gr-qc]
16. Denicol, G.S., Niemi, H., Molnar, E., Rischke, D.H.: Phys. Rev. D **85**, 114047 (2012). [arXiv:1202.4551 [nucl-th]]
17. Hiscock, W., Lindblom, L.: Ann. Phys. (N.Y.) **151**, 466 (1983)
18. Hiscock, W., Lindblom, L.: Phys. Rev. D **35**, 3723 (1987)
19. Olson, T.S.: Ann. Phys. **199**, 18 (1990). https://doi.org/10.1016/0003-4916(90)90366-V

20. Bemfica, F.S., Disconzi, M.M., Noronha, J.: Phys. Rev. Lett. **122**(22), 221602 (2019). https://doi.org/10.1103/PhysRevLett.122.221602, arXiv:1901.06701 [gr-qc]
21. Bemfica, F.S., Disconzi, M.M., Hoang, V., Noronha, J., Radosz, M.: arXiv:2005.11632 [hep-th]
22. Denicol, G.S., Koide, T., Rischke, D.H.: Phys. Rev. Lett. **105**, 162501 (2010). https://doi.org/10.1103/PhysRevLett.105.162501, arXiv:1004.5013 [nucl-th]
23. Denicol, G.S.: J. Phys. G **41**(12), 124004 (2014)
24. Kovtun, P., Son, D.T., Starinets, A.O.: Phys. Rev. Lett. **94**, 111601 (2005). https://doi.org/10.1103/PhysRevLett.94.111601. [arXiv:hep-th/0405231 [hep-th]]
25. Jackson, J.D.: Classical Electrodynamics. Wiley, New York (1999)
26. Brillouin, L.: Wave Propagation and Group Velocity. Academic Press, London (1960)

Analytical Solutions and Transient Dynamics

In the previous chapter, we have discussed the properties of relativistic fluid-dynamical theories in the linear regime, describing how perturbations of an equilibrium state evolve in space and time. In the linear regime, the equations can be solved analytically and their properties can be studied in detail. It was demonstrated that the relativistic generalization of Navier–Stokes theory is intrinsically unstable, since any perturbation around a moving homogeneous fluid grows exponentially with time (on time scales of the order $\sim \eta / (\varepsilon + P_0)$). On the other hand, transient theories of fluid dynamics can be designed to be linearly stable. If only the effects of shear viscosity are considered (i.e., if the bulk viscosity $\zeta = 0$), the theory will be linearly stable as long as the shear relaxation time, τ_π, satisfies the condition (2.109), i.e.,

$$\left(1 - c_s^2\right) \tau_\pi \geq \tau_{\text{eff}} \equiv \frac{4}{3} \frac{\eta}{\varepsilon + P_0} \ .$$

This condition was also shown to be related to the causality of the theory—it guarantees that no perturbations can travel faster than light.

In order to render the theory causal and stable, significant modifications had to be made, as discussed in Chap. 1. The main point was to extend the set of dynamical variables to include not only the traditional hydrodynamic fields (net-charge density, energy density, and 4-velocity) but also the dissipative currents (shear-stress tensor, bulk-viscous pressure, and charge-diffusion current). This implies that the dissipative quantities must also have their initial value specified on a given space-like hypersurface in order to obtain solutions of the theory. Since the conservation laws couple the traditional fluid-dynamical variables to the dissipative fluxes, the solution for the hydrodynamic fields n, ε, and u_μ is expected to be sensitive to the choices made for the initial conditions of Π, n^μ, and $\pi^{\mu\nu}$. This new feature is related to the appearance of the so-called non-hydrodynamic modes in the linear analyses performed in the

© Springer Nature Switzerland AG 2021
G. S. Denicol and D. H. Rischke, *Microscopic Foundations of Relativistic Fluid Dynamics*,
Lecture Notes in Physics 990, https://doi.org/10.1007/978-3-030-82077-0_3

previous chapter. In equilibrium, this dependence is of course erased but one may ask whether there is some type of non-equilibrium regime in which such a dependence can be observed.

In this chapter, we investigate the properties of transient fluid dynamics in more complex settings, where nonlinear effects cannot be ignored. This study can be carried out completely in highly symmetrical flow configurations, where the symmetries of the problem are sufficient to solve for the fluid 4-velocity and the remaining components of the conserved currents can be calculated analytically or, at least, semi-analytically. Such solutions are extremely rare and were mainly developed as toy models for the fluid-dynamical expansion of the hot and dense matter produced in ultrarelativistic heavy-ion collisions. Recently, these solutions have played a key role in understanding fluid dynamics itself, as they have allowed the calculation of the gradient series to arbitrarily high order—making it possible to determine, for the first time, the radius of convergence of such series for expanding systems [1]. In this chapter, we discuss and derive these rare solutions of relativistic fluid dynamics and show how they can be used to study the asymptotic properties of transient fluid dynamics, to a level that is not possible in the linear regime.

This chapter is organized as follows. In Sect. 3.1, we describe the form of the fluid-dynamical equations of motion in a general coordinate system. In Sects. 3.2 and 3.3, we derive rare solutions of relativistic fluid dynamics that can be obtained in highly symmetric flow configurations: Bjorken-flow solutions [2] and Gubser-flow solutions [3], respectively.

3.1   Fluid Dynamics in a General Metric System

In Chap. 1, the fluid-dynamical equations were derived in flat space, assuming a Minkowski metric (with the "mostly minus" sign convention). For the purposes of this chapter, it is necessary to discuss how these equations are modified when describing fluids in curved space or in other coordinate systems. In these cases, the metric tensor may depend on the space-time coordinates and, therefore, does not necessarily commute with the coordinate derivative ∂_μ. In order to maintain the covariance of the equations of motion, they must be re-expressed in terms of *covariant* derivatives, D_μ, which are linear operators constructed in such a way as to commute with the metric tensor,

$$D_\mu g^{\alpha\beta} = 0 \, .$$

Covariant derivatives of tensors are constructed to be tensors, while coordinate derivatives of tensors are not. The covariant derivative basically follows the rules usually associated with differentiation: it is a linear operation, satisfying the distributive property

$$D_\mu \left(A^{\nu_1 \cdots \nu_n} + B^{\nu_1 \cdots \nu_n} \right) = D_\mu A^{\nu_1 \cdots \nu_n} + D_\mu B^{\nu_1 \cdots \nu_n} \, ,$$

and follows the Leibniz rule

$$D_\mu \left(A^{\alpha_1 \cdots \alpha_n} B^{\beta_1 \cdots \beta_n} \right) = B^{\beta_1 \cdots \beta_n} D_\mu A^{\alpha_1 \cdots \alpha_n} + A^{\alpha_1 \cdots \alpha_n} D_\mu B^{\beta_1 \cdots \beta_n} .$$

Above, A and B are arbitrary tensors.

For the purposes of this chapter, it will be sufficient to know how to apply this operator to tensors of rank 0, 1, and 2. First, the covariant derivative of an arbitrary scalar field, V, is simply given by the usual coordinate derivative of V,

$$D_\mu V = \partial_\mu V . \tag{3.1}$$

On the other hand, the covariant derivative of an arbitrary 4-vector field, V^μ, and of an arbitrary second-rank tensor, $V^{\mu\nu}$, are given in terms of directional derivatives and the metric connections (Christoffel symbols) $\Gamma^\nu_{\mu\lambda}$ as

$$D_\mu V^\nu = \partial_\mu V^\nu + \Gamma^\nu_{\lambda\mu} V^\lambda , \tag{3.2}$$

$$D_\mu V^{\alpha\beta} = \partial_\mu V^{\alpha\beta} + \Gamma^\alpha_{\lambda\mu} V^{\lambda\beta} + \Gamma^\beta_{\lambda\mu} V^{\alpha\lambda} . \tag{3.3}$$

For a covariant vector as well as a mixed and fully covariant rank-2 tensor, these relations read

$$D_\mu V_\nu = \partial_\mu V_\nu - \Gamma^\lambda_{\mu\nu} V_\lambda , \tag{3.4}$$

$$D_\mu V^\alpha_\beta = \partial_\mu V^\alpha_\beta + \Gamma^\alpha_{\lambda\mu} V^\lambda_\beta - \Gamma^\lambda_{\beta\mu} V^\alpha_\lambda , \tag{3.5}$$

$$D_\mu V_{\alpha\beta} = \partial_\mu V_{\alpha\beta} - \Gamma^\lambda_{\alpha\mu} V_{\lambda\beta} - \Gamma^\lambda_{\beta\mu} V_{\alpha\lambda} . \tag{3.6}$$

The Christoffel symbol $\Gamma^\nu_{\mu\lambda}$ can be expressed in terms of coordinate derivatives of the metric tensor in the following way,

$$\Gamma^\nu_{\mu\lambda} = \frac{1}{2} g^{\nu\sigma} \left(\partial_\mu g_{\sigma\lambda} + \partial_\lambda g_{\sigma\mu} - \partial_\sigma g_{\mu\lambda} \right) . \tag{3.7}$$

Note that, using Eqs. (3.2) and (3.3) and the identity $\Gamma^\mu_{\mu\nu} = \partial_\nu \ln \sqrt{-g}$, where $g \equiv \det(g_{\mu\nu})$ is the determinant of the metric tensor, one can also derive simplified forms for the 4-divergence of a 4-vector field and a second-rank tensor,

$$D_\mu V^\mu = \frac{1}{\sqrt{-g}} \partial_\mu \left(\sqrt{-g} V^\mu \right) , \tag{3.8}$$

$$D_\mu V^{\mu\nu} = \frac{1}{\sqrt{-g}} \partial_\mu \left(\sqrt{-g} V^{\mu\nu} \right) + \Gamma^\nu_{\lambda\mu} V^{\mu\lambda} . \tag{3.9}$$

The conservation laws for net-charge (or particle number, in some specific cases) and energy-momentum can be written in the following covariant form,

$$D_\mu N^\mu = 0 , \tag{3.10}$$

$$D_\mu T^{\mu\nu} = 0 . \tag{3.11}$$

As before, we can separate the conservation laws into components that are parallel to the fluid 4-velocity, u^μ, and orthogonal to it. For a general metric, this procedure leads to the following equations of motion, cf. Eqs. (1.72)–(1.74),

$$D_\tau n + n\theta + D_\mu n^\mu = 0 \,, \tag{3.12}$$

$$D_\tau \varepsilon + (\varepsilon + P_0 + \Pi)\theta - \pi^{\mu\nu}\sigma_{\mu\nu} = 0 \,, \tag{3.13}$$

$$(\varepsilon + P_0 + \Pi)D_\tau u^\mu - \Delta_\lambda^\mu D^\lambda (P_0 + \Pi) + \Delta_\lambda^\mu D_\nu \pi^{\lambda\nu} = 0 \,, \tag{3.14}$$

where we now define $D_\tau \equiv u^\mu D_\mu$ as the comoving *covariant* derivative. Also, similar to the definitions employed in Chap. 1, we introduce $\theta = D_\mu u^\mu$ as the local expansion rate and $\sigma_{\mu\nu} = \Delta_{\mu\nu}^{\alpha\beta}D_\alpha u_\beta$ as the shear tensor—now both defined in terms of the covariant derivative.

The arguments developed by Israel and Stewart, as discussed in Chap. 1, to derive a transient theory of fluid dynamics using the second law of the thermodynamics can be implemented in the same way as before—one is just required to replace the coordinate derivatives by covariant ones. In the end, this leads to equations of motion of the same form as before, which are relaxation-type equations, as those obtained in the last section, Eqs. (1.131). For the purposes of this chapter, we consider a more general form of these equations of motion [4,5], derived from kinetic theory, cf. Chap. 5, Eqs. (5.168)–(5.170):

$$\begin{aligned}
\tau_\Pi D_\tau \Pi + \Pi &= -\zeta\theta - \ell_{\Pi n}D_\mu n^\mu - \tau_{\Pi n}n^\mu\nabla_\mu P_0 - \delta_{\Pi\Pi}\Pi\theta \\
&\quad - \lambda_{\Pi n}n^\mu\nabla_\mu\alpha_0 + \lambda_{\Pi\pi}\pi^{\mu\nu}\sigma_{\mu\nu} \,,
\end{aligned} \tag{3.15}$$

$$\begin{aligned}
\tau_n \Delta_\nu^\mu D_\tau n^\nu + n^\mu &= \varkappa\nabla^\mu\alpha_0 - \delta_{nn}n^\mu\theta - \ell_{n\Pi}\nabla^\mu\Pi + \ell_{n\pi}\Delta^{\mu\nu}\nabla_\lambda\pi_\nu^\lambda \\
&\quad + \tau_{n\Pi}\Pi\nabla^\mu P_0 - \tau_{n\pi}\pi^{\mu\nu}\nabla_\nu P_0 + \lambda_{n\Pi}\Pi\nabla^\mu\alpha_0 \\
&\quad - \lambda_{n\pi}\pi^{\mu\nu}\nabla_\nu\alpha_0 + \tau_n n_\nu\omega^{\mu\nu} + \lambda_{nn}n_\nu\sigma^{\mu\nu} \,,
\end{aligned} \tag{3.16}$$

$$\begin{aligned}
\tau_\pi \Delta_{\alpha\beta}^{\mu\nu}D_\tau\pi^{\alpha\beta} + \pi^{\mu\nu} &= 2\eta\sigma^{\mu\nu} - \delta_{\pi\pi}\pi^{\mu\nu}\theta - \tau_{\pi n}n^{\langle\mu}\nabla^{\nu\rangle}P_0 \\
&\quad + \ell_{\pi n}\nabla^{\langle\mu}n^{\nu\rangle} + \lambda_{\pi n}n^{\langle\mu}\nabla^{\nu\rangle}\alpha_0 + \tau_\pi\pi_\lambda^{\langle\mu}\omega^{\nu\rangle\lambda} \\
&\quad - \tau_{\pi\pi}\pi^{\lambda\langle\mu}\sigma_\lambda^{\nu\rangle} + \lambda_{\pi\Pi}\Pi\sigma^{\mu\nu} \,,
\end{aligned} \tag{3.17}$$

where we now generalize the nabla operator to be the projection of the covariant derivative, $\nabla^\mu \equiv \Delta^{\mu\nu}D_\nu$. We also introduced several new transport coefficients, which did not appear in the derivation of transient fluid dynamics explained in Chap. 1, but will emerge from more general derivations that use the Boltzmann equation as the starting point, as will be explained in Chaps. 5 and 8.

Note that the procedure employed to define the fluid 4-velocity (using, e.g., the Landau or Eckart picture, see Sect. 1.3.3) and its fictitious local-equilibrium state (matching conditions, see Sect. 1.3.1) do not need to be modified when using an arbitrary set of coordinates. One should just note that the local rest frame defined in one set of coordinates may correspond to a moving frame in another set of coordinates (this is the case for hyperbolic and Cartesian coordinates, which will be discussed in the next section). Therefore, as one changes the reference coordinate system, one may also be modifying the corresponding definition of the local rest frame.

In the following, we discuss solutions of relativistic fluid dynamics that can be obtained with the assistance of two coordinate systems: (i) hyperbolic or *Milne* coordinates, which are useful in deriving the so-called Bjorken-flow solutions [2], and (ii) Gubser coordinates, which are useful in deriving the so-called Gubser-flow solutions [3].

3.2 Bjorken Flow

In certain highly symmetric flow configurations, it is possible to exactly determine the flow velocity. In these cases, it becomes possible to solve the fluid-dynamical equations analytically (even in the transient regime) and study at least some properties of their solution, beyond the linear regime. In this section, we explain the most well-known of such configurations, the Bjorken flow [2].

3.2.1 Coordinates and Kinematic Properties

We start by considering a coordinate system with a space-time described by the metric [2]

$$ds^2 = g_{\mu\nu}dx^\mu dx^\nu = d\tau^2 - \left(dx^2 + dy^2 + \tau^2 d\eta_s^2\right) , \qquad (3.18)$$

where we introduced the Milne coordinates τ and η_s,

$$\tau \equiv \sqrt{t^2 - z^2} , \qquad \eta_s \equiv \frac{1}{2} \ln \frac{t + z}{t - z} = \text{Artanh} \frac{z}{t} , \qquad (3.19)$$

which replace the Cartesian coordinates t and z, i.e.,

$$t = \tau \cosh \eta_s , \quad z = \tau \sinh \eta_s . \qquad (3.20)$$

The time variable τ is invariant under Lorentz boosts in the z-direction, while the coordinate η_s is just shifted by the rapidity of such a boost. As it will turn out, in the Bjorken-flow scenario each fluid element moves with the velocity $v \equiv z/t$. Then, τ is equal to the proper time in the rest frame of this fluid element, while η_s is the rapidity corresponding to its velocity. Rapidity is additive under Lorentz boosts, which explains the abovementionend shift of η_s under such boosts. Note that in Milne coordinates the initial condition for the fluid variables is typically specified on a hypersurface of constant $\tau = \tau_0$, which corresponds to a hyperbola in the (t, z)-plane.

Here, we shall assume a fluid which is *symmetric* under reflections $\eta_s \to -\eta_s$, *homogeneous* (translationally invariant) along the η_s-direction, as well as *homogeneous and isotropic* (translationally invariant as well as rotationally invariant) in the (x, y)-plane. We note that the assumption of homogeneity in η_s-direction renders the fluid *boost-invariant*. Thus, all fluid-dynamical fields will depend solely on the time variable τ, without displaying any dependence on x, y, and η_s. This means that all spatial derivatives, i.e., derivatives with respect to x, y, and η_s, vanish. It further implies that the spatial part of any 4-vector must vanish—there cannot be any preferred direction neither in the transverse (x, y)-plane nor along the longitudinal η_s-axis.

As we can see from Eq. (3.18), the metric tensor is diagonal,

$$g_{\mu\nu} = \text{diag}\left(1, -1, -1, -\tau^2\right) \quad \Longrightarrow \quad g^{\mu\nu} = \text{diag}\left(1, -1, -1, -\frac{1}{\tau^2}\right), \quad (3.21)$$

and depends solely on the new time coordinate, τ. Therefore, there are only a few components of the Christoffel symbol (3.7) related to this metric tensor that are nonzero,

$$\Gamma^\tau_{\eta_s\eta_s} = \tau, \quad \Gamma^{\eta_s}_{\tau\eta_s} = \Gamma^{\eta_s}_{\eta_s\tau} = \frac{1}{\tau}. \quad (3.22)$$

Also, the determinant of this metric can be easily computed to be $g = -\tau^2$ and, consequently,

$$\sqrt{-g} = \tau. \quad (3.23)$$

The imposed assumptions imply that a solution to Eq. (3.14) is $u^\mu = (1, 0, 0, 0)^T$ (in hyperbolic coordinates). For such a velocity field,

$$\Delta^{\mu\nu} = g^{\mu\nu} - u^\mu u^\nu = \text{diag}\left(0, -1, -1, -\frac{1}{\tau^2}\right), \quad (3.24)$$

cf. Eq. (3.21). In order to prove that the solution u^μ to Eq. (3.14) has this simple form, let us first consider an arbitrary *scalar* function V. The spatially projected covariant derivative of V is zero,

$$\Delta^\mu_\nu D^\nu V = \Delta^\mu_\nu \partial^\nu V = \Delta^\tau_\tau \partial^\tau V = 0. \quad (3.25)$$

Here, we used the assumption that the system is spatially homogeneous, such that derivatives with respect to x, y, and η_s vanish, and we employed Eqs. (3.1) and (3.24). As a conclusion, the second term in Eq. (3.14) vanishes identically.

Now consider the shear-stress tensor $\pi^{\mu\nu}$. The third term in Eq. (3.14) is the spatially projected covariant derivative of the 4-divergence of the shear-stress tensor. Using the assumption of spatial homogeneity all derivatives except those with respect to τ vanish

$$\Delta^\mu_\lambda D_\nu \pi^{\nu\lambda} = \Delta^\mu_\lambda \left[\frac{1}{\tau}\partial_\tau\left(\tau\pi^{\tau\lambda}\right) + \Gamma^\lambda_{\rho\nu}\pi^{\nu\rho}\right]$$

$$= \Delta^\mu_\lambda \Gamma^\lambda_{\rho\nu}\pi^{\nu\rho} = \Delta^\mu_\tau \Gamma^\tau_{\eta_s\eta_s}\pi^{\eta_s\eta_s} = 0, \quad (3.26)$$

where we used Eq. (3.22) and the fact that, since $\pi^{\mu\nu}$ is orthogonal to u^μ, $u_\mu \pi^{\mu\nu} = 0$, we have that $\pi^{\tau\mu} = \pi^{\mu\tau} = 0$ for any $\mu = \tau$, x, y, or η_s. The final term vanishes because of Eq. (3.24).

We have now proved that the last two terms in Eq. (3.14) vanish. This equation then simplifies to $D_\tau u^\mu = 0$, but since u^μ can only have a time component (see discussion above), this actually means that

$$D_\tau u^\tau = \partial_\tau u^\tau + \Gamma^\tau_{\lambda\tau} u^\lambda = \partial_\tau u^\tau = 0 \, , \tag{3.27}$$

where we have used Eqs. (3.2) and (3.22). Obviously, this equation is solved by $u^\mu = (1, 0, 0, 0)^T$, as claimed. Note that a similar reasoning also applies in Cartesian coordinates, but the solution in that case is just the well-known global-equilibrium solution.

We can derive an immediate consequence of the solution $u^\mu = (1, 0, 0, 0)^T$ (in hyperbolic coordinates). Namely, if we rewrite this solution in Cartesian coordinates (for the moment denoted with primed variables), and remember that the 4-velocity u'^μ can be written as the derivative of the space-time coordinate vector x'^μ with respect to an evolution parameter, which we choose to be the variable τ, $u'^\mu \equiv dx'^\mu/d\tau$, we then compute with Eq. (3.20)

$$u'^\mu \equiv \frac{dx'^\mu}{d\tau} = \left(\frac{dt}{d\tau}, 0, 0, \frac{dz}{d\tau} \right)^T = (\cosh \eta_s, 0, 0, \sinh \eta_s)^T \equiv \gamma \, (1, 0, 0, v)^T \, , \tag{3.28}$$

with the velocity $v \equiv z/t$ in z-direction and the Lorentz-gamma factor $\gamma = (1 - v^2)^{-1/2}$. This proves the abovementioned claim that the velocity of each fluid element in the Bjorken-flow scenario is $v = z/t$.

Therefore, a fluid expanding in the longitudinal direction in Cartesian coordinates (with velocity $\sim z/t$) can be described as a homogeneous fluid, which just changes in time in hyperbolic coordinates—the expansion of the system being hidden in the metric and in the covariant derivatives. This allows us to solve certain relatively complicated fluid-dynamical problems using rather simple techniques and, in some cases, the problems can be solved even analytically. This is the case for the theoretical description of ultrarelativistic heavy-ion collisions, which are always better described in hyperbolic coordinates, because they better capture the strong initial longitudinal expansion which develops in such systems [6].

Even though the fluid velocity field can be locally static in hyperbolic coordinates, the fluid itself cannot be static. The time dependence of the metric tensor (3.18) will naturally induce a nonzero expansion rate,

$$\theta = \frac{1}{\sqrt{-g}} \partial_\mu (\sqrt{-g} \, u^\mu) = \frac{1}{\tau} \partial_\tau \tau = \frac{1}{\tau} \, , \tag{3.29}$$

where we have used Eq. (3.8).

The shear tensor $\sigma_{\mu\nu}$ is also nonzero (due to the intrinsic anisotropy of the hyperbolic coordinate system), but only its spatial diagonal elements have a nonzero value,

$$\sigma_{\mu\nu} = \Delta_{\mu\nu}^{\alpha\beta} D_\alpha u_\beta = -\Gamma_{\langle\mu\nu\rangle}^\tau = \text{diag}\left(0, \frac{1}{3\tau}, \frac{1}{3\tau}, -\frac{2}{3}\tau\right), \qquad (3.30)$$

where we have used Eqs. (3.4), (3.21), and (3.22), and $u^\mu = (1, 0, 0, 0)^T$. Both θ and the transverse components of $\sigma_{\mu\nu}$ are inversely proportional to the time coordinate, $\sim 1/\tau$, and, hence, become large at early *hyperbolic* times. At very late hyperbolic times, $\tau \gg 1$, these quantities vanish and the fluid's state resembles that of global equilibrium.

In the Bjorken-flow scenario, the vorticity tensor vanishes,

$$\begin{aligned}
\omega^{\mu\nu} &= \frac{1}{2}\Delta^{\mu\alpha}\Delta^{\nu\beta}\left(D_\alpha u_\beta - D_\beta u_\alpha\right) \\
&= \frac{1}{2}\Delta^{\mu\alpha}\Delta^{\nu\beta}\left(\partial_\alpha u_\beta - \Gamma_{\alpha\beta}^\tau - \partial_\beta u_\alpha + \Gamma_{\beta\alpha}^\tau\right) = 0,
\end{aligned} \qquad (3.31)$$

where we used Eq. (3.4) and the fact that $u^\mu = (1, 0, 0, 0)^T$ is constant.

According to the irreducible decomposition of the gradient field of the fluid 4-velocity [7],

$$D_\mu u_\nu = u_\mu D_\tau u_\nu + \omega_{\mu\nu} + \sigma_{\mu\nu} + \frac{\theta}{3}\Delta_{\mu\nu}, \qquad (3.32)$$

the kinematics of the Bjorken-flow velocity field is completely determined by Eqs. (3.29), (3.30), and (3.31), as well as the fact that the velocity field is acceleration-free, $D_\tau u_\mu = 0$.

Furthermore, any 4-vector V^μ that is orthogonal to u^μ must always be zero in this setting, since the orthogonality to the 4-velocity also implies that the 4-vector has a vanishing zeroth component, $u_\mu V^\mu = 0 \iff V^\tau = 0$. Thus, a diffusion 4-current cannot exist, $n^\mu \equiv 0$.

Finally, we can further restrict the form of the shear-stress tensor. From its orthogonality to u^μ, we have already concluded that $\pi^{\tau\mu} = \pi^{\mu\tau} = 0$. As a further constraint, we can impose the condition that it must display the same symmetries exhibited by the shear tensor and, for this reason, it is convenient to parametrize it in the following way

$$\pi^{\mu\nu} = \text{diag}\left(0, \frac{\bar\pi}{2}, \frac{\bar\pi}{2}, -\frac{\bar\pi}{\tau^2}\right). \qquad (3.33)$$

This parametrization also guarantees that $\pi^{\mu\nu}$ remains traceless ($\pi_\mu^\mu = 0$). We then see that, in this geometry, the shear-stress tensor is described solely by one degree of freedom, the function $\bar\pi(\tau)$.

3.2.2 Fluid-Dynamical Equations

With the assumed symmetries, we were already able to determine the velocity field of this system in the preceding subsection—it will always be of the form $u^\mu = (1, 0, 0, 0)^T$ in hyperbolic coordinates, regardless of the properties of the fluid being described. Now, using all the simplifying implications derived above, we can determine the equations of motion satisfied by the energy and particle (or net-charge) densities and the dissipative currents. First, we note that Eq. (3.13) can be rewritten in the form of an *ordinary* first-order differential equation for the energy density,

$$u_\nu D_\mu T^{\mu\nu} = \frac{d\varepsilon}{d\tau} + \frac{1}{\tau}\left(\varepsilon + P_0 + \Pi - \bar{\pi}\right) = 0\,, \tag{3.34}$$

where we used Eq. (3.29) as well as $\pi^{\mu\nu}\sigma_{\mu\nu} = \bar{\pi}/\tau$, cf. Eqs. (3.30) and (3.33). Due to the vanishing of the diffusion current, the equation of motion (3.12) for the particle (or net-charge) density can also be simplified as

$$D_\mu N^\mu = \frac{dn}{d\tau} + \frac{n}{\tau} = 0\,. \tag{3.35}$$

Thus, in Bjorken flow, the dynamics of the particle density decouples from that of the energy-momentum tensor. Similar to what occurred with the fluid 4-velocity, it is straightforward to show that the solution for the particle density is universal and independent of the properties of the fluid. Equation (3.35) can be solved analytically,

$$n(\tau) = n(\tau_0)\frac{\tau_0}{\tau} \quad\Longrightarrow\quad n(t, z) = \frac{n(\tau_0)\tau_0}{\sqrt{t^2 - z^2}}\,, \tag{3.36}$$

where $n(\tau_0)$ is the value of the particle density at some specific (hyperbolic) time, τ_0. Both quantities, τ_0 and $n(\tau_0)$, are free parameters of the solution. We also showed the solution in Cartesian coordinates, which clearly are not spatially homogeneous (and are valid within the forward lightcone, for any $t > z$).

Finally, the Israel–Stewart equations (3.15) and (3.17) also become ordinary first-order differential equations for Π and $\bar{\pi}$. Using Eqs. (3.21), (3.22), (3.24), (3.29), (3.30), (3.31), and (3.33), we obtain after some calculation [8]

$$\tau_\Pi \frac{d\Pi}{d\tau} + \Pi + \delta_{\Pi\Pi}\frac{\Pi}{\tau} = -\frac{\zeta}{\tau} + \lambda_{\Pi\pi}\frac{\bar{\pi}}{\tau}\,, \tag{3.37}$$

$$\tau_\pi \frac{d\bar{\pi}}{d\tau} + \bar{\pi} + \left(\delta_{\pi\pi} + \frac{\tau_{\pi\pi}}{3}\right)\frac{\bar{\pi}}{\tau} = \frac{4\eta}{3\tau} + \lambda_{\pi\Pi}\frac{2\Pi}{3\tau}\,, \tag{3.38}$$

where we have dropped many terms because the particle-diffusion 4-current vanishes. Thus, also Eq. (3.16) does not need to be solved. An equation of state, $P_0(\beta_0, \alpha_0)$, and the functional dependence of transport coefficients on the thermodynamical variables, $\eta(\beta_0, \alpha_0)$, $\zeta(\beta_0, \alpha_0)$, $\tau_\pi(\beta_0, \alpha_0)$, $\tau_\Pi(\beta_0, \alpha_0)$, $\delta_{\Pi\Pi}(\beta_0, \alpha_0)$, $\delta_{\pi\pi}(\beta_0, \alpha_0)$, $\tau_{\pi\pi}(\beta_0, \alpha_0)$, $\lambda_{\pi\Pi}(\beta_0, \alpha_0)$, and $\lambda_{\Pi\pi}(\beta_0, \alpha_0)$ must be known in order to solve these equations. A value for the dissipative fields, Π and $\bar{\pi}$, at the initial time τ_0 must also be specified. Once such information is provided, it is straightforward to solve the fluid-dynamical equations for the Bjorken flow numerically. For some types of fluids, it will be possible to solve them even analytically, as will be demonstrated in the next sections.

Note that Navier–Stokes theory can be simply recovered by taking the limit of vanishing relaxation times, i.e., $\tau_\pi \to 0$ and $\tau_\Pi \to 0$, and further imposing $\delta_{\Pi\Pi} \to 0$, $\delta_{\pi\pi} + \tau_{\pi\pi}/3 \to 0$, $\lambda_{\pi\Pi} \to 0$, and $\lambda_{\Pi\pi} \to 0$. This leads to the following constitutive relations for the dissipative currents,

$$\Pi_{NS} = -\frac{\zeta}{\tau} , \quad \bar{\pi}_{NS} = \frac{4\eta}{3\tau} . \tag{3.39}$$

3.2.3   Ideal-Fluid Limit

For the sake of simplicity, let us first consider the example of an ideal gas of massless particles in the *ideal*-fluid limit. This limit is obtained by taking all the transport coefficients in Eqs. (3.37) and (3.38) to be identically zero, leading to vanishing dissipative contributions, $\bar{\pi} = \Pi = 0$. The assumption that the fluid is an ideal gas made of massless particles fixes the equation of state, which is then given by $P_0 = \varepsilon/3$ (which assures that the trace of the energy-momentum tensor vanishes). Therefore, the equation of motion (3.34) can be expressed solely in terms of the energy density,

$$\frac{d\varepsilon}{d\tau} + \frac{4\varepsilon}{3\tau} = 0 . \tag{3.40}$$

This equation can be solved analytically and the solution is a power law,

$$\varepsilon(\tau) = \varepsilon(\tau_0) \left(\frac{\tau_0}{\tau}\right)^{4/3} . \tag{3.41}$$

Therefore, the energy density decreases as the hyperbolic time increases. At first glance, it may appear strange that the energy density of a homogeneous system may simply decrease with time. However, this is how the longitudinal expansion of the system in Cartesian coordinates manifests itself in the hyperbolic coordinate system.

In order to obtain the solutions for temperature $T \equiv 1/\beta_0$ and chemical potential $\mu \equiv T\alpha_0$, one must provide the functions $n(T, \mu)$ and $\varepsilon(T, \mu)$. For an ideal gas of classical and massless particles, we have

$$n = \frac{d_{\text{dof}}}{\pi^2} \exp\left(\frac{\mu}{T}\right) T^3 , \quad \varepsilon = 3\frac{d_{\text{dof}}}{\pi^2} \exp\left(\frac{\mu}{T}\right) T^4 , \tag{3.42}$$

where d_{dof} is the degeneracy factor of the microscopic degrees of freedom. We thus obtain

$$T(\tau) = \frac{\varepsilon(\tau)}{3n(\tau)} = T(\tau_0) \left(\frac{\tau_0}{\tau}\right)^{1/3} , \quad \frac{\mu(\tau)}{T(\tau)} = \frac{\mu(\tau_0)}{T(\tau_0)} . \tag{3.43}$$

This type of gas will have a temperature that decreases with time as $\tau^{-1/3}$, but will have a constant ratio of chemical potential to temperature. This is the type of behavior expected in general for conformal fluids, which share the same equation of state as illustrated above. Note that a gas of massless particles is not necessarily a conformal fluid, since the cross section of the particles introduces additional scales that break conformal invariance (a cross section that is proportional to $1/T^2$ will not break conformal invariance). Nevertheless, since the behavior of the cross section will only affect the transport coefficients of the fluid, it does not matter for the solutions in the ideal-fluid regime. In the following, we will discuss the effect of dissipation, i.e., nonzero transport coefficients, on the above solutions.

3.2.4 Relativistic Navier–Stokes Theory

In Chap. 2, we have shown that the relativistic generalization of Navier–Stokes theory suffers from intrinsic instabilities, due to the parabolic structure of its equations of motion, and cannot be considered as a viable theory of relativistic fluid dynamics. Nevertheless, it is still useful to study its solutions in the few cases where they exist. In the Bjorken-flow scenario, presented so far in this section, such solutions (even analytical ones) can be obtained since the 4-velocity field is determined *a priori*, at all space-time points, by the symmetries imposed on the fluid. In this sense, the parabolic nature of the theory is overcome by the stringent symmetries satisfied by the system.

In the Navier–Stokes limit, combining Eqs. (3.34) and (3.39) the equation of motion (3.34) satisfied by the energy density simply becomes

$$\frac{d\varepsilon}{d\tau} + \frac{\varepsilon + P_0}{\tau} - \frac{1}{\tau^2}\left(\zeta + \frac{4}{3}\eta\right) = 0 \ . \tag{3.44}$$

In order to solve this equation, we must first specify the properties of the fluid, here given by its equation of state, $P_0(T, \mu)$, the bulk viscosity, $\zeta(T, \mu)$, and the shear viscosity, $\eta(T, \mu)$. Once this is done, Eq. (3.44) can be solved numerically without the need of any sophisticated algorithm.

In order to provide a more thorough understanding of the theory, we also seek for analytical solutions of this equation. As already indicated, this will be possible for fluids which display sufficiently simple thermodynamic and transport properties. With this in mind, we consider once more a fluid composed of an ideal gas of massless particles. The assumption that the particles are massless leads to a vanishing bulk viscosity ($\zeta = 0$) and to an equation of state $P_0(\varepsilon) = \varepsilon/3$, with $\varepsilon(T, \mu)$ given by Eq. (3.42).

Then, the only quantity that remains to be specified is the shear-viscosity coefficient. In this section, we shall consider several parametrizations for this transport coefficient in order to gain some intuition on how the solutions for the dissipative currents depend on it.

3.2.4.1 The Case of Constant τ_η

First, we consider a simple Ansatz for the shear viscosity, namely where its ratio with the enthalpy, $\varepsilon + P_0$, is a constant [9],

$$\frac{\eta}{\varepsilon + P_0} \equiv \tau_\eta = \text{const. .} \tag{3.45}$$

Note that τ_η is a time scale related to the viscous damping of the transverse modes (or shear modes) of the theory, as discussed for perturbations of a fluid at rest in the Chap. 2, Sect. 2.4. It is also the time scale that determines the exponential growth of the unstable modes of Navier–Stokes theory, which appear when considering perturbations in a moving fluid.

The equation of motion (3.44) for $\varepsilon(\tau)$ then considerably simplifies

$$\frac{d\varepsilon}{d\tau} + \frac{4\varepsilon}{3\tau}\left(1 - \frac{4\tau_\eta}{3\tau}\right) = 0, \tag{3.46}$$

and has the following analytical solution,

$$\varepsilon(\tau) = \varepsilon(\tau_0)\left(\frac{\tau_0}{\tau}\right)^{\frac{4}{3}}\exp\left[-\frac{16}{9}\left(\frac{\tau_\eta}{\tau} - \frac{\tau_\eta}{\tau_0}\right)\right]$$

$$= \varepsilon_{\text{ideal}}(\tau)\exp\left[-\frac{16}{9}\left(\frac{\tau_\eta}{\tau} - \frac{\tau_\eta}{\tau_0}\right)\right], \tag{3.47}$$

where $\varepsilon_{\text{ideal}}(\tau)$ is the solution (3.41) for the energy density in the ideal-fluid limit. This solution explicitly demonstrates that the introduction of dissipative effects makes the energy density decrease slower with time, when compared to an ideal fluid. This behavior is actually expected and happens because of the entropy production due to the shear viscosity, which serves to heat up the system. The constant time scale τ_η determines how much slower the energy density of the system will actually decrease with time. Naturally, by taking the limit $\tau_\eta \to 0$, we recover the ideal-fluid solution obtained in the previous section. It is interesting to note that, in the limit of asymptotically large times compared to the microscopic time scale τ_η, $\tau \gg \tau_\eta$, we have

$$\varepsilon(\tau \gg \tau_\eta) \longrightarrow \varepsilon_{\text{ideal}}(\tau)\exp\left(\frac{16}{9}\frac{\tau_\eta}{\tau_0}\right). \tag{3.48}$$

Thus, the energy density in Navier–Stokes theory will be a factor $\exp\left[16\tau_\eta/(9\tau_0)\right]$ larger than for the ideal fluid.

We can also compute the corresponding solutions for the temperature and chemical potential. As already emphasized, for this purpose we need to specify an equation of state. For the simple equation of state of a gas of massless, classical particles, Eq. (3.42), we have

$$T(\tau) = \frac{\varepsilon(\tau)}{3n(\tau)} = T(\tau_0)\left(\frac{\tau_0}{\tau}\right)^{1/3}\exp\left[-\frac{16}{9}\left(\frac{\tau_\eta}{\tau} - \frac{\tau_\eta}{\tau_0}\right)\right], \tag{3.49}$$

$$\frac{\mu(\tau)}{T(\tau)} = \frac{\mu(\tau_0)}{T(\tau_0)} + \frac{16}{3}\left(\frac{\tau_\eta}{\tau} - \frac{\tau_\eta}{\tau_0}\right). \tag{3.50}$$

Here, we used the fact that the solution (3.36) for the density remains the same as in the ideal-fluid limit. Thus, one of the possible effects of dissipation is that the ratio of chemical potential to temperature is no longer a constant, but develops a time dependence.

3.2.4.2 The Case of Hard Spheres

Another, more realistic and yet still simple choice for the shear-viscosity coefficient can be obtained from the Boltzmann equation for a system of massless hard spheres [10]. In this case, the shear viscosity is found to be of the form (the derivation of this expression will be discussed in detail in Chap. 8),

$$\eta \sim \frac{T}{\sigma} \sim \frac{P_0}{n\sigma} , \tag{3.51}$$

where T is the temperature and σ is the total cross section, here assumed to be constant (hard-sphere approximation). In order to obtain the last expression, we used that the pressure of an ideal gas satisfies the relation, $P_0 = nT$. As already discussed, for a gas composed of massless particles, we can also use that $\varepsilon = 3P_0$. Inspired by these results, we consider another Ansatz for the shear viscosity,

$$\eta = \delta \frac{P_0}{n\sigma} \quad \Longleftarrow \quad \tau_\eta = \frac{\delta}{4n\sigma} , \tag{3.52}$$

where δ is a constant, which can be calculated from the Boltzmann equation ($\delta \approx 1.26$), but shall be considered as a free parameter in our following analysis.

Using Eq. (3.52), the equation of motion (3.34) for the energy density now becomes (the bulk viscosity vanishes for a gas of massless particles)

$$\frac{d\varepsilon}{d\tau} + \frac{4\varepsilon}{3\tau} \left(1 - \frac{\delta}{3n\tau\sigma} \right) = 0 . \tag{3.53}$$

This equation can also be solved analytically, since we already know that the quantity $n(\tau)\tau$ is a constant from the analytical solution (3.36) for the particle density (which remains valid in Navier–Stokes theory). We then further rewrite this equation of motion as

$$\frac{\partial \varepsilon}{\partial \tau} + \frac{4\varepsilon}{3\tau} (1 - \delta_{\text{eff}}) = 0 , \tag{3.54}$$

where we defined the constant dimensionless parameter

$$\delta_{\text{eff}} \equiv \frac{\delta}{3n(\tau_0)\tau_0\sigma} . \tag{3.55}$$

The solution for the energy density is then once more in the form of a power law (similar to the solution for an ideal fluid), but with a different exponent,

$$\varepsilon(\tau) = \varepsilon(\tau_0) \left(\frac{\tau_0}{\tau} \right)^{4(1-\delta_{\text{eff}})/3} = \varepsilon_{\text{ideal}}(\tau) \left(\frac{\tau}{\tau_0} \right)^{4\delta_{\text{eff}}/3} . \tag{3.56}$$

As expected, we see that the ideal-fluid limit is recovered by taking the viscosity to zero, $\delta_{\text{eff}} \to 0$ (taking the cross section to infinity). We also see that, as occurred with the previously discussed case, cf. Eq. (3.47), the energy density decreases slower with time due to the presence of dissipation. As already stated, this is a general feature of expanding dissipative fluids, since the entropy production due to friction among fluid elements heats the fluid. Nevertheless, we see that the form of this solution is very different from the previous one, indicating the influence that the transport properties of the fluid can have on its dynamics. For the other choice of shear viscosity, Eq. (3.45), this reduction was given by an exponential envelope function, while it now happens due to the reduction of the exponent in the power law. Furthermore, we see that the difference between the dissipative solution (3.56) and the ideal-fluid one actually increases with time. The energy density at late times will be dramatically different than the one found for ideal fluids (while in the other case, cf. Eq. (3.48), this difference becomes constant at late times).

Again, we compute the corresponding solutions for the temperature and chemical potential, assuming the equation of state (3.42). We obtain

$$T(\tau) = \frac{\varepsilon(\tau)}{3n(\tau)} = T(\tau_0) \left(\frac{\tau_0}{\tau}\right)^{\frac{1}{3}} \left(\frac{\tau}{\tau_0}\right)^{4\delta_{\text{eff}}/3} , \tag{3.57}$$

$$\frac{\mu(\tau)}{T(\tau)} = \frac{\mu(\tau_0)}{T(\tau_0)} + 4\delta_{\text{eff}} \ln\left(\frac{\tau_0}{\tau}\right) . \tag{3.58}$$

Once more, the ratio $\mu(\tau)/T(\tau)$ is found to depend on time, but with a different dependence when compared to the case of constant τ_η, cf. Eq. (3.50). This is yet another example for the sensitivity of the solutions of fluid dynamics to the choice of transport coefficients.

3.2.4.3 The Case of Constant η/s_0

Finally, we shall consider a parametrization of shear viscosity that is widely employed in simulations of ultrarelativistic heavy-ion collisions [11,12], where it is assumed that its ratio with the entropy density is a constant, i.e.,

$$\eta/s_0 \sim \text{const.}$$

This ansatz is valid for conformal fluids, where all quantities scale with the temperature, $\varepsilon \sim P_0 \sim T^4$ and $s_0 \sim \eta \sim T^3$. We study the solution of the fluid-dynamical equations for conformal fluids, where such scaling behavior occurs, in the limit of vanishing chemical potential. In this setting, it is more convenient to rewrite Eq. (3.46) as an equation of motion for the temperature, T,

$$\frac{dT}{d\tau} + \frac{T}{3\tau} = \frac{4}{9\tau^2} \frac{\eta}{s_0} , \tag{3.59}$$

where we used the thermodynamic relation, $\varepsilon + P_0 = T s_0$ (valid at vanishing chemical potential). Equation (3.59) can be solved analytically,

$$T(\tau) = T(\tau_0) \left(\frac{\tau_0}{\tau}\right)^{\frac{1}{3}} + \frac{2}{3} \frac{\eta}{s_0} \frac{1}{\tau} \left[\left(\frac{\tau}{\tau_0}\right)^{2/3} - 1\right] . \tag{3.60}$$

At asymptotically long times, $\tau \gg \tau_0$, the solution assumes the simple form

$$T(\tau) \sim \left[T(\tau_0) + \frac{2}{3\tau_0} \frac{\eta}{s_0}\right] \left(\frac{\tau_0}{\tau}\right)^{\frac{1}{3}} ,$$

which demonstrates that a dissipative system will have larger temperatures at late times when compared to an ideal fluid. As already mentioned, this happens because of the entropy production that occurs in the presence of dissipation and serves to heat the fluid.

In the next section, we investigate how these solutions change for fluids described by transient theories of fluid dynamics. In particular, we aim to verify if there is a relation among these solutions.

3.2.5 Transient Theory of Fluid Dynamics

We now aim to solve Eqs. (3.34) and (3.38), taking into account the effect of a nonzero relaxation time τ_π. As in the previous sections, we assume that the system is made of massless particles, thus neglecting all effects from bulk-viscous pressure and with an equation of state of the form given by Eq. (3.42),

$$\varepsilon = 3P_0 , \qquad P_0(T, \mu) \sim \exp\left(\frac{\mu}{T}\right) T^4 . \tag{3.61}$$

Even in this simplified scenario, we still have to specify the shear viscosity, the shear relaxation time, and the linear combination of transport coefficients $\delta_{\pi\pi} + \tau_{\pi\pi}/3$. We will further simplify the discussion by setting $\tau_{\pi\pi} = 0$ [13].

For a massless gas, it is reasonable to assume that the relaxation time is proportional to the dissipative time scale τ_η,

$$\tau_\pi \equiv b\tau_\eta = b \frac{\eta}{\varepsilon + P_0} , \tag{3.62}$$

where b is a dimensionless parameter that will be assumed to be constant (for a massless gas, this assumption holds for any choice of cross section as shown in Refs. [5,14,15] and will be discussed later in Chap. 5). Note that linear stability of the theory requires that $b \geq 2$ (taking $c_s^2 = 1/3$), cf. Eq. (2.111), when only effects of shear viscosity are considered.

Furthermore, in the massless limit, one can also obtain a general expression for $\delta_{\pi\pi}$, which is given solely in terms of the shear relaxation time,

$$\delta_{\pi\pi} = \frac{4}{3}\tau_\pi . \tag{3.63}$$

This result is valid for any conformal fluid [16] or for any gas of massless particles [5,14,15] and shall be used in the calculations performed next.

Now, the dynamical evolution of the shear-stress tensor will no longer be neglected and the dynamical evolution of the fluid is described by the following set of *coupled* ordinary differential equations,

$$\frac{d\varepsilon}{d\tau} + \frac{4\varepsilon}{3\tau} - \frac{\bar{\pi}}{\tau} = 0 , \tag{3.64}$$

$$\tau_\pi \frac{d\bar{\pi}}{d\tau} + \bar{\pi} + \frac{4\tau_\pi}{3\tau}\bar{\pi} = \frac{4\eta}{3\tau} . \tag{3.65}$$

It is convenient to rewrite these equations in terms of the dimensionless field $\chi \equiv \bar{\pi}/(\varepsilon + P_0) = 3\bar{\pi}/(4\varepsilon)$, leading to

$$\frac{d\varepsilon}{d\tau} + \frac{4\varepsilon}{3\tau}(1 - \chi) = 0 , \tag{3.66}$$

$$\frac{d\chi}{d\tau} + \frac{4\chi^2}{3\tau} + \frac{\chi}{\tau_\pi} = \frac{4}{3\tau}\frac{\tau_\eta}{\tau_\pi} . \tag{3.67}$$

It is also convenient to derive from Eq. (3.64) the equation of motion satisfied by the temperature,

$$\frac{dT}{d\tau} + \frac{4T}{3\tau}\left(\frac{1}{4} - \chi\right) = 0 , \tag{3.68}$$

where we used $\varepsilon = 3P_0 = 3nT$ (valid for classical gases) and the solution derived for n in Eq. (3.36).

A fluid-dynamical description is traditionally characterized by the existence of constitutive relations between dissipative currents and gradients of conserved quantities [17]. It is not clear how this phenomena occurs in a transient theory of fluid dynamics, where the initial state of the dissipative currents in principle affects the dynamical evolution of the fluid. However, one expects that, at sufficiently late times (times that are much larger than the relaxation time), transient effects will disappear and the dissipative corrections can be expressed in terms of constitutive relations. We shall discuss in detail how this occurs in the following, using cases where the equations of motion can be explicitly solved.

3.2.5.1 The Case of Constant τ_η

We now follow the same examples considered for Navier–Stokes theory. First, we assume that τ_η is constant, which, consequently, leads to a constant shear relaxation time, $\tau_\pi = b\tau_\eta$. In this case, we can further re-express the equations of motion (3.66) and (3.67) in terms of the dimensionless time variable

$$\hat{\tau} \equiv \frac{\tau}{\tau_\pi} ,$$

leading to the following coupled equations [9],

$$\frac{1}{\varepsilon\hat{\tau}^{4/3}} \frac{d(\varepsilon\hat{\tau}^{4/3})}{d\hat{\tau}} = \frac{4}{3}\frac{\chi}{\hat{\tau}} , \tag{3.69}$$

$$\frac{d\chi}{d\hat{\tau}} + \frac{4\chi^2}{3\hat{\tau}} + \chi - \frac{3}{4}\frac{a^2}{\hat{\tau}} = 0 , \tag{3.70}$$

where we further defined the dimensionless and constant variable

$$a^2 \equiv \frac{16}{9}\frac{\tau_\eta}{\tau_\pi} = \frac{16}{9b} . \tag{3.71}$$

Causality and stability around equilibrium at the linearized level holds when $\tau_\pi \geq 2\tau_\eta$ (which corresponds to $a^2 \leq 8/9$) [18], as we discussed in Chap. 2. It is important to remark that, in Bjorken flow, the variable $\hat{\tau}$ defined above is of the order of the inverse Knudsen number,

$$\hat{\tau} \sim \mathrm{Kn}^{-1} \equiv \frac{L}{\lambda} ,$$

since $\hat{\tau}$ corresponds to a ratio of a macroscopic scale L (here given by τ) to a microscopic scale λ (here given by $\tau_\pi \sim \tau_\eta$) [19].

Equation (3.70) is a Riccati equation that can be solved independently of Eq. (3.69)—a direct consequence of the assumption of a constant τ_η. First-order nonlinear ordinary differential equations of Riccati type can always be written as second-order *linear* differential equations and, as a matter of fact, this can be done in the present case using a new variable $y(\hat{\tau})$ defined via

$$\frac{1}{y}\frac{dy}{d\hat{\tau}} = \frac{4}{3}\frac{\chi}{\hat{\tau}} . \tag{3.72}$$

Inserting this into Eq. (3.70) yields

$$\frac{d^2y}{d\hat{\tau}^2} + \left(1 + \frac{1}{\hat{\tau}}\right)\frac{dy}{d\hat{\tau}} - \frac{a^2}{\hat{\tau}^2}y = 0 . \tag{3.73}$$

This linear differential equation can be solved and the general solution is

$$y(\hat{\tau}) = A\sqrt{\hat{\tau}}e^{-\hat{\tau}/2}\left\{B\left[K_{a-\frac{1}{2}}\left(\frac{\hat{\tau}}{2}\right) - K_{a+\frac{1}{2}}\left(\frac{\hat{\tau}}{2}\right)\right] + I_{a-\frac{1}{2}}\left(\frac{\hat{\tau}}{2}\right) + I_{a+\frac{1}{2}}\left(\frac{\hat{\tau}}{2}\right)\right\},$$
(3.74)

where A and B are constants and $I_n(z)$ and $K_n(z)$ are the modified Bessel functions of the first and second kind, respectively. The fact that Eq. (3.74) is a solution to Eq. (3.73) can be readily checked with the well-known relations

$$K'_\nu(z) = \frac{\nu}{z}K_\nu(z) - K_{\nu+1}(z) = -\frac{\nu}{z}K_\nu(z) - K_{\nu-1}(z),$$
(3.75)

$$I'_\nu(z) = \frac{\nu}{z}I_\nu(z) + I_{\nu+1}(z) = -\frac{\nu}{z}I_\nu(z) + I_{\nu-1}(z).$$
(3.76)

Inserting Eq. (3.74) into Eq. (3.72) and again using the relations (3.75) and (3.76), the following general analytical solution for χ is found,

$$\chi(\hat{\tau}) = \frac{3a}{4}\frac{B\left[K_{a-\frac{1}{2}}\left(\frac{\hat{\tau}}{2}\right) + K_{a+\frac{1}{2}}\left(\frac{\hat{\tau}}{2}\right)\right] + I_{a-\frac{1}{2}}\left(\frac{\hat{\tau}}{2}\right) - I_{a+\frac{1}{2}}\left(\frac{\hat{\tau}}{2}\right)}{B\left[K_{a-\frac{1}{2}}\left(\frac{\hat{\tau}}{2}\right) - K_{a+\frac{1}{2}}\left(\frac{\hat{\tau}}{2}\right)\right] + I_{a-\frac{1}{2}}\left(\frac{\hat{\tau}}{2}\right) + I_{a+\frac{1}{2}}\left(\frac{\hat{\tau}}{2}\right)}.$$
(3.77)

Comparing Eqs. (3.69) and (3.72), we observe that the energy density is actually $\varepsilon(\hat{\tau}) = \hat{\tau}^{-4/3}y(\hat{\tau})$, which allows to eliminate the integration constant A in terms of the initial energy density,

$$\varepsilon(\hat{\tau}) = \varepsilon\left(\hat{\tau}_0\right)e^{-(\hat{\tau}-\hat{\tau}_0)/2}\left(\frac{\hat{\tau}_0}{\hat{\tau}}\right)^{\frac{5}{6}}$$

$$\times\frac{B\left[K_{a-\frac{1}{2}}\left(\frac{\hat{\tau}}{2}\right) - K_{a+\frac{1}{2}}\left(\frac{\hat{\tau}}{2}\right)\right] + I_{a-\frac{1}{2}}\left(\frac{\hat{\tau}}{2}\right) + I_{a+\frac{1}{2}}\left(\frac{\hat{\tau}}{2}\right)}{B\left[K_{a-\frac{1}{2}}\left(\frac{\hat{\tau}_0}{2}\right) - K_{a+\frac{1}{2}}\left(\frac{\hat{\tau}_0}{2}\right)\right] + I_{a-\frac{1}{2}}\left(\frac{\hat{\tau}_0}{2}\right) + I_{a+\frac{1}{2}}\left(\frac{\hat{\tau}_0}{2}\right)}.$$
(3.78)

We note that, because $K_{\nu+1}(x) \geq K_\nu(x)$, we must impose the constraint $B \leq 0$ in order to guarantee that the energy density is always positive definite for any value of $\hat{\tau}_0$.

Equations (3.78) and (3.77) define the general analytical solution of the transient fluid-dynamical equations in Bjorken flow with a constant relaxation time. This solution is very different from the one found in Navier–Stokes theory. As was anticipated, it can be divided into two regimes: a transient regime, where the dynamics of the dissipative currents strongly depends on the initial value of χ (which determines the integration constant B), and an asymptotic late-time regime, where constitutive relations between dissipative currents and gradients start to appear.

It is straightforward to see from the analytical solution (3.77) that it completely loses the information about the initial conditions for χ (encoded in the parameter B) at late times. This happens because the Bessel functions display the following asymptotic form for sufficiently large values of their argument: $K_\nu(x) \sim e^{-x}/\sqrt{x}$

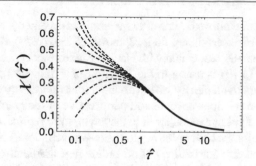

Fig. 3.1 Analytical non-equilibrium attractor (3.79) in solid red for $a^2 = 16/45$. The dashed curves correspond to the solution (3.77) for different initial conditions, which collapse onto the attractor before local equilibrium is reached (where χ vanishes). Figure from Ref. [9]

and $I_\nu(x) \sim e^x / \sqrt{x}$. Therefore, the terms containing $K_\nu (\hat{\tau}/2)$ become exponentially smaller compared to the terms containing $I_\nu (\hat{\tau}/2)$ as time increases. At a sufficiently long time, the solution (3.77) can be approximated as

$$\chi(\hat{\tau}) \to \chi_{\text{att}}(\hat{\tau}) = \frac{3a}{4} \frac{I_{a-\frac{1}{2}}\left(\frac{\hat{\tau}}{2}\right) - I_{a+\frac{1}{2}}\left(\frac{\hat{\tau}}{2}\right)}{I_{a-\frac{1}{2}}\left(\frac{\hat{\tau}}{2}\right) + I_{a+\frac{1}{2}}\left(\frac{\hat{\tau}}{2}\right)} . \tag{3.79}$$

This regime appears to have the structure usually associated with a fluid-dynamical (non-transient) picture, where the shear-stress tensor does not depend on its initial value and is simply expressed in terms of gradients of velocity ($\sigma^{\mu\nu} \sim \theta \sim 1/\tau$). This approximate late-time expression for the analytic solution has the non-trivial feature of also being an exact solution of the theory (the solution corresponding to the choice $B = 0$). For this reason, it is considered to be a so-called *attractor solution* of the hydrodynamic equations, an idea initially proposed in Ref. [20]. The typical attractor behavior is illustrated in Fig. 3.1, for the case where $a = \sqrt{16/45}$. One can see that all solutions of the equations converge to the attractor solution at late times, they are "attracted" by that solution. Equation (3.79) was the first analytical expression for an attractor of transient fluid dynamics and was derived in Ref. [9].

In Fig. 3.1, we clearly see that the dynamics for $\hat{\tau} < 1$ is transient and depends on the initial condition assumed for χ. Meanwhile, at times $\hat{\tau} > 1$, the solution for χ appears to lose any information about the initial state and becomes universal. We note that the analytical solutions for χ display an interesting feature at asymptotically early times, $\hat{\tau} \to 0$. In this limit, because of the behavior of the Bessel functions for vanishing argument, χ tends to two distinct values: $3a/4$, for $B = 0$, and $-3a/4$, for any $B \neq 0$. Therefore, the attractor is the *only* solution that goes to $3a/4$ at $\hat{\tau} = 0$. Indeed, this limiting behavior of the attractor in Bjorken flow has been used in previous works as a way to define it [20], even in situations where analytical solutions are not possible. In this case, one may find the attractor numerically by identifying it as the solution that obeys this boundary condition.

We note that, while *universal* constitutive relations do emerge at late times in the solutions of Israel–Stewart theory derived above, cf. Eq. (3.79), they do not appear in the form traditionally associated with Navier–Stokes theory, i.e., as $\chi(\hat{\tau}) \sim 1/\hat{\tau}$. The fact that Eq. (3.79) is a non-trivial function of $\hat{\tau}$ indicates that some sort of resummation occurs dynamically when solving a transient theory of fluid dynamics and the constitutive relations become non-perturbative in the gradients.

Thus, transient fluid dynamics, even in the asymptotic regime, cannot be strictly considered to be of a specific order in gradients—in principle, solutions of Israel–Stewart theory contain contributions of *all* orders in a gradient expansion. We can visualize this best by considering a late-time expansion of χ, Eq. (3.79), in powers of $1/\hat{\tau}$ (which corresponds to an expansion in powers of gradients in Bjorken flow),

$$\chi(\hat{\tau}) = \frac{3}{4}a^2 \sum_{n=0}^{\infty} \frac{c_n}{\hat{\tau}^n} . \tag{3.80}$$

The corresponding coefficients of the series, c_n, are found by inserting this equation into Eq. (3.70) and comparing coefficients of the same order in $1/\hat{\tau}$. This way, one obtains $c_0 = 0$, $c_1 = 1$, while all c_{n+1} with $n \geq 1$ fulfill the recursion relation

$$c_{n+1} = n\,c_n - a^2 \sum_{m=0}^{n} c_{n-m}c_m . \tag{3.81}$$

The asymptotic limit of solutions of transient fluid dynamics actually manages to dynamically re-sum all these gradients into just one expression, the solution found in Eq. (3.79). The connection to Navier–Stokes theory comes from the truncation of the series (3.80): if $\hat{\tau} \gg 1$, i.e., if the Knudsen number is sufficiently small, $\mathrm{Kn} \sim \hat{\tau}^{-1} \ll 1$, it is expected that the series can be truncated at a very low order. If only the leading non-vanishing term of the expansion is included, the constitutive relation that emerges from this procedure is exactly the one predicted from Navier–Stokes theory,

$$\chi(\hat{\tau}) \approx \frac{3}{4}\frac{a^2}{\hat{\tau}} = \frac{4}{3b}\frac{1}{\hat{\tau}} = \frac{4}{3}\frac{\tau_\eta}{\tau} . \tag{3.82}$$

The most surprising feature of the series above is that it has a vanishing radius of convergence, at least for parameter choices that guarantee that causality and stability are fulfilled, i.e., for $a^2 \leq 8/9$. The gradient series diverges since for large n the first term in Eq. (3.81) dominates, leading to factorial growth, $c_n \sim n!$. This can be also inferred from the properties of the modified Bessel functions. Nevertheless, the low-order truncations of the series (up to third order) still provide a reasonable approximation for the late-time solution [9].

3.2.5.2 The Case of Hard Spheres

We now consider the case where the fluid is a gas of massless hard spheres, for which the shear viscosity has the form of Eq. (3.52),

$$\eta = \delta \frac{P_0}{n\sigma} = \delta_{\text{eff}} \varepsilon \tau , \qquad (3.83)$$

where δ_{eff} is defined in Eq. (3.55). In order to write the second equality, we used the analytic solution (3.36) for $n(\tau)$. This implies that the relaxation time is *not* constant and is given by

$$\tau_\pi = \frac{3}{4} b \delta_{\text{eff}} \tau . \qquad (3.84)$$

In this setting, the equation of motion (3.67) for χ becomes [10]

$$\frac{d\chi}{d\bar{\tau}} + \chi^2 + \frac{\chi}{b\delta_{\text{eff}}} - \frac{1}{b} = 0 , \qquad (3.85)$$

where we defined the new dimensionless time variable $\bar{\tau} = (4/3) \ln (\tau/\tau_0)$. This fluid-dynamical equation can be solved analytically and the solution in this case is

$$\chi(\bar{\tau}) = \frac{\sqrt{1 + 4b\delta_{\text{eff}}^2}}{2b\delta_{\text{eff}}} \tanh \left[\frac{\sqrt{1 + 4b\delta_{\text{eff}}^2}}{2b\delta_{\text{eff}}} (\bar{\tau} + C) \right] - \frac{1}{2b\delta_{\text{eff}}} , \qquad (3.86)$$

where C is a free parameter, which can be fixed by specifying an initial condition for χ. In the long-time limit, the solutions once more become *universal* and independent of the initial state,

$$\chi(\bar{\tau}) \to \chi_{\text{att}}(\bar{\tau}) = \frac{\sqrt{1 + 4b\delta_{\text{eff}}^2} - 1}{2b\delta_{\text{eff}}} . \qquad (3.87)$$

This asymptotic solution is actually constant. As before, this asymptotic expression for the analytic solution has the non-trivial feature of being an exact solution of theory, corresponding to the choice $C \to \infty$. We thus identify it to be an attractor solution of the hydrodynamic equations. Once more, we note that $\chi_{\text{att}}(\bar{\tau})$ is different than the solution expected from the Navier–Stokes limit,

$$\chi_{\text{NS}} = \frac{3\bar{\pi}_{\text{NS}}}{4\varepsilon} = \frac{\eta}{\varepsilon\tau} = \delta_{\text{eff}} ,$$

where we have used the definition of χ, as well as Eqs. (3.39) and (3.83). Note that the Navier–Stokes limit of χ is also a constant.

The asymptotic solution (3.87) for χ can be better understood and interpreted when expressed in terms of the Knudsen number $\text{Kn} \equiv \tau_\pi/\tau$. We then see that, for this choice of transport coefficients, the Knudsen number is actually constant throughout the evolution of the system,

$$\text{Kn} \equiv \frac{\tau_\pi}{\tau} = \frac{3}{4} b \delta_{\text{eff}} .$$

Therefore, we can express the attractor solution as a function of the Knudsen number,

$$\chi_{\text{att}}(\text{Kn}) = 3 \frac{\sqrt{1 + \frac{4}{b}\left(\frac{4}{3}\text{Kn}\right)^2} - 1}{8\,\text{Kn}} .$$

This solution shows in a rather clear manner that transient theories of fluid dynamics (here in the form of Israel–Stewart theory) contain many orders of Knudsen number and that their solutions can be thought to contain many orders of powers of gradients, which are dynamically resummed. In this case, even at arbitrarily late times, the solution will never assume a Navier–Stokes form. Only if the Knudsen number is small, the attractor can be approximated by a Navier–Stokes form, which is the leading term in a gradient expansion,

$$\chi_{\text{att}}(\text{Kn}) = \frac{4}{3b}\,\text{Kn} + \mathcal{O}\left(\text{Kn}^2\right) .$$

Also in this case, the gradient expansion can be calculated explicitly

$$\chi_{\text{att}}(\text{Kn}) = \frac{3}{2} \sum_{n=1}^{\infty} \binom{1/2}{n} \frac{4^{3n-1}}{3^{2n} b^n} (\text{Kn})^{2n-1} .$$

and has a finite radius of convergence given by

$$\text{Kn} < \frac{3}{8}\sqrt{b} .$$

This is the only known example of a convergent gradient expansion in Bjorken flow (for *causal* theories of fluid dynamics).

When discussing Navier–Stokes theory, we also considered an example where the ratio of the shear viscosity over the entropy density is a constant, i.e., $\eta/s_0 = \text{const.}$. In this case, Eqs. (3.66) and (3.67) can still be solved numerically without the need of any sophisticated algorithm (e.g., the Runge–Kutta algorithm suffices), but deriving an analytic solution is no longer possible. Nevertheless, the solutions of Israel–Stewart theory have been studied extensively in the literature for such a choice of shear viscosity and shall not be analyzed in detail here. It is sufficient to state that it has the following general properties: (i) it has a transient part and an asymptotic part, with the latter behaving as an attractor solution and (ii) the expansion in powers of gradients of the attractor solution is a divergent series. As a matter of fact, the first example of a divergent gradient expansion for an expanding system was found for a conformal fluid in Ref. [20]. We now discuss the gradient expansion of Israel–Stewart theory in Bjorken flow in a more general setting, imposing less specific assumptions on the form of the transport coefficients.

3.2.6 Gradient Expansion of Transient Fluid Dynamics

It is interesting to determine the asymptotic gradient-expansion solution of a transient theory of fluid dynamics. This problem is usually very complicated and difficult to implement to arbitrarily high orders. However, the highly symmetric configuration of Bjorken flow makes it possible to solve this problem—something extremely rare and a great advantage of such a flow geometry. We have already performed this task for the two solutions of Israel–Stewart theory found in the previous two sections. Now, we construct the gradient series in a more general setting, assuming that the relaxation time is an arbitrary function of temperature, $\tau_\pi(T)$.

We shall carry out this procedure in a rigorous manner. First, we convert the original problem into a perturbation-theory problem by introducing a dimensionless parameter ϵ into the differential equation satisfied by χ as follows

$$\epsilon \frac{d\chi}{d\tau} + \epsilon \frac{4\chi^2}{3\tau} + \frac{\chi}{\tau_\pi} = \epsilon \frac{4}{3\tau} \frac{\tau_\eta}{\tau_\pi} . \tag{3.88}$$

Now, in this problem the shear-stress tensor and the temperature are functions of τ and the new variable ϵ, i.e., $T = T(\tau, \epsilon)$ and $\chi = \chi(\tau, \epsilon)$. The parameter ϵ was introduced in every term which contains a derivative or $1/\tau$ and, hence, it becomes a book-keeping parameter to count orders or powers of gradients. Therefore, an expansion in powers of ϵ will naturally lead to an expansion in powers of gradients. In Bjorken flow, it is straightforward to deduce (cf., e.g., Eqs. (3.29) and (3.30)) that powers of gradients correspond to inverse powers of the time coordinate τ [9] and a gradient expansion will naturally become an expansion in powers of $1/\tau$—this fact was already used in the previous section and will be confirmed below.

Next, we look for a solution for χ that can be represented as a power series in ϵ

$$\chi = \sum_{n=0}^{\infty} \chi_n(\tau, T)\epsilon^n . \tag{3.89}$$

This reduces the problem to solving an infinite number of simpler equations (in this case, algebraic equations), which are given by recurrence relations. At the end of the calculation, one sets $\epsilon = 1$ to recover the original problem. We note that such a perturbative procedure is only useful when the first few terms of the series contain at least some basic properties of the exact solution, i.e., when a low-order truncation of the series can capture basic trends of the solution. This does not necessarily mean that the series must converge—we already saw in the previous sections that divergent series will occur. Furthermore, in practice convergent series quite often do not offer useful representations of functions since they may be slowly convergent and require a large number of terms to provide a good approximation in a given domain [21]. On the other hand, truncations of divergent series are known to provide very good approximations of certain functions (as in the case of the error function).

Naturally, this procedure will not lead to a general solution of transient fluid dynamics, since the equations obtained in this type of perturbative approach do not

contain any free parameter associated with the initial condition for χ (the initial
condition for the temperature remains a free parameter, even in the perturbative
problem). In this sense, what will be obtained with this approach is just one solution for χ that cannot be adjusted to an arbitrary boundary condition. However, this
solution is expected to have physical meaning, reflecting the long-time, slow evolution of the system when all transient initial-state dynamics is lost and the system
enters a universal fluid-dynamical regime (as we already demonstrated with the two
examples discussed in the previous section).

We now continue the perturbative calculation, substituting the proposed series
(3.89) in powers of ϵ into the equation of motion (3.88) for χ. One then obtains the
following result:

$$\sum_{n=0}^{\infty} \frac{d\chi_n}{d\tau} \epsilon^{n+1} + \frac{4}{3\tau} \sum_{n=0}^{\infty} \sum_{m=0}^{\infty} \chi_n \chi_m \epsilon^{n+m+1} + \frac{1}{\tau_\pi} \sum_{n=0}^{\infty} \chi_n \epsilon^n = \frac{4}{3\tau} \frac{\tau_\eta}{\tau_\pi} \epsilon . \quad (3.90)$$

We now use the assumption that the relaxation time τ_π is proportional to τ_η, $\tau_\pi = b\tau_\eta$,
see Eq. (3.62). Note that even with this assumption the relaxation time is still a
general function of the temperature—the temperature dependence now occurs via
$\tau_\eta = \tau_\eta(T)$. Then, the equation simplifies to

$$\sum_{n=0}^{\infty} \frac{d\chi_n}{d\tau} \epsilon^{n+1} + \frac{4}{3\tau} \sum_{n=0}^{\infty} \sum_{m=0}^{\infty} \chi_n \chi_m \epsilon^{n+m+1} + \frac{1}{\tau_\pi} \sum_{n=0}^{\infty} \chi_n \epsilon^n = \frac{4}{3\tau b} \epsilon . \quad (3.91)$$

The problem in solving this equation, i.e., in collecting all terms that are of the
same power in ϵ, is that the term $d\chi_n/d\tau$ also has an ϵ-dependence that must be
considered when grouping the terms. This is actually the main difficulty in calculating
the gradient expansion and solving this problem in a general flow setting—a task that
has not been acomplished yet. In Bjorken flow, it is possible to solve this complication
by noticing that the series coefficient χ_n depends on τ only through two different
variables, $\chi_n = \chi_n(\tau, T(\tau))$—a direct consequence of the assumption $\tau_\pi = \tau_\pi(T)$.
Therefore, its derivative can be mathematically re-expressed in the following way

$$\frac{d\chi_n}{d\tau} = \left.\frac{\partial \chi_n}{\partial \tau}\right|_T + \left.\frac{\partial \chi_n}{\partial \ln T}\right|_\tau \frac{d \ln T}{d\tau} . \quad (3.92)$$

The relaxation time will carry the implicit dependence on the temperature that must
be resummed. From the conservation law, Eq. (3.68), it is clear that the temperature
derivative couples to χ and, thus, carries an ϵ-dependence

$$\frac{d \ln T}{d\tau} = -\frac{1}{3\tau} (1 - 4\chi) = -\frac{1}{3\tau} + \frac{4}{3\tau} \sum_{n=0}^{\infty} \chi_n (T) \epsilon^n , \quad (3.93)$$

which is the *only* ϵ–dependence in $d\chi_n/d\tau$. Therefore, we can rewrite Eq. (3.91) as

$$\sum_{n=0}^{\infty} \left(\left.\frac{\partial \chi_n}{\partial \tau}\right|_T - \frac{1}{3\tau} \left.\frac{\partial \chi_n}{\partial \ln T}\right|_\tau \right) \epsilon^{n+1} + \frac{4}{3\tau} \sum_{n=0}^{\infty} \sum_{m=0}^{\infty} \left(\chi_n + \left.\frac{\partial \chi_n}{\partial \ln T}\right|_\tau \right) \chi_m \epsilon^{n+m+1}$$

$$+ \frac{1}{\tau_\pi} \sum_{n=0}^{\infty} \chi_n \epsilon^n - \frac{4}{3b\tau} \epsilon = 0 \,, \tag{3.94}$$

Equating the terms that are of the same power in ϵ, one obtains algebraic equations that determine the expansion coefficients. The first two coefficients of the series can be calculated explicitly and are

$$\chi_0 = 0 \,, \qquad \chi_1 = \frac{4}{3b} \frac{\tau_\pi}{\tau} \,, \tag{3.95}$$

while the remaining coefficients are calculated using the recurrence relation

$$\chi_{n+1} = \frac{\tau_\pi}{3\tau} \left.\frac{\partial \chi_n}{\partial \ln T}\right|_\tau - \tau_\pi \left.\frac{\partial \chi_n}{\partial \tau}\right|_T - \frac{4\tau_\pi}{3\tau} \sum_{m=0}^{n} \left(\chi_{n-m} + \left.\frac{\partial \chi_{n-m}}{\partial \ln T}\right|_\tau \right) \chi_m \,.$$

For the sake of completeness, we explicitly calculate the next two terms in the series,

$$\chi_2 = \frac{4}{3b} \left(1 + \frac{1}{3} \frac{\partial \ln \tau_\pi}{\partial \ln T} \right) \frac{1}{\hat{\tau}^2} \,,$$

$$\chi_3 = \frac{4}{3b} \left[2 - \frac{16}{9b} + \frac{4}{3} \left(1 - \frac{4}{3b} \right) \frac{\partial \ln \tau_\pi}{\partial \ln T} \right.$$

$$\left. + \frac{2}{9} \left(\frac{\partial \ln \tau_\pi}{\partial \ln T} \right)^2 + \frac{1}{9} \frac{\partial^2 \ln \tau_\pi}{\partial (\ln T)^2} \right] \frac{1}{\hat{\tau}^3} \,. \tag{3.96}$$

Above, we re-expressed the expansion coefficient in terms of the dimensionless time variable $\hat{\tau} = \tau/\tau_\pi$—which is a proxy for the inverse Knudsen number in Bjorken flow, with τ_π representing a microscopic scale and τ a macroscopic scale. We thus infer that the perturbative expansion constructed is actually an expansion in powers of the Knudsen number.

After setting $\epsilon = 1$, it is straightforward to see that the zeroth-order solution of the expansion corresponds to the ideal-fluid limit ($\chi = 0$) and the first-order solution corresponds to the Navier–Stokes limit ($\chi = \frac{4}{3b} \frac{\tau_\pi}{\tau}$), as expected from a gradient expansion. Including the second-order and third-order terms will lead to the Burnett and super-Burnett theories, respectively.

For the sake of illustration, let us consider the following parametrization for the temperature dependence of τ_π,

$$\tau_\pi \sim T^\Delta \implies \ln \tau_\pi \sim \Delta \ln T \,, \tag{3.97}$$

where Δ is a constant parameter. In this case, all higher-order derivatives of $\ln \tau_\pi$ in $\ln T$ actually disappear, leading to the following solutions for the second and third expansion coefficients,

$$\chi_2 = \frac{4}{3b}\left(1 + \frac{1}{3}\Delta\right)\frac{1}{\hat{\tau}^2},$$

$$\chi_3 = \frac{4}{3b}\left[2 - \frac{16}{9b} + \frac{4}{3}\left(1 - \frac{4}{3b}\right)\Delta + \frac{2}{9}\Delta^2\right]\frac{1}{\hat{\tau}^3}. \qquad (3.98)$$

Thus, the solution for the gradient expansion, including terms up to order $\mathcal{O}_3$, becomes (setting $\epsilon = 1$)

$$\chi = \frac{4}{3b}\left\{\frac{1}{\hat{\tau}} + \left(1 + \frac{1}{3}\Delta\right)\frac{1}{\hat{\tau}^2} + \left[2 - \frac{16}{9b} + \frac{4}{3}\left(1 - \frac{4}{3b}\right)\Delta + \frac{2}{9}\Delta^2\right]\frac{1}{\hat{\tau}^3}\right\} + \mathcal{O}_4.$$

It is convenient to remove the explicit time dependence from the expansion coefficients and rewrite the gradient expansion (3.89) in the form

$$\chi = \sum_{n=0}^{\infty}\frac{\hat{\chi}_n(T)}{\hat{\tau}^n}, \qquad (3.99)$$

where we introduced the rescaled variable, $\hat{\chi}_n(T) \equiv \chi_n(\tau, T)/\hat{\tau}^n$. Now $\hat{\chi}_n$ only depends on time indirectly, via the temperature (if the parametrization (3.97) is assumed, the rescaled expansion coefficient, $\hat{\chi}_n$, does not depend on time at all). One can either restart the whole calculation of the recurrence relations assuming the expansion (3.99) above or simply calculate $\hat{\chi}_n$ using the solutions of χ_n—both methods will of course yield the same result.

The magnitude of the coefficients $\hat{\chi}_n$ at arbitrarily high order was calculated in Ref. [22] for a few choices of Δ. It was found that the rescaled expansion coefficients display factorial growth for all cases considered. This implies that the series has zero radius of convergence for several temperature dependences of the relaxation time. At this stage, the only example of a convergent series was the one shown in the previous section, when the relaxation time was assumed to be proportional to the mean free path.

Now that it is known that the gradient series diverges, one may ask if a different mathematical representation, which does not rely on the assumption of small gradients, can be formulated to describe the hydrodynamic regime. The first step in this direction was made in Ref. [20] with the proposal that hydrodynamic behavior may be meaningfully defined even far from equilibrium as long as a late-time "attractor" structure is present. In this context, Navier–Stokes theory already played the role of an attractor since the system, regardless of its initial conditions, approaches this limit when sufficiently close to equilibrium. The novelty of the proposal of Ref. [20] is that this may occur in the far-from-equilibrium regime as well. Several works have since then investigated this in Bjorken flow [9,23–29], but the question of what happens in systems with less symmetries still remains [26].

3.3   Gubser Flow

Describing the longitudinal expansion of a fluid by using a coordinate system in which the former becomes spatially homogeneous (leading to a trivial flow profile) is what allowed us to obtain rare and unique analytical solutions of fluid dynamics and, also, to calculate the gradient expansion to arbitrarily high orders. As implied, such type of highly symmetric flow configurations, which can be described solely in terms of a coordinate system, are extremely rare, but also extremely useful. They were originally found with the aim of studying the hot and dense matter formed in ultrarelativistic heavy-ion collisions, but have been extremely useful in studying the properties of hydrodynamic solutions and/or series, as discussed in the previous section. Next, we shall discuss a generalization of Bjorken flow, valid only for conformal fluids, which also includes the effects of radial expansion. This solution was derived by Steven Gubser [3] and, thus, is commonly referred to as Gubser flow.

As a starting point, we once more consider fluids whose dynamics is boost-invariant in the longitudinal direction (z-axis) and, thus, is naturally described in terms of the hyperbolic coordinates τ, η_s (instead of the Cartesian coordinates t and z). Furthermore, as in Bjorken flow, we impose that the system is invariant under reflections of the η_s-axis. The main difference to Bjorken flow is that we no longer impose that the fluid is homogeneous (invariant under translations) in the transverse (x, y)-plane, but we still demand that it remains isotropic (invariant under rotations) in that plane, i.e., azimuthally symmetric. Then, the fluid can exhibit a non-trivial radial dependence.

In more mathematical terms [3], it is imposed that the fluid flow is invariant under $SO(3) \otimes SO(1, 1) \otimes Z_2$. The Z_2-symmetry corresponds to invariance under $\eta_s \rightarrow -\eta_s$, while $SO(1, 1)$ denotes invariance under boosts along the longitudinal axis. These two symmetries are the same that we already imposed in Bjorken flow. The remaining symmetries are more abstract: the $SO(3)$ group is a subgroup of the $SO(4, 2)$ conformal group which describes the symmetry of the solution under rotations in the transverse plane and two operations constructed using special conformal transformations that *replace* translation invariance in the transverse plane. For more details regarding the generators of the $SO(3)$ symmetry group of this solution, see Ref. [3].

Conformal invariance is a crucial assumption in the construction of such solutions and, thus, restricts the types of fluid that obey this solution (this is in contrast to Bjorken flow, where *any* fluid could satisfy the Bjorken-flow solution). It is important to emphasize that conformal invariance imposes certain constraints on the thermodynamic and transport properties of a fluid: such systems are restricted to have a vanishing bulk-viscous pressure and $\varepsilon = 3P_0$ (leading to a *traceless* energy-momentum tensor), must have a *constant* ratio of shear viscosity to entropy density, $\eta \sim s_0 \sim T^3$, the shear relaxation time is always proportional to the inverse temperature, $\tau_\pi \sim T^{-1}$, and $\delta_{\pi\pi} = 4\tau_\pi/3$. Similar to what happened in Bjorken flow, the symmetries imposed on the system guarantee that there are neither particle nor net-charge diffusion currents and, thus, these will be omitted from the following discussion.

3.3.1   Coordinates and Kinematical Properties

Imposing azimuthal symmetry in the transverse plane, the metric measure in Milne coordinates, Eq. (3.18), reads

$$ds^2 = d\tau^2 - (dr^2 + r^2 d\phi^2) - \tau^2 d\eta_s^2 \,, \tag{3.100}$$

where $r \equiv \sqrt{x^2 + y^2}$ is the radial distance from the center, and $\phi \equiv \arctan(y/x)$ the azimuthal angle in the transverse plane. We now replace τ and r by the new, so-called *Gubser coordinates* ρ and θ via the transformation

$$\sinh \rho \equiv -\frac{1 - (q\tau)^2 + (qr)^2}{2q\tau} \,, \qquad \tan \theta \equiv \frac{2qr}{1 + (q\tau)^2 - (qr)^2} \,, \tag{3.101}$$

where the parameter q is arbitrary. It will turn out that it determines quantitative properties of the flow profile. Note that the variable θ has nothing to do with the polar angle in spherical coordinates. In the new coordinates, the metric measure (3.100) reads

$$ds^2 = \tau^2 \left(d\rho^2 - \cosh^2\rho \, d\theta^2 - \cosh^2\rho \sin^2\theta \, d\phi^2 - d\eta_s^2 \right) \,, \tag{3.102}$$

such that the metric tensor becomes

$$g_{\mu\nu} = \mathrm{diag}(\tau^2, -\tau^2 \cosh^2\rho, -\tau^2 \cosh^2\rho \sin^2\theta, -\tau^2) \,. \tag{3.103}$$

Similar to the case of Bjorken flow, the flow profile in Gubser coordinates $(\rho, \theta, \phi, \eta_s)$ is particularly simple,

$$u^\mu = (\tau^{-1}, 0, 0, 0)^T \implies u_\mu = g_{\mu\nu}u^\nu = (\tau, 0, 0, 0) \,. \tag{3.104}$$

From the transformation law for a covariant vector A_μ under coordinate transformations,

$$A'_\mu = A_\nu \frac{\partial x^\nu}{\partial x'^\mu} \,, \tag{3.105}$$

we then compute from Eq. (3.101)

$$\frac{\partial \rho}{\partial \tau} = \frac{1}{\tau}(qr \sin\theta + \cos\theta) = \frac{1}{\tau} \frac{1}{\sqrt{1 - v_\perp^2(\tau, r)}} \,, \tag{3.106}$$

$$\frac{\partial \rho}{\partial r} = -q \sin\theta = -\frac{1}{\tau} \frac{v_\perp(\tau, r)}{\sqrt{1 - v_\perp^2(\tau, r)}} \,, \tag{3.107}$$

where we defined the transverse 3-velocity

$$v_\perp(\tau, r) \equiv \frac{2q^2 \tau r}{1 + (q\tau)^2 + (qr)^2} \,. \tag{3.108}$$

With the corresponding rapidity $\eta_\perp \equiv \mathrm{Artanh}\, v_\perp$, we obtain from Eq. (3.105) (with $A'_\mu \to u'_\mu = (u_\tau, u_r, u_\phi, u_{\eta_s})$ being the covariant 4-velocity vector in (τ, r, ϕ, η_s) coordinates) [3],

$$u_\tau = \cosh[\eta_\perp(\tau, r)]\,, \quad u_r = -\sinh[\eta_\perp(\tau, r)]\,, \quad u_\phi = u_{\eta_s} = 0\,. \qquad (3.109)$$

In the following, this solution will be referred to as *Gubser flow*.

Note that for any nonzero τ, the value of ρ decreases with r, while for a fixed r the value of ρ increases with τ. Thus, when $\rho \ll 0$ one probes regions where $r \gg 1$, and when $\rho \gg 1$ one has $\tau \gg 1$. In this sense, we expect that physically meaningful solutions for the temperature to behave as $\lim_{\rho \to \pm\infty} T(\rho) = 0$. Similar arguments also hold for the shear-stress tensor.

Conformal invariance further dictates that the fluid-dynamical evolution (or any other evolution, for that matter) is invariant under a *Weyl rescaling* of the measure,

$$ds^2 \longrightarrow d\hat{s}^2 = ds^2/\tau^2\,. \qquad (3.110)$$

The Weyl rescaling of the metric considerably simplifies the problem from a mathematical point of view, since it will render the fluid homogeneous in the coordinates $(\rho, \theta, \phi, \eta_s)$,

$$\hat{u}^\mu = (1, 0, 0, 0)^T\,. \qquad (3.111)$$

This is analogous to the case of Bjorken flow. Note that, once a Weyl rescaling is performed, all fields become dimensionless. Because of the metric rescaling and the coordinate transformation $x^\mu = (\tau, r, \phi, \eta)^T \to \hat{x}^\mu = (\rho, \theta, \phi, \eta_s)^T$, the dimensionless dynamical variables in the rescaled system (denoted with a *hat*: $\hat{T}$, $\hat{u}_\nu$, and $\hat{\pi}_{\mu\nu}$) are related to those in hyperbolic coordinates (T, u_ν, and $\pi_{\mu\nu}$) as follows [3],

$$u_\mu(\tau, r) = \hat{u}_\nu \tau \frac{\partial \hat{x}^\nu}{\partial x^\mu}\,, \qquad (3.112)$$

$$T(\tau, r) = \frac{\hat{T}}{\tau}\,, \qquad (3.113)$$

$$\pi_{\mu\nu}(\tau, r) = \hat{\pi}_{\alpha\beta} \frac{1}{\tau^2} \frac{\partial \hat{x}^\alpha}{\partial x^\mu} \frac{\partial \hat{x}^\beta}{\partial x^\nu}\,. \qquad (3.114)$$

The factors of τ in the transformation rules above come from the properties of these fields under Weyl transformations [3], and restore the dimension of each field. For instance, since $\pi_{\mu\nu} \to \Omega^2 \pi_{\mu\nu}$ under Weyl rescaling, $g_{\mu\nu} \to \Omega^{-2} g_{\mu\nu}$, with $\Omega = \tau$, there is a factor of $1/\tau^2$ in Eq. (3.114).

Combining Eqs. (3.102) and (3.110), we thus consider a *homogeneous* system of massless particles that are expanding in (ordinary) space-time described by the measure [3],

$$d\hat{s}^2 = \hat{g}_{\mu\nu} dx^\mu dx^\nu = d\rho^2 - \left(\cosh^2\rho\, d\theta^2 + \cosh^2\rho \sin^2\theta\, d\phi^2 + d\eta_s^2\right)\,, \qquad (3.115)$$

i.e., the rescaled metric tensor is

$$\hat{g}_{\mu\nu} = \text{diag}\left(1, -\cosh^2\rho, -\cosh^2\rho\sin^2\theta, -1\right) . \tag{3.116}$$

The nonzero Christoffel symbols (3.7) pertaining to this metric are

$$\hat{\Gamma}^{\rho}_{\theta\theta} = \cosh\rho\sinh\rho , \quad \hat{\Gamma}^{\rho}_{\phi\phi} = \sin^2\theta\cosh\rho\sinh\rho ,$$

$$\hat{\Gamma}^{\theta}_{\rho\theta} = \hat{\Gamma}^{\psi}_{\rho\phi} = \tanh\rho , \quad \hat{\Gamma}^{\theta}_{\phi\phi} = -\sin\theta\cos\theta , \quad \hat{\Gamma}^{\phi}_{\theta\phi} = \cot\theta , \tag{3.117}$$

and its determinant is

$$\sqrt{-\hat{g}} = \sin\theta\cosh^2\rho . \tag{3.118}$$

Since the system is homogeneous, all fields depend only on the time-like variable ρ, without displaying any dependence on θ, ϕ, and η_s. If we transform back to Minkowski space-time in hyperbolic coordinates (τ, r, ϕ, η_s), where the line element is given by Eq. (3.100), this homogeneous system is mapped into a longitudinally boost-invariant fluid, whose expansion in the transverse plane is radially symmetric.

After the Weyl rescaling (3.110) of the measure, the 4-velocity field simplifies to Eq. (3.111), which automatically satisfies the momentum-conservation equation (for conformal fluids). Nevertheless, in this curved space-time, the fluid still has a nonzero expansion rate and shear tensor. Using Eqs. (3.4) and (3.8), together with Eqs. (3.117) and (3.118), we obtain

$$\hat{\theta} \equiv \hat{D}_\mu \hat{u}^\mu = 2\tanh\rho , \tag{3.119}$$

$$\hat{\sigma}_{\mu\nu} \equiv -\hat{\Gamma}^{\rho}_{\langle\mu\nu\rangle} = \text{diag}\left(0, -\frac{1}{3}\cosh^2\rho, -\frac{1}{3}\cosh^2\rho\sin^2\theta, \frac{2}{3}\right)\tanh\rho ,$$

$$\implies \hat{\sigma}^{\mu}_{\nu} = \text{diag}\left(0, \frac{1}{3}, \frac{1}{3}, -\frac{2}{3}\right)\tanh\rho . \tag{3.120}$$

Similar to Bjorken flow, the vorticity tensor vanishes for this flow profile,

$$\hat{\omega}_{\mu\nu} \equiv \frac{1}{2}\hat{\Delta}^{\alpha}_{\mu}\hat{\Delta}^{\beta}_{\nu}\left(\hat{D}_\alpha\hat{u}_\beta - \hat{D}_\beta\hat{u}_\alpha\right) = 0 , \tag{3.121}$$

and plays no role for the results of this section.

The shear-stress tensor should have the same form as the shear tensor. It must be diagonal and can be parametrized in the following way (identical to what occurred in the Bjorken-flow scenario),

$$\hat{\pi}^{\mu}_{\nu} = \text{diag}\left(0, \frac{\hat{\pi}}{2}, \frac{\hat{\pi}}{2}, -\hat{\pi}\right) . \tag{3.122}$$

3.3.2 Fluid-Dynamical Equations of Motion

The equation of motion for the energy density $\hat{\varepsilon}(\rho)$ becomes

$$\hat{u}_\nu \hat{D}_\mu \hat{T}^{\mu\nu} = \frac{d\hat{\varepsilon}}{d\rho} + \frac{8}{3}\hat{\varepsilon} \tanh \rho - \hat{\pi} \tanh \rho = 0 , \tag{3.123}$$

where we used that $\hat{\pi}^{\mu\nu}\hat{\sigma}_{\mu\nu} = \hat{\pi} \tanh \rho$. The equation of motion (3.17) for $\hat{\pi}(\rho)$ is

$$\hat{\tau}_\pi \frac{d\hat{\pi}}{d\rho} + \frac{8\hat{\tau}_\pi}{3}\hat{\pi} \tanh \rho + \hat{\pi} = \frac{4\hat{\eta}}{3} \tanh \rho , \tag{3.124}$$

with the Weyl-rescaled shear-viscosity coefficient $\hat{\eta}$ and shear relaxation time $\hat{\tau}_\pi$. It is convenient to express these equations in terms of the temperature, $\hat{T}$ (the system is conformal, hence $\hat{\varepsilon} \sim \hat{T}^4$), and the variable $\hat{\chi}$ defined in the same way as in the discussion of the Bjorken-flow scenario,

$$\hat{\chi} \equiv \frac{\hat{\pi}}{\hat{\varepsilon} + \hat{P}_0} = \frac{3\hat{\pi}}{4\hat{\varepsilon}} . \tag{3.125}$$

This variable is convenient since it is invariant under Weyl transformations. We then rewrite the equations of motion in the following form [30]

$$\frac{1}{\hat{T}} \frac{d\hat{T}}{d\rho} + \frac{2}{3} \tanh \rho - \frac{\hat{\chi}}{3} \tanh \rho = 0 , \tag{3.126}$$

$$\hat{\tau}_\pi \frac{d\hat{\chi}}{d\rho} + \hat{\chi} + \frac{4}{3}\hat{\tau}_\pi \hat{\chi}^2 \tanh \rho = \frac{4}{3}\hat{\tau}_\eta \tanh \rho . \tag{3.127}$$

We follow a similar notation as employed in the previous chapters and define $\hat{\tau}_\eta \equiv \hat{\eta}/(\hat{\varepsilon} + \hat{P}_0)$.

Therefore, we have once more reduced the complicated nonlinear partial differential equations of fluid dynamics to simple ordinary coupled differential equations for $\hat{T}$ and $\hat{\chi}$. These equations can be solved numerically using simple algorithms, as was first done in Ref. [30]. These equations and their limiting forms will be investigated in the remainder of this chapter.

3.3.3 Ideal-Fluid Limit

For the sake of simplicity, let us first consider the example of an ideal gas of massless particles in the ideal-fluid limit. In practice, the ideal-fluid solution of the equations of motion is obtained by setting $\chi = 0$, leading to the following equation of motion for the temperature,

$$\frac{1}{\hat{T}} \frac{d\hat{T}}{d\rho} + \frac{2}{3} \tanh \rho = 0 . \tag{3.128}$$

The solution for this equation can be readily found,

$$\hat{T}(\rho) = \frac{\hat{T}(0)}{\cosh^{2/3}\rho} .$$
(3.129)

Using Eq. (3.113), we conclude that the temperature in the original hyperbolic coordinates is given by

$$T_{\text{ideal}}(\tau, r) = \frac{\hat{T}(0)(2q\tau)^{2/3}}{\tau\left\{\left[1 + (q\tau)^2 + (qr)^2\right]^2 - 4(q\tau)^2(qr)^2\right\}^{1/3}} ,$$
(3.130)

and, at the time $\tau_0 = 1/q$, one finds $T_{\text{ideal}}(\tau_0, 0) = \hat{T}(0) q$.

3.3.4   Relativistic Navier–Stokes Theory

The relativistic Navier–Stokes approximation consists in setting the relaxation time to zero, i.e., $\hat{\tau}_\pi = 0$, while keeping $\hat{\eta}/\hat{s}_0$ fixed in Eq. (3.127). In this case, the dissipative current $\hat{\chi}$ satisfies the constitutive relation

$$\hat{\chi} = \frac{4}{3}\hat{\tau}_\eta \tanh\rho = \frac{4}{3}\frac{\hat{\eta}}{\hat{s}_0}\frac{\tanh\rho}{\hat{T}} ,$$
(3.131)

where we used the thermodynamic relation $\hat{\varepsilon} + \hat{P}_0 = \hat{T}\hat{s}_0$, valid in the absence of any conserved charge. The equation of motion (3.126) for the temperature $\hat{T}$ then becomes

$$\frac{d\hat{T}}{d\rho} + \frac{2}{3}\hat{T}\tanh\rho - \frac{4}{9}\frac{\hat{\eta}}{\hat{s}_0}\tanh^2\rho = 0 .$$
(3.132)

The analytical solution, first derived in Ref. [3], is

$$\hat{T}_{\text{NS}}(\rho) = \frac{\hat{T}_{\text{NS}}(0)}{\cosh^{2/3}\rho} + \frac{4}{27}\frac{\hat{\eta}}{\hat{s}_0}\frac{\sinh^3\rho}{\cosh^{2/3}\rho}\,_2F_1\left(\frac{3}{2},\frac{7}{6};\frac{5}{2};-\sinh^2\rho\right) ,$$
(3.133)

where $_2F_1$ is a hypergeometric function. From the equation of motion (3.132), the condition $\lim_{\rho\to\pm\infty} d\hat{T}/d\rho = 0$ shows that $\lim_{\rho\to\pm\infty}\hat{T}_{\text{NS}}(\rho) = \pm 2\hat{\eta}/3\hat{s}_0$. In this case, when $\hat{\eta}/\hat{s}_0 \neq 0$, for any given τ, there is always a value of r beyond which the temperature switches sign and becomes negative (which is very different than the ideal-fluid case, where $\lim_{\rho\to\pm\infty}\hat{T}_{\text{ideal}} = 0$). This unphysical feature may be connected with the well-known acausality issue of the relativistic Navier–Stokes equations, discussed in Chap. 2. We shall see below that once the relaxation-time coefficient is taken into account one can find a solution where $\hat{T}$ is positive definite and $\lim_{\rho\to-\infty}\hat{T}(\rho) = 0$.

3.3.5 Transient Theory of Fluid Dynamics

In general $\hat{\chi}$ obeys the ordinary nonlinear differential equation (3.127), which requires an independent initial condition. First we check whether it is possible to find solutions that are well-behaved in the limits $\rho \to \pm\infty$. For example, we first note that if one imposes $\lim_{\rho \to \pm\infty} \hat{T}(\rho) = 0$ and, simultaneously, $\lim_{\rho \to \pm\infty} d\hat{\chi}/d\rho = 0$, one can find that $\lim_{\rho \to \pm\infty} |\hat{\chi}(\rho)| = \sqrt{1/b}$, where we assumed the following parametrization for the relaxation time,

$$\hat{\tau}_\pi = b\hat{\tau}_\eta = b\frac{\hat{\eta}}{\hat{s}_0\hat{T}} \sim \frac{1}{\hat{T}}, \tag{3.134}$$

which has already been employed above. Therefore, solutions with $\lim_{\rho \to \pm\infty} \hat{T}(\rho) = 0$ are possible and do occur in practice, as long as b is finite—something ensured by causality.

There is a limit in which one can find analytical solutions for $\hat{T}$ and $\hat{\chi}$. This will become possible when the fluid is very viscous or when the temperature is very small, i.e., when $\hat{\eta}/\left(\hat{T}\hat{s}_0\right) \gg 1$. In this case, the term $\hat{\chi}$ becomes negligible in comparison to all the other terms in Eq. (3.127), which are all linear in $\hat{\eta}/\left(\hat{T}\hat{s}_0\right)$, and the equation can be approximated as

$$\frac{d\hat{\chi}}{d\rho} + \frac{4}{3}\hat{\chi}^2 \tanh \rho = \frac{4}{3b} \tanh \rho . \tag{3.135}$$

The solution of this equation is

$$\hat{\chi}(\rho) = \sqrt{\frac{1}{b}} \tanh\left[\sqrt{\frac{1}{b}}\left(\frac{4}{3}\ln\cosh\rho - \hat{\chi}_0 b\right)\right], \tag{3.136}$$

where $\hat{\chi}_0$ is an integration constant. Substituting this into Eq. (3.126), we obtain an analytic solution for the temperature,

$$\hat{T}(\rho) = \hat{T}_1 \frac{\exp\left(b\hat{\chi}_0/2\right)}{\cosh^{2/3}\rho} \cosh^{1/4}\left[\sqrt{\frac{1}{b}}\left(\frac{4}{3}\ln\cosh\rho - \hat{\chi}_0 b\right)\right], \tag{3.137}$$

where $\hat{T}_1$ is an integration constant. The analytical solutions (3.136) and (3.137) are even in ρ, $\hat{T}$ is positive definite, and $\lim_{\rho \to \pm\infty} \hat{T}(\rho) = 0$ if $4b > 1$. Moreover, note that, as long as $b > 1$, $\hat{\chi}$ is always smaller than 1 for any value of ρ, i.e., the dissipative correction to the energy-momentum tensor is always smaller than the ideal-fluid contribution.

We show in Fig. 3.2 a comparison between $\hat{T}$ and $\hat{\chi}$ computed for an ideal fluid, Navier–Stokes theory, and Israel–Stewart theory, for $\hat{\eta}/\hat{s}_0 = 0.2$, a value usually used in heavy-ion collisions, and $b = 5$, which is the typical value obtained from

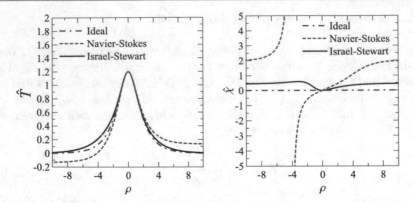

Fig. 3.2 Comparison between the solutions for $\hat{T}$ and $\hat{\chi}$ for $\hat{\eta}/\hat{s}_0 = 0.2$, $b = 5$, and $\hat{T}(0) = 1.2$ found using different versions of the relativistic fluid-dynamical equations of motion. The solid black lines denote solutions of Israel–Stewart theory, Navier–Stokes results are in dashed blue, while the dashed-dotted red curves correspond to the ideal-fluid case. Figure taken from Ref. [30]

approximations of the Boltzmann equation [5]. We have chosen the initial conditions for the equations such that $\hat{T}(0) = 1.2$, for all the cases, and, for the Israel–Stewart case, $\hat{\chi}(0) = 0$. The solution for Israel–Stewart theory is shown in solid black, the Navier–Stokes results in dashed blue, and the ideal fluid corresponds to the dashed-dotted red curve. One can see that the solution of Israel–Stewart theory for $\hat{T}$ is positive definite and $\lim_{\rho \to \pm\infty} \hat{T}(\rho) = 0$. Moreover, viscous effects break the parity of the solutions with respect to $\rho \to -\rho$. Note that, as mentioned before, $\hat{\chi}$ goes to $\sqrt{1/b}$ when $\rho \to \pm\infty$ in Israel–Stewart theory, while for Navier–Stokes theory this quantity diverges at $\rho \approx -4.19$, which is the value of ρ at which $\hat{T}_{NS} = 0$ and the temperature changes sign. It is important to remark how much the solutions of Israel–Stewart theory can differ from those of Navier–Stokes theory. In some regimes, we can see that they are not even qualitatively similar.

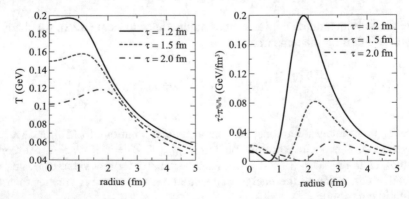

Fig. 3.3 Temperature and $\tau^2 \pi^{\eta_s \eta_s}$ profiles for $\tau = 1.2$ fm (solid black curves), $\tau = 1.5$ fm (dashed blue curves), and $\tau = 2$ fm (dashed-dotted red curves) with $q = 1$ fm^{-1}, $\hat{\eta}/\hat{s}_0 = 0.2$, $b = 5$, and $\hat{T}(0) = 1.2$. Figure taken from Ref. [30]

In order to study the space-time dependence of the solutions we define $q = 1\,\mathrm{fm}^{-1}$ so that $\rho = 0$ corresponds to $\tau = 1\,\mathrm{fm}$ and $r = 0$. Therefore, in standard hyperbolic coordinates, $T\,(r = 0, \tau_0 = 1\,\mathrm{fm}) = 1.2\,\mathrm{fm}^{-1}$ and $\chi\,(r = 0, \tau_0 = 1\,\mathrm{fm}) = 0$. In Fig. 3.3 we show a comparison between the temperature profiles for Israel–Stewart theory at the times $\tau = 1.2, 1.5, 2\,\mathrm{fm}$, with $\hat{\eta}/\hat{s}_0 = 0.2$, $b = 5$. Also, in the same figure, we show $\tau^2 \pi^{\eta_s \eta_s}$ as a function of the radius for the same times. The other components of the shear-stress tensor can be obtained using Eq. (3.114), but do not provide any new information.

3.3.6 Gradient Expansion of Transient Fluid Dynamics

We now discuss the gradient expansion of asymptotic solutions of Israel–Stewart theory in the context of Gubser flow. Here, we shall see once more that such a complicated expansion can be carried out to arbitrarily high orders. We follow the procedure outlined in Refs. [9,31], as has been applied to Bjorken-flow solutions in the previous section. The idea is to convert the original problem into a perturbation-theory problem by introducing a dimensionless parameter ϵ into the differential equation (3.127) satisfied by $\hat{\chi}$ as follows

$$\epsilon \frac{d\hat{\chi}}{d\rho} + \frac{\hat{\chi}}{\hat{\tau}_\pi} + \frac{4}{3}\epsilon \hat{\chi}^2 \tanh \rho = \frac{4}{3b}\epsilon \tanh \rho \,. \tag{3.138}$$

As explained before, the shear-stress tensor and the temperature then become functions of ρ and the new variable ϵ, i.e., $\hat{T} = \hat{T}(\rho, \epsilon)$ and $\hat{\pi} = \hat{\pi}(\rho, \epsilon)$. The parameter ϵ was introduced in every term which contains a derivative or $\tanh \rho$ and, hence, it becomes a book-keeping parameter to count orders or powers of gradients. Note that while in Bjorken flow it was straightforward to deduce that powers of gradients correspond to inverse powers of the time coordinate τ [9], here the situation is not that simple and the corresponding terms are obtained from the perturbative procedure itself.

Next, we look for a solution for $\hat{\chi}$ that can be represented as a power series in ϵ

$$\hat{\chi} = \sum_{n=0}^{\infty} \hat{\chi}_n(\rho)\epsilon^n \,. \tag{3.139}$$

This reduces the problem to solving an infinite number of recurrence relations. At the end of the calculation, one sets $\epsilon = 1$ to recover the parameters of the original problem.

We now continue the perturbative calculation, substituting the proposed series solution (3.139) into the equation of motion (3.138) for $\hat{\chi}$. One then obtains the following result:

$$\sum_{n=0}^{\infty} \frac{d\hat{\chi}_n}{d\rho}\epsilon^{n+1} + \frac{1}{\hat{\tau}_\pi}\sum_{n=0}^{\infty} \hat{\chi}_n \epsilon^n + \frac{4}{3}\sum_{n=0}^{\infty}\sum_{m=0}^{\infty} \hat{\chi}_n \hat{\chi}_m \epsilon^{n+m+1} \tanh \rho = \frac{4}{3b}\epsilon \tanh \rho \,. \tag{3.140}$$

The problem in solving this equation, i.e., in collecting all terms that are of the same power in ϵ, is that the term $d\hat{\chi}_n/d\rho$ also has an ϵ-dependence that must be considered when grouping the terms. This can be taken into account in the same way as in the previous section. The coefficient $\hat{\chi}_n$ depends on ρ through two different variables, $\hat{\chi}_n = \hat{\chi}_n\left(\rho, \hat{T}(\rho)\right)$ and, thus, its derivative can be mathematically re-expressed in the following way

$$\frac{d\hat{\chi}_n}{d\rho} = \left.\frac{\partial\hat{\chi}_n}{\partial\rho}\right|_{\hat{T}} + \left.\frac{\partial\hat{\chi}_n}{\partial\hat{T}}\right|_{\rho}\frac{d\hat{T}}{d\rho} . \tag{3.141}$$

Above, the derivatives are taken as if ρ and $\hat{T}$ were two independent variables. It is the temperature derivative that carries the ϵ-dependence and, thus, using Eq. (3.126) we can rewrite Eq. (3.140) as

$$\sum_{n=0}^{\infty}\left(\left.\frac{\partial\hat{\chi}_n}{\partial\rho}\right|_{\hat{T}} - \frac{2}{3}\hat{T}\tanh\rho\left.\frac{\partial\hat{\chi}_n}{\partial\hat{T}}\right|_{\rho}\right)\epsilon^{n+1} + \frac{\hat{T}}{3}\tanh\rho\sum_{n=0}^{\infty}\sum_{m=0}^{\infty}\hat{\chi}_m\left.\frac{\partial\hat{\chi}_n}{\partial\hat{T}}\right|_{\rho}\epsilon^{n+m+1}$$

$$+ \frac{1}{\hat{\tau}_\pi}\sum_{n=0}^{\infty}\hat{\chi}_n\epsilon^n + \frac{4}{3}\sum_{n=0}^{\infty}\sum_{m=0}^{\infty}\hat{\chi}_n\hat{\chi}_m\epsilon^{n+m+1}\tanh\rho - \frac{4}{3b}\epsilon\tanh\rho = 0 . \tag{3.142}$$

Now that the terms are properly organized in powers of ϵ we can group the terms together that are of the same order and obtain the set of recurrence relations that must be solved to obtain $\hat{\chi}_n$. The zeroth-order term must satisfy

$$\hat{\chi}_0 = 0 , \tag{3.143}$$

which describes a system that is in local equilibrium, as expected. Collecting the terms that are of first order in ϵ one obtains

$$\hat{\chi}_1(\rho) = \frac{4}{3b}\hat{\tau}_\pi\tanh\rho = \frac{4}{3}\hat{\tau}_\eta\tanh\rho , \tag{3.144}$$

which corresponds to the constitutive relation employed in relativistic Navier–Stokes theory, cf. Eq. (3.131). This expression also reflects the fact that the shear-stress tensor in Navier–Stokes theory is linear in the Knudsen number $\mathrm{Kn} \sim \hat{\tau}_\pi\sqrt{\hat{\sigma}_{\mu\nu}\hat{\sigma}^{\mu\nu}}$ for Gubser flow. Finally, the terms that are of second order or higher in ϵ, ($n \geq 2$), satisfy the equation

$$\hat{\chi}_{n+1} = \frac{\hat{\tau}_\pi}{3}\tanh\rho\left[2\left.\frac{\partial\hat{\chi}_n}{\partial\ln\hat{T}}\right|_{\rho} - \sum_{m=0}^{n}\left(4\hat{\chi}_{n-m} + \left.\frac{\partial\hat{\chi}_{n-m}}{\partial\ln\hat{T}}\right|_{\rho}\right)\hat{\chi}_m\right] - \hat{\tau}_\pi\left.\frac{\partial\hat{\chi}_n}{\partial\rho}\right|_{\hat{T}} . \tag{3.145}$$

Since we already know $\hat{\chi}_0$ and $\hat{\chi}_1$, we can calculate all the $\hat{\chi}_n$'s that follow. We note that $\hat{\tau}_\pi \sim \hat{T}^{-1}$, therefore $d \ln \hat{\tau}_\pi / d \ln \hat{T} = -1$. For the sake of completeness, we also give the result up to third order,

$$\hat{\chi} = \frac{4}{3b} \epsilon \hat{\tau}_\pi \tanh \rho - \frac{4}{3b} \left(\epsilon \hat{\tau}_\pi \right)^2 \left(1 - \frac{1}{3} \tanh^2 \rho \right)$$
$$+ \frac{8}{9b} \left(\epsilon \hat{\tau}_\pi \right)^3 \left[\tanh \rho \left(1 + \frac{1}{3} \tanh^2 \rho \right) - \frac{2}{b} \tanh^3 \rho \right] + \mathcal{O} \left(\epsilon^4 \right) . \quad (3.146)$$

Therefore, we can write the solution as a series in powers of $\epsilon \hat{\tau}_\pi$, with all temperature contributions to the shear-stress tensor contained in the relaxation time. Also, we note that while the Navier–Stokes result depends solely on the combination $\hat{\tau}_\pi \tanh \rho$, the same is not true for the higher-order terms in the gradient expansion, which depend separately on $\hat{\tau}_\pi$ and $\tanh \rho$.

These equations have the form that is traditionally associated with a gradient expansion: the zeroth- and first-order truncations obtained in this section correspond to well-known results, ideal hydrodynamics and Navier–Stokes theory, respectively. Both of these examples were already studied extensively in the literature [3].

3.3.7   Divergence of the Gradient Expansion

The recurrence relations derived above were computed up to $n = 100$ in Ref. [31], assuming $b = 5$. We reproduce these results in the left panel of Fig. 3.4 for $\rho = 0.1$, $\rho = 1$, and $\rho = 10$. In all these plots, the coefficients were rescaled by the relaxation time in the following manner, $\pi_n \equiv \hat{\chi}_n / \hat{\tau}_\pi^n$, in such a way to remove any dependence of the result on the value of the shear viscosity. In the right panel of Fig. 3.4, we also show the corresponding result when $\rho = 0$ and $\rho \to \infty$. We do not plot results for negative values of ρ since the modulus of each coefficient does not depend on the sign of ρ. Both figures clearly display the factorial growth of the coefficients of the series, indicating that the gradient expansion has zero radius of convergence. We also remark that the results vary very little when ρ is changed.

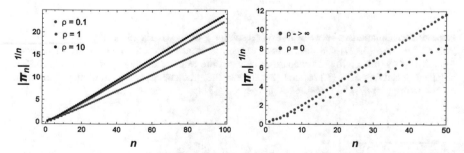

Fig. 3.4 *Left:* $|\pi_n|^{1/n}$ as a function of n for $\rho = 0.1$ (red), $\rho = 1$ (blue), and $\rho = 10$ (black). *Right:* $|\pi_n|^{1/n}$ as a function of n for $\rho = 0$ and $\rho \to \infty$. Figure taken from Ref. [31]

It is interesting to see that when $\rho = 0$ (i.e., when the shear tensor is zero) the gradient expansion still diverges. Moreover, since the first-order term in the expansion is proportional to $\tanh \rho$, this term vanishes when $\rho = 0$ and, as a matter of fact, one can show that all odd terms in the series vanish in this case. However, the coefficients $\pi_{2n}(0)$ do not vanish and these terms alone display factorial growth.

3.3.8 Domain of Applicability of the Gradient Expansion

As already mentioned, low-order truncations of a divergent expansion can still be used to provide reasonable approximations for solutions of the theory, at least in some domains. In Fig. 3.5, we compare several truncations of this divergent series with exact solutions obtained by numerically solving the Israel–Stewart equations (3.126) and (3.127), for two values of $\hat{\eta}/\hat{s}_0$: $1/(4\pi)$ and 1. The initial conditions for the numerical problem where chosen to be $\hat{T}(\rho_0) = 0.0057$ and $\hat{\chi}(\rho_0) = 0.4$, with $\rho_0 = -30$. We remark that the results do not depend strongly on the value chosen for $\hat{\chi}(\rho_0)$, but they do depend on the choice of $\hat{T}(\rho_0)$.[1] In this comparison, we take the exact temperature profile $\hat{T}(\rho)$ and insert it into the corresponding constitutive relations obtained for $\hat{\chi}$ using the gradient expansion.

We find that, when the viscosity is small ($\hat{\eta}/\hat{s}_0 = 1/4\pi$), the first- and second-order truncations of the gradient expansion provide a good approximation to the exact solution in a wide region around $\rho = 0$. We remark that the best approximation to the exact solution, for this value of viscosity, is obtained by truncating the expansion

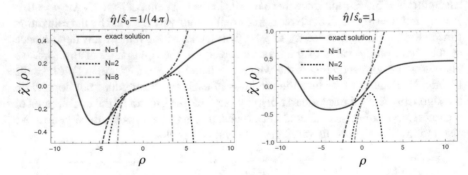

Fig. 3.5 *Left*: Comparison between the exact result for $\hat{\chi}$ defined in Eq. (3.125) obtained by solving Eqs. (3.126) and (3.127) and the gradient expansion truncated at different orders. In this plot $\hat{\eta}/\hat{s}_0 = 1/(4\pi)$ (black). *Right*: Comparison between the exact result for $\hat{\chi}$ defined in Eq. (3.125) obtained by solving Eqs. (3.126) and (3.127) and the gradient expansion truncated at different orders. In this plot $\hat{\eta}/\hat{s}_0 = 1$. Figure taken from Ref. [31]

---

[1] It is straightforward to see from the equations of motion for $\hat{T}$ and $\hat{\chi}$ that rescaling the initial value of the temperature is equivalent to changing the shear viscosity of the system. This happens in such a way that reducing the initial temperature of the system corresponds to effectively increasing the value of $\hat{\eta}/\hat{s}_0$.

at second order, as also happened when performing this analysis assuming Bjorken flow [9]. However, the divergent nature of the series is manifest by the fact that the eighth-order truncation is significantly worse than the lower orders, indicating that the optimal truncation of the series is indeed at a lower order.

Meanwhile, for a larger value of viscosity, $\hat{\eta}/\hat{s}_0 = 1$, none of the different truncations of the gradient series is able to provide a reasonable description of the exact solution. In fact, in this case one even finds that $\hat{\chi}(0)$ deviates significantly from zero in the exact solution—a result that is very hard to describe using the gradient expansion, since it implies that the Navier-Stokes limit is not approached at all even when $\rho \approx 0$. In general, our results suggests that truncations of the gradient expansion can provide a good description of solutions of Israel–Stewart theory around $\rho = 0$, though how far in ρ this occurs (or if this occurs at all) clearly depends on the value of $\hat{\eta}/\hat{s}_0$.

Furthermore, we remark that the gradient expansion cannot describe the quantitative and qualitative behavior of the solution when $|\rho|$ is very large. In fact, it is known from Ref. [30], and was also discussed in the previous sections, that the exact solution for $\hat{\chi}$ in Israel–Stewart theory asymptotes to a constant when $|\rho| \gg 1$—a result that can never be obtained within a gradient expansion. However, this is not the limit where one would expect the gradient expansion to be useful since in this case all powers of $\tanh \rho$ become of the same order.

3.3.9 Slow-Roll Expansion

It is now convenient to introduce another perturbative solution of transient fluid dynamics: the slow-roll expansion [20]. The slow-roll expansion in Gubser flow was first determined in Ref. [31] and is defined by the following perturbative problem

$$\epsilon c \frac{d\hat{\chi}}{d\rho} + \hat{T}\hat{\chi} + \frac{4}{3}c\hat{\chi}^2 \tanh \rho = \frac{4c}{3b} \tanh \rho \,, \qquad (3.147)$$

where now, in contrast to Eq. (3.138), the book-keeping parameter ϵ multiplies only the derivative term of the equation. For the sake of convenience, we introduced the parameter c, defined by the relation $\hat{\tau}_\pi \equiv c/\hat{T}$. Using the parametrization introduced in Eq. (3.134), one can express this new variable in terms of the shear viscosity-to-entropy density ratio as $c = b\hat{\eta}/\hat{s}_0$. As before, we look for perturbative solutions of the form

$$\hat{\chi} = \sum_{n=0}^{\infty} \bar{\pi}_n \epsilon^n \,, \qquad (3.148)$$

and, at the end of the calculation, set $\epsilon = 1$, recovering, in principle, a solution of the original equation of motion. Here, we labeled the expansion coefficients as $\bar{\pi}_n$, so that they are not confused with the expansion coefficients of the gradient series (3.139), labeled as $\hat{\chi}_n$. Substituting this expansion into Eq. (3.147), we obtain the following equation

$$c \sum_{n=0}^{\infty} \frac{d\bar{\pi}_n}{d\rho} \epsilon^{n+1} + \hat{T} \sum_{n=0}^{\infty} \bar{\pi}_n \epsilon^n + \frac{4}{3} c \sum_{n=0}^{\infty} \sum_{m=0}^{\infty} \bar{\pi}_n \bar{\pi}_m \epsilon^{n+m} \tanh \rho = \frac{4c}{3b} \tanh \rho \ .$$

(3.149)

Analogously to the calculation involving the gradient expansion, we write the derivative of $\bar{\pi}_n$ as

$$\frac{d\bar{\pi}_n}{d\rho} = \left. \frac{\partial \bar{\pi}_n}{\partial \rho} \right|_{\hat{T}} + \left. \frac{\partial \bar{\pi}_n}{\partial \hat{T}} \right|_{\rho} \frac{d\hat{T}}{d\rho} \ ,$$

which allows us to properly collect all powers of ϵ that are contained in the temperature derivative. Using Eq. (3.126), this leads to the following equation,

$$c \sum_{n=0}^{\infty} \left(\left. \frac{\partial \bar{\pi}_n}{\partial \rho} \right|_{\hat{T}} - \frac{2}{3} \tanh \rho \ \hat{T} \left. \frac{\partial \bar{\pi}_n}{\partial \hat{T}} \right|_{\rho} \right) \epsilon^{n+1} + \frac{c}{3} \tanh \rho \sum_{n=0}^{\infty} \sum_{m=0}^{\infty} \bar{\pi}_m \hat{T} \left. \frac{\partial \bar{\pi}_n}{\partial \hat{T}} \right|_{\rho} \epsilon^{n+m+1}$$

$$+ \hat{T} \sum_{n=0}^{\infty} \bar{\pi}_n \epsilon^n + \frac{4}{3} c \sum_{n=0}^{\infty} \sum_{m=0}^{\infty} \bar{\pi}_n \bar{\pi}_m \epsilon^{n+m} \tanh \rho = \frac{4c}{3b} \tanh \rho \ . \qquad (3.150)$$

The zeroth-order solution is the Gubser-flow generalization of the well-known result first derived for Bjorken flow in Ref. [20],

$$\frac{4}{3} \bar{\pi}_0^2 \hat{\tau}_\pi \tanh \rho + \bar{\pi}_0 = \frac{4}{3b} \hat{\tau}_\pi \tanh \rho \ ,$$

$$\Longrightarrow \bar{\pi}_0^{\pm}(\rho) = \frac{-3 \pm \sqrt{9 + \frac{1}{b} \left(8\hat{\tau}_\pi \tanh \rho \right)^2}}{8\hat{\tau}_\pi \tanh \rho} \ .$$

(3.151)

The remaining terms of the expansion must be calculated using *one* of the solutions above. The most natural choice is to choose the one which converges to the Navier–Stokes solution in the limit $\rho \to 0$, i.e., we consider only the solution $\bar{\pi}_0^+$. In fact, if one expands Eq. (3.151) in powers of $\hat{\tau}_\pi$ one finds

$$\bar{\pi}_0^+(\rho) = \frac{4}{3b} \hat{\tau}_\pi \tanh \rho - \frac{64}{27b^2} \left(\hat{\tau}_\pi \tanh \rho \right)^3 + \mathcal{O}(\hat{\tau}_\pi^5) \ . \qquad (3.152)$$

Thus, we see that the zeroth-order term in the slow-roll expansion recovers the Navier–Stokes constitutive relation (3.131), i.e., it matches the gradient expansion truncated at first order. However, it is important to notice that the higher-order terms generated by Taylor-expanding $\bar{\pi}_0^+$ differ from the higher-order terms of the gradient expansion.

Also, it is interesting to note that the zeroth-order slow-roll term $\bar{\pi}_0^+$ is solely a function of the Knudsen number $\hat{\tau}_\pi \tanh \rho \sim \hat{\tau}_\pi \sqrt{\sigma_{\mu\nu} \sigma^{\mu\nu}}$ for Gubser flow, as also happened with the first-order truncation of the gradient expansion. Nevertheless, we shall explicitly show in the following that the higher-order terms of the slow-roll series in Gubser flow are functions of both $\hat{\tau}_\pi$ and $\tanh \rho$ separately, as also occurred in the gradient expansion.

Using Eq. (3.151), higher-order solutions are obtained by solving the recurrence relation

$$
\frac{\partial \bar{\pi}_n}{\partial \rho}\bigg|_{\hat{T}} - \frac{2\hat{T}}{3} \tanh \rho \left. \frac{\partial \bar{\pi}_n}{\partial \hat{T}}\right|_\rho + \frac{\hat{T}}{3} \tanh \rho \sum_{m=0}^{n} \bar{\pi}_m \left. \frac{\partial \bar{\pi}_{n-m}}{\partial \hat{T}}\right|_\rho
$$

$$
+ \frac{\hat{T}}{c} \bar{\pi}_{n+1} + \frac{4}{3} \sum_{m=0}^{n+1} \bar{\pi}_{n+1-m} \bar{\pi}_m \tanh \rho = 0 , \tag{3.153}
$$

which can be simplified to

$$
\frac{1}{\hat{\tau}_\pi} \left(1 + \frac{8}{3} \hat{\tau}_\pi \tanh \rho \, \bar{\pi}_0^+ \right) \bar{\pi}_{n+1} = -\frac{2}{3} \tanh \rho \, \hat{\tau}_\pi \left. \frac{\partial \bar{\pi}_n}{\partial \hat{\tau}_\pi}\right|_\rho - \left. \frac{\partial \bar{\pi}_n}{\partial \rho}\right|_{\hat{\tau}_\pi}
$$

$$
+ \frac{1}{3} \tanh \rho \sum_{m=0}^{n} \bar{\pi}_m \, \hat{\tau}_\pi \left. \frac{\partial \bar{\pi}_{n-m}}{\partial \hat{\tau}_\pi}\right|_\rho - \frac{4}{3} \sum_{m=1}^{n} \bar{\pi}_{n+1-m} \bar{\pi}_m \tanh \rho . \tag{3.154}
$$

Solving this recurrence relation for $n = 0$, one can determine the first-order solution as

$$
\bar{\pi}_1(\rho) = \frac{3b\bar{\pi}_0^+}{\hat{\tau}_\pi \tanh \rho} \frac{\left(1 + \bar{\pi}_0^+\right) \left(\hat{\tau}_\pi \tanh \rho\right)^2 - 3\hat{\tau}_\pi^2}{9b + \left(8\hat{\tau}_\pi \tanh \rho\right)^2} . \tag{3.155}
$$

This result clearly demonstrates that $\bar{\pi}_1$ depends on both $\hat{\tau}_\pi$ and $\tanh \rho$ separately, a fact that remains true for all higher-order coefficients of the slow-roll expansion.

Furthermore, expanding now the series truncated at first order, i.e., $\bar{\pi} = \bar{\pi}_0^+ + \bar{\pi}_1$, in powers of $\hat{\tau}_\pi$ leads to

$$
\bar{\pi}(\rho) = \frac{4}{3b} \hat{\tau}_\pi \tanh \rho - \frac{4}{3b} \hat{\tau}_\pi^2 \left(1 - \frac{1}{3} \tanh^2 \rho \right) - \frac{16}{9b^2} (\hat{\tau}_\pi \tanh \rho)^3 + \mathcal{O}(\hat{\tau}_\pi^4) , \tag{3.156}
$$

which shows that the first-order truncation of the slow-roll series is able to recover the result obtained from the gradient expansion truncated at second order, see Eq. (3.146). In general, a Taylor expansion of the slow-roll expansion truncated at the Nth term will reduce to the correct expression for the gradient expansion truncated at order $N + 1$. In this sense, the slow-roll expansion can be seen as a type of reorganization of the gradient series, which contains an infinite resummation of gradients at a given order.

3.3.10 Divergence of the Slow-Roll Series

In Ref. [31], Eq. (3.154) was solved numerically to determine the behavior of the slow-roll series at higher orders. In the left panel of Fig. 3.6, we show how the

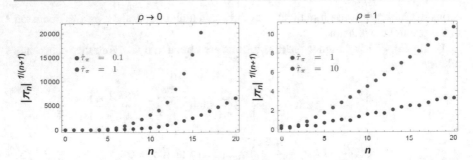

Fig. 3.6 (Color online) *Left*: Large-order behavior of the slow-roll series when $\rho \to 0$ and $\hat{\tau}_\pi = 0.1, 1$. *Right*: Large-order behavior of the slow-roll series when $\rho = 1$ and $\hat{\tau}_\pi = 1, 10$. Figure taken from Ref. [31]

coefficients of the series change with n when $\rho \to 0$ and for the relaxation times $\hat{\tau}_\pi = 0.1$ and 1. This plot indicates that, in this limit, the magnitude of the terms grows larger than $n!$ when n is large. To illustrate the fact that the large-order behavior of the slow-roll series now depends on two parameters (i.e., ρ and $\hat{\tau}_\pi$), we show in the right panel of Fig. 3.6 what happens when $\rho = 1$. Even though the series still appears to diverge, in this case $|\bar{\pi}_n|^{1/(n+1)}$ only grows linearly with n (the same qualitative result appears for other values of $\hat{\tau}_\pi$ and also when ρ is negative). Therefore, in Israel–Stewart theory both the gradient and the slow-roll expansions generally diverge in Gubser flow, just as it occurred in Bjorken flow.

The only exception occurs when $|\rho| \to \infty$. Since the temperature vanishes in this limit one must also take $\hat{\tau}_\pi \to \infty$, which implies that $\bar{\pi}_0^{\pm} \to \pm\text{sgn}\,\rho/\sqrt{b}$ and Eq. (3.154) gives $\bar{\pi}_{n>0} = 0$. This shows that the slow-roll expansion in fact converges in this limit to $\bar{\pi}_0^{\pm}$. As already discussed in this chapter, solutions of the Israel–Stewart equation (3.127) for the shear-stress tensor display the same behavior and, thus, one can see that the slow-roll series necessarily converges to solutions of the Israel–Stewart equation when $|\rho| \to \infty$.

A comparison between different truncations of the slow-roll series for $\hat{\chi}$ with the exact solution of the Israel–Stewart equations can be found in Ref. [31]. Overall, low-order truncations of this series provide a better description of the exact solution of Israel–Stewart theory, when compared to low-order truncations of the gradient expansion, but they are still not able to describe such solutions when $\hat{\eta}/\hat{s}_0$ is not small.

3.3.11 Attractor Solution

In this section, we investigate the attractor solution of the Israel–Stewart equations for a system expanding according to Gubser flow, which was first studied in Ref. [31]. We shall interpret the attractor solution as a resummed slow-roll expansion. The analysis presented in this section aims to explore two fundamental aspects of the attractor: its possible functional dependence on the Knudsen number and its approximate description using the slow-roll series truncated at higher orders.

In the previous section, we found that the slow-roll series converges when $|\rho| \to \infty$ and we can use this fact to numerically construct a resummed version of the slow-roll solution. We found that there are two possible convergent solutions for $\hat{\chi}$ when $|\rho| \to \infty$: $1/\sqrt{b}$ and $-1/\sqrt{b}$. It is natural to expect that one of these boundary conditions will lead to the exact solution related to the slow-roll expansion. In practice, we observe that the vast majority of solutions of Israel–Stewart theory actually converges to $1/\sqrt{b}$ when $\rho \to -\infty$ (for $\rho \to \infty$, all known numerical solutions of Israel–Stewart theory also converge to $1/\sqrt{b}$). So far, the only case in which we are able to obtain a solution that is equal to $-1/\sqrt{b}$, when $\rho \to -\infty$, is when we give exactly this boundary condition at a very small value of ρ. Any small deviation from $-1/\sqrt{b}$ will make the solution tend to $1/\sqrt{b}$ when we decrease the value of ρ. This behavior suggests that the boundary condition $\hat{\chi}(-\infty) = -1/\sqrt{b}$ defines a unique solution of the equation and that such unique solution can be identified as the resummed result for the slow-roll series. On the other hand, the other boundary condition at $\rho = -\infty$, $\hat{\chi}(-\infty) = 1/\sqrt{b}$, is satisfied by an infinite number of solutions and cannot be used to define any specific solutions of the equations.

This is illustrated in Fig. 3.7 for two vastly different values of $\hat{\eta}/\hat{s}_0$. In these plots, the solid red curve depicts the solution of the Israel–Stewart equations assuming $\hat{T}(-30) = 9.222 \times 10^{-8}$ and $\hat{\chi}(-30) = -1/\sqrt{b}$ for $b = 5$. The dashed curves are computed keeping the initial condition for the temperature fixed while considering very small variations of $\hat{\chi}(-30)$, of at most 1% around $-1/\sqrt{b}$. We see that any small variation of the value of $\hat{\chi}$ at $\rho = -30$ makes the solution converge to $1/\sqrt{b}$ when ρ is decreased. This only does not happen when we fix $\hat{\chi}$ to be exactly $-1/\sqrt{b}$ (red curve). Furthermore, one can see that all solutions converge extremely rapidly to the solution where $\lim_{\rho \to -\infty} \hat{\chi}(\rho) = -1/\sqrt{b}$.

In Fig. 3.7, we also see that the profile of the attractor depends on the value of $\hat{\eta}/\hat{s}_0$, becoming closer to a step function as $\hat{\eta}/\hat{s}_0$ is increased even further. Such a behavior is very hard to describe using the slow-roll series, as one can see in the left panel of Fig. 3.8. In this figure, we compare the attractor solutions (solid red) with truncations of the slow-roll series. For $\hat{\eta}/\hat{s}_0 = 1/(4\pi)$ we used the second-order truncation while for the extremely large value of $\hat{\eta}/\hat{s}_0 = 1000$ we only took the zeroth-order term in the series. When $\hat{\eta}/\hat{s}_0$ is small, the truncated slow-roll series provides an excellent description of the attractor while for very large values of $\hat{\eta}/\hat{s}_0$ this perturbative approach provides a poor description, even though still qualitatively accurate, of the attractor away from the asymptotic regime, as expected.

Now we explore a different feature, so far exclusive to Gubser flow, which is the fact that the attractor solution per se depends on the temperature. This can be seen in the right panel of Fig. 3.8 where we now fix $\hat{\chi}(-30) = -1/\sqrt{5}$ (attractor solutions) and consider three very different values for the initial temperature $\hat{T}(-30) = 9.222 \times 10^{-8}$, $\hat{T}(-30) = 9.222 \times 10^{-10}$, and $\hat{T}(-30) = 9.222 \times 10^{-12}$ and $\hat{\eta}/\hat{s}_0 = 1/(4\pi)$. These variations in initial temperature lead to very different values for the temperature at $\rho = 0$ in the attractor solutions. One obtains three different profiles for the attractor, which are also compared to their corresponding

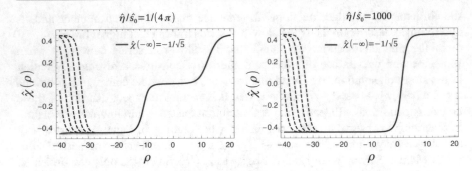

Fig. 3.7 *Left*: Attractor solution of the Israel–Stewart equations for Gubser flow (solid red) compared to other solutions of the equations (dashed black curves) for $\hat{\eta}/\hat{s}_0 = 1/(4\pi)$, $b = 5$. *Right*: Attractor solution of the Israel–Stewart equations for Gubser flow (solid red) compared to other solutions of the equations (dashed black curves) for a very large value of $\hat{\eta}/\hat{s}_0 = 1000$, $b = 5$. Figure taken from Ref. [31]

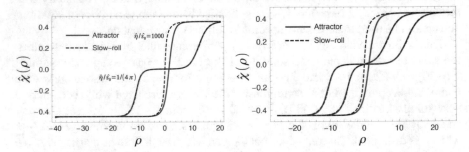

Fig. 3.8 *Left*: Comparison between the attractor solutions of Israel–Stewart equations undergoing Gubser flow, for $\hat{\eta}/\hat{s}_0 = 1/(4\pi)$ and $\hat{\eta}/\hat{s}_0 = 1000$, and the corresponding approximations using the slow-roll series (dashed). *Right*: Comparison between the attractor solutions of Israel–Stewart equations undergoing Gubser flow, computed using different initial conditions for the temperature, and the corresponding approximations using the slow-roll series (dashed). In this plot $\hat{\eta}/\hat{s}_0 = 1/(4\pi)$. Figure taken from Ref. [31]

slow-roll series truncated at second order for the first two solutions and at zeroth order for the last one. Even though $\hat{\eta}/\hat{s}_0$ is small, by significantly decreasing the initial value of the temperature one can again recover the large-$\hat{\tau}_\pi(0)$ regime, where the slow-roll series does not work well.

3.4   Summary

In this chapter, we investigated the properties of transient fluid dynamics in the nonlinear regime. This task was performed using highly symmetrical flow configurations, where the symmetries of the problem are sufficiently restrictive to completely determine the fluid 4-velocity. Such solutions are extremely rare. Two of these were explicitly derived in this chapter, the Bjorken and the Gubser flow. We further demonstrated how the remaining components of the conserved currents can be calculated analytically or, at least, semi-analytically, using a convenient set of coordinates. We then discussed and derived these rare solutions of relativistic fluid dynamics for simple fluid properties and showed how they can be used to study the asymptotic properties of transient fluid dynamics, to a level that is not possible in the linear regime.

We also developed a perturbative scheme to construct asymptotic solutions, such as the gradient and slow-roll expansions, of Israel–Stewart theory undergoing Bjorken and Gubser flow. We then determined the large-order behavior of these perturbative expansions for these highly symmetric flow configurations. We demonstrated numerically that the expansion coefficients in both cases grow factorially, indicating that these series have zero radius of convergence. Even though these series appear to diverge, their low-order truncations can still offer a reasonable description of exact solutions of Israel–Stewart theory at large times in Bjorken flow, or near the origin in Gubser flow. We finally found attractor solutions of transient fluid dynamics and showed how they are related to resummations of the gradient expansion or the slow-roll series. Since the gradient expansion appears to diverge in most of the known cases, the attractor regime of transient fluid dynamics is interpreted as a new hydrodynamic regime.

References

1. Florkowski, W., Heller, M.P., Spalinski, M.: Rept. Prog. Phys. **81**(4), 046001 (2018). https://doi.org/10.1088/1361-6633/aaa091, arXiv:1707.02282 [hep-ph]
2. Bjorken, J.D.: Phys. Rev. D **27**, 140 (1983)
3. Gubser, S.S: Phys. Rev. D **82**, 085027 (2010); Gubser, S.S., Yarom, A.: Nucl. Phys. B **846**, 469 (2011)
4. Denicol, G.S., Koide, T., Rischke, D.H.: Phys. Rev. Lett. **105**, 162501 (2010). https://doi.org/10.1103/PhysRevLett.105.162501, arXiv:1004.5013 [nucl-th]
5. Denicol, G.S., Niemi, H., Molnar, E., Rischke, D.H.: Phys. Rev. D **85**, 114047 (2012)
6. Ollitrault, J.Y.: Eur. J. Phys. **29**, 275-302 (2008). https://doi.org/10.1088/0143-0807/29/2/010, arXiv:0708.2433 [nucl-th]
7. Rezzolla, L., Zanotti, O.: Relativistic Hydrodynamics. Oxford University Press, Oxford (2013)
8. Denicol, G.S., Florkowski, W., Ryblewski, R., Strickland, M.: Phys. Rev. C **90**(4), 044905 (2014). https://doi.org/10.1103/PhysRevC.90.044905, arXiv:1407.4767 [hep-ph]
9. Denicol, G.S., Noronha, J.: Phys. Rev. D **97**(5), 056021 (2018)
10. Denicol, G.S., Noronha, J.: Phys. Rev. Lett. **124**(15), 152301 (2020). https://doi.org/10.1103/PhysRevLett.124.152301, arXiv:1908.09957 [nucl-th]
11. Heinz, U., Snellings, R.: Ann. Rev. Nucl. Part. Sci. **63**, 123–151 (2013). https://doi.org/10.1146/annurev-nucl-102212-170540, arXiv:1301.2826 [nucl-th]
12. Gale, C., Jeon, S., Schenke, B.: Int. J. Mod. Phys. A **28**, 1340011 (2013). https://doi.org/10.1142/S0217751X13400113, arXiv:1301.5893 [nucl-th]
13. The case $\tau_{\pi\pi} \neq 0$ can also be analytically solved, see Ref. [Denicol, G.S., Noronha, J.: Phys. Rev. D **97**(5), 056021 (2018)]
14. Denicol, G.S., Molnar, E., Niemi, H., Rischke, D.H.: Eur. Phys. J. A **48**, 170 (2012)
15. Denicol, G.S., Jeon, S., Gale, C.: Phys. Rev. C **90**(2), 024912 (2014)
16. Baier, R., Romatschke, P., Son, D.T., Starinets, A.O., Stephanov, M.A.: JHEP **04**, 100 (2008). https://doi.org/10.1088/1126-6708/2008/04/100, arXiv:0712.2451 [hep-th]

17. Chapman, S., Cowling, T.G.: The Mathematical Theory of Non-Uniform Gases. Cambridge University Press, Cambridge, England (1952)
18. Pu, S., Koide, T., Rischke, D.H.: Phys. Rev. D **81**, 114039 (2010)
19. Denicol, G.S., Noronha, J.: arXiv:1608.07869 [nucl-th]
20. Heller, M.P., Spalinski, M.: Phys. Rev. Lett. **115**(7), 072501 (2015). https://doi.org/10.1103/PhysRevLett.115.072501
21. Bender, C.M., Orszag, S.A.: Advanced Mathematical Methods for Scientists and Engineers: Asymptotic Methods and Perturbation Theory. Springer, New York (1999)
22. Heller, M.P., Svensson, V.: Phys. Rev. D **98**(5), 054016 (2018)
23. Romatschke, P.: Phys. Rev. Lett. **120**(1), 012301 (2018). https://doi.org/10.1103/PhysRevLett.120.012301, arXiv:1704.08699 [hep-th]
24. Spalinski, M.: Phys. Lett. B **776**, 468 (2018). https://doi.org/10.1016/j.physletb.2017.11.059, arXiv:1708.01921 [hep-th]
25. Strickland, M., Noronha, J., Denicol, G.: Phys. Rev. D **97**(3), 036020 (2018). https://doi.org/10.1103/PhysRevD.97.036020, arXiv:1709.06644 [nucl-th]
26. Romatschke, P.: JHEP **1712**, 079 (2017). https://doi.org/10.1007/JHEP12(2017)079, arXiv:1710.03234 [hep-th]
27. Florkowski, W., Maksymiuk, E., Ryblewski, R.: Phys. Rev. C **97**(2), 024915 (2018). https://doi.org/10.1103/PhysRevC.97.024915, arXiv:1710.07095 [hep-ph]
28. Casalderrey-Solana, J., Gushterov, N.I., Meiring, B.: arXiv: 1712.02772 [hep-th]
29. Almaalol, D., Strickland, M.: arXiv:1801.10173 [hep-ph]
30. Marrochio, H., Noronha, J., Denicol, G.S., Luzum, M., Jeon, S., Gale, C.: Phys. Rev. C **91**(1), 014903 (2015)
31. Denicol, G.S., Noronha, J.: Phys. Rev. D **99**(11), 116004 (2019). https://doi.org/10.1103/PhysRevD.99.116004

Microscopic Origin of Transport Coefficients: Linear-Response Theory

4

In this chapter, we discuss a general method to compute, from quantum field theory, all linear equations of motion and transport coefficients (such as the shear viscosity and its corresponding relaxation time) associated with the dissipative behavior of fluids. In particular, we explain how the linearized equations of motion of any dissipative current are determined by the analytical structure of the associated retarded Green's function and show that the transient dynamics of a fluid is determined by the slowest microscopic and not by the fastest fluid-dynamical time scale. We explicitly prove that the fluid-dynamical relaxation times are given by the inverse of the pole of the retarded Green's function nearest to the origin in the complex ω-plane and that, if the relaxation time is sent to zero, or equivalently, the pole to infinity, the dissipative currents approach the values given by the standard gradient expansion. The discussion presented in this chapter is based on Ref. [1].

This chapter is organized as follows. In Sect. 4.1 we define our notation. In Sect. 4.2, we show the equivalence of the gradient expansion in coordinate space with a Taylor series in momentum space. In Sect. 4.3, we derive the main results of this chapter and establish the connection between the analytical structure of retarded Green's functions and the linear transport coefficients in fluid dynamics. We show in Sects. 4.4.1 and 4.4.2 that the linear transport coefficients computed using either the linearized Boltzmann equation or via the disturbances in the space-time metric follow the general method and formulae derived in Sect. 4.3. In Sect. 4.5, we compare our results to previous derivations of relaxation-type equations for the dissipative currents based on the gradient expansion. We conclude this chapter with a summary of our results in Sect. 4.6.

© Springer Nature Switzerland AG 2021
G. S. Denicol and D. H. Rischke, *Microscopic Foundations of Relativistic Fluid Dynamics*,
Lecture Notes in Physics 990, https://doi.org/10.1007/978-3-030-82077-0_4

4.1   Preliminaries

Let us consider a general linear relation between a dissipative current $J(X)$ and a thermodynamical force $F(X)$,

$$J(X) = \int d^4X' \, G_R(X - X') \, F(X') \,. \tag{4.1}$$

Any translationally invariant theory that has a linear relation between J and F can always be written in the form of Eq. (4.1). In fluid dynamics, J could, e.g., be the shear-stress tensor $\pi^{\mu\nu}$ and F the shear tensor $\sigma^{\mu\nu}$, or J could be the diffusion current n^μ of a particle or charge and $F \sim \partial^\mu n$, where n is the associated particle or charge density.

In this chapter, we define the Fourier transformation in the following way:

$$\tilde{A}(Q) = \int d^4X \exp\left(iQ \cdot X\right) A(X) \,, \tag{4.2}$$

$$A(X) = \int \frac{d^4Q}{(2\pi)^4} \exp\left(-iQ \cdot X\right) \tilde{A}(Q) \,. \tag{4.3}$$

Here, $Q \equiv q^\mu = (\omega, \mathbf{q})^T$ is the momentum 4-vector, and $Q \cdot X \equiv q^\mu x_\mu$ is the scalar 4-product of the momentum 4-vector Q with the coordinate 4-vector $X \equiv x^\mu = (t, \mathbf{x})^T$. Using this convention, we can rewrite Eq. (4.1) in terms of the Fourier transforms of the retarded Green's function and of the thermodynamic force, $\tilde{G}_R(Q)$ and $\tilde{F}(Q)$, respectively, which then gives

$$\tilde{J}(Q) = \tilde{G}_R(Q)\tilde{F}(Q) \,. \tag{4.4}$$

We only consider systems where the microscopic dynamics is invariant under time reversal and, thus, Re $\tilde{G}_R$ is an even function of ω, while Im $\tilde{G}_R$ is an odd function of ω. Note that Eq. (4.4) implies that the current J can also be expressed as an integral over Q,

$$J(X) = \int \frac{d^4Q}{(2\pi)^4} \exp\left(-iQ \cdot X\right) \tilde{G}_R(Q)\tilde{F}(Q) \,. \tag{4.5}$$

Thus, $\tilde{G}_R$ at *all* frequencies can contribute to the dynamics of J. Equations (4.1), (4.4), and (4.5) are, of course, equivalent and contain all the information about the underlying microscopic theory that can be obtained through a linear analysis. In this chapter, we show how the analytical structure of the retarded Green's function $\tilde{G}_R$ determines the equation of motion for the current J.

4.2 Equivalence Between Gradient Expansion and Taylor Series

In this section, we show that the gradient expansion in space-time is actually equivalent to a Taylor series in 4-momentum space. For the sake of simplicity, we suppress any dependence on spatial coordinates or, equivalently, on 3-momentum, retaining only the dependence on time t and (complex) frequency ω. The coordinate, or 3-momentum dependence, respectively, will be restored later. We implicitly work in the rest frame of the fluid, such that the equations of motion do not appear to be relativistically covariant, however, covariance can be restored by a proper Lorentz boost.

Let us assume that $\tilde{G}_R(\omega)$ is analytic in the whole complex ω-plane. This can be considered as the limiting case when $\tilde{G}_R(\omega)$ has singularities, but all of them are pushed to infinity by some suitable limiting procedure. In this situation, a Taylor expansion of $\tilde{G}_R(\omega)$ around the origin has infinite convergence radius and thus provides a valid representation of $\tilde{G}_R(\omega)$ in the whole complex plane,

$$\tilde{G}_R(\omega) = \tilde{G}_R(0) + \partial_\omega \tilde{G}_R(\omega)\Big|_{\omega=0} \omega + \frac{1}{2}\partial_\omega^2 \tilde{G}_R(\omega)\Big|_{\omega=0} \omega^2 + O\left(\omega^3\right) . \quad (4.6)$$

Using Eqs. (4.3) and (4.6), as well as the Fourier representation of Dirac's delta function

$$\delta\left(t - t'\right) = \int \frac{d\omega}{2\pi} \exp\left[-i\omega\left(t - t'\right)\right] , \quad (4.7)$$

it is straightforward to obtain the general form of $G_R(t - t')$,

$$G_R(t - t') = \tilde{G}_R(0)\delta\left(t - t'\right) + i\partial_\omega \tilde{G}_R(\omega)\Big|_{\omega=0} \partial_t\delta\left(t - t'\right)$$
$$-\frac{1}{2}\partial_\omega^2 \tilde{G}_R(\omega)\Big|_{\omega=0} \partial_t^2\delta\left(t - t'\right) + O\left(\partial_t^3\right) . \quad (4.8)$$

Substituting Eq. (4.8) into Eq. (4.1), we obtain the following equation of motion for the dissipative current:

$$J(t) = \bar{D}_0 F(t) + \bar{D}_1 \partial_t F(t) + \bar{D}_2 \partial_t^2 F(t) + O\left(\partial_t^3 F\right) , \quad (4.9)$$

where we introduced the coefficients

$$\bar{D}_0 = \tilde{G}_R(0) ,$$
$$\bar{D}_1 = i\partial_\omega \tilde{G}_R(\omega)\Big|_{\omega=0} , \quad (4.10)$$
$$\bar{D}_2 = -\frac{1}{2}\partial_\omega^2 \tilde{G}_R(\omega)\Big|_{\omega=0} ,$$

$$\vdots$$

This is nothing but the so-called gradient expansion, where the current J is expressed in terms of the thermodynamic force F and its derivatives. Note that the standard gradient expansion does not involve time derivatives of the thermodynamic force. This is not a problem, since one can always replace time derivatives with spatial gradients using the conservation equations of fluid dynamics.

As explained in Sect. 1.3.5, this derivative expansion is controlled by a small parameter called the Knudsen number, $\mathrm{Kn} \equiv \lambda/L$, which is the ratio between a microscopic length scale λ and the macroscopic length scale L over which fluid-dynamical variables vary. The thermodynamic force, F, is already proportional to a gradient of a macroscopic variable, and thus $F \sim L^{-1}$. Every additional derivative ∂_t brings in another inverse power of L, $\partial_t^n F \sim L^{-(n+1)}$. The microscopic scale λ is contained in $\tilde{G}_R$ and its derivatives with respect to ω. Thus, up to some overall power of λ (which restores the correct scaling dimension), $\tilde{G}_R(0) \sim \lambda$, and each additional derivative ∂_ω brings in another power of λ, such that $\bar{D}_n \sim \lambda^{n+1}$. Therefore, the terms $\bar{D}_0 F$, $\bar{D}_1 \partial_t F$, and $\bar{D}_2 \partial_t^2 F$ in Eq. (4.9) are of order Kn, Kn^2, and Kn^3, respectively. The gradient expansion of F, Eq. (4.9), can be truncated at a given order in Kn, and one obtains a closed macroscopic theory for the dissipative current J.

4.3   The Role of the Analytical Structure of $\tilde{G}_R(\omega)$

In the previous section, it was shown how to relate the gradient expansion of the thermodynamical force F with the Taylor expansion of the Green's function $\tilde{G}_R$. The viability of the latter required the assumption that the singularities of $\tilde{G}_R$ are pushed to infinity by some suitable limiting procedure. For instance, if $\tilde{F}(\omega)$ has only support in a region of small $|\omega|$, which is well-separated from the singularities of $\tilde{G}_R$, one can devise a limiting procedure that effectively pushes these singularities to infinity. However, a priori it is not at all clear that $\tilde{F}(\omega)$ has vanishing support in the region of the complex ω-plane where $\tilde{G}_R(\omega)$ has singularities.

Therefore, we have to consider the case that $\tilde{G}_R(\omega)$ has some singularities in the complex ω-plane. This fact necessarily restricts the convergence radius of the Taylor expansion. If the singularities are simple poles, it is better to use a Laurent expansion around these poles.

4.3.1 $\tilde{G}_R(\omega)$ with One Pole

In order to illustrate this, we consider a retarded Green's function, $\tilde{G}_R(\omega)$, with a single simple pole at ω_0; cf. Fig. 4.1 for an illustration in the complex plane. A function with a single pole can always be expressed in the following form:

$$\tilde{G}_R(\omega) = \frac{f(\omega)}{\omega - \omega_0} , \qquad (4.11)$$

Fig. 4.1 Analytic structure
of the retarded Green's
function with a single
singularity at ω_0. The dashed
line illustrates the radius of
convergence of the Taylor
expansion around the origin.
Figure from Ref. [1]

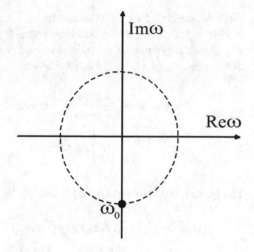

where $f(\omega)$ is an analytic function in the complex plane. In order for $\operatorname{Re} \tilde{G}_R$ to be
an even function of real ω and $\operatorname{Im} \tilde{G}_R$ to be an odd function of real ω, we have to
require that $\omega_0 \equiv -i\zeta$, where ζ is positive and real. We also have to require that
$\operatorname{Re} f(\omega)$ is odd in ω, while $\operatorname{Im} f(\omega)$ is even in ω. Since J is not a conserved quantity,
we exclude the case where the pole is at the origin, $\zeta = 0$.

Since $\tilde{G}_R(\omega)$ has a pole at ω_0, the Taylor series around $\omega = 0$ has a radius of
convergence $|\omega_0|$ and, consequently, this expansion is not able to describe $\tilde{G}_R(\omega)$
beyond the pole. In such cases, the Laurent expansion for $\tilde{G}_R(\omega)$ around the pole ω_0
should be used,

$$
\tilde{G}_R(\omega) = \frac{f(\omega_0)}{\omega - \omega_0} + \partial_\omega f(\omega)|_{\omega=\omega_0}
$$
$$
+ \frac{1}{2}\partial_\omega^2 f(\omega)|_{\omega=\omega_0} (\omega - \omega_0) + O\left[(\omega - \omega_0)^2\right] . \qquad (4.12)
$$

The series of positive powers in $\omega - \omega_0$ can be rearranged into a series of positive
powers in ω. The coefficients of the latter series can be most conveniently expressed
by matching the Laurent expansion to the Taylor expansion around $\omega = 0$. To this
end, we expand $f(\omega_0)(\omega - \omega_0)^{-1}$ around the origin. Then we match Eq. (4.6) with
Eq. (4.12). In this way, we obtain, e.g., for the coefficients of the constant and linear
terms in ω

$$
\tilde{G}_R(0) = -\frac{f(\omega_0)}{\omega_0} + \partial_\omega f(\omega)|_{\omega=\omega_0} - \frac{1}{2}\partial_\omega^2 f(\omega)|_{\omega=\omega_0} \omega_0 + \cdots ,
$$
$$
\partial_\omega \tilde{G}_R(\omega)\Big|_{\omega=0} = -\frac{f(\omega_0)}{\omega_0^2} + \frac{1}{2}\partial_\omega^2 f(\omega)|_{\omega=\omega_0} + \cdots . \qquad (4.13)
$$

We now observe that *all* constant terms in ω in Eq. (4.12) can be expressed as $\tilde{G}_R(0) + f(\omega_0)/\omega_0$, while *all* linear terms in ω can be written as $\partial_\omega \tilde{G}_R(\omega)|_{\omega=0} + f(\omega_0)/\omega_0^2$. Higher-order terms in ω can be expressed in a similar fashion. Thus, we can rewrite Eq. (4.12) in the following way:

$$\tilde{G}_R(\omega) = \frac{f(\omega_0)}{\omega - \omega_0} + \left[\tilde{G}_R(0) + \frac{f(\omega_0)}{\omega_0} \right]$$
$$+ \left[\partial_\omega \left. \tilde{G}_R(\omega) \right|_{\omega=0} + \frac{f(\omega_0)}{\omega_0^2} \right] \omega + O(\omega^2) . \qquad (4.14)$$

The Fourier transform of $\tilde{G}_R(\omega)$ is straightforwardly obtained and has the form

$$G_R(t - t') = -if(\omega_0) \exp\left[-i\omega_0(t - t')\right] \theta(t - t')$$
$$+ \left[\tilde{G}_R(0) + \frac{f(\omega_0)}{\omega_0} \right] \delta(t - t')$$
$$+ i \left[\partial_\omega \left. \tilde{G}_R(\omega) \right|_{\omega=0} + \frac{f(\omega_0)}{\omega_0^2} \right] \partial_t \delta(t - t') + O(\partial_t^2) . \quad (4.15)$$

One readily convinces oneself by explicit calculation that this retarded Green's function is the solution of the following differential equation:

$$\partial_t G_R(t - t') + i\omega_0 G_R(t - t') = i\omega_0 \tilde{G}_R(0) \delta(t - t')$$
$$- \left[\omega_0 \partial_\omega \left. \tilde{G}_R(\omega) \right|_{\omega=0} - \tilde{G}_R(0) \right] \partial_t \delta(t - t') + O(\partial_t^2) . \quad (4.16)$$

Dividing Eq. (4.16) by $i\omega_0$, multiplying by $F(t')$, and integrating over t', one now obtains an *equation of motion* for the current J defined in Eq. (4.1), instead of a simple algebraic identity as in the gradient expansion, cf. Eq. (4.9). This equation of motion reads

$$\tau_R \partial_t J + J = D_0 F + D_1 \partial_t F + O\left(\partial_t^2 F\right) , \qquad (4.17)$$

where the coefficients are

$$\tau_R = \frac{1}{i\omega_0} = \frac{1}{\zeta} ,$$
$$D_0 = \tilde{G}_R(0) ,$$
$$D_1 = i \partial_\omega \left. \tilde{G}_R(\omega) \right|_{\omega=0} + D_0 \tau_R . \qquad (4.18)$$

Note that, in case $\tilde{G}_R(\omega)$ has a single simple pole, Eq. (4.17) is *exact*.

It is clear that Eq. (4.17) is nothing but a relaxation equation which is similar in structure to that occurring in the transient theories for non-relativistic fluid dynamics proposed by Grad [2] and extended to relativistic fluids by Israel and Stewart in Refs. [3–5]. The appearance of the time derivative of J is due to the existence of the pole in the retarded Green's function. Also, the transport coefficient τ_R, usually known as the relaxation-time coefficient, is directly related to the singularity, $\omega_0 \equiv -i\zeta$, of $\tilde{G}_R(\omega)$. Since $\zeta > 0$, τ_R is real and positive, as expected. It is interesting to note that $D_0 \equiv \bar{D}_0$, with $\bar{D}_0$ from Eq. (4.10), while D_1 is *not* identical to $\bar{D}_1$; cf. see Eqs. (4.10) and (4.18).

It should be noted that the right-hand side of Eq. (4.17) can be truncated in the same manner as Eq. (4.9). Truncating after the first order in gradients, the dissipative current J satisfies

$$\tau_R \partial_t J + J = D_0 F + D_1 \partial_t F . \tag{4.19}$$

The truncation assumes that Kn $\ll 1$, with Kn being the appropriate Knudsen number. The particular truncation in Eq. (4.19) neglects terms of order $O\left(\mathrm{Kn}^3\right)$.

As was mentioned in the beginning of this section, the Taylor expansion around $\omega = 0$ is valid in a radius $|\omega_0|$ around the origin; see Fig. 4.1. Thus, when the pole is pushed to infinity, $|\omega_0| \to \infty$, the radius of convergence of the Taylor series becomes infinite and we should recover the gradient expansion. In fact, taking the limit $|\omega_0| \to \infty$, that is, $\tau_R \to 0$, cf. Eqs. (4.18), one recovers Eq. (4.9), with *identical* coefficients. The coefficient D_0 was already seen to be identical to $\bar{D}_0$, while D_1 agrees with $\bar{D}_1$ only in the limit of vanishing relaxation time [6].

In the case where the thermodynamic force varies slowly on the time scale given by τ_R, eventually, i.e., for times $t \gg \tau_R$, the dissipative current J will follow the time dependence imposed by the right-hand side of Eq. (4.19). In other words, the transient term $\tau_R \partial_t J$ in Eq. (4.19) will become small. Then it is permissible to replace J in this term by the right-hand side of Eq. (4.19), and we obtain up to terms of order $O(\mathrm{Kn}^3)$

$$J \simeq D_0 F + D_1 \partial_t F - \tau_R D_0 \partial_t F \equiv \bar{D}_0 F + \bar{D}_1 \partial_t F , \tag{4.20}$$

i.e., we recover the result (4.9) given by the gradient expansion. In this sense, the gradient expansion is the asymptotic solution of Eq. (4.17) for times $t \gg \tau_R$.

For non-relativistic systems, when the transient dynamics can be neglected at all times, it is known that the first-order truncation of the gradient expansion can actually serve not only as an asymptotic solution but also as an effective theory to describe the system. For instance, substituting the first-order result into the conservation equations one obtains the non-relativistic diffusion equation and the non-relativistic Navier–Stokes equation. However, for relativistic theories, this is not possible because of the violation of causality. As was mentioned in Sect. 2.1, because of Eq. (2.109), for any nonzero value of the transport coefficients (shear viscosity, bulk viscosity, etc.) it is not possible to take the (acausal) limit of vanishing relaxation time, if one wants to obtain causal and (linearly) stable relativistic fluid-dynamical equations of motion.

4.3.2 $\tilde{G}_R(\omega)$ with Two Poles

In order to better understand the consequences induced by the retarded Green's function's non-trivial analytic structure, it is useful to analyze in detail the case where $\tilde{G}_R(\omega)$ has two poles, ω_1 and ω_2,

$$\tilde{G}_R(\omega) = \frac{f_1(\omega)}{\omega - \omega_1} + \frac{f_2(\omega)}{\omega - \omega_2} . \tag{4.21}$$

We employ exactly the same steps as before and expand *each* term of $\tilde{G}_R(\omega)$ in a Laurent series around its respective pole. The result is

$$\tilde{G}_R(\omega) = \frac{f_1(\omega_1)}{\omega - \omega_1} + \frac{f_2(\omega_2)}{\omega - \omega_2} + \partial_\omega f_1(\omega)|_{\omega=\omega_1}$$

$$+ \partial_\omega f_2(\omega)|_{\omega=\omega_2} + \frac{1}{2}\partial_\omega^2 f_1(\omega)|_{\omega=\omega_1} (\omega - \omega_1) \tag{4.22}$$

$$+ \frac{1}{2}\partial_\omega^2 f_2(\omega)|_{\omega=\omega_2} (\omega - \omega_2) + O\big[(\omega - \omega_i)^2\big] .$$

As before, cf. Eq. (4.14), we can match this expansion to the Taylor expansion near the origin. This enables us to rewrite $\tilde{G}_R(\omega)$ as

$$\tilde{G}_R(\omega) = \frac{f_1(\omega_1)}{\omega - \omega_1} + \frac{f_2(\omega_2)}{\omega - \omega_2} + \left[\tilde{G}_R(0) + \frac{f_1(\omega_1)}{\omega_1} + \frac{f_2(\omega_2)}{\omega_2} \right]$$

$$+ \left[\partial_\omega \tilde{G}_R(\omega) \Big|_{\omega=0} + \frac{f_1(\omega_1)}{\omega_1^2} + \frac{f_2(\omega_2)}{\omega_2^2} \right] \omega \tag{4.23}$$

$$+ \left[\frac{1}{2} \partial_\omega^2 \tilde{G}_R(\omega) \Big|_{\omega=0} + \frac{f_1(\omega_1)}{\omega_1^3} + \frac{f_2(\omega_2)}{\omega_2^3} \right] \omega^2 + O(\omega^3) .$$

With the last expression, we can determine the Green's function $G_R(t - t')$ and also its equation of motion. Similar to the previous section, cf. Eq. (4.15), it is straightforward to show that

$$G_R(t - t') = -i \left\{ f_1(\omega_1) \exp\big[-i\omega_1(t - t')\big] + f_2(\omega_2) \exp\big[-i\omega_2(t - t')\big] \right\} \theta(t - t')$$

$$+ \left[\tilde{G}_R(0) + \frac{f_1(\omega_1)}{\omega_1} + \frac{f_2(\omega_2)}{\omega_2} \right] \delta(t - t')$$

$$+ i \left[\partial_\omega \tilde{G}_R(\omega) \Big|_{\omega=0} + \frac{f_1(\omega_1)}{\omega_1^2} + \frac{f_2(\omega_2)}{\omega_2^2} \right] \partial_t \delta(t - t') \tag{4.24}$$

$$- \left[\frac{1}{2} \partial_\omega^2 \tilde{G}_R(\omega) \Big|_{\omega=0} + \frac{f_1(\omega_1)}{\omega_1^3} + \frac{f_2(\omega_2)}{\omega_2^3} \right] \partial_t^2 \delta(t - t') + O(\partial_t^3) .$$

An explicit calculation shows that the equation satisfied by G_R is

$$
\begin{aligned}
-\partial_t^2 G_R(t - t') = {} & i\,(\omega_1 + \omega_2)\,\partial_t G_R(t - t') \\
& - \omega_1\omega_2 G_R(t - t') + \omega_1\omega_2 \tilde{G}_R(0)\delta(t - t') \\
& + i\left[\omega_1\omega_2\partial_\omega\,\tilde{G}_R(\omega)\Big|_{\omega=0} - (\omega_1 + \omega_2)\,\tilde{G}_R(0)\right]\partial_t\delta(t - t') \\
& - \left[\frac{1}{2}\,\omega_1\,\omega_2\,\partial_\omega^2\,\tilde{G}_R(\omega)\Big|_{\omega=0} - (\omega_1 + \omega_2)\,\partial_\omega\,\tilde{G}_R(\omega)\Big|_{\omega=0}\right. \\
& \left. + \tilde{G}_R(0)\right]\partial_t^2\,\delta(t - t') + O(\partial_t^3)\,.
\end{aligned}
\tag{4.25}
$$

The main difference to the previous case is that, due to the existence of a second pole, the Green's function now satisfies a *second*-order differential equation, instead of a first-order one. As will be shown later, the order of the differential equation satisfied by $\tilde{G}_R(\omega)$ is equal to the number of its singularities. Dividing by $-\omega_1\omega_2$, multiplying the equation by $F(t')$, and integrating over t', we can determine the equation of motion for J,

$$
\chi_2\partial_t^2 J + \chi_1\partial_t J + J = D_0 F(t) + D_1\partial_t F(t) + D_2\partial_t^2 F(t) + O\left[(\lambda/L)^4\right]\,. \tag{4.26}
$$

Here, we introduced the following transport coefficients:

$$
\begin{aligned}
\chi_2 &= -\frac{1}{\omega_1\omega_2}\,, \\
\chi_1 &= \frac{1}{i\omega_1} + \frac{1}{i\omega_2}\,, \\
D_0 &= \tilde{G}_R(0)\,, \\
D_1 &= i\partial_\omega\,\tilde{G}_R(\omega)\Big|_{\omega=0} + D_0\chi_1\,, \\
D_2 &= -\frac{1}{2}\partial_\omega^2\,\tilde{G}_R(\omega)\Big|_{\omega=0} + D_1\chi_1 + D_0\left(\chi_2 - \chi_1^2\right)\,. \tag{4.27}
\end{aligned}
$$

Note that χ_2 and χ_1 have contributions from *both* poles.

Next, we shall investigate under which circumstances a relaxation equation for J can be obtained. Due to time-reversal invariance, the two poles of $\tilde{G}_R(\omega)$ can appear in two ways: (i) both poles are on the imaginary axis, in which case we assume, without any loss of generality, that $|\omega_2| > |\omega_1|$; (ii) both poles have the same imaginary part, but opposite real parts, being symmetric with respect to the imaginary axis. In this case, $|\omega_1| = |\omega_2|$; cf. Fig. 4.2. Both cases reflect distinct physical scenarios.

Let us consider the case where the thermodynamic force is turned off, and the current J is left to relax to equilibrium. This process is governed by the equation

$$
\chi_2\partial_t^2 J + \chi_1\partial_t J + J = 0\,. \tag{4.28}
$$

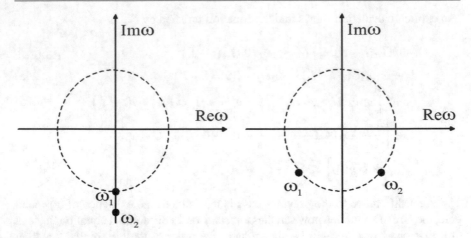

Fig. 4.2 Analytic structure of the retarded Green's function with two singularities, ω_1 and ω_2. Case (i), where both poles are on the imaginary axis, is illustrated on the left. Case (ii), where both poles are symmetric around the imaginary axis, is illustrated on the right. The dashed line illustrates the radius of convergence of the Taylor expansion around the origin. Figure from Ref. [1]

This is the equation of motion of a damped harmonic oscillator,

$$\ddot{x} + 2\gamma\,\dot{x} + \omega_0^2 x = 0 . \tag{4.29}$$

The harmonic oscillator is overdamped, if $\gamma > \omega_0$, and underdamped, if $\gamma < \omega_0$. Identifying the coefficients, we find

$$\gamma = \frac{\chi_1}{2\,\chi_2} , \quad \omega_0^2 = \frac{1}{\chi_2} . \tag{4.30}$$

Therefore, if $\chi_1^2 > 4\chi_2$, the dissipative current J relaxes to equilibrium without oscillating, while for $\chi_1^2 < 4\chi_2$, it relaxes in an oscillatory fashion. Using the definitions (4.27), we see that in the overdamped case, $4\omega_1\omega_2 > (\omega_1 + \omega_2)^2$, while in the underdamped case, $4\omega_1\omega_2 < (\omega_1 + \omega_2)^2$.

In case (i), both poles are purely imaginary, $\omega_i \equiv -i\zeta_i$, with $\zeta_i > 0$, $i = 1, 2$. Since always $4\zeta_1\zeta_2 < (\zeta_1 + \zeta_2)^2$, we are in the overdamped limit. In this case, a relaxation equation is obtained by applying the limiting procedure $\chi_2 \to 0$. That is, the second pole is pushed to infinity, in which case the relaxation time is given by the inverse of the first pole,

$$\tau_R \equiv \chi_1 = \frac{1}{i\omega_1} .$$

Naturally, this theory is only valid for frequencies that are small compared to the second pole.

In case (ii), $|\omega_1| = |\omega_2|$ and $\omega_1 = -\omega_2^*$. We always have $|\omega_1|^2 > (\text{Im}\,\omega_1)^2$, which can be written in the form $4\omega_1\omega_2 = -4|\omega_1|^2 < (\omega_1 - \omega_1^*)^2 = (\omega_1 + \omega_2)^2$. Thus, we are in the underdamped case. In this case, due to the symmetries of the retarded Green's function, one *cannot* disregard one of the poles while keeping the other. Thus, the term including the second time derivative of J must be included (otherwise, there would be no oscillation) and the full equation of motion must be solved. It is important to remark that, in this case, the coefficient χ_1 cannot be interpreted as a relaxation time.

Note that the exponential decay of $J(t)$ in case (i) can also be seen when inserting the Green's function $\tilde{G}_R(\omega)$ into Eq. (4.5) and performing the ω-integral via contour integration, picking up the poles via the residue theorem. In case (ii), one gets an exponentially damped factor from the imaginary part of the pole, $\sim \exp(-|\mathrm{Im}\,\omega_i|\,t)$, and an oscillatory factor from the real part, $\sim \exp(i\,\mathrm{Re}\,\omega_i\,t)$. If $|\mathrm{Im}\,\omega_i| \gg |\mathrm{Re}\,\omega_i|$, the damping is much stronger than the oscillation, and already within a single oscillation, the dissipative current has decayed a couple of e-folds toward its stationary solution. On the other hand, if $|\mathrm{Im}\,\omega_i| \ll |\mathrm{Re}\,\omega_i|$, the current will oscillate many times before a substantial decay occurs. The oscillatory behavior of the dissipative current in case (ii) was already noticed in Ref. [7].

It is clear from this analysis that the derivation of relaxation equations for systems with more than one pole is not possible if the first pole does not lie on the imaginary axis. With this in mind, we can finally consider the general case of an arbitrary number of poles. The result will be qualitatively similar to what was found above.

4.3.3 $\tilde{G}_R(Q)$ with N Poles

Now we assume that $\tilde{G}_R(Q)$ has N poles in the complex ω-plane, $\omega_1(\mathbf{q}), \ldots,$ $\omega_N(\mathbf{q})$. We furthermore restore the $\mathbf{q}$-dependence that was neglected in the previous sections, and assume that the Green's function is analytic in $\mathbf{q}$. Since J is not a conserved quantity, we can safely assume that all poles remain within a finite distance from the origin even when $\mathbf{q} \to 0$. Here, we consider the case of a finite (but arbitrarily large) number of poles. Also, in many cases $\tilde{G}_R(Q)$ contains branch cuts. We assume that these branch cuts are located further from the origin of the complex frequency plane than these N poles and thus will not be considered any further in this chapter. The retarded Green's function can be written in the general form

$$\tilde{G}_R(Q) = \sum_{i=1}^{N} \frac{f_i(Q)}{\omega - \omega_i(\mathbf{q})} = \frac{\Xi(Q)}{\left[\omega - \omega_1(\mathbf{q})\right] \cdots \left[\omega - \omega_N(\mathbf{q})\right]}, \tag{4.31}$$

where the $f_i(Q)$ and $\Xi(Q)$ are analytic functions in the complex ω-plane. We now use the identity

$$1 + \Phi_1(\mathbf{q})(-i\omega) + \cdots + \Phi_N(\mathbf{q})(-i\omega)^N = (-1)^N \frac{\left[\omega - \omega_1(\mathbf{q})\right] \cdots \left[\omega - \omega_N(\mathbf{q})\right]}{\omega_1(\mathbf{q}) \cdots \omega_N(\mathbf{q})}, \tag{4.32}$$

where

$$\Phi_m(\mathbf{q}) = (-i)^m \sum_{1 \le i_1 \cdots < i_m \le N} \frac{1}{\omega_{i_1}(\mathbf{q}) \cdots \omega_{i_m}(\mathbf{q})} \, . \tag{4.33}$$

From this expression, we can find the following equation for $\tilde{G}_R(Q)$:

$$\left[1 + \Phi_1(\mathbf{q})(-i\omega) + \cdots + \Phi_N(\mathbf{q})(-i\omega)^N \right] \tilde{G}_R(Q) = \frac{(-1)^N \, \Xi(Q)}{\omega_1(\mathbf{q}) \cdots \omega_N(\mathbf{q})} \, . \tag{4.34}$$

After taking the Fourier transform and expanding the functions $\Phi_m(\mathbf{q})$ in a Taylor series around $\mathbf{q} = 0$, we obtain a differential equation satisfied by G_R. It is clear that this will be a linear differential equation of order N in time. The equation of motion for J can be obtained in the same way as before. The result is

$$\chi_N \partial_t^N J + \cdots + \chi_1 \partial_t J + J = D_0 F + \cdots + D_N \partial_t^N F + O\left(\partial_t^{N+1} F, \partial_\mathbf{x}\right) \, . \tag{4.35}$$

Here, we omitted all the terms involving spatial derivatives. The coefficients χ_m are

$$\chi_m = \Phi_m(\mathbf{0}) = (-i)^m \sum_{1 \le i_1 \cdots < i_m \le N} \frac{1}{\omega_{i_1}(\mathbf{0}) \cdots \omega_{i_m}(\mathbf{0})} \, . \tag{4.36}$$

As before, the coefficients D_0, D_1, and D_2 can be expressed as

$$D_0 = \tilde{G}_R(\omega, \mathbf{0})\Big|_{\omega=0} \, ,$$

$$D_1 = i\partial_\omega \, \tilde{G}_R(\omega, \mathbf{0})\Big|_{\omega=0} + D_0 \chi_1 \, ,$$

$$D_2 = -\frac{1}{2}\partial_\omega^2 \, \tilde{G}_R(\omega, \mathbf{0})\Big|_{\omega=0} + D_1 \chi_1 + D_0 \left(\chi_2 - \chi_1^2\right) \, , \tag{4.37}$$

cf. Eq. (4.27). In general, the coefficients D_k have the following form:

$$D_k = i^k \frac{(-1)^N}{k!} \frac{\partial_\omega^k \, \Xi(\omega, \mathbf{0})|_{\omega=0}}{\omega_1(\mathbf{0}) \cdots \omega_N(\mathbf{0})} \, . \tag{4.38}$$

We remark that up to now we have not employed any approximation besides the initial assumptions regarding the singularities of $\tilde{G}_R$. As mentioned before, if there is a clear separation of scales, it is possible to simplify the equation of motion. Since $D_n \sim \lambda^{n+1}$, assuming that $\lambda/L \ll 1$, we can safely assume $D_0 F \gg D_1 \partial_t F \gg D_2 \partial_t^2 F$ and truncate the right-hand side of Eq. (4.35).

As discussed in the previous section, a relaxation equation can be obtained only for the cases where the pole nearest to the origin (in the following referred to as the "first pole") lies on the imaginary axis. In this case, the limiting procedure which pushes all other poles to infinity, $\chi_i \to 0$, $i \ge 2$, can be applied without breaking any symmetries of the retarded Green's function. Assuming that this can be done, we obtain the following equation of motion for J:

$$\tau_R \partial_t J + J = D_0 F + D_1 \partial_t F + D_2 \partial_t^2 F + O\left(D_3 \partial_t^3 F, \partial_x\right) , \tag{4.39}$$

where

$$\tau_R = \frac{1}{i\omega_1(0)},$$

$$D_0 = \tilde{G}_R(\omega, \mathbf{0})\Big|_{\omega=0} ,$$

$$D_1 = i\partial_\omega \tilde{G}_R(\omega, \mathbf{0})\Big|_{\omega=0} + D_0 \tau_R ,$$

$$D_2 = -\frac{1}{2}\partial_\omega^2 \tilde{G}_R(\omega, \mathbf{0})\Big|_{\omega=0} + D_1 \tau_R - D_0 \tau_R^2 . \tag{4.40}$$

On the other hand, if the first pole and, for reasons of symmetry, its counterpart on the other side of the imaginary axis have nonzero real parts, the dissipative current will oscillate and the equation of motion cannot be reduced to a simple relaxation equation, even in the small-frequency domain.

4.4   Applications

4.4.1   The Linearized Boltzmann Equation

The discussion presented above is valid for both weakly and strongly coupled theories. The prime example of a weakly coupled theory is given by the Boltzmann equation. In this section, we calculate the shear viscosity and relaxation-time coefficients for a weakly coupled gas via the Boltzmann equation following the general method presented above. This calculation was first presented in Ref. [1].

We start from the relativistic Boltzmann equation

$$k^\mu \partial_\mu f_{\mathbf{k}} = C[f] , \tag{4.41}$$

where $k^\mu \equiv K = (k_0, \mathbf{k})^T$, $k_0 = \sqrt{\mathbf{k}^2 + m^2}$, and m is the particle mass. For the collision term $C[f]$, we consider only binary elastic collisions with incoming momenta k, k', and outgoing momenta p, p',

$$C[f] = \frac{1}{\nu} \int dK' dP dP' W_{\mathbf{k}\mathbf{k}' \to \mathbf{p}\mathbf{p}'} \left(f_{\mathbf{p}} f_{\mathbf{p}'} \tilde{f}_{\mathbf{k}} \tilde{f}_{\mathbf{k}'} - f_{\mathbf{k}} f_{\mathbf{k}'} \tilde{f}_{\mathbf{p}} \tilde{f}_{\mathbf{p}'} \right) , \tag{4.42}$$

where ν is a symmetry factor ($= 2$ for identical particles), $W_{\mathbf{k}\mathbf{k}' \to \mathbf{p}\mathbf{p}'}$ is the Lorentz-invariant transition rate, and $dK \equiv d_{\mathrm{dof}} d^3\mathbf{k}/\left[(2\pi)^3 k^0\right]$ is the Lorentz-invariant momentum-space volume, with d_{dof} being the number of internal degrees of freedom. We introduced the notation

$$\tilde{f}_{\mathbf{k}} \equiv 1 - a f_{\mathbf{k}} , \tag{4.43}$$

where $a = 1$ ($a = -1$) for fermions (bosons) and $a = 0$ for a classical gas. In the following, we restrict our consideration to the latter case, i.e., we take $a = 0$.

We now linearize the Boltzmann equation (4.41) around the classical equilibrium state, i.e., we define

$$\delta f_{\mathbf{k}} \equiv f_{\mathbf{k}} - f_{0\mathbf{k}} , \tag{4.44}$$

where

$$f_{0\mathbf{k}} \equiv \exp(y_{0\mathbf{k}}) , \qquad y_{0\mathbf{k}} \equiv \alpha_0 - \beta_0 E_{\mathbf{k}} , \tag{4.45}$$

with the inverse temperature $\beta_0 \equiv 1/T$, the thermal potential $\alpha_0 \equiv \mu/T$, and the energy in the local rest frame $E_{\mathbf{k}} \equiv u_\mu k^\mu$. As in the preceding chapters, u^μ is the fluid 4-velocity.

The linearized Boltzmann equation can be written as

$$\delta \dot{f}_{\mathbf{k}} + E_{\mathbf{k}}^{-1} k^\mu \nabla_\mu \delta f_{\mathbf{k}} - \hat{C}(X, K)\delta f_{\mathbf{k}} = S(X, K) , \tag{4.46}$$

where, as in Chap. 1, we used the notation $\dot{A} \equiv u^\mu \partial_\mu A$, $\nabla_\mu \equiv \Delta_{\mu\nu}\partial^\nu$, with $\Delta^{\mu\nu} = g^{\mu\nu} - u^\mu u^\nu$ being the 3-space projector orthogonal to u^μ, and

$$
\begin{aligned}
S(X, K) \equiv f_{0\mathbf{k}} \Bigg[-\dot{\alpha}_0 + E_{\mathbf{k}}\dot{\beta}_0 + k^\mu \left(\beta_0 \dot{u}_\mu + \nabla_\mu \beta_0 - E_{\mathbf{k}}^{-1}\nabla_\mu \alpha_0 \right) \\
+ \frac{\beta_0}{3} E_{\mathbf{k}}^{-1} \left(m^2 - E_{\mathbf{k}}^2 \right) \theta + \beta_0 E_{\mathbf{k}}^{-1} k^{\langle\mu} k^{\nu\rangle} \sigma_{\mu\nu} \Bigg] .
\end{aligned}
\tag{4.47}
$$

Here, as in Chap. 1, the expansion scalar is $\theta \equiv \partial_\mu u^\mu$ and the shear tensor is $\sigma^{\mu\nu} \equiv \nabla^{\langle\mu} u^{\nu\rangle}$, where $A^{\langle\mu\nu\rangle} = \Delta^{\mu\nu\alpha\beta} A_{\alpha\beta}$, with $\Delta^{\mu\nu\alpha\beta}$ being the rank-4 symmetric traceless projection operator from Eq. (1.55). In Eq. (4.46), we also introduced the collision operator

$$\hat{C}(X, K)\delta f_{\mathbf{k}} = \frac{1}{\nu E_{\mathbf{k}}} \int dK' dP dP' W_{\mathbf{k}\mathbf{k}' \to \mathbf{p}\mathbf{p}'} f_{0\mathbf{k}} f_{0\mathbf{k}'} \left(\frac{\delta f_{\mathbf{p}'}}{f_{0\mathbf{p}'}} + \frac{\delta f_{\mathbf{p}}}{f_{0\mathbf{p}}} - \frac{\delta f_{\mathbf{k}'}}{f_{0\mathbf{k}'}} - \frac{\delta f_{\mathbf{k}}}{f_{0\mathbf{k}}} \right) , \tag{4.48}$$

which we obtain from linearizing the collision integral (4.42) around the classical equilibrium state. In deriving Eq. (4.48), we made use of the fact that due to the principle of detailed balance, the collision integral vanishes in equilibrium, $C[f_0] = 0$, and that $f_{0\mathbf{k}} f_{0\mathbf{k}'} \equiv f_{0\mathbf{p}} f_{0\mathbf{p}'}$ due to 4-momentum conservation in binary collisions.

We now consider Eq. (4.46) in the local rest frame, where $u^\mu = (1, 0, 0, 0)^T$. We assume a situation where the fluid does not accelerate, $\dot{u}^\mu = 0$, and does not expand, $\theta = 0$, and where temperature and chemical potential are constant. With these assumptions, the source term (4.47) reduces to

$$S(X, K) = \beta_0 f_{0\mathbf{k}} E_{\mathbf{k}}^{-1} k^{\langle\mu} k^{\nu\rangle} \sigma_{\mu\nu}(X) , \tag{4.49}$$

and the collision operator no longer depends on X, $\hat{C}(X, K) \equiv \hat{C}(K)$. The Boltzmann equation (4.46) takes the form

$$\partial_t \delta f_{\mathbf{k}} + \mathbf{v} \cdot \nabla \delta f_{\mathbf{k}} - \hat{C}(K) \delta f_{\mathbf{k}} = S(X, K) , \tag{4.50}$$

where $\mathbf{v} \equiv \mathbf{k}/E_{\mathbf{k}}$.

We solve the inhomogeneous, linear, partial integro-differential equation (4.50) for $\delta f_{\mathbf{k}}$ in 4-momentum space. Taking the Fourier transform, we obtain

$$-i\omega \delta \tilde{f}_{\mathbf{k}}(Q) + i\mathbf{v} \cdot \mathbf{q}\, \delta \tilde{f}_{\mathbf{k}}(Q) - \hat{C}(K)\delta \tilde{f}_{\mathbf{k}}(Q) = \tilde{S}(Q, K) . \tag{4.51}$$

From now on, it is important not to confuse the momenta of the particles, K, K', P, and P', with the variable Q from the Fourier transformation.

The formal solution of Eq. (4.51) can be expressed in the following form:

$$\delta \tilde{f}_{\mathbf{k}}(Q) = \frac{1}{-i\omega + i\mathbf{v} \cdot \mathbf{q} - \hat{C}(K)} \tilde{S}(Q, K) . \tag{4.52}$$

As will be shown in Chap. 5, Eq. (5.28), the shear-stress tensor $\pi^{\mu\nu}$ can be expressed in terms of $\delta f_{\mathbf{k}}$ as [8]

$$\pi^{\mu\nu} = \int dK\, k^{\langle \mu} k^{\nu \rangle} \delta f_{\mathbf{k}} . \tag{4.53}$$

Taking the Fourier transform of Eq. (4.53) and using Eqs. (4.49) and (4.52), we obtain

$$\tilde{\pi}^{\mu\nu}(Q) = \int dK\, k^{\langle \mu} k^{\nu \rangle} \frac{1}{-i\omega + i\mathbf{v} \cdot \mathbf{q} - \hat{C}(K)} \beta_0 f_{0\mathbf{k}} E_{\mathbf{k}}^{-1} k^{\langle \alpha} k^{\beta \rangle} \tilde{\sigma}_{\alpha\beta}(Q) . \tag{4.54}$$

Note that Eq. (4.54) has the following form:

$$\tilde{\pi}^{\mu\nu}(Q) = \tilde{G}_R^{\mu\nu\alpha\beta}(Q)\tilde{\sigma}_{\alpha\beta}(Q) , \tag{4.55}$$

where we introduced

$$\tilde{G}_R^{\mu\nu\alpha\beta}(Q) = \int dK\, k^{\langle \mu} k^{\nu \rangle} \frac{1}{-i\omega + i\mathbf{v} \cdot \mathbf{q} - \hat{C}(K)} \beta_0 f_{0\mathbf{k}} E_{\mathbf{k}}^{-1} k^{\langle \alpha} k^{\beta \rangle} . \tag{4.56}$$

As already mentioned, we want to calculate the two main transport coefficients that describe the linearized dynamics of the shear-stress tensor: the shear relaxation time, τ_π, and the shear-viscosity coefficient, η. Let us start by defining the function

$$B^{\alpha\beta}(Q, K) \equiv \frac{1}{-i\omega + i\mathbf{v} \cdot \mathbf{q} - \hat{C}(K)} \beta_0 f_{0\mathbf{k}} E_{\mathbf{k}}^{-1} k^{\langle \alpha} k^{\beta \rangle} , \tag{4.57}$$

i.e., by definition $B^{\alpha\beta}(Q, K)$ satisfies

$$\left[-i\omega + i\mathbf{v} \cdot \mathbf{q} - \hat{C}(K) \right] B^{\alpha\beta}(Q, K) = \beta_0 E_{\mathbf{k}}^{-1} k^{\langle\alpha} k^{\beta\rangle} f_{0\mathbf{k}} \ . \tag{4.58}$$

In general, $B^{\alpha\beta}(Q, K)$ is a function of the 4-momenta Q and K. However, from Eqs. (4.36) and (4.37) we already know that in order to calculate the relaxation time and the viscosity coefficient, it is sufficient to consider the case $\mathbf{q} = 0$. Then $B^{\alpha\beta}(Q, K) = B^{\alpha\beta}(\omega, K)$. Since the tensor structure on both sides of Eq. (4.58) must be the same, we make the Ansatz

$$B^{\alpha\beta}(\omega, K) = f_{0\mathbf{k}} k^{\langle\alpha} k^{\beta\rangle} \sum_{n=0}^{\infty} a_n(\omega) E_{\mathbf{k}}^n \ . \tag{4.59}$$

For the sake of convenience, we have factored out an equilibrium distribution function $f_{0\mathbf{k}}$ and, since the remaining K-dependence of $B^{\alpha\beta}(\omega, K)$ can only be a dependence on the scalar quantity $E_{\mathbf{k}} = u_\mu K^\mu$, we expanded this dependence in terms of a power series in $E_{\mathbf{k}}$, with yet to be determined coefficients $a_n(\omega)$. Substituting Eq. (4.59) into Eq. (4.56), it follows that

$$\tilde{G}_R^{\mu\nu\alpha\beta}(\omega, \mathbf{0}) = \sum_{n=0}^{\infty} a_n(\omega) \int dK \, k^{\langle\mu} k^{\nu\rangle} k^{\langle\alpha} k^{\beta\rangle} E_{\mathbf{k}}^n f_{0\mathbf{k}}$$

$$= 2\Delta^{\mu\nu\alpha\beta} \sum_{n=0}^{\infty} I_{n+4,2} \, a_n(\omega) \ , \tag{4.60}$$

where we used

$$\int dK \, k^{\langle\mu} k^{\nu\rangle} k^{\langle\alpha} k^{\beta\rangle} E_{\mathbf{k}}^n f_{0\mathbf{k}} = \frac{2}{5!!} \Delta^{\mu\nu\alpha\beta} \int dK \, E_{\mathbf{k}}^n f_{0\mathbf{k}} \left(m^2 - E_{\mathbf{k}}^2 \right)^2 \ , \tag{4.61}$$

which can be shown using the orthogonality relation (6.7) proved in App. 6.7 (see also Ref. [9]). In Eq. (4.60), we also introduced the thermodynamic function

$$I_{nq} \equiv \frac{1}{(2q + 1)!!} \int dK \, f_{0\mathbf{k}} E_{\mathbf{k}}^{n-2q} \left(E_{\mathbf{k}}^2 - m^2 \right)^q \ . \tag{4.62}$$

Thus, Eq. (4.55) can be cast into the more convenient form

$$\tilde{\pi}^{\mu\nu}(\omega, \mathbf{0}) = 2\tilde{G}_R(\omega, \mathbf{0}) \tilde{\sigma}^{\mu\nu}(\omega, \mathbf{0}) \ , \tag{4.63}$$

where we introduced the retarded Green's function

$$\tilde{G}_R(\omega, \mathbf{0}) = \sum_{n=0}^{\infty} I_{n+4,2} \, a_n(\omega) \ . \tag{4.64}$$

On account of the Navier–Stokes relation $\pi^{\mu\nu} = 2\eta\sigma^{\mu\nu}$, we pulled out a factor 2 from $\tilde{G}_R(\omega, \mathbf{0})$ in Eq. (4.63). As shown in the previous section, Eqs. (4.36) and (4.37), the shear relaxation time is determined by the first pole ω_1 of $\tilde{G}_R(\omega, \mathbf{0})$,

$$\tau_\pi = \frac{1}{i\omega_1(0)} , \qquad (4.65)$$

while the shear-viscosity coefficient is determined by the retarded Green's function $\tilde{G}_R(\omega, \mathbf{0})$ at the origin,

$$\eta = \tilde{G}_R(\omega, \mathbf{0})\Big|_{\omega=0} . \qquad (4.66)$$

Thus, the problem of finding the linear transport coefficients has been reduced to determining the analytic properties of the coefficient functions $a_n(\omega)$.

Equations such as Eq. (4.58) appear quite often in problems involving the extraction of transport coefficients from the Boltzmann equation. We now proceed to solve this equation by substituting the expansion (4.59) into Eq. (4.58), multiplying by $E_{\mathbf{k}}^m k^{\langle\mu} k^{\nu\rangle}$, and integrating over dK. Then one obtains (setting $\mathbf{q} = 0$)

$$\sum_{n=0}^{\infty} a_n(\omega) \int dK \, E_{\mathbf{k}}^m k^{\langle\mu} k^{\nu\rangle} \left[-i\omega - \hat{C}(K)\right] f_{0\mathbf{k}} E_{\mathbf{k}}^n k^{\langle\alpha} k^{\beta\rangle}$$

$$= \beta_0 \int dK \, E_{\mathbf{k}}^{m-1} k^{\langle\mu} k^{\nu\rangle} k^{\langle\alpha} k^{\beta\rangle} f_{0\mathbf{k}} . \qquad (4.67)$$

The most complicated term in this equation that we need to compute is the one involving the collision operator $\hat{C}(K)$. First, we use the definition (4.48) of the operator $\hat{C}(K)$, but now acting on $f_{0\mathbf{k}} E_{\mathbf{k}}^n k^{\langle\alpha} k^{\beta\rangle}$ instead of $\delta f_{\mathbf{k}}$. Then, the respective term in Eq. (4.67) has the form

$$\int dK \, E_{\mathbf{k}}^m k^{\langle\mu} k^{\nu\rangle} \hat{C}(K) f_{0\mathbf{k}} E_{\mathbf{k}}^n k^{\langle\alpha} k^{\beta\rangle}$$

$$= \frac{1}{\nu} \int dK dK' dP dP' W_{\mathbf{kk'}\to\mathbf{pp'}} E_{\mathbf{k}}^{m-1} f_{0\mathbf{k}} f_{0\mathbf{k'}} k^{\langle\mu} k^{\nu\rangle}$$

$$\times \left(E_{\mathbf{p'}}^n p'^{\langle\alpha} p'^{\beta\rangle} + E_{\mathbf{p}}^n p^{\langle\alpha} p^{\beta\rangle} - E_{\mathbf{k'}}^n k'^{\langle\alpha} k'^{\beta\rangle} - E_{\mathbf{k}}^n k^{\langle\alpha} k^{\beta\rangle} \right) . \qquad (4.68)$$

The tensor structure of this expression implies that it must be of the form

$$\int dK \, E_{\mathbf{k}}^m k^{\langle\mu} k^{\nu\rangle} \hat{C}(K) f_{0\mathbf{k}} E_{\mathbf{k}}^n k^{\langle\alpha} k^{\beta\rangle} \equiv -2\mathcal{A}^{mn} \Delta^{\mu\nu\alpha\beta} , \qquad (4.69)$$

where we defined the matrix

$$\mathcal{A}^{mn} \equiv -\frac{1}{10} \int dK \, E_{\mathbf{k}}^m k_{\langle\alpha} k_{\beta\rangle} \hat{C}(K) f_{0\mathbf{k}} E_{\mathbf{k}}^n k^{\langle\alpha} k^{\beta\rangle} . \qquad (4.70)$$

Using Eqs. (4.61), (4.62), and (4.69), Eq. (4.67) implies

$$\sum_{n=0}^{\infty} \left(-i\omega \mathcal{D}^{mn} + \mathcal{A}^{mn} \right) a_n(\omega) = \beta_0 I_{m+3,2} , \qquad (4.71)$$

where we defined the matrix

$$\mathcal{D}^{mn} = \frac{1}{5!!} \int dK f_{0\mathbf{k}} E_{\mathbf{k}}^{m+n} \left(m^2 - E_{\mathbf{k}}^2 \right)^2 \equiv I_{n+m+4,2} . \qquad (4.72)$$

Provided the matrix $(-i\omega \mathcal{D} + \mathcal{A})$ is invertible, the solution for $a_n(\omega)$ is then

$$a_m(\omega) = \beta_0 \sum_{n=0}^{\infty} \left[(-i\omega \mathcal{D} + \mathcal{A})^{-1} \right]^{mn} I_{n+3,2} , \qquad (4.73)$$

and Eq. (4.64) becomes

$$\tilde{G}_R(\omega, \mathbf{0}) = \beta_0 \sum_{m=0}^{\infty} \sum_{n=0}^{\infty} I_{m+4,2} \left[(-i\omega \mathcal{D} + \mathcal{A})^{-1} \right]^{mn} I_{n+3,2} . \qquad (4.74)$$

The poles of this function can be obtained from the roots of the determinant $\det (-i\omega \mathcal{D} + \mathcal{A})$, which is given by the product of the eigenvalues λ_n of the operator $-i\omega \mathcal{D} + \mathcal{A}$. In other words, a pole of $\tilde{G}_R(\omega, \mathbf{0})$ is given by a vanishing eigenvalue λ_n. Note that because $\mathcal{D}$ and $\mathcal{A}$ are real matrices, all the poles are on the imaginary axis. Thus, the truncation of the equation of motion to a relaxation-type form is possible, if the separation between the poles is large enough. The shear-viscosity coefficient is given by

$$\eta \equiv \tilde{G}_R(\omega, \mathbf{0}) \Big|_{\omega=0} = \beta_0 \sum_{m=0}^{\infty} \sum_{n=0}^{\infty} I_{m+4,2} (\mathcal{A}^{-1})^{mn} I_{n+3,2} . \qquad (4.75)$$

Thus, in order to find the relaxation times and viscosity coefficients from the linearized Boltzmann equation, one has to invert and compute eigenvalues of infinite matrices. In practice, however, one never deals with infinite matrices because the series in Eq. (4.59) is always truncated at some value of n; see, e.g., the discussion of Chapman–Enskog theory in Sect. 5.2 and Ref. [9].

Let us consider the simplest possible case and take only one term in the expansion (4.59). We remark that this corresponds to using the 14-moment approximation in the method of moments; see Chap. 6. Then, $\tilde{G}_R(\omega, \mathbf{0})$ has the following simple form:

$$\tilde{G}_R(\omega, \mathbf{0}) = \frac{i\beta_0 I_{32}}{\omega + i\mathcal{A}^{00}/I_{42}} , \qquad (4.76)$$

where we used $\mathcal{D}^{00} = I_{42}$. In this case, the retarded Green's function has only one pole, $\omega_0 = -i\mathcal{A}^{00}/I_{42}$. The relaxation time is obtained as

$$\tau_\pi = \frac{1}{i\omega_0} = \frac{I_{42}}{\mathcal{A}^{00}} \,. \tag{4.77}$$

On the other hand, the shear viscosity is given by $\eta = \tilde{G}_R(\omega, \mathbf{0})\big|_{\omega=0}$ and becomes

$$\eta = \beta_0 \frac{I_{42} I_{32}}{\mathcal{A}^{00}} \,. \tag{4.78}$$

In the massless limit, for a gas of hard spheres, one determines $\mathcal{A}^{00}$ to be (calculations of integrals of the collision operator will be performed explicitly in Chap. 8; see also Ref. [10])

$$\mathcal{A}^{00} = \frac{3}{5} I_{42} n_0 \sigma \,, \tag{4.79}$$

where σ is the total cross section and n_0 is the particle number density. Then,

$$\eta = \frac{4}{3\beta_0 \sigma} \,, \tag{4.80}$$

$$\tau_\pi = \frac{5}{3n_0 \sigma} \,, \tag{4.81}$$

where we used that, in the massless limit,

$$I_{42} = 4\frac{P_0}{\beta_0^2} \,, \qquad I_{32} = \frac{4}{5}\frac{P_0}{\beta_0} \,, \tag{4.82}$$

where $P_0 \equiv n_0/\beta_0$ is the thermodynamic pressure of a classical gas. Also, in this particular example, one can show that the ratio η/τ_π is independent of the cross section,

$$\frac{\eta}{\tau_\pi} = \beta_0 I_{32} = \frac{4}{5} P_0 = \frac{4}{5}\frac{n_0}{\beta_0} \,. \tag{4.83}$$

These are exactly the results obtained in Refs. [8], from kinetic theory, and in Refs. [11–13], via the projection-operator method. This demonstrates that the relaxation time in Israel–Stewart theories, which determines the time scale of the transient dynamics of the dissipative currents, is indeed a microscopic and not a fluid-dynamical time scale. It is determined by the interparticle scattering rate, and not by the time scales of fluid dynamics located near the origin of the complex ω-plane. We shall demonstrate in Sect. 4.5 that attempts to extract the value of the relaxation time from the dynamics on fluid-dynamical time scales in general fail to give the correct expression.

It is important to remark that by including more terms in the expansion (4.59), we obtain a retarded Green's function with more poles. Furthermore, the expression for the first pole and, consequently, the relaxation time, will also be modified. All other transport coefficients will also receive corrections.

4.4.2   Linear-Response Theory and Metric Perturbations

We now apply our formalism to the case studied in Ref. [14], where the transport coefficients are determined from perturbations $h_{\mu\nu}$ of the metric tensor

$$g_{\mu\nu} = \eta_{\mu\nu} + h_{\mu\nu} . \tag{4.84}$$

This method can be equally applied at strong and weak couplings. The variation of the energy-momentum tensor $T^{\mu\nu}$ due to the metric perturbations is [15]

$$\delta T^{\mu\nu}(X) = \frac{1}{2} \int\limits_{-\infty}^{\infty} d^4 X'\, G_R^{\mu\nu\alpha\beta}(X - X')\, h_{\alpha\beta}(X') , \tag{4.85}$$

where $G_R^{\mu\nu\alpha\beta}(X - X')$ is the retarded Green's function.

We consider a fluid with vanishing bulk viscosity, for which the energy-momentum tensor assumes the form

$$T^{\mu\nu} = \varepsilon\, u^\mu u^\nu - \Delta^{\mu\nu} P_0 + \pi^{\mu\nu} . \tag{4.86}$$

We now assume that, in the absence of metric perturbations, $h^{\mu\nu} = 0$, this fluid is at rest in an equilibrium state, $\varepsilon = \text{const.}$, $P_0 = \text{const.}$, $u^\mu = (1, 0, 0, 0)^T$, i.e., considering the energy-momentum tensor as a function of the metric tensor, $T^{\mu\nu} = T^{\mu\nu}(g_{\alpha\beta})$, for the unperturbed state we have

$$T^{\mu\nu}(\eta_{\alpha\beta}) = \varepsilon\, u^\mu u^\nu - \Delta_0^{\mu\nu} P_0 , \tag{4.87}$$

where $\Delta_0^{\mu\nu} \equiv \eta^{\mu\nu} - u^\mu u^\nu$. The metric perturbation will induce a change in the energy-momentum tensor,

$$\delta T^{\mu\nu} \equiv T^{\mu\nu}(\eta_{\alpha\beta} + h_{\alpha\beta}) - T^{\mu\nu}(\eta_{\alpha\beta}) = \frac{\delta T^{\mu\nu}}{\delta g_{\alpha\beta}} h_{\alpha\beta} + O(h^2) . \tag{4.88}$$

We now restrict the consideration to the special metric perturbation $h_{xy} = h_{xy}(t, z)$, with all other components of the metric tensor left unperturbed [14,16]. This is a tensor perturbation and thus cannot change the scalar and vector quantities ε, P_0, and u^μ. Using Eqs. (4.86) and (4.88), we obtain for the xy-components of the perturbation $\delta T^{\mu\nu}$

$$\delta T^{xy} = -P_0\, h^{xy} + \delta\pi^{xy} , \tag{4.89}$$

where $\delta\pi^{xy} \equiv (\delta\pi^{\mu\nu}/\delta g_{xy}) h_{xy}$ is the xy-component of the shear-stress tensor generated by the metric perturbations.

Furthermore, for the simple metric perturbation $h_{xy} = h_{xy}(t, z)$ Eq. (4.85) becomes [15],

$$\delta T^{xy}(t, z) = \int\limits_{-\infty}^{\infty} dt'\, dz'\, G_R^{xyxy}\left(t - t';\, z - z'\right) h_{xy}\left(t', z'\right) . \tag{4.90}$$

Inserting Eq. (4.89), we arrive at

$$\delta \pi^{xy} = P_0\, h^{xy} + \int\limits_{-\infty}^{\infty} dt'\, dz'\, G_R^{xyxy}\left(t - t';\, z - z'\right) h_{xy}\left(t', z'\right) . \tag{4.91}$$

Taking the Fourier transform, we obtain

$$\delta \tilde{\pi}^{xy}(Q) = \tilde{G}_R(Q)\, \tilde{h}_{xy}(Q) , \tag{4.92}$$

where $\tilde{G}_R(Q) = -P_0 + \tilde{G}_R^{xyxy}(Q)$. Note that $h^{xy} = -h_{xy}$, which follows from Eq. (4.84) and the fact that $g^{\mu\nu}$ is the inverse of $g_{\mu\nu}$, $g^{\mu\lambda} g_{\lambda\nu} = \delta_\nu^\mu$. Since the pressure has no dependence on Q, it is clear that $\tilde{G}_R(Q)$ has the same analytic structure as $\tilde{G}_R^{xyxy}(Q)$. Note also that static, homogeneous perturbations do not produce any shear stress, such that

$$\delta \tilde{\pi}^{xy}(\omega, \mathbf{0})\big|_{\omega=0} = \tilde{G}_R(\omega, \mathbf{0})\, \tilde{h}_{xy}(\omega, \mathbf{0})\big|_{\omega=0} = 0 \implies \tilde{G}_R(\omega, \mathbf{0})\big|_{\omega=0} = 0 . \tag{4.93}$$

Assuming that $\tilde{G}_R^{xyxy}$ has N poles, $\omega_i(\mathbf{q})$, and that the first pole is purely imaginary, we can apply Eq. (4.39) and obtain the equation of motion for $\delta \pi^{xy}$,

$$\tau_\pi \partial_t \delta \pi^{xy} + \delta \pi^{xy} = D_0 h_{xy} + D_1 \partial_t h_{xy} + D_2 \partial_t^2 h_{xy} + O\left(\partial_t^3 h_{xy}, \partial_z^2 h_{xy}\right) . \tag{4.94}$$

Note that, in case the first pole is not purely imaginary, a simple relaxation equation would not be able to describe the transient dynamics, even for long times. In Eq. (4.94), we followed Eq. (4.40) and introduced the following transport coefficients:

$$\tau_\pi = \frac{1}{i\omega_1(0)} , \tag{4.95}$$

$$D_0 = \tilde{G}_R(\omega, \mathbf{0})\big|_{\omega=0} = 0 , \tag{4.96}$$

$$D_1 = i\partial_\omega\, \tilde{G}_R(\omega, \mathbf{0})\big|_{\omega=0} + \tau_\pi D_0 = i\partial_\omega\, \tilde{G}_R(\omega, \mathbf{0})\big|_{\omega=0} \equiv \eta , \tag{4.97}$$

$$D_2 = -\frac{1}{2}\partial_\omega^2\, \tilde{G}_R(\omega, \mathbf{0})\big|_{\omega=0} + D_1 \tau_\pi - D_0 \tau_\pi^2$$

$$\equiv -\frac{1}{2}\partial_\omega^2\, \tilde{G}_R(\omega, \mathbf{0})\big|_{\omega=0} + \eta \tau_\pi . \tag{4.98}$$

For Eq. (4.96), we have employed Eq. (4.93). In order to understand why D_1 is identical to the shear-viscosity coefficient, cf. Eq. (4.97), we first note that we have already computed the shear tensor in a general metric and in a static background in Chap. 3; cf. Eq. (3.30). This calculation also applies here, but the Christoffel symbols (3.7) for the metric (4.84) are of course different. Computing to first order in the metric perturbation $h_{\mu\nu}$, we obtain

$$
\begin{aligned}
\sigma_{xy} &= -\Gamma^t_{\langle xy \rangle} = -\Delta^{\mu\nu}_{xy} \Gamma^t_{\mu\nu} \\
&= -\frac{1}{2} \left(\Delta^\mu_x \Delta^\nu_y - \frac{1}{3} \Delta^{\mu\nu} \Delta_{xy} \right) \left(\partial_\mu h_{t\nu} + \partial_\nu h_{t\mu} - \partial_t h_{\mu\nu} \right) \\
&= \frac{1}{2} \partial_t h_{xy} ,
\end{aligned}
\tag{4.99}
$$

where we used the fact that the background is static, $u^\mu = (1, 0, 0, 0)^T$, and only the xy-component of the metric perturbation is nonzero. Inserting Eq. (4.99) into Eq. (4.94), we conclude that, in order to get the Navier–Stokes term $2\eta\sigma_{xy}$, we need to identify $D_1 \equiv \eta$. This is different from the Boltzmann case, where the coefficient of the shear tensor was D_0, and not D_1. On the other hand, the relaxation time (4.95) is still given by the inverse of the first pole.

4.5   Discussion

As discussed in this chapter, the gradient expansion corresponds to a Taylor expansion of the retarded Green's function around the origin of the complex ω-plane, and thus *per se* cannot account for relaxation-type dynamics, since it does not account for poles of the Green's function in the complex ω-plane, which, as we have seen, are necessary to obtain relaxation-type equations for the dissipative currents. Nevertheless, such equations have been derived in Ref. [14] based on the gradient expansion. We give a brief account in terms of our notation of the strategy employed in that work to achieve this.

The starting point is the gradient expansion (4.9), assuming that a Knudsen-number counting as explained at the end of Sect. 4.2 is applicable. The gradient expansion (4.9) is not an equation of motion for the dissipative current, but one can construct one by taking the first-order solution, $J = \bar{D}_0 F + O(\mathrm{Kn}^2)$, and then replacing the first time derivative of F on the right-hand side by a time derivative of J,

$$
J = \bar{D}_0 F + \bar{D}_1 \partial_t \left(\frac{J}{\bar{D}_0} \right) + O(\mathrm{Kn}^3) \quad \Longrightarrow \quad \bar{\tau}_R \partial_t J + J \simeq \bar{D}_0 F , \tag{4.100}
$$

where $\bar{\tau}_R \equiv -\bar{D}_1/\bar{D}_0$ has the dimension of time and where we have neglected the time derivative of $\bar{D}_0$ (this coefficient depends in principle on the thermodynamic variables α_0, β_0, the time derivatives of which can be computed using the conservation equations for energy and particle or charge number). By construction, this is a relaxation-type equation of motion for the dissipative current J, with a relaxation time $\bar{\tau}_R$.

From our previous discussion, however, it is clear that this need not be the correct equation of motion for the dissipative current. If the poles of the retarded Green function $\tilde{G}_R(\omega)$ associated with the dissipative current J are off the imaginary axis, the equation of motion for J is never of relaxation type, and the above way to construct one is misleading, since it fails to capture the correct physics. Only if the first pole of $\tilde{G}_R(\omega)$ lies on the imaginary axis and is sufficiently separated from the other singularities, one can obtain a relaxation-type equation for J. In this case, however, the true relaxation time is $\tau_R = 1/[i\omega_1(\mathbf{0})]$, and not $\bar{\tau}_R = -\bar{D}_1/\bar{D}_0$.

There is, however, a particular case, where $\tau_R = \bar{\tau}_R$, namely when $D_1 = 0$, i.e., when $\Xi(\omega, \mathbf{0}) = $ const.; cf. Eqs. (4.38). This is most easily seen in the one-pole case, where, cf. Eqs. (4.10) and (4.18),

$$0 = i \left. \partial_\omega \tilde{G}_R(\omega, \mathbf{0}) \right|_{\omega=0} + D_0 \tau_R \equiv \bar{D}_1 + \bar{D}_0 \tau_R , \qquad (4.101)$$

i.e., $\tau_R = -\bar{D}_1/\bar{D}_0 \equiv \bar{\tau}_R$.

This argument can also be applied to the case discussed in Sect. 4.4.2, where the thermodynamic force is not given by F but $\partial_t F$. Then, $D_0 = 0$, and $D_1 = \bar{D}_1 = \eta$ is the transport coefficient. In this case, $\bar{\tau}_\pi = -\bar{D}_2/\bar{D}_1$, and equivalency to the true relaxation time requires that $D_2 = 0$, i.e.,

$$0 = -\frac{1}{2} \left. \partial_\omega^2 \tilde{G}_R(\omega, \mathbf{0}) \right|_{\omega=0} + D_1 \tau_\pi \equiv \bar{D}_2 + \bar{D}_1 \tau_\pi , \qquad (4.102)$$

or

$$\eta \tau_\pi = \frac{1}{2} \left. \partial_\omega^2 \tilde{G}_R(\omega, \mathbf{0}) \right|_{\omega=0} . \qquad (4.103)$$

This equation is rather similar to the one given in Refs. [14, 16], $\eta \bar{\tau}_\pi = [\partial_\omega^2 \tilde{G}_R(0) - \partial_{q_z}^2 \tilde{G}_R(0)]/2$ [17]. Here, the additional derivatives with respect to momentum enter because second-order time derivatives also arise from space-time curvature in Eq. (4.9) (they were not explicitly denoted in that equation); cf. Ref. [14]. These are not subjected to the construction of a relaxation-type equation for J as explained above. In turn, $\eta \bar{\tau}_\pi$ receives an additional contribution from second-order spatial derivatives; for details, see Ref. [14].

Regardless of these considerations, the true relaxation time is always given by the first pole of the retarded Green function. In general, the location of this pole cannot be found from a *truncated* Taylor expansion around the origin.

A calculation of the shear-viscosity relaxation-time coefficient has also been performed in Refs. [11, 12, 18–21] and within the Mori-Zwanzig formalism [22–25]. The implications of our findings to those works will not be discussed in this book.

4.6   Summary

In this chapter, we have derived equations of motion for the dissipative currents, assuming these currents to be linearly related to the thermodynamic forces. We have shown how these equations of motion are determined by the analytic structure of the associated retarded Green's function in the complex ω-plane. We have demonstrated that the standard gradient expansion is equivalent to a Taylor expansion of the retarded Green's function around the origin in the complex ω-plane. This Taylor series is convergent only when all singularities of the retarded Green's function are pushed to infinity. In general, however, these singularities appear at finite values of $|\omega|$, which consequently severely restricts the applicability of the gradient expansion.

We have furthermore demonstrated that if the retarded Green's function has simple poles in the complex ω-plane, the dissipative current obeys a differential equation with source terms which are the thermodynamic force and gradients thereof. This is different from the gradient expansion where the current is directly proportional to the thermodynamic force and its gradients.

In general, the equation of motion for the dissipative current is not a relaxation-type equation. However, in the limit where all singularities of the retarded Green's function except the pole nearest to the origin are pushed to infinity, it is possible to approximate the dynamical equation satisfied by J as a relaxation-type equation, similar to the ones appearing in Israel–Stewart theory. This is only possible if the first pole is purely imaginary. The relaxation time is equal to minus the inverse of the imaginary part of the pole. The gradient expansion constitutes the asymptotic solution of the resulting relaxation-type equation, and can be obtained by taking the relaxation time to zero or, equivalently, pushing the first pole to infinity. This is consistent with the above statement that the gradient expansion arises from a Taylor expansion of the retarded Green's function around the origin.

In relativistic systems, in order to have stable and causal equations of motion for the dissipative currents, one cannot take the relaxation time to be arbitrarily small; see Sect. 2.1 and Refs. [26,27]. Thus, one cannot push the first pole of the retarded Green's function to infinity. This is the reason why one cannot use the gradient expansion to obtain stable and causal equations of motion for the dissipative currents.

As an example, in this chapter we have studied the Boltzmann equation for a classical gas and demonstrated that the retarded Green's function associated with the shear-stress tensor has (in principle, infinitely many) simple poles along the imaginary axis. Therefore, it is possible to reduce the equation of motion for the shear-stress tensor to a relaxation-type equation. Our results are consistent with those of Refs. [8,11–13], when one truncates the collision operator at the lowest order. This convincingly demonstrates that the time scale of transient dynamics determined by the relaxation time is of microscopic, and not of fluid-dynamical origin: it is the slowest microscopic time scale, not the fastest fluid-dynamical time scale as implicitly assumed in attempts to extract the relaxation time by expanding the retarded Green's function around the origin. Consequently, the expression for the relaxation time derived in Refs. [14,16,28] by using an expansion around the origin is, in general, different from the one derived from the first pole of the retarded Green's function.

In fact, as we have demonstrated in this work, when the retarded Green's function has simple poles off the imaginary axis in the complex ω-plane, the true dynamics of the system at long wavelengths and low frequencies is not even of relaxation type. Strongly coupled theories, like those emerging from the AdS/CFT correspondence [15], belong to this class.

References

1. Denicol, G.S., Noronha, J., Niemi, H., Rischke, D.H.: Phys. Rev. D **83**, 074019 (2011). https://doi.org/10.1103/PhysRevD.83.074019, arXiv:1102.4780 [hep-th]
2. Grad, H.: Commun. Pure Appl. Math. **2**, 331 (1949)
3. Israel, W.: Ann. Phys. (N.Y.) **100**, 310 (1976)
4. Stewart, J.M.: Proc. Roy. Soc. A **357**, 59 (1977)
5. Israel, W., Stewart, J.M.: Ann. Phys. (N.Y.) **118**, 341 (1979)
6. Even though we only show terms up to order $O\left(\text{Kn}^3\right)$ in Eq. (4.19), this equivalence can be easily confirmed to all orders
7. Kovtun, P.K., Starinets, A.O.: Phys. Rev. D **72**, 086009 (2005)
8. Denicol, G.S., Koide, T., Rischke, D.H.: Phys. Rev. Lett. **105**, 162501 (2010). https://doi.org/10.1103/PhysRevLett.105.162501, arXiv:1004.5013 [nucl-th]
9. de Groot, S.R., van Leeuwen, W.A., van Weert, Ch.G.: Relativistic Kinetic Theory—Principles and Applications. North-Holland (1980)
10. Denicol, G.S., Niemi, H., Molnar, E., Rischke, D.H.: Phys. Rev. D **85**, 114047 (2012). https://doi.org/10.1103/PhysRevD.85.114047, arXiv:1202.4551 [nucl-th]
11. Koide, T., Kodama, T.: Phys. Rev. E **78**, 051107 (2008)
12. Koide, T., Nakano, E., Kodama, T.: Phys. Rev. Lett. **103**, 052301 (2009)
13. Denicol, G.S., Huang, X.G., Koide, T., Rischke, D.H.: Phys. Lett. B **708**, 174-178 (2012) https://doi.org/10.1016/j.physletb.2012.01.018 [arXiv:1003.0780 [hep-th]]
14. Baier, R., Romatschke, P., Son, D.T., Starinets, A.O., Stephanov, M.A.: JHEP **0804**, 100 (2008)
15. Son, D.T., Starinets, A.O.: Ann. Rev. Nucl. Part. Sci. **57**, 95–118 (2007)
16. Moore, G.D., Sohrabi, K.A.: arXiv:1007.5333 [hep-ph]
17. Note that this result was derived using a different metric signature, i.e., $\eta_{\mu\nu} = \text{diag}\,(-, +, +, +)$. See Ref. [Baier, R., Romatschke, P., Son, D.T., Starinets, A.O., Stephanov, M.A.: JHEP **0804**, 100 (2008)] for details
18. Koide, T.: Phys. Rev. E **75**, 060103 (2007)
19. Koide, T., Conf, A.I.P.: Proc. **1312**, 27 (2010)
20. Huang, X.-G., Kodama, T., Koide, T., Rischke, D.H.: Phys. Rev. C **83**, 024906 (2011)
21. Huang, X.-G., Koide, T.: arXiv:1105.2483 [hep-th]
22. Nakajima, S.: Prog. Theor. Phys. **20**, 948 (1958)
23. Zwanzig, R.: J. Chem. Phys. **33**, 1338 (1960)
24. Mori, H.: Prog. Theor. Phys. **33**, 423 (1965)
25. Zwanzig, R.: Nonequilibrium Statistical Mechanics. Oxford University, New York (2004)
26. Denicol, G.S., Kodama, T., Koide, T., Mota, P.: J. Phys. G **35**, 115102 (2008)
27. Pu, S., Koide, T., Rischke, D.H.: Phys. Rev. D **81**, 114039 (2010)
28. Lu, E., Moore, G.D.: Phys. Rev. C **83**, 044901 (2011)

Fluid Dynamics from Kinetic Theory: Traditional Approaches

5

In Chap. 1, we have investigated how the equations of relativistic dissipative fluid dynamics can be derived phenomenologically, in Chap. 2 we have discussed their basic linear properties around equilibrium, and in Chap. 3 we have constructed explicit analytic solutions for rapidly expanding systems. In Chap. 4, we analyzed how the transport coefficients of dissipative fluid dynamics emerge from an underlying microscopic theory, using linear-response theory. In particular, we discussed in what ways transient fluid dynamics differs significantly from the traditional derivation of fluid dynamics based on the gradient expansion. This directly affects the definition of new transport coefficients introduced in transient fluid dynamics (or second-order theories), such as the relaxation times.

In this chapter, we continue this discussion from the perspective of relativistic dilute gases. In this case, the derivation of relativistic dissipative fluid dynamics from a microscopic theory, here the relativistic Boltzmann equation, can be carried out in great detail, going beyond the linear regime. This issue will be the main topic for the remainder of this book. In this chapter, we start addressing this topic by discussing the two most widespread methods usually employed to derive relativistic fluid dynamics from the Boltzmann equation: the Chapman–Enskog expansion [1] and the method of moments as proposed by Israel and Stewart [2–4].

This chapter is organized as follows: In Sect. 5.1, we briefly introduce the Boltzmann equation and discuss how the fluid-dynamical degrees of freedom are defined in this setting. In Sect. 5.2, we introduce Chapman–Enskog theory and derive the fluid-dynamical equations following this procedure. In Sect. 5.3, we discuss Israel's and Stewart's original derivation of transient fluid dynamics and the differences to Chapman and Enskog's approach.

© Springer Nature Switzerland AG 2021
G. S. Denicol and D. H. Rischke, *Microscopic Foundations of Relativistic Fluid Dynamics*,
Lecture Notes in Physics 990, https://doi.org/10.1007/978-3-030-82077-0_5

5.1   Matching Fluid-Dynamical with Kinetic Degrees of Freedom

We start with the relativistic Boltzmann equation (4.41) for the single-particle distribution function $f_{\mathbf{k}}$, which we quote again for the sake of convenience,

$$k^{\mu}\partial_{\mu} f_{\mathbf{k}} = C[f] \; . \tag{5.1}$$

The notation was already explained in Sect. 4.4.1, i.e., the 4-momentum of a particle is $k^{\mu} = (k^0, \mathbf{k})^T$, with $k^0 = \sqrt{\mathbf{k}^2 + m^2}$ being the on-shell energy and m its mass. We take the same collision term as in Eq. (4.42), i.e., we consider only binary elastic collisions,

$$C[f] = \frac{1}{\nu} \int dK' dP dP' W_{\mathbf{kk'} \to \mathbf{pp'}} \left(f_{\mathbf{p}} f_{\mathbf{p'}} \tilde{f}_{\mathbf{k}} \tilde{f}_{\mathbf{k'}} - f_{\mathbf{k}} f_{\mathbf{k'}} \tilde{f}_{\mathbf{p}} \tilde{f}_{\mathbf{p'}} \right) , \tag{5.2}$$

where ν is a symmetry factor ($= 2$ for identical particles), $W_{\mathbf{kk'} \to \mathbf{pp'}}$ is the Lorentz-invariant transition rate, and $dK \equiv d_{\mathrm{dof}} d^3 \mathbf{k} / [(2\pi)^3 k^0]$ is the Lorentz-invariant momentum-space volume, with d_{dof} being the number of internal degrees of freedom. We introduced the notation

$$\tilde{f}_{\mathbf{k}} \equiv 1 - a f_{\mathbf{k}} , \tag{5.3}$$

where $a = 1$ ($a = -1$) for fermions (bosons) and $a = 0$ for a classical gas.

In kinetic theory, the conserved particle current N^{μ} and the energy-momentum tensor $T^{\mu\nu}$ are expressed as moments of the single-particle distribution function $f_{\mathbf{k}}$,

$$N^{\mu} = \langle k^{\mu} \rangle , \tag{5.4}$$

$$T^{\mu\nu} = \langle k^{\mu} k^{\nu} \rangle , \tag{5.5}$$

where the angular brackets are defined as

$$\langle \cdots \rangle \equiv \int dK \; (\cdots) \, f_{\mathbf{k}} , \tag{5.6}$$

i.e., a momentum-space average with the distribution function $f_{\mathbf{k}}$ as weight factor.

5.1.1   Macroscopic Conservation Laws

The macroscopic conservation laws are expressed in terms of continuity equations. They can be obtained using basic properties of the collision operator, as will be derived in this subsection. To this end, it is convenient to consider the following set of integrals of the collision operator,

$$\int dK \, G_{\mathbf{k}} C[f] = \frac{1}{\nu} \int dK dK' dP dP' W_{\mathbf{kk'} \to \mathbf{pp'}} G_{\mathbf{k}} \left(f_{\mathbf{p}} f_{\mathbf{p'}} \tilde{f}_{\mathbf{k}} \tilde{f}_{\mathbf{k'}} - f_{\mathbf{k}} f_{\mathbf{k'}} \tilde{f}_{\mathbf{p}} \tilde{f}_{\mathbf{p'}} \right) , \tag{5.7}$$

where $G_{\mathbf{k}}$ is an arbitrary function of the 3-momentum $\mathbf{k}$. The function $G_{\mathbf{k}}$ can be a Lorentz-tensor of arbitrary rank.

We now use a set of properties satisfied by the transition rate, $W_{\mathbf{kk'}\to\mathbf{pp'}}$. First, we note that $W_{\mathbf{kk'}\to\mathbf{pp'}}$ is invariant under the exchange $\mathbf{k} \leftrightarrow \mathbf{k'}$ or $\mathbf{p} \leftrightarrow \mathbf{p'}$, i.e.,

$$W_{\mathbf{kk'}\to\mathbf{pp'}} = W_{\mathbf{k'k}\to\mathbf{pp'}} = W_{\mathbf{k'k}\to\mathbf{p'p}} \;. \tag{5.8}$$

In other words, the transition probability per unit of time from an initial state (particles before the collision) to a final state (particles after the collision) cannot depend on which particle carries the incoming momentum $\mathbf{k}$ or $\mathbf{k'}$ or the outgoing momentum $\mathbf{p}$ or $\mathbf{p'}$. This property allows us to rewrite the collision integral (5.7) in the following form

$$\int dK\, G_{\mathbf{k}} C[f]$$
$$= \frac{1}{\nu} \int dK dK' dP dP'\, W_{\mathbf{kk'}\to\mathbf{pp'}} \frac{G_{\mathbf{k}} + G_{\mathbf{k'}}}{2} \left(f_{\mathbf{p}} f_{\mathbf{p'}} \tilde{f}_{\mathbf{k}} \tilde{f}_{\mathbf{k'}} - f_{\mathbf{k}} f_{\mathbf{k'}} \tilde{f}_{\mathbf{p}} \tilde{f}_{\mathbf{p'}} \right) \;. \tag{5.9}$$

Next, we note that time-reversal symmetry further imposes that the transition rate is invariant under the exchange $\mathbf{kk'} \leftrightarrow \mathbf{pp'}$, i.e.,

$$W_{\mathbf{kk'}\to\mathbf{pp'}} = W_{\mathbf{pp'}\to\mathbf{kk'}} \;. \tag{5.10}$$

That is, collisions that change the momenta of the particles in the direction $(\mathbf{k}, \mathbf{k'}) \to (\mathbf{p}, \mathbf{p'})$ are as likely to happen as those that change the momenta in the direction $(\mathbf{p}, \mathbf{p'}) \to (\mathbf{k}, \mathbf{k'})$. If we impose this fundamental symmetry, we derive the following relation,

$$\int dK\, G_{\mathbf{k}} C[f] = \frac{1}{4\nu} \int dK dK' dP dP'\, W_{\mathbf{kk'}\to\mathbf{pp'}} \left(G_{\mathbf{k}} + G_{\mathbf{k'}} - G_{\mathbf{p}} - G_{\mathbf{p'}} \right)$$
$$\times \left(f_{\mathbf{p}} f_{\mathbf{p'}} \tilde{f}_{\mathbf{k}} \tilde{f}_{\mathbf{k'}} - f_{\mathbf{k}} f_{\mathbf{k'}} \tilde{f}_{\mathbf{p}} \tilde{f}_{\mathbf{p'}} \right) \;. \tag{5.11}$$

Expression (5.11) for the collision integral (5.7) is very useful since it clearly shows that it vanishes if $G_{\mathbf{k}}$ is a quantity that is conserved in microscopic collisions between the particles. Such quantities are well known in physics and are the 4-momentum, k^{μ}, and the charge or, since we only consider binary elastic collisions, the particle number. Thus, taking $G_{\mathbf{k}} = 1$ or $G_{\mathbf{k}} = k^{\mu}$, we obtain

$$\int dK\, C[f] = 0 \;, \tag{5.12}$$

$$\int dK\, k^{\mu} C[f] = 0 \;. \tag{5.13}$$

Since there are no other quantities which are conserved in particle collisions, no other integral of the collision term vanishes—all other integrals will lead to finite contributions.

If we integrate the Boltzmann equation over 4-momentum K and use property (5.12) of the collision operator, we arrive at the continuity equation related to particle-number conservation (or net-charge conservation, if we consider processes that change the particle number),

$$\int dK \, k^\mu \partial_\mu f_{\mathbf{k}} = \int dK \, C[f] = 0 \quad \Longleftrightarrow \quad \partial_\mu \langle k^\mu \rangle = \partial_\mu N^\mu = 0 . \qquad (5.14)$$

Similarly, if we multiply the Boltzmann equation by k^ν, integrate over K, and use property (5.13) of the collision operator, we arrive at the continuity equation expressing energy-momentum conservation,

$$\int dK \, k^\mu k^\nu \partial_\mu f_{\mathbf{k}} = \int dK \, k^\nu C[f] = 0 \quad \Longleftrightarrow \quad \partial_\mu \langle k^\mu k^\nu \rangle = \partial_\mu T^{\mu\nu} = 0 . \qquad (5.15)$$

As already mentioned, any other integral of the collision operator will not vanish and hence, no additional conservation laws can be derived from the Boltzmann equation. We note that, for spin-zero particles, angular-momentum conservation is a consequence of energy-momentum conservation, as long as the energy-momentum tensor is symmetric—a property that is fulfilled by definition, cf. Eq. (5.5).

5.1.2   Fluid-Dynamical Variables and Matching Conditions

The particle current N^μ and the energy-momentum tensor $T^{\mu\nu}$ can be tensor-decomposed with respect to the fluid 4-velocity u^μ. We introduce u^μ as the time-like, normalized ($u_\mu u^\mu = 1$) eigenvector of the energy-momentum tensor,

$$T^{\mu\nu} u_\nu \equiv \varepsilon u^\mu , \qquad (5.16)$$

where the eigenvalue ε is the energy density. This choice of fluid velocity is traditionally referred to as the Landau frame [5], see Chap. 1, Sect. 1.3.3. Next, we divide the momentum of the particles k^μ into two parts: one parallel and one orthogonal to u^μ,

$$k^\mu = E_{\mathbf{k}} u^\mu + k^{\langle \mu \rangle} , \qquad (5.17)$$

where we defined the scalar

$$E_{\mathbf{k}} \equiv k^\mu u_\mu \qquad (5.18)$$

and used the notation $A^{\langle \mu \rangle} = \Delta^\mu_\nu A^\nu$, with $\Delta^{\mu\nu}$ being the projection operator onto the 3-space orthogonal to u^μ given by Eq. (1.22).

Then, the tensor decomposition of N^μ and $T^{\mu\nu}$ reads

$$N^\mu = n u^\mu + n^\mu , \tag{5.19}$$

$$T^{\mu\nu} = \varepsilon\, u^\mu u^\nu - \Delta^{\mu\nu}\, (P_0 + \Pi) + \pi^{\mu\nu} , \tag{5.20}$$

where the particle density n, the particle-diffusion current n^μ, the energy density ε, the shear-stress tensor $\pi^{\mu\nu}$, and the sum of thermodynamic pressure P_0 and bulk-viscous pressure Π are defined by

$$n \equiv \langle E_\mathbf{k} \rangle , \quad n^\mu \equiv \left\langle k^{\langle\mu\rangle} \right\rangle , \quad \varepsilon \equiv \left\langle E_\mathbf{k}^2 \right\rangle ,$$

$$\pi^{\mu\nu} \equiv \left\langle k^{\langle\mu} k^{\nu\rangle} \right\rangle , \quad P_0 + \Pi \equiv -\frac{1}{3}\left\langle \Delta^{\mu\nu} k_\mu k_\nu \right\rangle , \tag{5.21}$$

where $A^{\langle\mu\nu\rangle} \equiv \Delta^{\mu\nu}_{\alpha\beta} A^{\alpha\beta}$, with $\Delta^{\mu\nu}_{\alpha\beta}$ as defined in Eq. (1.55).

Next, we introduce the local-equilibrium distribution function as

$$f_{0\mathbf{k}} = [\exp(\beta_0 E_\mathbf{k} - \alpha_0) + a]^{-1} , \tag{5.22}$$

where $\beta_0 \equiv 1/T$ is the inverse temperature and $\alpha_0 = \mu/T$ is the ratio of the chemical potential to temperature, respectively. The values of α_0 and β_0 are defined by the matching conditions, as explained in Sect. 1.3.1,

$$n \equiv n_0 = \langle E_\mathbf{k} \rangle_0 , \quad \varepsilon \equiv \varepsilon_0 = \left\langle E_\mathbf{k}^2 \right\rangle_0 , \tag{5.23}$$

where we use the following notation for the momentum-space average with the local-equilibrium distribution function as weight factor

$$\langle \cdots \rangle_0 \equiv \int dK\, (\cdots)\, f_{0\mathbf{k}} . \tag{5.24}$$

Then, the separation between thermodynamic pressure and bulk-viscous pressure is achieved by

$$P_0 = -\frac{1}{3}\left\langle \Delta^{\mu\nu} k_\mu k_\nu \right\rangle_0 \tag{5.25}$$

and

$$\Pi = -\frac{1}{3}\left\langle \Delta^{\mu\nu} k_\mu k_\nu \right\rangle_\delta , \tag{5.26}$$

where

$$\langle \cdots \rangle_\delta = \langle \cdots \rangle - \langle \cdots \rangle_0 . \tag{5.27}$$

In equilibrium, the bulk-viscous pressure vanishes by definition, $\Pi_0 \equiv 0$. The particle- or charge-diffusion current n^μ and the shear-stress tensor $\pi^{\mu\nu}$ vanish as well, $n_0^\mu \equiv \langle k^{\langle\mu\rangle} \rangle_0 = 0$, $\pi_0^{\mu\nu} = \langle k^{\langle\mu} k^{\nu\rangle} \rangle_0 = 0$. This is obvious from the symmetries of the equilibrium distribution $f_{0\mathbf{k}}$. This, in turn, implies that

$$n^\mu = \left\langle k^{\langle\mu\rangle} \right\rangle_\delta \,, \qquad \pi^{\mu\nu} = \left\langle k^{\langle\mu} k^{\nu\rangle} \right\rangle_\delta \,. \tag{5.28}$$

As shown in Sect. 1.3.3, the conservation laws provide equations of motion only for n, ε, and u^μ, and hence one still needs to derive the equations of motion for the dissipative corrections Π, n^μ, and $\pi^{\mu\nu}$. In kinetic theory, this task can be performed rigorously, and it is possible to derive the equations of motion and transport coefficients even in the nonlinear regime. Nevertheless, there are several methods that can be used to determine the required equations and transport coefficients. In the following, we review the two most traditional methods: Chapman–Enskog theory [1] and the method of moments proposed by Israel and Stewart [2–4].

5.2 Chapman–Enskog Theory

The Chapman–Enskog expansion [1] is the most traditional formalism used to derive fluid dynamics from the Boltzmann equation. It was originally developed for non-relativistic systems, but Israel proved that it could be used with almost no modifications to describe relativistic systems as well [6,7].

The Chapman–Enskog formalism corresponds to the microscopic implementation of the gradient expansion, already discussed in previous chapters. It assumes that the single-particle distribution function depends only on the five primary fluid-dynamical variables, i.e., temperature, chemical potential, and the three independent components of the fluid-velocity field, as well as their gradients. The corrections to the local distribution function are then systematically arranged in terms of an expansion in powers of the Knudsen number. As is well known and will be shown in this section, the first-order truncation of the expansion leads to Navier–Stokes theory. Keeping second- and higher-order terms one obtains the Burnett and super-Burnett equations, respectively [8]. In principle, one can construct the solution to an arbitrarily high order in Knudsen number.

As first pointed out by Grad, the Chapman–Enskog expansion is an asymptotic series [9]. Also, in the relativistic case, the Chapman–Enskog expansion leads to unstable equations of motion [10] and, therefore, has very little use in the description of realistic systems. Despite this major drawback, the Chapman–Enskog expansion is an important development in kinetic theory and its results are useful to understand the asymptotic behavior of the Boltzmann equation.

The first step is to rewrite the Boltzmann equation (5.1) by decomposing the 4-derivative ∂_μ into its time-like and space-like parts,

$$\partial_\mu = u_\mu u^\nu \partial_\nu + \Delta_\mu^\nu \partial_\nu \equiv u_\mu D + \nabla_\mu \,, \tag{5.29}$$

where we defined the comoving derivative

$$D \equiv u^\nu \partial_\nu \,, \tag{5.30}$$

which is equal to the ordinary time derivative in the local rest frame of the fluid, $D_{LR} \equiv \partial/\partial t$, and the 3-space gradient

$$\nabla_\mu \equiv \Delta_\mu^\nu \partial_\nu \,, \tag{5.31}$$

which is equal to the ordinary spatial gradient in the local rest frame of the fluid, $\nabla_{LR,\mu} \equiv (0, \nabla)$. The Boltzmann equation (5.1) then reads

$$D f_{\mathbf{k}} + \frac{1}{E_{\mathbf{k}}} k^\mu \nabla_\mu f_{\mathbf{k}} = \frac{1}{E_{\mathbf{k}}} C[f] \,. \tag{5.32}$$

In the local rest frame of the system, the gradient ∇_μ determines an inverse macroscopic distance scale, L^{-1}, over which the single-particle distribution function (and its momentum integrals) varies in space. Similarly, the covariant derivative D defines an inverse macroscopic time scale, $\bar{\tau}^{-1}$. It is convenient to redefine these derivatives as $\nabla_\mu \equiv L^{-1} \hat{\nabla}_\mu$ and $D \equiv \bar{\tau}^{-1} \hat{D}$, where $\hat{\nabla}_\mu$ and $\hat{D}$ are unitless derivatives of order one. Then, multiplying the Boltzmann equation by the mean free path λ of the particles, we obtain the dimensionless equation of motion

$$\overline{\mathrm{Kn}} \, \hat{D} f_{\mathbf{k}} + \frac{\mathrm{Kn}}{E_{\mathbf{k}}} k^\mu \hat{\nabla}_\mu f_{\mathbf{k}} = \frac{\lambda}{E_{\mathbf{k}}} C[f] \,, \tag{5.33}$$

where $\mathrm{Kn} = \lambda/L$ is the usual Knudsen number. For the sake of completeness, we also introduced another type of Knudsen number, $\overline{\mathrm{Kn}} = \lambda/\bar{\tau}$, which characterizes the macroscopic time variations relative to the mean free path.

In Chapman–Enskog theory, a solution of the Boltzmann equation is obtained by expanding the single-particle distribution function in powers of Kn,

$$f_{\mathbf{k}} = f_{\mathbf{k}}^{(0)} + \mathrm{Kn} \, f_{\mathbf{k}}^{(1)} + \mathrm{Kn}^2 \, f_{\mathbf{k}}^{(2)} + \cdots \,. \tag{5.34}$$

If the Knudsen number is small, it should be possible to truncate this expansion and find an approximate expression for $f_{\mathbf{k}}$. This solution is found using perturbation theory, by substituting Eq. (5.34) into Eq. (5.33) and solving it order by order in Knudsen number.

First, we substitute expression (5.34) into the collision term (5.2), obtaining

$$C[f] = C^{(0)} + \mathrm{Kn} \, C^{(1)} + \mathrm{Kn}^2 \, C^{(2)} + \cdots \,, \tag{5.35}$$

where the first two terms of the expansion are

$$C^{(0)} \equiv \frac{1}{\nu} \int dK' dP dP' \, W_{\mathbf{k}\mathbf{k}' \to \mathbf{p}\mathbf{p}'} \left(f_{\mathbf{p}}^{(0)} f_{\mathbf{p}'}^{(0)} \tilde{f}_{\mathbf{k}}^{(0)} \tilde{f}_{\mathbf{k}'}^{(0)} - f_{\mathbf{k}}^{(0)} f_{\mathbf{k}'}^{(0)} \tilde{f}_{\mathbf{p}}^{(0)} \tilde{f}_{\mathbf{p}'}^{(0)} \right) \,, \tag{5.36}$$

$$C^{(1)} \equiv \frac{1}{\nu} \int dK' dP dP' W_{\mathbf{kk'} \to \mathbf{pp'}}$$

$$\times \left[f_{\mathbf{p}}^{(0)} f_{\mathbf{p'}}^{(0)} \tilde{f}_{\mathbf{k}}^{(0)} \tilde{f}_{\mathbf{k'}}^{(0)} \left(\frac{f_{\mathbf{p}}^{(1)}}{f_{\mathbf{p}}^{(0)}} + \frac{f_{\mathbf{p'}}^{(1)}}{f_{\mathbf{p'}}^{(0)}} - a \frac{f_{\mathbf{k}}^{(1)}}{\tilde{f}_{\mathbf{k}}^{(0)}} - a \frac{f_{\mathbf{k'}}^{(1)}}{\tilde{f}_{\mathbf{k'}}^{(0)}} \right) \right.$$

$$\left. - f_{\mathbf{k}}^{(0)} f_{\mathbf{k'}}^{(0)} \tilde{f}_{\mathbf{p}}^{(0)} \tilde{f}_{\mathbf{p'}}^{(0)} \left(\frac{f_{\mathbf{k}}^{(1)}}{f_{\mathbf{k}}^{(0)}} + \frac{f_{\mathbf{k'}}^{(1)}}{f_{\mathbf{k'}}^{(0)}} - a \frac{f_{\mathbf{p}}^{(1)}}{\tilde{f}_{\mathbf{p}}^{(0)}} - a \frac{f_{\mathbf{p'}}^{(1)}}{\tilde{f}_{\mathbf{p'}}^{(0)}} \right) \right] . \tag{5.37}$$

Note that we introduced two types of Knudsen numbers: one related to temporal variations of integrals of $f_{\mathbf{k}}$, $\overline{\mathrm{Kn}}$, and another related to spatial variations of integrals of $f_{\mathbf{k}}$, Kn. In general, these two quantities do not need to be equal or even related. On the other hand, in the fluid-dynamical limit, it is not absurd to expect these two quantities to become related or, at least, to be of the same order of magnitude. In the Chapman–Enskog expansion, one goes a step further and assumes that Kn and $\overline{\mathrm{Kn}}$ are exactly the same, $\mathrm{Kn} = \overline{\mathrm{Kn}}$. Only with this assumption it becomes possible to solve the Boltzmann equation perturbatively in powers of Kn, as was initially proposed. We shall see later that this assumption will ensure that time-like gradients can always be replaced by space-like gradients and, consequently, one can always arrange the solution of the single-particle distribution function in powers of space-like gradients.

Therefore, using $\mathrm{Kn} = \overline{\mathrm{Kn}}$, and inserting the expansions (5.34) and (5.35) into the Boltzmann equation (5.33), one obtains by comparing order by order in Kn the following solution,

$$0 = C^{(0)} , \tag{5.38}$$

$$\left[\hat{D} f_{\mathbf{k}} \right]^{(1)} + \frac{1}{E_{\mathbf{k}}} k^\mu \hat{\nabla}_\mu f_{\mathbf{k}}^{(0)} = \frac{\lambda}{E_{\mathbf{k}}} C^{(1)} , \tag{5.39}$$

$$\left[\hat{D} f_{\mathbf{k}} \right]^{(2)} + \frac{1}{E_{\mathbf{k}}} k^\mu \hat{\nabla}_\mu f_{\mathbf{k}}^{(1)} = \frac{\lambda}{E_{\mathbf{k}}} C^{(2)} . \tag{5.40}$$

Equation (5.38) can be used to solve for $f_{\mathbf{k}}^{(0)}$. Once $f_{\mathbf{k}}^{(0)}$ is known, Eq. (5.39) can be used to solve for $f_{\mathbf{k}}^{(1)}$ and so on. In principle, one could go on indefinitely and construct the solution $f_{\mathbf{k}}$ to any order in Knudsen number. However, after solving for the correction of first order in Knudsen number, the calculations start to become cumbersome.

Note that in Eqs. (5.38)–(5.40), we wrote $\left[\hat{D} f_{\mathbf{k}} \right]^{(n)}$ to indicate terms of order Kn^n from the time derivative. This needs to be distinguished from $\hat{D} f_{\mathbf{k}}^{(n)}$ because, as we shall see below, the latter can have contributions of all orders in Kn, i.e.,

$$\hat{D} f_{\mathbf{k}}^{(n)} = \left[\hat{D} f_{\mathbf{k}}^{(n)} \right]^{(1)} + \mathrm{Kn} \left[\hat{D} f_{\mathbf{k}}^{(n)} \right]^{(2)} + \cdots , \tag{5.41}$$

where $\left[\hat{D} f_{\mathbf{k}}^{(n)}\right]^{(m)}$ is the coefficient of order Kn^{n+m} in the Boltzmann equation (5.33). Therefore,

$$\left[\hat{D} f_{\mathbf{k}}\right]^{(0)} = 0 \,, \tag{5.42}$$

$$\left[\hat{D} f_{\mathbf{k}}\right]^{(1)} = \left[\hat{D} f_{\mathbf{k}}^{(0)}\right]^{(1)} \,, \tag{5.43}$$

$$\left[\hat{D} f_{\mathbf{k}}\right]^{(2)} = \left[\hat{D} f_{\mathbf{k}}^{(0)}\right]^{(2)} + \left[\hat{D} f_{\mathbf{k}}^{(1)}\right]^{(2)} \,, \tag{5.44}$$

and so on.

Let us now clarify why time derivatives contain also higher orders in the Knudsen number. To this end, we rewrite the conservation laws Eqs. (1.72)–(1.74), under the assumption $\text{Kn} = \overline{\text{Kn}}$ as

$$\hat{D}\varepsilon_0 + (\varepsilon_0 + P_0 + \Pi)\,\hat{\theta} - \pi^{\alpha\beta}\hat{\sigma}_{\alpha\beta} = 0 \,, \tag{5.45}$$

$$(\varepsilon_0 + P_0 + \Pi)\,\hat{D}u^\mu - \hat{\nabla}^\mu (P_0 + \Pi) - \pi^{\mu\beta}\hat{D}u_\beta + \Delta_\alpha^\mu \hat{\nabla}_\beta \pi^{\alpha\beta} = 0 \,, \tag{5.46}$$

$$\hat{D}n_0 + n_0\hat{\theta} - n^\mu \hat{D}u_\mu + \hat{\nabla}_\mu n^\mu = 0 \,. \tag{5.47}$$

Here, we introduced $\theta \equiv L^{-1}\hat{\theta}$ and $\sigma^{\mu\nu} \equiv L^{-1}\hat{\sigma}^{\mu\nu}$, with the expansion scalar θ and the shear tensor $\sigma^{\mu\nu}$ defined in Eqs. (1.30) and (1.75). In order to obtain Eqs. (5.46), (5.47) from Eqs. (1.73), (1.74) we also rewrote $\partial_\mu n^\mu$ and $\Delta_\alpha^\mu \partial_\beta \pi^{\alpha\beta}$ using Eq. (5.29) and the orthogonality relations $n^\mu u_\mu = 0$, $u_\beta \pi^{\alpha\beta} = 0$.

Let us now define the thermodynamic functions

$$I_{nq} \equiv \frac{1}{(2q+1)!!} \int dK \, f_{0\mathbf{k}} E_{\mathbf{k}}^{n-2q} \left(E_{\mathbf{k}}^2 - m^2\right)^q \,, \tag{5.48}$$

$$J_{nq} \equiv \frac{1}{(2q+1)!!} \int dK \, f_{0\mathbf{k}} \tilde{f}_{0\mathbf{k}} E_{\mathbf{k}}^{n-2q} \left(E_{\mathbf{k}}^2 - m^2\right)^q \,. \tag{5.49}$$

Note that the function I_{nq} was already introduced in Eq. (4.62) in Chap. 4. From Eqs. (5.23)–(5.25) we observe that

$$n_0 = I_{10} \,, \quad \varepsilon_0 = I_{20} \,, \quad P_0 = I_{21} \,. \tag{5.50}$$

From the definition of $f_{0\mathbf{k}}$, Eq. (5.22), we derive the identity

$$dI_{nq} = J_{nq} \, d\alpha_0 - J_{n+1,q} \, d\beta_0 \,. \tag{5.51}$$

Using Eq. (5.22) in the definitions (5.48), (5.49) and integrating by parts also yields the identity

$$\beta_0 J_{nq} = I_{n-1,q-1} + (n-2q) I_{n-1,q} \,, \tag{5.52}$$

from which we deduce

$$I_{10} = \beta_0 J_{21} , \quad I_{20} + I_{21} = \beta_0 J_{31} . \tag{5.53}$$

From the definition (5.49), one also readily proves the identity

$$J_{nq} \equiv (2q + 3)J_{n,q+1} + m^2 J_{n-2,q} , \tag{5.54}$$

which also holds for the integrals I_{nq}. For further use, we also define the thermodynamic functions

$$G_{nm} \equiv J_{n0}J_{m0} - J_{n-1,0}J_{m+1,0} , \quad D_{nq} \equiv J_{n+1,q}J_{n-1,q} - J_{nq}^2 . \tag{5.55}$$

From Eqs. (5.50) and (5.51), we then derive the following thermodynamic relations,

$$d\alpha_0 = -\frac{J_{20}}{D_{20}}d\varepsilon_0 + \frac{J_{30}}{D_{20}}dn_0 , \tag{5.56}$$

$$d\beta_0 = -\frac{J_{10}}{D_{20}}d\varepsilon_0 + \frac{J_{20}}{D_{20}}dn_0 . \tag{5.57}$$

Rewriting the total derivative into a (dimensionless time-like derivative) and using the hydrodynamic equations of motion (5.45) and (5.47), we finally obtain equations of motion for α_0 and β_0,

$$\hat{D}\alpha_0 = \frac{1}{D_{20}} \left\{ [(\varepsilon_0 + P_0) J_{20} - n_0 J_{30}] \hat{\theta} + J_{20} \left(\Pi\hat{\theta} - \pi^{\alpha\beta}\hat{\sigma}_{\alpha\beta} \right) \right.$$
$$\left. - J_{30} \left(\hat{\nabla}_\mu n^\mu - n^\mu \hat{D}u_\mu \right) \right\} , \tag{5.58}$$

$$\hat{D}\beta_0 = \frac{1}{D_{20}} \left\{ [(\varepsilon_0 + P_0) J_{10} - n_0 J_{20}] \hat{\theta} + J_{10} \left(\Pi\hat{\theta} - \pi^{\alpha\beta}\hat{\sigma}_{\alpha\beta} \right) \right.$$
$$\left. - J_{20} \left(\hat{\nabla}_\mu n^\mu - n^\mu \hat{D}u_\mu \right) \right\} . \tag{5.59}$$

We see that the time-like derivatives are proportional to terms of first order in gradients (the terms $\sim \hat{\theta}$) or, equivalently, proportional to one power of the Knudsen number, respectively, but that also terms enter which are proportional to gradients times dissipative quantities (Π, n^μ, $\pi^{\mu\nu}$), which (according to Navier–Stokes theory) are themselves at least of first order in the Knudsen number. Therefore, time derivatives of a quantity of given order in Kn in general involve also higher orders in Kn, as written formally in Eq. (5.41) for $f_{\mathbf{k}}^{(n)}$.

5.2.1 Solving the Chapman–Enskog Expansion: Zeroth- and First-Order Solutions

The solution of Eq. (5.38) is well known, and is given by the local-equilibrium single-particle distribution function (5.22),

$$f_{\mathbf{k}}^{(0)} = f_{0\mathbf{k}} = [\exp{(\beta_0 E_{\mathbf{k}} - \alpha_0)} + a]^{-1}, \tag{5.60}$$

where α_0, β_0, and u^μ are functions of space and time (global equilibrium corresponds to the particular case where there is no space-time dependence). Therefore, the zeroth-order truncation of the gradient expansion leads to the equations of ideal fluid dynamics, with all conserved currents given by their respective equilibrium values. In this case, we have already demonstrated in Chap. 1 that the continuity equations describing net-charge and energy-momentum conservation are sufficient to describe the dynamics of the fluid, since the pressure is determined by an equation of state (here, the equation of state is the one of a dilute gas, given by $P_0 = I_{21}(\alpha_0, \beta_0)$).

However, ideal fluid dynamics does not arise from a solution of the Boltzmann equation since $f_{0\mathbf{k}}$ alone does not satisfy this equation: the collision term vanishes, Eq. (5.38), but the left-hand side of the Boltzmann equation obviously does not vanish if α_0, β_0, and u^μ are functions of space and time. On the other hand, the above result implies that ideal fluid dynamics can be interpreted as the zeroth-order truncation of an expansion in powers of Knudsen number. In this sense, it should be understood solely as an approximate solution, that will never be exactly realized but, in practice, may lead to a reliable description of several physical systems.

The local-equilibrium variables α_0 and β_0 are defined by the matching conditions (5.23), while the fluid 4-velocity is defined by the choice of frame (in our case the Landau frame, Eq. (5.16)). In order to satisfy these constraints, for all $n \geq 1$ the corrections $f_{\mathbf{k}}^{(n)}$ must satisfy

$$\int dK \, E_{\mathbf{k}} \, f_{\mathbf{k}}^{(n)} = 0, \tag{5.61}$$

$$\int dK \, E_{\mathbf{k}}^2 \, f_{\mathbf{k}}^{(n)} = 0, \tag{5.62}$$

$$\int dK \, E_{\mathbf{k}} k^{\langle \mu \rangle} \, f_{\mathbf{k}}^{(n)} = 0. \tag{5.63}$$

The last condition can be obtained by projecting Eq. (5.16) onto Δ_μ^α and ensures that there is no flow of energy relative to u^μ. These conditions guarantee that, to any order of approximation, the solution depends solely on α_0, β_0, and u^μ, and their gradients. Also, they remove the freedom that we could add the solution of the homogeneous Boltzmann equation to the solution of the perturbative series (5.34), i.e., a global-equilibrium distribution function with another temperature, chemical potential, and velocity.

In the following, we shall construct the solution of the Chapman–Enskog expansion to first order in Knudsen number, i.e., the solution to Eq. (5.39). Using Eqs. (5.43) and (5.60), as well as simplifying Eq. (5.37) using Eq. (5.60), we must solve

$$\left[\hat{D}f_{0\mathbf{k}}\right]^{(1)} + \frac{1}{E_{\mathbf{k}}} k^{\mu}\hat{\nabla}_{\mu} f_{0\mathbf{k}} = -\lambda \hat{C} f_{\mathbf{k}}^{(1)} , \tag{5.64}$$

where we defined the *linear* collision operator acting on $f_{\mathbf{k}}^{(1)}$ as

$$\hat{C} f_{\mathbf{k}}^{(1)} \equiv \frac{1}{\nu E_{\mathbf{k}}} \int dK' dP dP' W_{\mathbf{kk}'\to\mathbf{pp}'} f_{0\mathbf{k}} f_{0\mathbf{k}'} \tilde{f}_{0\mathbf{p}} \tilde{f}_{0\mathbf{p}'}$$

$$\times \left(\frac{f_{\mathbf{k}}^{(1)}}{f_{0\mathbf{k}}\tilde{f}_{0\mathbf{k}}} + \frac{f_{\mathbf{k}'}^{(1)}}{f_{0\mathbf{k}'}\tilde{f}_{0\mathbf{k}'}} - \frac{f_{\mathbf{p}}^{(1)}}{f_{0\mathbf{p}}\tilde{f}_{0\mathbf{p}}} - \frac{f_{\mathbf{p}'}^{(1)}}{f_{0\mathbf{p}'}\tilde{f}_{0\mathbf{p}'}}\right) . \tag{5.65}$$

The time-like and space-like derivatives of $f_{0\mathbf{k}}$ appearing on the left-hand side of Eq. (5.64) are straightforward to calculate using Eq. (5.60),

$$\hat{D} f_{0\mathbf{k}} = -f_{0\mathbf{k}} \tilde{f}_{0\mathbf{k}} \left(E_{\mathbf{k}}\hat{D}\beta_0 + \beta_0 k^{\langle\nu\rangle}\hat{D}u_{\nu} - \hat{D}\alpha_0\right) , \tag{5.66}$$

$$\hat{\nabla}_{\mu} f_{0\mathbf{k}} = -f_{0\mathbf{k}} \tilde{f}_{0\mathbf{k}} \left(E_{\mathbf{k}}\hat{\nabla}_{\mu}\beta_0 + \beta_0 k^{\langle\nu\rangle}\hat{\nabla}_{\mu}u_{\nu} - \hat{\nabla}_{\mu}\alpha_0\right) , \tag{5.67}$$

where we used $u^{\nu}\hat{D}u_{\nu} = u^{\nu}\hat{\nabla}_{\mu}u_{\nu} = 0$.

All time-like derivatives of α_0, β_0, and u^{μ} that appear in Eq. (5.66) can be replaced by space-like gradients using the conservation equation (5.46), as well as Eqs. (5.58), (5.59), respectively. However, note that the bulk-viscous pressure, particle-diffusion 4-current, and shear-stress tensor do not have zeroth-order contributions in the Knudsen number, because they vanish in equilibrium,

$$(\Pi)^{(0)} = -\frac{1}{3}\left\langle\Delta^{\mu\nu}k_{\mu}k_{\nu}\right\rangle_0 - P_0 = 0 , \tag{5.68}$$

$$(n^{\mu})^{(0)} = \left\langle k^{\langle\mu\rangle}\right\rangle_0 = 0 , \tag{5.69}$$

$$(\pi^{\mu\nu})^{(0)} = \left\langle k^{\langle\mu} k^{\nu\rangle}\right\rangle_0 = 0 , \tag{5.70}$$

i.e., the dissipative currents are *at least* of first order in Knudsen number. Therefore, substituting these results into Eqs. (5.46), (5.58), and (5.59), one obtains

$$\hat{D}\alpha_0 = \frac{(\varepsilon_0 + P_0) J_{20} - n_0 J_{30}}{D_{20}}\hat{\theta} + O\left(\text{Kn}^2\right) , \tag{5.71}$$

$$\hat{D}\beta_0 = \frac{(\varepsilon_0 + P_0) J_{10} - n_0 J_{20}}{D_{20}}\hat{\theta} + O\left(\text{Kn}^2\right) , \tag{5.72}$$

$$\hat{D}u^{\mu} = \frac{1}{\varepsilon_0 + P_0}\hat{\nabla}^{\mu}P_0 + O\left(\text{Kn}^2\right). \tag{5.73}$$

Therefore, to first order in Knudsen number, the comoving time derivatives of α_0, β_0, and u^μ are linearly proportional to their spatial gradients. This whole procedure illustrates the subtleties involved in replacing time derivatives by spatial ones in Chapman–Enskog theory. To first order, the procedure is fairly simple, but it can get rather complicated when one goes to higher orders, since it is extremely difficult to establish *a priori* what is the general structure of the contributions of higher order in Knudsen number.

Then, Eq. (5.64) can be reduced to

$$f_{0\mathbf{k}} \tilde{f}_{0\mathbf{k}} \left(A_{\mathbf{k}} \hat{\theta} + B_{\mathbf{k}} k^{\langle \mu \rangle} \hat{\nabla}_\mu \alpha_0 + \frac{\beta_0}{E_{\mathbf{k}}} k^{\langle \mu} k^{\nu \rangle} \hat{\sigma}_{\mu\nu} \right) = \lambda \hat{C} f_{\mathbf{k}}^{(1)} , \qquad (5.74)$$

where we made use of the thermodynamic relation (1.97) and the fact that $k^{\langle \mu \rangle} k^{\langle \nu \rangle} = k^{\langle \mu} k^{\nu \rangle} + \Delta^{\mu\nu} \left(\Delta^{\alpha\beta} k_\alpha k_\beta \right) /3$. We furthermore introduced the scalar functions

$$A_{\mathbf{k}} \equiv \frac{(\varepsilon_0 + P_0) J_{10} - n_0 J_{20}}{D_{20}} E_{\mathbf{k}} - \frac{(\varepsilon_0 + P_0) J_{20} - n_0 J_{30}}{D_{20}} + \frac{\beta_0}{3E_{\mathbf{k}}} \Delta^{\alpha\beta} k_\alpha k_\beta ,$$
$$(5.75)$$

$$B_{\mathbf{k}} \equiv h_0^{-1} - \frac{1}{E_{\mathbf{k}}} . \qquad (5.76)$$

We note that the linear operator $\hat{C}$, defined in Eq. (5.65), has five degenerate eigenfunctions (1, $E_{\mathbf{k}}$, and $k^{\langle \mu \rangle}$, multiplied by $f_{0\mathbf{k}} \tilde{f}_{0\mathbf{k}}$) with eigenvalues equal to zero

$$\hat{C}(f_{0\mathbf{k}} \tilde{f}_{0\mathbf{k}} \, 1) = 0 , \quad \hat{C}(f_{0\mathbf{k}} \tilde{f}_{0\mathbf{k}} \, E_{\mathbf{k}}) = 0 , \quad \hat{C}(f_{0\mathbf{k}} \tilde{f}_{0\mathbf{k}} \, k^{\langle \mu \rangle}) = 0 . \qquad (5.77)$$

These correspond to quantities which are conserved in microscopic collisions.

Nevertheless, solving Eq. (5.74) is still a complicated task because one needs to invert the collision operator $\hat{C}$ in order to obtain $f_{\mathbf{k}}^{(1)}$. Still, since the equation is linear in $f_{\mathbf{k}}^{(1)}$ and the left-hand side is linear in the gradients $\hat{\theta}$, $\hat{\nabla}_\mu \alpha_0$, and $\hat{\sigma}_{\mu\nu}$, one already knows that the general solution for $f_{\mathbf{k}}^{(1)}$ must have the following form,

$$\frac{f_{\mathbf{k}}^{(1)}}{f_{0\mathbf{k}} \tilde{f}_{0\mathbf{k}}} = \varphi_{\mathbf{k}}^{\mathrm{s}} \hat{\theta} + \varphi_{\mathbf{k}}^{\mathrm{v}} \beta_0 k^{\langle \mu \rangle} \hat{\nabla}_\mu \alpha_0 + \varphi_{\mathbf{k}}^{\mathrm{t}} \beta_0^2 k^{\langle \mu} k^{\nu \rangle} \hat{\sigma}_{\mu\nu} + \varphi_{\mathbf{k}}^{\mathrm{hom}} . \qquad (5.78)$$

The functions $\varphi_{\mathbf{k}}^i$, $i = $s,v,t, were constructed to be dimensionless and depend on momentum only through $E_{\mathbf{k}}$, i.e., $\varphi_{\mathbf{k}}^i = \varphi_{\mathbf{k}}^i (E_{\mathbf{k}})$. The function $\varphi_{\mathbf{k}}^{\mathrm{hom}}$ is the homogeneous solution,

$$\hat{C} \varphi_{\mathbf{k}}^{\mathrm{hom}} = 0 , \qquad (5.79)$$

which is constructed as a linear combination of the eigenfunctions 1, $E_{\mathbf{k}}$, and $k^{\langle \mu \rangle}$,

$$\varphi_{\mathbf{k}}^{\mathrm{hom}} = a_0 + a_1 E_{\mathbf{k}} + a_{2\mu} k^{\langle \mu \rangle} . \qquad (5.80)$$

The coefficients a_0, a_1, and $a_{2\mu}$ must be determined using the matching conditions (5.61)–(5.63).

For the following, it is advantageous to define

$$\alpha_r^s = \int dK \, E_{\mathbf{k}}^r A_{\mathbf{k}} f_{0\mathbf{k}} \tilde{f}_{0\mathbf{k}}$$

$$= \frac{(\varepsilon_0 + P_0) J_{10} - n_0 J_{20}}{D_{20}} J_{r+1,0} - \frac{(\varepsilon_0 + P_0) J_{20} - n_0 J_{30}}{D_{20}} J_{r,0} - \beta_0 J_{r+1,1} \,,$$

$$(5.81)$$

$$\alpha_r^v = \frac{1}{3} \int dK \, E_{\mathbf{k}}^r B_{\mathbf{k}} \left(\Delta^{\alpha\beta} k_\alpha k_\beta \right) f_{0\mathbf{k}} \tilde{f}_{0\mathbf{k}} = J_{r+1,1} - h_0^{-1} J_{r+2,1} \,, \tag{5.82}$$

$$\alpha_r^t = \frac{2}{15} \beta_0 \int dK \, E_{\mathbf{k}}^{r-1} \left(\Delta^{\alpha\beta} k_\alpha k_\beta \right)^2 f_{0\mathbf{k}} \tilde{f}_{0\mathbf{k}} = 2\beta_0 J_{r+3,2} \,, \tag{5.83}$$

where we used Eqs. (5.49), (5.75), and (5.76). We note that, in the massless limit, α_r^s vanishes for any value of r. This can be proven noting that, for $m = 0$, $I_{n0} = 3I_{n1}$, $J_{n0} = 3J_{n1}$, cf. Eqs. (5.48), (5.49), and using the relations (1.97), (5.51), and (5.53).

Then, inserting the Ansatz (5.78) into Eq. (5.74), multiplying by an arbitrary power of energy, $E_{\mathbf{k}}^r$, and integrating over momentum, dK, one obtains the equation satisfied by $\varphi_{\mathbf{k}}^s$,

$$\alpha_r^s = \frac{\lambda}{\nu} \int dK dK' dP dP' \, W_{\mathbf{k}\mathbf{k}' \to \mathbf{p}\mathbf{p}'} f_{0\mathbf{k}} f_{0\mathbf{k}'} \tilde{f}_{0\mathbf{p}} \tilde{f}_{0\mathbf{p}'} E_{\mathbf{k}}^{r-1} \left(\varphi_{\mathbf{k}}^s + \varphi_{\mathbf{k}'}^s - \varphi_{\mathbf{p}}^s - \varphi_{\mathbf{p}'}^s \right) \,. \tag{5.84}$$

Here, we have used orthogonality relations on the left- and right-hand sides. On the left-hand side, we have used the relation

$$\int dK \, \mathrm{F}(E_{\mathbf{k}}) k^{\langle \mu_1} \cdots k^{\mu_m \rangle} k_{\langle \nu_1} \cdots k_{\nu_n \rangle}$$

$$= \frac{n! \, \delta_{mn}}{(2n+1)!!} \Delta_{\nu_1 \cdots \nu_n}^{\mu_1 \cdots \mu_m} \int dK \, \mathrm{F}(E_{\mathbf{k}}) \left(\Delta^{\alpha\beta} k_\alpha k_\beta \right)^n \,, \tag{5.85}$$

for an arbitrary function $\mathrm{F}(E_{\mathbf{k}})$, which depends only on $E_{\mathbf{k}} = k^\mu u_\mu$. The proof of this relation can be found in Appendix 6.7. To obtain the right-hand side of Eq. (5.84), we have used the fact that the tensor structure of a quantity

$$(\mathcal{A}_r)_{\nu_1 \cdots \nu_n}^{\mu_1 \cdots \mu_m} \equiv \frac{1}{\nu} \int dK dK' dP dP' \, W_{\mathbf{k}\mathbf{k}' \to \mathbf{p}\mathbf{p}'} f_{0\mathbf{k}} f_{0\mathbf{k}'} \tilde{f}_{0\mathbf{p}} \tilde{f}_{0\mathbf{p}'} E_{\mathbf{k}}^{r-1} k^{\langle \mu_1} \cdots k^{\mu_m \rangle}$$

$$\times \left(\mathrm{H}_{\mathbf{k}} k_{\langle \nu_1} \cdots k_{\nu_n \rangle} + \mathrm{H}_{\mathbf{k}'} k'_{\langle \nu_1} \cdots k'_{\nu_n \rangle} - \mathrm{H}_{\mathbf{p}} p_{\langle \nu_1} \cdots p_{\nu_n \rangle} - \mathrm{H}_{\mathbf{p}'} p'_{\langle \nu_1} \cdots p'_{\nu_n \rangle} \right) \,, \tag{5.86}$$

where $\mathrm{H}_{\mathbf{k}}$ is an arbitrary function of $E_{\mathbf{k}}$, can only be of a form which satisfies

$$(\mathcal{A}_r)_{\nu_1 \cdots \nu_n}^{\mu_1 \cdots \mu_m} = \delta_{mn} \mathcal{A}_r^{(n)} \Delta_{\nu_1 \cdots \nu_n}^{\mu_1 \cdots \mu_m} \,, \tag{5.87}$$

with

$$
\mathcal{A}_r^{(n)} = \frac{1}{\nu\,(2n+1)} \int dK dK' dP dP'\, W_{\mathbf{kk'}\to\mathbf{pp'}} f_{0\mathbf{k}} f_{0\mathbf{k'}} \tilde{f}_{0\mathbf{p}} \tilde{f}_{0\mathbf{p'}} E_{\mathbf{k}}^{r-1} k^{\langle\mu_1} \cdots k^{\mu_n\rangle}
$$
$$
\times \left(H_{\mathbf{k}}\, k_{\langle\mu_1} \cdots k_{\mu_n\rangle} + H_{\mathbf{k'}}\, k'_{\langle\mu_1} \cdots k'_{\mu_n\rangle} - H_{\mathbf{p}}\, p_{\langle\mu_1} \cdots p_{\mu_n\rangle} - H_{\mathbf{p'}}\, p'_{\langle\mu_1} \cdots p'_{\mu_n\rangle} \right),
$$

$$(5.88)$$

cf. Eqs. (6.45) and (6.47) in Chap. 6, where these relations are explicitly proven.

Similarly, multiplying Eq. (5.74) by $E_{\mathbf{k}}^r k^{\langle\nu\rangle}$ and integrating over dK, one obtains the equation for $\varphi_{\mathbf{k}}^{\mathrm{v}}$,

$$
\alpha_r^{\mathrm{v}} = \frac{\beta_0 \lambda}{3\nu} \int dK dK' dP dP'\, W_{\mathbf{kk'}\to\mathbf{pp'}} f_{0\mathbf{k}} f_{0\mathbf{k'}} \tilde{f}_{0\mathbf{p}} \tilde{f}_{0\mathbf{p'}}
$$
$$
\times E_{\mathbf{k}}^{r-1} k_{\langle\mu\rangle} \left(\varphi_{\mathbf{k}}^{\mathrm{v}} k^{\langle\mu\rangle} + \varphi_{\mathbf{k'}}^{\mathrm{v}} k'^{\langle\mu\rangle} - \varphi_{\mathbf{p}}^{\mathrm{v}} p^{\langle\mu\rangle} - \varphi_{\mathbf{p'}}^{\mathrm{v}} p'^{\langle\mu\rangle} \right).
$$

$$(5.89)$$

Finally, multiplying by $E_{\mathbf{k}}^r k^{\langle\alpha} k^{\beta\rangle}$ and once more integrating over dK, one obtains the equation for $\varphi_{\mathbf{k}}^{\mathrm{t}}$,

$$
\alpha_r^{\mathrm{t}} = \frac{\beta_0^2 \lambda}{5\nu} \int dK dK' dP dP'\, W_{\mathbf{kk'}\to\mathbf{pp'}} f_{0\mathbf{k}} f_{0\mathbf{k'}} \tilde{f}_{0\mathbf{p}} \tilde{f}_{0\mathbf{p'}}
$$
$$
\times E_{\mathbf{k}}^{r-1} k_{\langle\mu} k_{\nu\rangle} \left(\varphi_{\mathbf{k}}^{\mathrm{t}} k^{\langle\mu} k^{\nu\rangle} + \varphi_{\mathbf{k'}}^{\mathrm{t}} k'^{\langle\mu} k'^{\nu\rangle} - \varphi_{\mathbf{p}}^{\mathrm{t}} p^{\langle\mu} p^{\nu\rangle} - \varphi_{\mathbf{p'}}^{\mathrm{t}} p'^{\langle\mu} p'^{\nu\rangle} \right).
$$

$$(5.90)$$

The next step is to expand the functions $\varphi_{\mathbf{k}}^i$ in terms of a complete basis formed of $E_{\mathbf{k}}$. The most common approach is to simply take a power series,

$$
\varphi_{\mathbf{k}}^i = \sum_{n=0}^{N^i} \epsilon_n^i E_{\mathbf{k}}^n,
$$

$$(5.91)$$

where for practical purposes the series is truncated at some power $N^i < \infty$. Note that for the scalar contribution ($i = \mathrm{s}$), the terms $n = 0, 1$ of the expansion can be incorporated into the homogeneous solution, while for the vector contribution ($i = \mathrm{v}$) the same occurs for the $n = 0$ term of the expansion. Overall, this reduces the problem of solving for the remaining coefficients ϵ_n^i. Inserting the expansion (5.91) into Eqs. (5.84), (5.89), and (5.90) leads to the following linear algebraic relations for the ϵ_n^i,

$$
\alpha_r^i = \lambda \sum_{n=0}^{N^i} \mathcal{A}_{rn}^i \epsilon_n^i, \quad i = \mathrm{s}, \mathrm{v}, \mathrm{t}.
$$

$$(5.92)$$

The dimension of the matrices $\mathcal{A}_{rn}^i$ is determined by N^i; the more terms one includes in the expansion (5.91), the higher the dimension of the matrix becomes. In practice, one can never include an infinite number of terms in the series, but one can check its convergence and truncate at that N^i for which $\varphi_{\mathbf{k}}^i$ reaches the required accuracy (up to a given value of momentum).

The matrices $\mathcal{A}_{rn}^i$ are defined as

$$
\mathcal{A}_{rn}^s = \frac{1}{\nu} \int dK dK' dP dP' \, W_{\mathbf{k}\mathbf{k}' \to \mathbf{p}\mathbf{p}'} f_{0\mathbf{k}} f_{0\mathbf{k}'} \tilde{f}_{0\mathbf{p}} \tilde{f}_{0\mathbf{p}'} E_{\mathbf{k}}^{r-1} \left(E_{\mathbf{k}}^n + E_{\mathbf{k}'}^n - E_{\mathbf{p}}^n - E_{\mathbf{p}'}^n \right) ,
$$

$$
\mathcal{A}_{rn}^v = \frac{\beta_0}{3\nu} \int dK dK' dP dP' \, W_{\mathbf{k}\mathbf{k}' \to \mathbf{p}\mathbf{p}'} f_{0\mathbf{k}} f_{0\mathbf{k}'} \tilde{f}_{0\mathbf{p}} \tilde{f}_{0\mathbf{p}'}
$$
$$
\times E_{\mathbf{k}}^{r-1} k_{\langle \mu \rangle} \left(E_{\mathbf{k}}^n k^{\langle \mu \rangle} + E_{\mathbf{k}'}^n k'^{\langle \mu \rangle} - E_{\mathbf{p}}^n p^{\langle \mu \rangle} - E_{\mathbf{p}'}^n p'^{\langle \mu \rangle} \right) , \tag{5.93}
$$

$$
\mathcal{A}_{rn}^t = \frac{\beta_0^2}{5\nu} \int dK dK' dP dP' \, W_{\mathbf{k}\mathbf{k}' \to \mathbf{p}\mathbf{p}'} f_{0\mathbf{k}} f_{0\mathbf{k}'} \tilde{f}_{0\mathbf{p}} \tilde{f}_{0\mathbf{p}'}
$$
$$
\times E_{\mathbf{k}}^{r-1} k_{\langle \mu} k_{\nu \rangle} \left(E_{\mathbf{k}}^n k^{\langle \mu} k^{\nu \rangle} + E_{\mathbf{k}'}^n k'^{\langle \mu} k'^{\nu \rangle} - E_{\mathbf{p}}^n p^{\langle \mu} p^{\nu \rangle} - E_{\mathbf{p}'}^n p'^{\langle \mu} p'^{\nu \rangle} \right) .
$$

5.2.2 Minimal Truncation Scheme

The conservation of the number of particles (in binary collisions), energy, and momentum make the following components of the collision matrices $\mathcal{A}^s$ and $\mathcal{A}^v$ vanish: $\mathcal{A}_{r0}^s$, $\mathcal{A}_{r1}^s$, $\mathcal{A}_{1n}^s$, $\mathcal{A}_{2n}^s$, $\mathcal{A}_{r0}^v$, and $\mathcal{A}_{1n}^v$. Hence, these matrices have some rows and columns with all their entries being zero. Furthermore, from the definitions (5.81), (5.82), the identities (5.50), (5.51), and (5.53), as well as the thermodynamical relation (1.97) one can prove that the components α_1^s, α_2^s, and α_1^v are also zero and, consequently, Eq. (5.92) becomes a trivial identity for $r = 1, 2$ ($i =$ s) and $r = 1$ ($i =$ v). When inverting Eq. (5.92), these trivial lines and columns must be removed, making it possible to solve for all, except three, expansion coefficients ϵ_n^i. The three coefficients that are undetermined in Eq. (5.92) must be obtained from the matching conditions—they actually correspond to terms that can be incorporated into the homogeneous solution specified in Eq. (5.80). For this reason, when truncating the expansion (5.91), one must have at least $N^s \geq 2$ and $N^v \geq 1$.

In order to better understand this, let us assume the simplest possible truncation scheme for Eq. (5.91), i.e., $N^s = 2$, $N^v = 1$, and $N^t = 0$. Then, the non-trivial lines of Eq. (5.92) (obtained by removing all contributions related to the homogeneous solution) lead to the following equations for ϵ_2^s, ϵ_1^v, and ϵ_0^t,

$$
\epsilon_2^s = \frac{\alpha_0^s}{\lambda \mathcal{A}_{02}^s} , \quad \epsilon_1^v = \frac{\alpha_0^v}{\lambda \mathcal{A}_{01}^v} , \quad \epsilon_0^t = \frac{\alpha_0^t}{\lambda \mathcal{A}_{00}^t} , \tag{5.94}
$$

leaving ϵ_0^s, ϵ_1^s, and ϵ_0^v still undetermined. On the other hand, the matching conditions (5.61)–(5.63) provide the following additional constraints that must be satisfied by $\varphi_{\mathbf{k}}^s$ and $\varphi_{\mathbf{k}}^v$

$$\int dK \, E_{\mathbf{k}} \varphi_{\mathbf{k}}^{s} f_{0\mathbf{k}} \tilde{f}_{0\mathbf{k}} = 0 \,, \tag{5.95}$$

$$\int dK \, E_{\mathbf{k}}^{2} \varphi_{\mathbf{k}}^{s} f_{0\mathbf{k}} \tilde{f}_{0\mathbf{k}} = 0 \,, \tag{5.96}$$

$$\int dK \, E_{\mathbf{k}} \left(\Delta_{\mu\nu} k^{\mu} k^{\nu} \right) \varphi_{\mathbf{k}}^{v} f_{0\mathbf{k}} \tilde{f}_{0\mathbf{k}} = 0 \,. \tag{5.97}$$

These equations relate the so far undetermined coefficients to ϵ_2^s and ϵ_1^v,

$$\begin{pmatrix} J_{10} & J_{20} \\ J_{20} & J_{30} \end{pmatrix} \begin{pmatrix} \epsilon_0^s \\ \epsilon_1^s \end{pmatrix} = - \begin{pmatrix} J_{30} \\ J_{40} \end{pmatrix} \epsilon_2^s \,, \tag{5.98}$$

$$J_{31} \epsilon_0^v = -J_{41} \epsilon_1^v \,. \tag{5.99}$$

With Eq. (5.94), the solution is

$$\epsilon_0^s = \frac{D_{30}}{D_{20}} \frac{\alpha_0^s}{\lambda \mathcal{A}_{02}^s} \,, \tag{5.100}$$

$$\epsilon_1^s = \frac{G_{23}}{D_{20}} \frac{\alpha_0^s}{\lambda \mathcal{A}_{02}^s} \,, \tag{5.101}$$

$$\epsilon_0^v = -\frac{J_{41}}{J_{31}} \frac{\alpha_0^v}{\lambda \mathcal{A}_{01}^v} \,, \tag{5.102}$$

where we used the thermodynamic functions (5.55).

With all the coefficients of the truncated expansion solved for, we can write down the first-order solution for the bulk-viscous pressure,

$$\begin{aligned}
(\Pi)^{(1)} &= -\frac{1}{3} \left\langle \Delta^{\mu\nu} k_\mu k_\nu \right\rangle^{(1)} = -\frac{1}{3} \int dK \left(\Delta^{\mu\nu} k_\mu k_\nu \right) f_{\mathbf{k}}^{(1)} \\
&= -\frac{1}{3} \int dK \left(\Delta^{\mu\nu} k_\mu k_\nu \right) \varphi_{\mathbf{k}}^{s} f_{0\mathbf{k}} \tilde{f}_{0\mathbf{k}} \, \hat{\theta} \\
&= \left(\frac{J_{21} D_{30} + J_{31} G_{23}}{D_{20}} + J_{41} \right) \frac{\alpha_0^s}{\lambda \mathcal{A}_{02}^s} \, \hat{\theta} \,, \tag{5.103}
\end{aligned}$$

for the diffusion 4-current,

$$\begin{aligned}
(n^\mu)^{(1)} &= \left\langle k^{\langle \mu \rangle} \right\rangle^{(1)} = \int dK \, k^{\langle \mu \rangle} f_{\mathbf{k}}^{(1)} \\
&= \beta_0 \int dK \, k^{\langle \mu \rangle} k^{\langle \nu \rangle} \varphi_{\mathbf{k}}^{v} f_{0\mathbf{k}} \tilde{f}_{0\mathbf{k}} \, \hat{\nabla}_\nu \alpha_0 \\
&= \frac{D_{31}}{J_{31}} \frac{\beta_0 \alpha_0^v}{\lambda \mathcal{A}_{01}^v} \, \hat{\nabla}^\mu \alpha_0 \,, \tag{5.104}
\end{aligned}$$

and for the shear-stress tensor,

$$
\left(\pi^{\mu\nu}\right)^{(1)} = \left\langle k^{\langle\mu}k^{\nu\rangle}\right\rangle^{(1)} = \int dK\, k^{\langle\mu}k^{\nu\rangle}\, f_{\mathbf{k}}^{(1)}
$$
$$
= \beta_0^2 \int dK\, k^{\langle\mu}k^{\nu\rangle}k^{\langle\alpha}k^{\beta\rangle}\varphi_{\mathbf{k}}^{\mathrm{t}}\, f_{0\mathbf{k}}\tilde{f}_{0\mathbf{k}}\,\hat{\sigma}_{\alpha\beta} = 2J_{42}\frac{\beta_0^2\alpha_0^{\mathrm{t}}}{\lambda\mathscr{A}_{00}^{\mathrm{t}}}\,\hat{\sigma}^{\mu\nu}\,. \qquad (5.105)
$$

In order to obtain these results, we have again made use of the orthogonality condition (5.85).

Therefore, as already mentioned at the beginning of this section, in the Chapman–Enskog expansion relativistic Navier–Stokes theory appears as the first-order truncation of an expansion in powers of gradients of temperature, chemical potential, and velocity. With $\mathrm{Kn} = \lambda/L$, $\hat{\theta} = L\theta$, $\hat{\nabla}^\mu = L\nabla^\mu$, and $\hat{\sigma}^{\mu\nu} = L\sigma^{\mu\nu}$, we then identify by comparison with Eqs. (1.80), (1.81), and (1.82) the microscopic formulas for bulk-viscosity, particle-diffusion, and shear-viscosity coefficients as

$$
\zeta = -\left(\frac{J_{21}D_{30} + J_{31}G_{23}}{D_{20}} + J_{41}\right)\frac{\alpha_0^{\mathrm{s}}}{\mathscr{A}_{02}^{\mathrm{s}}}\,, \qquad (5.106)
$$

$$
\varkappa = \frac{D_{31}}{J_{31}}\frac{\beta_0\alpha_0^{\mathrm{v}}}{\mathscr{A}_{01}^{\mathrm{v}}}\,, \qquad (5.107)
$$

$$
\eta = J_{42}\frac{\beta_0^2\alpha_0^{\mathrm{t}}}{\mathscr{A}_{00}^{\mathrm{t}}}\,, \qquad (5.108)
$$

respectively. We note that the solution found for the shear viscosity is equivalent to the one obtained in Sect. 4.4.1 using linear-response theory, cf. Eq. (4.78). In order to see this, note that $\alpha_0^{\mathrm{t}} \equiv 2\beta_0 J_{32}$, cf. Eq. (5.83), such that $\eta = 2\beta_0^3 J_{42}J_{32}/\mathscr{A}_{00}^{\mathrm{t}}$. Furthermore, in the classical limit, $\lim_{a\to 0} J_{nq} = I_{nq}$, cf. Eqs. (5.48) and (5.49). The equivalence is then established by noting that (in the classical limit) $\mathscr{A}_{00}^{\mathrm{t}} = 2\beta_0^2\mathscr{A}^{00}$, with $\mathscr{A}^{00}$ defined in Eq. (4.70). Since in that calculation an equivalent truncation scheme was employed when inverting the collision operator, the agreement is not too surprising. Numerically more precise expressions for these transport coefficients can be obtained by including more terms in the power series (5.92).

5.3   Israel–Stewart Theory

In this section, we explain Israel's and Stewart's 14-moment approximation as it was originally proposed in Ref. [4]. This method is conceptually different from the Chapman–Enskog theory and is not constructed from an expansion in a small parameter, such as the Knudsen number. The main assumption made in this approach is that, since a fluid-dynamical description requires only the conserved currents N^μ and $T^{\mu\nu}$ to specify the state of the fluid, the single-particle distribution function should also be well described by these fields. In practice, this is accomplished by expanding $f_{\mathbf{k}}$ in terms of its moments and truncate it in such a way that it only depends on N^μ and $T^{\mu\nu}$. We shall describe the details of this approach next.

5.3.1   14-Moment Approximation

The starting point is the Israel–Stewart Ansatz for the non-equilibrium single-particle distribution function

$$f_{\mathbf{k}} = \left[\exp\left(-y_{\mathbf{k}}\right) + a\right]^{-1} , \tag{5.109}$$

where $a = 1$ ($a = -1$) for fermions (bosons) and $a = 0$ for a classical gas. In the traditional Israel–Stewart approach, the parameter $y_{\mathbf{k}}$ is expanded in momentum space around its local-equilibrium value, $y_{0\mathbf{k}} = \alpha_0 - \beta_0 u_\mu k^\mu$, in terms of a series of (reducible) Lorentz-tensors formed from the particle 4-momentum k^μ, i.e., 1, k^μ, $k^\mu k^\nu$, …. Therefore,

$$\delta y_{\mathbf{k}} \equiv y_{\mathbf{k}} - y_{0\mathbf{k}} = \epsilon + k^\mu \epsilon_\mu + k^\mu k^\nu \epsilon_{\mu\nu} + k^\mu k^\nu k^\lambda \epsilon_{\mu\nu\lambda} + \cdots . \tag{5.110}$$

For small momenta, the non-equilibrium single-particle distribution function can be further expanded around the local-equilibrium state

$$f_{\mathbf{k}} = f_{0\mathbf{k}} + f_{0\mathbf{k}} \tilde{f}_{0\mathbf{k}} \delta y_{\mathbf{k}} + O\left(\delta y_{\mathbf{k}}^2\right) . \tag{5.111}$$

In kinetic theory, the particle 4-current and the energy-momentum tensor are given by Eqs. (5.4) and (5.5). Substituting Eqs. (5.110) and (5.111) into these equations, we express the conserved currents in terms of the expansion coefficients $\epsilon_{\mu_1 \cdots \mu_m}$,

$$N^\mu = I_0^\mu + \epsilon J_0^\mu + J_0^{\mu\nu} \epsilon_\nu + J_0^{\mu\nu\lambda} \epsilon_{\nu\lambda} + J_0^{\mu\nu\lambda\rho} \epsilon_{\nu\lambda\rho} + \cdots , \tag{5.112}$$

$$T^{\mu\nu} = I_0^{\mu\nu} + \epsilon J_0^{\mu\nu} + J_0^{\mu\nu\lambda} \epsilon_\lambda + J_0^{\mu\nu\lambda\rho} \epsilon_{\lambda\rho} + J_0^{\mu\nu\lambda\rho\sigma} \epsilon_{\lambda\rho\sigma} + \cdots , \tag{5.113}$$

where we introduced the tensors,

$$I_0^{\alpha_1 \cdots \alpha_n} \equiv \int dK \, k^{\alpha_1} \cdots k^{\alpha_n} f_{0\mathbf{k}} , \tag{5.114}$$

$$J_0^{\alpha_1 \cdots \alpha_n} \equiv \int dK \, k^{\alpha_1} \cdots k^{\alpha_n} f_{0\mathbf{k}} \tilde{f}_{0\mathbf{k}} . \tag{5.115}$$

The terms that are not multiplied by the expansion coefficients ϵ, ϵ^μ, and $\epsilon^{\mu\nu}$ are identified as the equilibrium currents (1.20),

$$N_{\text{ideal}}^\mu = I_0^\mu , \tag{5.116}$$

$$T_{\text{ideal}}^{\mu\nu} = I_0^{\mu\nu} . \tag{5.117}$$

Naturally, the remaining terms originate from the dissipative corrections to $f_{\mathbf{k}}$. The tensors (5.114), (5.115) depend only on the thermal potential, α_0, the inverse temperature, β_0, and the 4-velocity, u^μ. Therefore, their tensor structure must be constructed solely from combinations of u^μ and the metric tensor, $g^{\mu\nu}$. For the sake of convenience, we tensor decompose the moments (5.114), (5.115) in terms of the fluid velocity u^μ and the projection operator $\Delta^{\mu\nu}$. One then obtains

$$
\begin{aligned}
I_0^\mu &= I_{10} u^\mu \, , \\
I_0^{\mu\nu} &= I_{20} u^\mu u^\nu - I_{21} \Delta^{\mu\nu} \, , \\
J_0^\mu &= J_{10} u^\mu \, , \\
J_0^{\mu\nu} &= J_{20} u^\mu u^\nu - J_{21} \Delta^{\mu\nu} \, , \\
J_0^{\mu\nu\lambda} &= J_{30} u^\mu u^\nu u^\lambda - 3 J_{31} u^{(\mu} \Delta^{\nu\lambda)} \, , \\
J_0^{\mu\nu\lambda\rho} &= J_{40} u^\mu u^\nu u^\lambda u^\rho - 6 J_{41} u^{(\mu} u^\nu \Delta^{\lambda\rho)} + 3 J_{42} \Delta^{(\mu\nu} \Delta^{\lambda\rho)} \, ,
\end{aligned} \tag{5.118}
$$

where the thermodynamic integrals I_{nq} and J_{nq} were defined in Eqs. (5.48) and (5.49). Due to Eq. (5.50), Eqs. (5.116) and (5.117) reduce to what was derived in Chap. 1, Eq. (1.20). The parentheses around the Lorentz indices indicate symmetrization of the tensor with respect to all indices (where only independent terms are counted), whereas the prefactor counts the number of (independent) terms resulting from this symmetrization.

Obtaining the expansion coefficients $\epsilon^{\mu_1 \cdots \mu_m}$ is not a trivial task. In the Chapman–Enskog expansion, these were expressed in terms of gradients of α_0, β_0, and u^μ using perturbation theory, by assuming an expansion in powers of Knudsen number. Following the developments made by Grad in the non-relativistic regime, Israel and Stewart proposed a different approach. Instead of a gradient expansion, they suggested an *ad hoc* truncation of the expansion (5.110) at second order in momentum, i.e., one only keeps the tensors 1, k^μ, and $k^\mu k^\nu$ in the expansion,

$$
\delta y_{\mathbf{k}} \approx \epsilon + k^\mu \epsilon_\mu + k^\mu k^\nu \epsilon_{\mu\nu} \, . \tag{5.119}
$$

Without loss of generality, $\epsilon_{\mu\nu}$ can be assumed to be symmetric and traceless, i.e., $\epsilon^{\mu\nu} = \epsilon^{\nu\mu}$ and $\epsilon_\mu^\mu = 0$ (the symmetry under $\mu \leftrightarrow \nu$ is obvious, since any antisymmetric part would vanish in the contraction with $k^\mu k^\nu$, while the trace of $\epsilon_{\mu\nu}$ can always be absorbed in the scalar coefficient ϵ). This leaves us with 14 unknown degrees of freedom in the expansion coefficients ϵ, ϵ_μ, and $\epsilon_{\mu\nu}$. We note that the equilibrium variables introduced in $y_{0\mathbf{k}}$, i.e., α_0, β_0, and u^μ, are defined by the matching conditions (5.23) and the definition of the local rest frame, e.g. the Landau choice (5.16).

5.3.2   Matching Procedure

The 14 degrees of freedom of the *truncated* expansion can be uniquely related to the 14 components of the particle 4-current, N^μ, and the energy-momentum tensor, $T^{\mu\nu}$, the so-called *matching procedure*. This procedure will generate a single-particle distribution function that is completely determined by the components of the conserved currents. Israel and Stewart expected this to be a good approximation in the fluid-dynamical regime, where N^μ and $T^{\mu\nu}$ are considered to be sufficient to describe the state of the system.

The expansion coefficients can be solved using the constraints already derived in Eqs. (5.16), (5.21), and (5.23), that is

$$\Delta_{\mu\nu}N^\nu = n_\mu , \tag{5.120}$$

$$\Delta^{\mu\nu}_{\alpha\beta}T^{\alpha\beta} = \pi^{\mu\nu} , \tag{5.121}$$

$$-\frac{1}{3}\Delta_{\mu\nu}\left(T^{\mu\nu} - T^{\mu\nu}_{\text{ideal}}\right) = \Pi , \tag{5.122}$$

$$u_\mu\left(N^\mu - N^\mu_{\text{ideal}}\right) = 0 , \tag{5.123}$$

$$u_\nu\left(T^{\mu\nu} - T^{\mu\nu}_{\text{ideal}}\right) = 0 . \tag{5.124}$$

By solving this set of 14 linear equations, the expansion coefficients ϵ, ϵ_μ, and $\epsilon_{\mu\nu}$ can be expressed in terms of the 14 (independent) variables α_0, β_0, Π, u^μ, n^μ, and $\pi^{\mu\nu}$. The relations (5.120)–(5.122) define the dissipative currents, while the restrictions (5.123), (5.124) come from the matching conditions and the Landau choice for the local rest frame. If we were using the Eckart frame, for example, we would have to use, instead of Eqs. (5.123) and (5.124),

$$N^\mu - N^\mu_{\text{ideal}} = 0 , \tag{5.125}$$

$$u_\mu u_\nu\left(T^{\mu\nu} - T^{\mu\nu}_{\text{ideal}}\right) = 0 . \tag{5.126}$$

Combining Eqs. (5.116), (5.117), and (5.118), and truncating the expansion at terms quadratic in 4-momentum, the conditions (5.120), (5.121), and (5.122) imply that

$$n^\mu = -J_{21}\Delta^{\mu\nu}\epsilon_\nu - 2J_{31}\Delta^{\mu\nu}u^\lambda\epsilon_{\nu\lambda} , \tag{5.127}$$

$$\pi^{\mu\nu} = 2J_{42}\Delta^{\mu\nu}_{\lambda\rho}\epsilon^{\lambda\rho} , \tag{5.128}$$

$$\Pi = J_{21}\epsilon + J_{31}u^\lambda\epsilon_\lambda + \left(J_{41} + \frac{5}{3}J_{42}\right)u^\lambda u^\rho\epsilon_{\lambda\rho} , \tag{5.129}$$

where in the last equation we have used the fact that $\epsilon_{\lambda\rho}$ is traceless. The expressions (5.127)–(5.129) motivate the following Ansatz for the expansion coefficients,

$$\epsilon = E_0\Pi ,$$

$$\epsilon_\lambda = D_0\Pi u_\lambda + D_1 n_\lambda , \tag{5.130}$$

$$\epsilon_{\lambda\rho} = B_0\left(\Delta_{\lambda\rho} - 3u_\lambda u_\rho\right)\Pi + B_1 u_{(\lambda}n_{\rho)} + B_2\pi_{\lambda\rho} .$$

Then, in order to determine the single-particle distribution function (5.111), we have to compute the coefficients E_0, D_0, D_1, B_0, B_1, and B_2. This can be done by substituting the Ansatz (5.130) into Eqs. (5.123), (5.124), (5.127), (5.128), and (5.129), leading to

$$J_{21}D_1 + J_{31}B_1 = -1 \,, \tag{5.131}$$

$$2J_{42}B_2 = 1 \,, \tag{5.132}$$

$$J_{21}E_0 + J_{31}D_0 - (3J_{41} + 5J_{42})\,B_0 = 1 \,, \tag{5.133}$$

$$J_{10}E_0 + J_{20}D_0 - 3\,(J_{30} + J_{31})\,B_0 = 0 \,, \tag{5.134}$$

$$J_{31}D_1 + J_{41}B_1 = 0 \,, \tag{5.135}$$

$$J_{20}E_0 + J_{30}D_0 - 3\,(J_{40} + J_{41})\,B_0 = 0 \,. \tag{5.136}$$

Equations (5.131), (5.132), and (5.133) come directly from (5.127), (5.128), and (5.129), respectively. Equations (5.134), (5.135), and (5.136) are consequences of Eqs. (5.123) and (5.124). The solution of this set of equations is

$$\frac{E_0}{3B_0} = m^2 + 4\,\frac{J_{31}J_{30} - J_{41}J_{20}}{D_{20}} \equiv -C_1 \,,$$

$$\frac{D_0}{3B_0} = -4\,\frac{J_{31}J_{20} - J_{41}J_{10}}{D_{20}} \equiv -C_2 \,,$$

$$B_0 = -\frac{1}{3C_1J_{21} + 3C_2J_{31} + 3J_{41} + 5J_{42}} \,,$$

$$B_1 = \frac{J_{31}}{D_{31}} \,,$$

$$D_1 = -\frac{J_{41}}{D_{31}} \,,$$

$$B_2 = \frac{1}{2J_{42}} \,. \tag{5.137}$$

In the first two equations, we have used the identity (5.54).

5.3.3   Moment Equations

Now that the momentum distribution function is known, we can calculate the equations of motion satisfied by the dissipative currents. Israel and Stewart derived the equations of motion for Π, n^μ, and $\pi^{\mu\nu}$ from the second moment of the Boltzmann equation [2–4, 11]

$$\partial_\mu \left\langle k^\mu k^\nu k^\lambda \right\rangle = \int dK \; k^\nu k^\lambda C\,[f] \,. \tag{5.138}$$

In this case, the equations for Π, n^μ, and $\pi^{\mu\nu}$ are obtained from the projections $u_\nu u_\lambda \partial_\mu \left\langle k^\mu k^\nu k^\lambda \right\rangle$, $\Delta^\alpha_\lambda u_\nu \partial_\mu \left\langle k^\mu k^\nu k^\lambda \right\rangle$, and $\Delta^{\alpha\beta}_{\nu\lambda} \partial_\mu \left\langle k^\mu k^\nu k^\lambda \right\rangle$, respectively, together with the 14-moment approximation, Eqs. (5.111), (5.119), and (5.130). These equations determine the time evolution of Π, n^μ, and $\pi^{\mu\nu}$ through their (projected) comoving derivatives, $D\Pi$, $Dn^{\langle\mu\rangle} \equiv \Delta^\mu_\nu Dn^\nu$, and $D\pi^{\langle\mu\nu\rangle} \equiv \Delta^{\mu\nu}_{\alpha\beta} D\pi^{\alpha\beta}$, respectively.

However, extracting the equations of motion from the second moment of the Boltzmann equation is just a choice. The 14-moment approximation itself does not specify which moment of the Boltzmann equation should be chosen to close the conservation laws—once the 14-moment approximation is employed *any* moment of the Boltzmann equation will lead to a closed set of equations for the dissipative currents [12, 13]. This happens because the distribution function itself is already a unique function of the fluid-dynamical variables. In general, the form of the equations of motion will always be the same, but the transport coefficients appearing in the final equations will depend on the choice of the moment [13]. This happens because the 14-moment approximation is not a truncation in Knudsen number and there is no moment of the Boltzmann equation that carries the complete contribution to each term.

In the following, we will slightly deviate from Israel and Stewart and derive the equations satisfied by Π, n^μ, and $\pi^{\mu\nu}$ following Ref. [12], where it was shown that using the second moment of the Boltzmann equation to obtain the equations of motion for the dissipative currents introduces an unnecessary ambiguity in the derivation of fluid dynamics. This happens because the exact equations of motion for Π, n^μ, and $\pi^{\mu\nu}$ are known and can be derived directly from the Boltzmann equation. Therefore, instead of choosing an arbitrary moment of the Boltzmann equation to derive such equations, one should just use the exact equations of motion for the dissipative currents.

Following Ref. [12], the comoving derivatives of Π, n^μ, and $\pi^{\mu\nu}$ are calculated exactly using

$$DΠ = -\frac{1}{3}m^2 \int dK\, D\delta f_\mathbf{k} \, , \tag{5.139}$$

$$Dn^{\langle\mu\rangle} = \int dK\, k^{\langle\mu\rangle}\, D\delta f_\mathbf{k} \, , \tag{5.140}$$

$$D\pi^{\langle\mu\nu\rangle} = \int dK\, k^{\langle\mu} k^{\nu\rangle}\, D\delta f_\mathbf{k} \, . \tag{5.141}$$

Note that in the first equation, we used the matching condition $\varepsilon = \varepsilon_0$, such that $\langle E_\mathbf{k}^2 \rangle_\delta = 0$. Then, replacing $D\delta f_\mathbf{k}$ by using the Boltzmann equation (4.41) in the form

$$D\delta f_\mathbf{k} = -Df_{0\mathbf{k}} - \frac{1}{E_\mathbf{k}} k^{\langle\mu\rangle} \nabla_\mu f_\mathbf{k} + \frac{1}{E_\mathbf{k}} C[f] \, , \tag{5.142}$$

we obtain the *exact* equations

$$DΠ + C = -\frac{m^2}{3}\alpha_0^s \theta - \left(\frac{2}{3} - \frac{m^2}{3}\frac{G_{20}}{D_{20}}\right)\Pi\theta - \frac{m^2}{3}\frac{G_{20}}{D_{20}}\pi^{\mu\nu}\sigma_{\mu\nu} - \frac{m^2}{3}\frac{G_{30}}{D_{20}}\partial_\mu n^\mu$$

$$+ \frac{m^4}{9}\left\langle E_\mathbf{k}^{-2}\right\rangle_\delta \theta + \frac{m^2}{3}\left\langle E_\mathbf{k}^{-2}k^{\langle\mu}k^{\nu\rangle}\right\rangle_\delta \sigma_{\mu\nu} + \frac{m^2}{3}\nabla_\mu \left\langle E_\mathbf{k}^{-1}k^{\langle\mu\rangle}\right\rangle_\delta \, , \tag{5.143}$$

$$Dn^{\langle\mu\rangle} - C^\mu = \alpha_0^v \nabla^\mu \alpha_0 + n^v \omega_v^\mu - n^\mu \theta - \frac{3}{5} n^v \sigma_v^\mu$$

$$+ \frac{\beta_0 J_{21}}{\varepsilon_0 + P_0} \left(\Pi D u^\mu - \nabla^\mu \Pi - \pi^{\mu\nu} D u_\nu + \Delta_v^\mu \nabla_\lambda \pi^{\lambda\nu} \right)$$

$$- \frac{m^2}{3} \left\langle E_{\mathbf{k}}^{-2} k^{\langle\mu\rangle} \right\rangle_\delta \theta - \Delta_\lambda^\mu \nabla_v \left\langle E_{\mathbf{k}}^{-1} k^{\langle\lambda} k^{v\rangle} \right\rangle_\delta - \frac{2m^2}{5} \left\langle E_{\mathbf{k}}^{-2} k^{\langle v\rangle} \right\rangle_\delta \sigma_v^\mu$$

$$- \frac{m^2}{3} \nabla^\mu \left\langle F_{\mathbf{k}}^{-1} \right\rangle_\delta - \left\langle E_{\mathbf{k}}^{-2} k^{\langle\mu} k^v k^{\lambda\rangle} \right\rangle_\delta \sigma_{\lambda v} , \qquad (5.144)$$

$$D\pi^{\langle\mu\nu\rangle} - C^{\mu\nu} = \alpha_0^t \sigma^{\mu\nu} - \frac{4}{3} \pi^{\mu\nu} \theta - \frac{10}{7} \pi^{\lambda\langle\mu} \sigma_\lambda^{v\rangle} + 2\pi^{\lambda\langle\mu} \omega_\lambda^{v\rangle} + \frac{6}{5} \Pi \sigma^{\mu\nu}$$

$$- \frac{4m^2}{7} \Delta_{\alpha\beta}^{\mu\nu} \left\langle E_{\mathbf{k}}^{-2} k^{\langle\lambda} k^{\alpha\rangle} \right\rangle_\delta \sigma_\lambda^\beta - \frac{2m^4}{15} \left\langle E_{\mathbf{k}}^{-2} \right\rangle_\delta \sigma^{\mu\nu}$$

$$- \frac{2m^2}{5} \Delta_{\alpha\beta}^{\mu\nu} \nabla^\alpha \left\langle E_{\mathbf{k}}^{-1} k^{\langle\beta\rangle} \right\rangle_\delta - \frac{m^2}{3} \left\langle E_{\mathbf{k}}^{-2} k^{\langle\mu} k^{v\rangle} \right\rangle_\delta \theta$$

$$- \left\langle E_{\mathbf{k}}^{-2} k^{\langle\mu} k^v k^\lambda k^{\rho\rangle} \right\rangle_\delta \sigma_{\lambda\rho} - \Delta_{\alpha\beta}^{\mu\nu} \nabla_\lambda \left\langle E_{\mathbf{k}}^{-1} k^{\langle\alpha} k^\beta k^{\lambda\rangle} \right\rangle_\delta . \qquad (5.145)$$

The functions α_0^i were defined previously in the context of the Chapman–Enskog expansion in Eqs. (5.81)–(5.83). We also used the equations of motion (5.46), (5.58), and (5.59) (in their unscaled versions), as well as the thermodynamic identity (1.97). We also used the relations

$$k^{\langle\mu\rangle} k^{\langle\nu\rangle} = k^{\langle\mu} k^{v\rangle} + \frac{1}{3} \Delta^{\mu\nu} \left(\Delta^{\alpha\beta} k_\alpha k_\beta \right) , \qquad (5.146)$$

$$k^{\langle\mu\rangle} k^{\langle\nu\rangle} k^{\langle\lambda\rangle} = k^{\langle\mu} k^v k^{\lambda\rangle} + \frac{1}{5} \left(\Delta^{\alpha\beta} k_\alpha k_\beta \right)$$
$$\times \left(\Delta^{\mu\nu} k^{\langle\lambda\rangle} + \Delta^{\mu\lambda} k^{\langle\nu\rangle} + \Delta^{\nu\lambda} k^{\langle\mu\rangle} \right) , \qquad (5.147)$$

$$k^{\langle\mu\rangle} k^{\langle\nu\rangle} k^{\langle\lambda\rangle} k^{\langle\rho\rangle} = k^{\langle\mu} k^v k^\lambda k^{\rho\rangle} + \frac{1}{7} \left(\Delta^{\alpha\beta} k_\alpha k_\beta \right)$$
$$\times \left(\Delta^{\mu\nu} k^{\langle\lambda} k^{\rho\rangle} + \Delta^{\mu\lambda} k^{\langle\nu} k^{\rho\rangle} + \Delta^{\mu\rho} k^{\langle\nu} k^{\lambda\rangle} \right.$$
$$\left. + \Delta^{\nu\lambda} k^{\langle\mu} k^{\rho\rangle} + \Delta^{\nu\rho} k^{\langle\mu} k^{\lambda\rangle} + \Delta^{\lambda\rho} k^{\langle\mu} k^{v\rangle} \right)$$
$$+ \frac{1}{15} \left(\Delta^{\alpha\beta} k_\alpha k_\beta \right)^2 \left(\Delta^{\mu\nu} \Delta^{\lambda\rho} + \Delta^{\mu\lambda} \Delta^{\nu\rho} + \Delta^{\mu\rho} \Delta^{\nu\lambda} \right) , (5.148)$$

where $k^{\langle\mu} k^v k^{\lambda\rangle} = \Delta_{\alpha\beta\gamma}^{\mu\nu\lambda} k^\alpha k^\beta k^\gamma$, $k^{\langle\mu} k^v k^\lambda k^{\rho\rangle} = \Delta_{\alpha\beta\gamma\delta}^{\mu\nu\lambda\rho} k^\alpha k^\beta k^\gamma k^\delta$, with the projection operators $\Delta_{\alpha\beta\gamma}^{\mu\nu\lambda}$ and $\Delta_{\alpha\beta\gamma\delta}^{\mu\nu\lambda\rho}$ defined in Appendix 6.7. Finally, we also defined the collision integrals,

$$C = \frac{m^2}{3} \int dK \, E_{\mathbf{k}}^{-1} C[f] \, , \tag{5.149}$$

$$C^\mu = \int dK \, E_{\mathbf{k}}^{-1} k^{\langle \mu \rangle} C[f] \, , \tag{5.150}$$

$$C^{\mu\nu} = \int dK \, E_{\mathbf{k}}^{-1} k^{\langle \mu} k^{\nu \rangle} C[f] \, . \tag{5.151}$$

5.3.4 Calculation of the Collision Integrals

In Israel–Stewart theory, the distribution function was expanded in powers of $\delta y_{\mathbf{k}}$, retaining only the first-order correction, cf. Eq. (5.111). For the sake of consistency, the same approximation has to be made in the calculation of the collision terms (5.149)–(5.151). In the end, this is equivalent to using the linearized collision operator in the integrals above. Up to terms of order $O\left(\delta y_{\mathbf{k}}^2\right)$, the collision terms reduce to

$$
\begin{aligned}
C = \frac{m^2}{3\nu} \int dK dK' dP dP' \\
\times \frac{1}{E_{\mathbf{k}}} W_{\mathbf{kk'} \to \mathbf{pp'}} f_{0\mathbf{k}} f_{0\mathbf{k'}} \tilde{f}_{0\mathbf{p}} \tilde{f}_{0\mathbf{p'}} \left(y_{\mathbf{p}} + y_{\mathbf{p'}} - y_{\mathbf{k}} - y_{\mathbf{k'}} \right) \, ,
\end{aligned}
\tag{5.152}
$$

$$
\begin{aligned}
C^\mu = \frac{1}{\nu} \int dK dK' dP dP' \\
\times \frac{k^{\langle \mu \rangle}}{E_{\mathbf{k}}} W_{\mathbf{kk'} \to \mathbf{pp'}} f_{0\mathbf{k}} f_{0\mathbf{k'}} \tilde{f}_{0\mathbf{p}} \tilde{f}_{0\mathbf{p'}} \left(y_{\mathbf{p}} + y_{\mathbf{p'}} - y_{\mathbf{k}} - y_{\mathbf{k'}} \right) \, ,
\end{aligned}
\tag{5.153}
$$

$$
\begin{aligned}
C^{\mu\nu} = \frac{1}{\nu} \int dK dK' dP dP' \\
\times \frac{k^{\langle \mu} k^{\nu \rangle}}{E_{\mathbf{k}}} W_{\mathbf{kk'} \to \mathbf{pp'}} f_{0\mathbf{k}} f_{0\mathbf{k'}} \tilde{f}_{0\mathbf{p}} \tilde{f}_{0\mathbf{p'}} \left(y_{\mathbf{p}} + y_{\mathbf{p'}} - y_{\mathbf{k}} - y_{\mathbf{k'}} \right) \, .
\end{aligned}
\tag{5.154}
$$

Here, we have written $y_{\mathbf{k}}$ instead of $\delta y_{\mathbf{k}}$, since $y_{0\mathbf{k}} = \alpha_0 - \beta_0 E_{\mathbf{k}}$ consists of two collision invariants (1 and $E_{\mathbf{k}}$), and consequently $y_{0\mathbf{p}} + y_{0\mathbf{p'}} - y_{0\mathbf{k}} - y_{0\mathbf{k'}} = 0$. Inserting the moment expansion (5.119) of $\delta y_{\mathbf{k}}$ into the collision integrals, they further simplify to

$$C = -\frac{4}{9} m^2 \mathcal{A}_{02}^{\mathrm{s}} u_\alpha u_\beta \epsilon^{\alpha\beta} \, , \tag{5.155}$$

$$C^\mu = -2 \frac{\mathcal{A}_{01}^{\mathrm{v}}}{\beta_0} u_\alpha \Delta_\beta^\mu \epsilon^{\alpha\beta} \, , \tag{5.156}$$

$$C^{\mu\nu} = -\frac{\mathcal{A}_{00}^{\mathrm{t}}}{\beta_0^2} \Delta_{\alpha\beta}^{\mu\nu} \epsilon^{\alpha\beta} \, , \tag{5.157}$$

where we have used the definition (5.93) of the matrix elements $\mathcal{A}_{mn}^{i}$, $i =$s,v,t. We note that the only term from the moment expansion that contributes to the collision integral is $k_\alpha k_\beta \epsilon^{\alpha\beta}$. The remaining two terms, ϵ and $\epsilon_\mu k^\mu$, are proportional to collision invariants, and consequently, make the collision integrals vanish. Furthermore, we have made use of the orthogonality relation (5.87) and of the tracelessness of $\epsilon^{\alpha\beta}$. Then, using Eq. (5.130), which expresses the expansion coefficient $\epsilon^{\mu\nu}$ in terms of Π, n^μ, and $\pi^{\mu\nu}$, we obtain the final expressions for the collision integrals C, C^μ, and $C^{\mu\nu}$ in the 14-moment approximation

$$C = \frac{4}{3}m^2 B_0 \mathcal{A}_{02}^{s}\Pi \ , \tag{5.158}$$

$$C^\mu = -B_1 \frac{\mathcal{A}_{01}^{v}}{\beta_0} n^\mu \ , \tag{5.159}$$

$$C^{\mu\nu} = -B_2 \frac{\mathcal{A}_{00}^{t}}{\beta_0^2} \pi^{\mu\nu} \ . \tag{5.160}$$

The thermodynamic functions B_0, B_1, and B_2 were calculated in Eq. (5.137) while the matrices $\mathcal{A}_{nm}^{i}$ are defined in Eq. (5.93). The dependence of the collision integrals on the particle cross sections come exclusively from $\mathcal{A}_{nm}^{i}$.

In addition, the following terms appearing in the exact equations of motion can be computed using Eqs. (5.85), (5.119), and (5.130) as

$$\left\langle E_{\mathbf{k}}^{-1}\right\rangle_\delta = \left[J_{-1,0}E_0 + J_{00}D_0 - 3\left(J_{11} + J_{10}\right)B_0\right]\Pi \equiv \gamma_1^{\Pi}\Pi \ , \tag{5.161}$$

$$\left\langle E_{\mathbf{k}}^{-2}\right\rangle_\delta = \left[J_{-2,0}E_0 + J_{-1,0}D_0 - 3\left(J_{01} + J_{00}\right)B_0\right]\Pi \equiv \gamma_2^{\Pi}\Pi \ , \tag{5.162}$$

$$\left\langle E_{\mathbf{k}}^{-1}k^{\langle\mu\rangle}\right\rangle_\delta = -\left(J_{11}D_1 + J_{21}B_1\right)n^\mu \equiv \gamma_1^{n}n^\mu \ , \tag{5.163}$$

$$\left\langle E_{\mathbf{k}}^{-2}k^{\langle\mu\rangle}\right\rangle_\delta = -\left(J_{01}D_1 + J_{11}B_1\right)n^\mu \equiv \gamma_2^{n}n^\mu \ , \tag{5.164}$$

$$\left\langle E_{\mathbf{k}}^{-1}k^{\langle\mu}k^{\nu\rangle}\right\rangle_\delta = \frac{J_{32}}{J_{42}}\pi^{\mu\nu} \equiv \gamma_1^{\pi}\pi^{\mu\nu} \ , \tag{5.165}$$

$$\left\langle E_{\mathbf{k}}^{-2}k^{\langle\mu}k^{\nu\rangle}\right\rangle_\delta = \frac{J_{22}}{J_{42}}\pi^{\mu\nu} \equiv \gamma_2^{\pi}\pi^{\mu\nu} \ , \tag{5.166}$$

$$\left\langle E_{\mathbf{k}}^{-2}k^{\langle\mu}k^{\nu}k^{\lambda\rangle}\right\rangle_\delta = 0 \ , \quad \left\langle E_{\mathbf{k}}^{-2}k^{\langle\mu}k^{\nu}k^{\lambda}k^{\rho\rangle}\right\rangle_\delta = 0 \ , \tag{5.167}$$

where the coefficients γ_1^{Π}, γ_2^{Π}, γ_1^{n}, γ_2^{n}, γ_1^{π}, and γ_2^{π} are defined by the right-hand sides of these equations.

5.3.5   Hydrodynamic Equations of Motion

Implementing the simplifications (5.158)–(5.167) on account of the 14-moment approximation, the system of equations (5.143)–(5.145) will be closed in terms of the fluid-dynamical variables. Then, we obtain the following equations of motion,

$$\tau_\Pi \dot{\Pi} = -\Pi - \zeta\theta - \delta_{\Pi\Pi}\Pi\theta + \lambda_{\Pi\pi}\pi^{\mu\nu}\sigma_{\mu\nu}$$
$$- \ell_{\Pi n}\nabla_\mu n^\mu - \tau_{\Pi n}n^\mu Du_\mu - \lambda_{\Pi n}n^\mu \nabla_\mu \alpha_0 , \tag{5.168}$$

$$\tau_n \dot{n}^{\langle\mu\rangle} = -n^\mu + \varkappa\nabla^\mu\alpha_0 - \tau_n n_\nu \omega^{\nu\mu} - \delta_{nn}n^\mu\theta - \lambda_{nn}n^\nu\sigma_\nu^\mu + \ell_{n\pi}\Delta^{\mu\nu}\nabla_\lambda\pi_\nu^\lambda$$
$$- \tau_{n\pi}\pi_\nu^\mu Du^\nu - \lambda_{n\pi}\pi^{\mu\nu}\nabla_\nu\alpha_0 - \ell_{n\Pi}\nabla^\mu\Pi + \tau_{n\Pi}\Pi Du^\mu + \lambda_{n\Pi}\Pi\nabla^\mu\alpha_0 , \tag{5.169}$$

$$\tau_\pi \dot{\pi}^{\langle\mu\nu\rangle} = -\pi^{\mu\nu} + 2\eta\sigma^{\mu\nu} + 2\tau_\pi\pi_\alpha^{\langle\mu}\omega^{\nu\rangle\alpha} - \delta_{\pi\pi}\pi^{\mu\nu}\theta - \tau_{\pi\pi}\pi_\alpha^{\langle\mu}\sigma^{\nu\rangle\alpha}$$
$$- \tau_{\pi n}n^{\langle\mu}Du^{\nu\rangle} + \ell_{\pi n}\nabla^{\langle\mu}n^{\nu\rangle} + \lambda_{\pi n}n^{\langle\mu}\nabla^{\nu\rangle}\alpha_0 + \lambda_{\pi\Pi}\Pi\sigma^{\mu\nu} , \tag{5.170}$$

where we neglected terms of third order in dissipative currents or gradients. In total, these equations contain 25 transport coefficients. The bulk viscosity, the particle diffusion, and the shear viscosity are identified as

$$\zeta = \frac{\alpha_0^s}{4B_0\mathcal{A}_{02}^s} , \qquad \varkappa = \frac{\beta_0\alpha_0^v}{B_1\mathcal{A}_{01}^v} , \qquad \eta = \frac{\beta_0^2\alpha_0^t}{2B_2\mathcal{A}_{00}^t} . \tag{5.171}$$

Note that these expressions for ζ, $\varkappa$, and η are the same as those obtained in Chapman–Enskog theory, Eqs. (5.106)–(5.108), when the simplest truncation scheme possible is employed, $N^s = 2$, $N^v = 1$, and $N^t = 0$. (The proof that the bulk-viscosity coefficient agrees with Eq. (5.106) utilizes the relation (5.54).)

The relaxation times, which have no analogue in Navier–Stokes theory, are given by,

$$\tau_\Pi = \frac{3}{4\,m^2 B_0\mathcal{A}_{02}^s} , \qquad \tau_n = \frac{\beta_0}{B_1\mathcal{A}_{01}^v} , \qquad \tau_\pi = \frac{\beta_0^2}{B_2\mathcal{A}_{00}^t} . \tag{5.172}$$

Note that the ratios of the relaxation times to their corresponding viscosity and diffusion coefficients are actually thermodynamic functions and independent of the collision term,

$$\frac{\tau_\Pi}{\zeta} = \frac{3}{m^2\alpha_0^s} , \qquad \frac{\tau_n}{\varkappa} = \frac{1}{\alpha_0^v} , \qquad \frac{\tau_\pi}{\eta} = \frac{2}{\alpha_0^t} . \tag{5.173}$$

In particular, in the classical and massless limit $\tau_\pi/\eta = 1/(\beta_0 I_{32}) = 5/(\varepsilon_0 + P_0)$. This value respects the asymptotic causality condition discussed in Chap. 2 and was widely employed in Chap. 3 to derive the analytic solutions for Bjorken and Gubser flow.

The remaining transport coefficients related to the bulk-viscous pressure are

$$
\frac{\delta_{\Pi\Pi}}{\tau_\Pi} = \frac{2}{3} - \frac{m^2}{3}\frac{G_{20}}{D_{20}} - \frac{m^4}{9}\gamma_2^\Pi , \qquad \frac{\lambda_{\Pi\pi}}{\tau_\Pi} = \frac{m^2}{3}\left(\gamma_2^\pi - \frac{G_{20}}{D_{20}}\right) ,
$$

$$
\frac{\lambda_{\Pi n}}{\tau_\Pi} = -\frac{m^2}{3}\left(\frac{\partial\gamma_1^n}{\partial\alpha_0} + h_0^{-1}\frac{\partial\gamma_1^n}{\partial\beta_0}\right) , \qquad \frac{\ell_{\Pi n}}{\tau_\Pi} = \frac{m^2}{3}\left(\frac{G_{30}}{D_{20}} - \gamma_1^n\right) ,
$$

$$
\frac{\tau_{\Pi n}}{\tau_\Pi} = -\frac{m^2}{3}\left(\frac{G_{30}}{D_{20}} - \beta_0\frac{\partial\gamma_1^n}{\partial\beta_0}\right) , \tag{5.174}
$$

while those related to the particle-diffusion current are

$$
\frac{\delta_{nn}}{\tau_n} = 1 + \frac{m^2}{3}\gamma_2^n , \qquad \frac{\lambda_{nn}}{\tau_n} = \frac{3}{5} + \frac{2}{5}m^2\gamma_2^n ,
$$

$$
\frac{\ell_{n\pi}}{\tau_n} = \frac{\beta_0 J_{21}}{\varepsilon_0 + P_0} - \gamma_1^\pi , \qquad \frac{\tau_{n\pi}}{\tau_n} = \beta_0\frac{J_{21}}{\varepsilon_0 + P_0} - \beta_0\frac{\partial\gamma_1^\pi}{\partial\beta_0} ,
$$

$$
\frac{\lambda_{n\pi}}{\tau_n} = \frac{\partial\gamma_1^\pi}{\partial\alpha_0} + h_0^{-1}\frac{\partial\gamma_1^\pi}{\partial\beta_0} , \qquad \frac{\ell_{n\Pi}}{\tau_n} = \frac{\beta_0 J_{21}}{\varepsilon_0 + P_0} + \frac{m^2}{3}\gamma_1^\Pi ,
$$

$$
\frac{\tau_{n\Pi}}{\tau_n} = \frac{\beta_0 J_{21}}{\varepsilon_0 + P_0} + \frac{m^2}{3}\beta_0\frac{\partial\gamma_1^\Pi}{\partial\beta_0} , \qquad \frac{\lambda_{n\Pi}}{\tau_n} = -\frac{m^2}{3}\left(\frac{\partial\gamma_1^\Pi}{\partial\alpha_0} + h_0^{-1}\frac{\partial\gamma_1^\Pi}{\partial\beta_0}\right) , \tag{5.175}
$$

and, finally, those related to the shear-stress tensor are

$$
\frac{\delta_{\pi\pi}}{\tau_\pi} = \frac{4}{3} + \frac{m^2}{3}\gamma_2^\pi , \qquad\qquad \frac{\tau_{\pi\pi}}{\tau_\pi} = \frac{10}{7} + \frac{4}{7}m^2\gamma_2^\pi ,
$$

$$
\frac{\tau_{\pi n}}{\tau_\pi} = -\frac{2}{5}m^2\beta_0\frac{\partial\gamma_1^n}{\partial\beta_0} , \qquad\qquad \frac{\ell_{\pi n}}{\tau_\pi} = -\frac{2}{5}m^2\gamma_1^n ,
$$

$$
\frac{\lambda_{\pi n}}{\tau_\pi} = -\frac{2}{5}m^2\left(\frac{\partial\gamma_1^n}{\partial\alpha_0} + h_0^{-1}\frac{\partial\gamma_1^n}{\partial\beta_0}\right) , \qquad \frac{\lambda_{\pi\Pi}}{\tau_\pi} = \frac{6}{5} - \frac{2}{15}m^4\gamma_2^\Pi . \tag{5.176}
$$

We observe that, in the massless limit, $\delta_{\pi\pi}/\tau_\pi = 4/3$ and $\tau_{\pi\pi}/\tau_\pi = 10/7$, as mentioned in Chap. 3.

We note that equations (5.168), (5.169), and (5.170) are not identical to the original equations obtained by Israel and Stewart in Ref. [4]. The following terms did not appear in the original Israel–Stewart equations: in the equation for the bulk-viscous pressure the terms proportional to $\Pi\theta$, $\pi^{\mu\nu}\sigma_{\mu\nu}$, and $n_\mu\nabla^\mu\alpha_0$; in the equation for the diffusion 4-current the terms proportional to $n^\mu\theta$, $n^\nu\sigma_\nu^\mu$, $\pi^{\mu\nu}\nabla_\nu\alpha_0$, and $\Pi\nabla^\mu\alpha_0$; and in the equation for the shear-stress tensor the terms proportional to $\pi^{\mu\nu}\theta$, $\pi_\alpha^{\langle\mu}\sigma^{\nu\rangle\alpha}$, $n^{\langle\mu}\nabla^{\nu\rangle}\alpha_0$, and $\Pi\sigma^{\mu\nu}$. These missing terms made some believe that the formalism proposed by Israel and Stewart to derive fluid dynamics necessarily led to incomplete equations of motion. As we demonstrated in this chapter, this is certainly not the case. The issue was that such terms were originally dropped in their original work because they were considered to be unimportant for applications to cosmology, where the expansion rate of the fluid is usually quite small and the shear tensor is zero. Therefore, the disappearance of such terms is only a reflection of the power-counting scheme originally adopted by Israel–Stewart, it is not a feature of the formalism.

However, for the purposes of describing the quark-gluon plasma produced in heavy-ion collisions, such a power-counting scheme is simply not sufficient, and the terms that were originally dropped can be of relevance to the description of such a rapidly expanding system. The first to note that Israel–Stewart theory was not incomplete were the authors of Refs. [14–16], who actually wrote down the complete equations that follow from Israel–Stewart's derivation procedure, i.e., Eqs. (5.168)–(5.170).

Furthermore, we note that the transport coefficients derived in this section are slightly different from those derived by Israel and Stewart in Ref. [4]. The reason behind this is well known: we did not use the second moment of the Boltzmann equation to derive the equations of motion for the dissipative currents using the 14-moment approximation—we instead used the exact equation of motion for the dissipative currents. The moment of the Boltzmann equation employed to derive the equations of motion does not change the form of the equations of motion, but does affect the microscopic expressions obtained for the transport coefficients. We remark that this is basically a relativistic effect and, in the non-relativistic (or low-temperature) limit, the set of transport coefficients will have the same values regardless of the moment employed.

In contrast to Chapman–Enskog theory, Israel's and Stewart's framework allows for a rather simple derivation of fluid dynamics. This derivation leads to a fairly accurate theory of fluid dynamics for dilute gases, which takes into account the transient dynamics of the dissipative currents and includes several higher-order terms. Following the arguments constructed in Chap. 2, linearized Israel–Stewart theory can be shown to be stable under perturbations and also to respect causality, depending on the values taken for η/τ_π, $\varkappa/\tau_n$, and ζ/τ_Π. The microscopic expressions derived in this section for the viscosity coefficients and relaxation times are consistent with the causality and stability conditions obtained in Chap. 2. Naturally, improvements to this formalism are still required. Recent developments will be discussed in detail in the forthcoming chapters.

5.4   Summary

In this chapter, we have discussed in detail the derivation of relativistic dissipative fluid dynamics from kinetic theory following the two most traditional approaches: the Chapman–Enskog expansion and the method of moments, as originally proposed by Israel and Stewart (14-moment approximation).

In Sect. 5.1, we showed how the fluid-dynamical degrees of freedom can be matched to moments of the single-particle momentum distribution function. This is the first step required to derive fluid dynamics from the Boltzmann equation, and this section's results will also be employed in the following two chapters.

In Sect. 5.2, we discussed the Chapman–Enskog expansion. In this scheme, an asymptotic solution of the Boltzmann is constructed using perturbation theory, by expanding the single-particle distribution function in powers of the Knudsen number. This results in a gradient expansion, in which all moments of the distribution function depend solely on the five primary fluid-dynamical variables and their gradients. The corrections to the local distribution function are then systematically arranged in terms of an expansion in powers of the Knudsen number. We showed in this section that the zeroth order truncation of this expansion leads to ideal fluid dynamics and the first-order truncation to Navier–Stokes theory. We further obtained the microscopic expressions for the viscosity and diffusion coefficients. Keeping second- and higher-order terms one would obtain the Burnett and super-Burnett equations, but these higher-order solutions were not explicitly calculated here.

In Sect. 5.3, we discussed Israel's and Stewart's derivation of relativistic fluid dynamics. This procedure is based on a truncation of the moment expansion of the single-particle distribution function. Then, the distribution function is expressed solely in terms of 14 degrees of freedom, which can be matched to the 14 independent components of the conserved currents. This truncated version of the single-particle distribution function, the so-called 14-moment approximation, is substituted into the exact equations of motion for the dissipative currents, leading to a closed set of dynamical equations for these fields. The novel equations of motion obtained satisfy the causality condition obtained in Chap. 2 and, therefore, are stable.

In the next chapters, we shall see how the method of moments can be improved, in order to derive even more precise fluid-dynamical equations of motion.

References

1. Chapman, S., Cowling, T.G.: The Mathematical Theory of Non-Uniform Gases, 3rd edn. Cambridge University Press, New York (1974)
2. Israel, W.: Ann. Phys. (N.Y.) **100**, 310 (1976)
3. Stewart, J.M.: Proc. Roy. Soc. A **357**, 59 (1977)
4. Israel, W., Stewart, J.M.: Ann. Phys. (N.Y.) **118**, 341 (1979)
5. Landau, L.D., Lifshitz, E.M.: Fluid Mechanics. Pergamon, New York (1959)
6. Sasaki, M.: In: Kockel, B. et al. (eds.) Max-Planck-Festschrift. Veb Deutscher Verlag der Wissenschaften, Berlin (1958)
7. Israel, W.: J. Math. Phys. **4**, 1163 (1963)
8. Burnett, D.: Proc. Lond. Math. Soc. **39**, 385 (1935); Proc. Lond. Math. Soc. **40**, 382 (1936)
9. Grad, H.: Phys. Fluids **6**, 147 (1963)
10. Hiscock, W., Lindblom, L.: Phys. Rev. D **31**, 725 (1985)
11. de Groot, S.R., van Leeuwen, W.A., van Weert, Ch.G.: Relativistic Kinetic Theory—Principles and Applications. North-Holland (1980)
12. Denicol, G.S., Koide, T., Rischke, D.H.: Phys. Rev. Lett. **105** (2010)
13. Denicol, G.S., Molnar, E., Niemi, H., Rischke, D.H.: Eur. Phys. J. A **48**, 170 (2012)
14. Betz, B., Henkel, D., Rischke, D.H.: Prog. Part. Nucl. Phys. **62**, 556–561 (2009)
15. Betz, B., Henkel, D., Rischke, D.H.: J. Phys. G **36** (2009)
16. Betz, B., Denicol, G.S., Koide, T., Molnar, E., Niemi, H., Rischke, D.H.: EPJ Web Conf. **13**, 07005 (2011)

Method of Moments: Equilibrium Reference State

The two widespread methods to derive relativistic fluid dynamics from the Boltzmann equation, the Chapman–Enskog expansion [1] and the 14-moment approximation [2–4], both described in the previous chapter, have flaws. In the relativistic case, the Chapman–Enskog expansion leads to equations of fluid dynamics that are acausal and intrinsically unstable [5] and, consequently, it should not be applied to derive the equations of relativistic fluid dynamics from kinetic theory. The method of moments [2–4,6] is in principle free of such problems and leads to transient fluid-dynamical equations that can be constructed to be causal and stable [7–11], as shown in Chap. 2. However, in the form presented in the previous chapter, it still features an ambiguity as to how to close the system of fluid-dynamical equations of motion. In this chapter, we will discuss the necessary prerequisites to resolve this problem.

The method of moments is generally considered the method of choice for deriving the fluid-dynamical equations of motion from the Boltzmann equation. This approach was first formulated consistently by Grad [12] for non-relativistic systems and consists of expanding the non-equilibrium correction to the single-particle distribution function in terms of a complete set of Hermite polynomials [13]. The generalization of Grad's method of moments to relativistic systems is, nevertheless, not trivial and has been pursued by several authors [14–19]. The main challenge is to find a suitable set of orthogonal polynomials which could replace the Hermite polynomials in a relativistic formulation [3,6]. This issue was circumvented by Israel and Stewart by simply expanding the non-equilibrium correction to the single-particle distribution function in a Taylor series in 4-momentum. The main drawback of this approach is that, since the expansion is not realized in terms of an orthogonal basis, the expansion coefficients cannot be trivially determined in terms of moments of the single-particle distribution function. Israel and Stewart then introduced another approximation, the so-called 14-moment approximation [4], already discussed in Sect. 5.3 of Chap. 5. In this case, the Taylor expansion in momentum is truncated at second order, leaving

© Springer Nature Switzerland AG 2021

G. S. Denicol and D. H. Rischke, *Microscopic Foundations of Relativistic Fluid Dynamics*, Lecture Notes in Physics 990, https://doi.org/10.1007/978-3-030-82077-0_6

only 14 expansion coefficients to be determined. Israel and Stewart then introduced a set of 14 constraints, with which it was possible to express the expansion coefficients in terms of the conserved currents, N^μ and $T^{\mu\nu}$—the so-called matching procedure.

A more reliable relativistic formulation of the method of moments, which expands the single-particle distribution function in momentum using an orthogonal and complete basis, was formulated in Ref. [20]. The main goal of this chapter is to describe this expansion and to derive the equations of motion satisfied by the corresponding expansion coefficients, the so-called *irreducible moments*, which turn out to be moments of the deviation of the single-particle distribution function from a chosen reference state. As we shall discuss in Chap. 8, such a generalization of the method of moments will be essential in understanding how relativistic fluid dynamics actually emerges from the relativistic Boltzmann equation, what is its domain of applicability, and how the equations of motion can be systematically improved. In this chapter, the reference state is assumed to be in local thermodynamical equilibrium, described by the single-particle distribution function (5.60). In Chap. 9, we will show how to generalize this to a particular non-equilibrium reference state, exhibiting a spatial anisotropy.

This chapter is organized as follows. In Sect. 6.1, we demonstrate how to expand the single-particle distribution function in terms of a *complete, orthogonal* basis in momentum space. In contrast to Israel's and Stewart's non-orthogonal basis 1, k^μ, $k^\mu k^\nu$, ..., our approach uses *irreducible tensors* in 4-momentum k^μ, and is thus orthogonal. The coefficients of the irreducible tensors in the expansion of the single-particle distribution function are orthogonal polynomials of the rest-frame energy, multiplied by the above mentioned irreducible moments. Section 6.2 derives an infinite set of equations for these moments, which is still completely equivalent to the Boltzmann equation. Section 6.3 explicitly calculates the moments of the collision term, which appear in the equations of motion for the irreducible moments. In Sect. 6.4, we summarize this chapter. Several mathematical details are delegated to a set of appendices.

6.1   Moment Expansion

In principle, the momentum dependence of the non-equilibrium component of $f_{\mathbf{k}}$, $\delta f_{\mathbf{k}} \equiv f_{\mathbf{k}} - f_{0\mathbf{k}}$, should be obtained by solving the relativistic Boltzmann equation, which in general poses a very complicated task. In this context, the method of moments can be a convenient tool, since it allows us to obtain an approximate expression for $\delta f_{\mathbf{k}}$, which is able to capture some features of solutions of the Boltzmann equation. In this section, we explain and develop the moment expansion of the single-particle distribution function.

It is convenient to factorize the local-equilibrium distribution function $f_{0\mathbf{k}}$ from $f_{\mathbf{k}}$ in the following way

$$f_{\mathbf{k}} = f_{0\mathbf{k}} + \delta f_{\mathbf{k}} \equiv f_{0\mathbf{k}} \left(1 + \mathcal{G}_{\mathbf{k}} \phi_{\mathbf{k}} \right) . \tag{6.1}$$

Above, we introduced $\mathcal{G}_\mathbf{k}$ as an arbitrary function of $E_\mathbf{k} \equiv u_\mu k^\mu$, and $\phi_\mathbf{k}$ as an arbitrary function of the space-time coordinates x^μ and the on-shell 4-momentum $k^\mu = (k_0, \mathbf{k})^T$, $k_0 = \sqrt{\mathbf{k}^2 + m^2}$.

The next step is to expand $\phi_\mathbf{k}$ in terms of a complete basis of tensors formed of k^μ. One choice is to follow the approach proposed by Israel and Stewart [4] and expand $\phi_\mathbf{k}$ using the following basis,

$$1, \; k^\mu, \; k^\mu k^\nu, \; k^\mu k^\nu k^\lambda, \ldots . \tag{6.2}$$

In this case, the formal expression for $\phi_\mathbf{k}$ becomes

$$\phi_\mathbf{k} = \epsilon + \epsilon_\mu k^\mu + \epsilon_{\mu\nu} k^\mu k^\nu + O\left(k^\mu k^\nu k^\lambda\right) , \tag{6.3}$$

where the expansion coefficients $\epsilon_{\mu_1 \cdots \mu_m}$ carry all the space-time dependence of $\phi_\mathbf{k}$.

The main disadvantage of the moment expansion proposed by Israel and Stewart is that the expansion coefficients are simply unknown and can only be extracted approximately. This happens because the basis (6.2) is not orthogonal. In their original work, Israel and Stewart overcame this problem by truncating the expansion at second order in momentum and extracting the coefficients of the truncated expansion by matching them to certain moments of the single-particle distribution function. However, the coefficients extracted from such a matching procedure are not unique and can change according to the order in which the expansion is truncated.

This unpleasant aspect of Israel–Stewart theory can be easily avoided by using an orthogonal basis for the expansion. Here, we follow the approach developed in Ref. [20] and expand $\phi_\mathbf{k}$ using as a basis the *irreducible* tensors,

$$1, \; k^{\langle\mu\rangle}, \; k^{\langle\mu} k^{\nu\rangle}, \; k^{\langle\mu} k^\nu k^{\lambda\rangle}, \ldots , \tag{6.4}$$

and orthogonal polynomials in $E_\mathbf{k}$,

$$P_{\mathbf{k}n}^{(\ell)} = \sum_{r=0}^{n} a_{nr}^{(\ell)} E_\mathbf{k}^r . \tag{6.5}$$

Note that, in the local rest frame, the irreducible tensors (6.4) are (certain) polynomials formed of powers of the components of 3-momentum $\mathbf{k}$, while the orthogonal polynomials (6.5) are polynomials in energy k_0. The term "irreducible" refers to the fact that the tensors are irreducible with respect to the group of Lorentz transformations Λ which leave the fluid velocity invariant, $\Lambda_\nu^\mu u^\nu \equiv u^\mu$ (the so-called *little group* associated with u^μ). In the local rest frame, these are spatial rotations.

The irreducible tensors form a *complete and orthogonal* set and are defined by using the symmetrized and, for $m > 1$ traceless, projection orthogonal to u^μ of tensors constructed from k^μ. That is,

$$k^{\langle\mu_1} \cdots k^{\mu_m\rangle} \equiv \Delta_{\nu_1 \cdots \nu_m}^{\mu_1 \cdots \mu_m} k^{\nu_1} \cdots k^{\nu_m} , \tag{6.6}$$

where the projectors $\Delta_{\nu_1 \cdots \nu_m}^{\mu_1 \cdots \mu_m}$ are constructed in Appendix 6.5, Eq. (6.75), see also Refs. [6,21]. Naturally, these projectors are orthogonal to the fluid 4-velocity.

The irreducible tensors (6.4) satisfy an orthogonality condition,

$$
\int dK \, \mathrm{F}(E_{\mathbf{k}}) \, k^{\langle \mu_1} \cdots k^{\mu_m \rangle} k_{\langle \nu_1} \cdots k_{\nu_n \rangle}
$$
$$
= \frac{m! \, \delta_{mn}}{(2m+1)!!} \Delta_{\nu_1 \cdots \nu_m}^{\mu_1 \cdots \mu_m} \int dK \, \mathrm{F}(E_{\mathbf{k}}) \left(\Delta^{\alpha\beta} k_\alpha k_\beta \right)^m , \tag{6.7}
$$

for an arbitrary function $\mathrm{F}(E_{\mathbf{k}})$, which depends only on $E_{\mathbf{k}} = k^\mu u_\mu$, for the proof see Appendix 6.7. Likewise, the orthogonal polynomials (6.5) fulfill an orthogonality condition,

$$
\int dK \, \omega^{(\ell)} \, P_{\mathbf{k}m}^{(\ell)} P_{\mathbf{k}n}^{(\ell)} = \delta_{mn} , \tag{6.8}
$$

with the measure $\omega^{(\ell)}$ being given by

$$
\omega^{(\ell)} \equiv \frac{W^{(\ell)}}{(2\ell+1)!!} \left(\Delta^{\alpha\beta} k_\alpha k_\beta \right)^\ell \mathcal{G}_{\mathbf{k}} f_{0\mathbf{k}} , \tag{6.9}
$$

cf. Appendix 6.8 for further details.

We note that the basis constructed from irreducible tensors and orthogonal polynomials is completely equivalent to the basis used by Israel and Stewart, as long as $\mathcal{G}_{\mathbf{k}} = \tilde{f}_{0\mathbf{k}}$. However, the fact that it is orthogonal will make it more convenient to use. Next, we explain how the orthogonality relations (6.7), (6.8) satisfied by the basis elements allow us to calculate the expansion coefficients.

Using the basis introduced in Eqs. (6.4) and (6.5), the quantity $\phi_{\mathbf{k}}$ defined in Eq. (6.1) is expanded as

$$
\phi_{\mathbf{k}} = \sum_{\ell=0}^{\infty} \sum_{n=0}^{N_\ell} c_n^{\langle \mu_1 \cdots \mu_\ell \rangle} P_{\mathbf{k}n}^{(\ell)} \, k_{\langle \mu_1} \cdots k_{\mu_\ell \rangle} . \tag{6.10}
$$

The expansion coefficients $c_n^{\langle \mu_1 \cdots \mu_\ell \rangle}$ can be obtained by multiplying Eq. (6.10) by the corresponding basis element, $\mathcal{G}_{\mathbf{k}} f_{0\mathbf{k}} P_{\mathbf{k}n}^{(\ell)} k^{\langle \mu_1} \cdots k^{\mu_\ell \rangle}$, and integrating over momentum, dK. Together with the definition (6.1) of $\delta f_{\mathbf{k}}$, the orthogonality conditions (6.7), (6.8) then imply that

$$
c_n^{\langle \mu_1 \cdots \mu_\ell \rangle} = \frac{W^{(\ell)}}{\ell!} \int dK \, P_{\mathbf{k}n}^{(\ell)} k^{\langle \mu_1} \cdots k^{\mu_\ell \rangle} \delta f_{\mathbf{k}} . \tag{6.11}
$$

Plugging this into Eq. (6.10) and the result into Eq. (6.1), using Eq. (6.5) the full non-equilibrium single-particle distribution function can be expressed in the following way,

$$
f_{\mathbf{k}} = f_{0\mathbf{k}} + f_{0\mathbf{k}} \sum_{\ell=0}^{\infty} \sum_{n=0}^{N_\ell} \mathcal{G}_{\mathbf{k}} \mathcal{H}_{\mathbf{k}n}^{(\ell)} \rho_n^{\mu_1 \cdots \mu_\ell} k_{\langle \mu_1} \cdots k_{\mu_\ell \rangle} , \tag{6.12}
$$

where we introduced the function

$$\mathcal{H}_{\mathbf{k}n}^{(\ell)} \equiv \frac{W^{(\ell)}}{\ell!} \sum_{m=n}^{N_\ell} a_{mn}^{(\ell)} P_{\mathbf{k}m}^{(\ell)} , \tag{6.13}$$

and the irreducible moments $\rho_n^{\mu_1 \cdots \mu_\ell}$ of $\delta f_{\mathbf{k}}$,

$$\rho_n^{\mu_1 \cdots \mu_\ell} \equiv \int dK \, E_{\mathbf{k}}^n \, k^{\langle \mu_1} \cdots k^{\mu_\ell \rangle} \, \delta f_{\mathbf{k}} \equiv \left\langle E_{\mathbf{k}}^n \, k^{\langle \mu_1} \cdots k^{\mu_\ell \rangle} \right\rangle_\delta . \tag{6.14}$$

On the right-hand side, we made contact to the notation (5.27) introduced in Chap. 5.

We remark that our choice of matching conditions (5.23) and the Landau-frame definition (5.16) of the fluid velocity [22] imply that the following irreducible moments vanish

$$\rho_1 = \rho_2 = \rho_1^\mu = 0 . \tag{6.15}$$

The vanishing of the two scalar moments, ρ_1 and ρ_2, define the temperature and chemical potential of the fictitious local-equilibrium state characterized by the single-particle distribution function $f_{0\mathbf{k}}$. The rank-1 moment ρ_1^μ corresponds to energy diffusion, which vanishes by definition in the Landau frame.

Note that, so far, the function $\mathcal{G}_{\mathbf{k}}$ was assumed to be independent of the index ℓ. This was done just for the sake of simplicity and, in principle, nothing would prevent us from introducing a function $\mathcal{G}_{\mathbf{k}}^{(\ell)}$, if this provides a more useful prefactor in Eq. (6.1). The arbitrary function $\mathcal{G}_{\mathbf{k}}^{(\ell)}$ would determine the overall behavior of the polynomial basis employed to expand $\phi_{\mathbf{k}}$. Setting $\mathcal{G}_{\mathbf{k}}^{(\ell)} = \tilde{f}_{0\mathbf{k}}$ corresponds to employing the same basis used by Israel and Stewart, which, once truncated, is a good approximation to describe the function for small values of $\beta_0 E_{\mathbf{k}}$ (as long as there are no singularities). This basis is sufficient to derive fluid dynamics and was used for this purpose in previous work [20]. On the other hand, one can also set $\mathcal{G}_{\mathbf{k}}^{(\ell)} = \tilde{f}_{0\mathbf{k}} / (1 + \beta_0 E_{\mathbf{k}})^\ell$, which, at large $\beta_0 E_{\mathbf{k}}$, is equivalent to the basis $1/E_{\mathbf{k}}$, $1/E_{\mathbf{k}}^2$, $1/E_{\mathbf{k}}^3, \ldots$, while, at small $\beta_0 E_{\mathbf{k}}$, is equivalent to the usual basis 1, $E_{\mathbf{k}}$, $E_{\mathbf{k}}^2$, $E_{\mathbf{k}}^3, \ldots$. This (truncated) basis is expected to be viable when $\beta_0 E_{\mathbf{k}}$ is large and should provide a better estimate of the momentum dependence of $\phi_{\mathbf{k}}$. In summary, with a reasonable choice of $\mathcal{G}_{\mathbf{k}}^{(\ell)}$ one can describe the single-particle distribution function for several domains of $\beta_0 E_{\mathbf{k}}$.

For the particular choice $\mathcal{G}_k = \tilde{f}_{0\mathbf{k}}$ one can easily show that, substituting Eqs. (6.12) and (6.13) into Eq. (6.14) and using Eq. (6.5), the orthogonality condition (6.7), as well as the definition of the auxiliary thermodynamic integrals (6.30), all irreducible moments are linearly related to each other (for more details see, Eq. (72) of Ref. [23]),

$$\rho_i^{\mu_1 \cdots \mu_\ell} \equiv (-1)^\ell \, \ell! \sum_{n=0}^{N_\ell} \rho_n^{\mu_1 \cdots \mu_\ell} \gamma_{in}^{(\ell)} , \tag{6.16}$$

where

$$\gamma_{in}^{(\ell)} = \frac{W^{(\ell)}}{\ell!} \sum_{m=n}^{N_\ell} \sum_{r=0}^{m} a_{mn}^{(\ell)} a_{mr}^{(\ell)} J_{i+r+2\ell,\ell} \ . \tag{6.17}$$

Note that these relations are also valid for moments with *negative* i, hence it is possible to express the irreducible moments with negative powers of $E_{\mathbf{k}}$ in terms of the ones with positive i, for details see Eq. (65) of Ref. [20].

The moment expansion described in this section is a very powerful tool, with applications to all methods that derive fluid dynamics from the relativistic Boltzmann equation.

6.2   Equations of Motion for the Irreducible Moments

The time-evolution equations for the irreducible moments $\rho_r^{\mu_1 \cdots \mu_\ell}$ can be obtained directly from the Boltzmann equation by applying the comoving derivative to the definition (6.14), together with the symmetrized traceless projection,

$$\dot{\rho}_r^{\langle \mu_1 \cdots \mu_\ell \rangle} = \Delta_{\nu_1 \cdots \nu_\ell}^{\mu_1 \cdots \mu_\ell} \frac{d}{d\tau} \int dK E_{\mathbf{k}}^r k^{\langle \nu_1} \cdots k^{\nu_\ell \rangle} \delta f_{\mathbf{k}} \ , \tag{6.18}$$

where $\dot{A} \equiv u^\mu \partial_\mu A \equiv dA/d\tau$ and $\dot{\rho}_r^{\langle \mu_1 \cdots \mu_\ell \rangle} \equiv \Delta_{\nu_1 \cdots \nu_\ell}^{\mu_1 \cdots \mu_\ell} \dot{\rho}_r^{\nu_1 \cdots \nu_\ell}$. Using the Boltzmann equation (4.41) in the form (5.142),

$$\delta \dot{f}_{\mathbf{k}} = -\dot{f}_{0\mathbf{k}} - E_{\mathbf{k}}^{-1} k^\nu \nabla_\nu f_{0\mathbf{k}} - E_{\mathbf{k}}^{-1} k^\nu \nabla_\nu \delta f_{\mathbf{k}} + E_{\mathbf{k}}^{-1} C[f] \ , \tag{6.19}$$

where $\nabla_\mu \equiv \Delta_\mu^\nu \partial_\nu$, and substituting this expression into Eq. (6.18), one can obtain the *exact* equations for the comoving derivatives of $\rho_r^{\mu_1 \cdots \mu_\ell}$.

Using the power-counting scheme that will be developed in Sect. 8.1 of Chap. 8, we can show that, in order to derive the equations of motion for relativistic fluid dynamics, it is sufficient to know the time-evolution equations for the moments (6.14) up to rank two, i.e., for ρ_r, ρ_r^μ, and $\rho_r^{\mu\nu}$. Similar equations can also be derived for higher-rank irreducible moments, if needed. Thus, using Eqs. (6.18) and (6.19), we obtain after some lengthy calculation

$$\begin{aligned}
\dot{\rho}_r - C_{r-1} = {} & \alpha_r^{(0)} \theta - \frac{G_{2r}}{D_{20}} \Pi \theta + \frac{G_{2r}}{D_{20}} \pi^{\mu\nu} \sigma_{\mu\nu} + \frac{G_{3r}}{D_{20}} \partial_\mu n^\mu + (r-1) \rho_{r-2}^{\mu\nu} \sigma_{\mu\nu} \\
& + r \rho_{r-1}^\mu \dot{u}_\mu - \nabla_\mu \rho_{r-1}^\mu - \frac{1}{3} \left[(r+2) \rho_r - (r-1) m^2 \rho_{r-2} \right] \theta \ ,
\end{aligned} \tag{6.20}$$

$$\begin{aligned}
\dot{\rho}_r^{\langle \mu \rangle} - C_{r-1}^{\langle \mu \rangle} = {} & \alpha_r^{(1)} I^\mu + \rho_r^\nu \omega_\nu^\mu + \frac{1}{3} \left[(r-1) m^2 \rho_{r-2}^\mu - (r+3) \rho_r^\mu \right] \theta \\
& + \frac{1}{5} \left[(2r-2) m^2 \rho_{r-2}^\nu - (2r+3) \rho_r^\nu \right] \sigma_\nu^\mu + \frac{1}{3} \left[m^2 r \rho_{r-1} - (r+3) \rho_{r+1} \right] \dot{u}^\mu \\
& + \frac{\beta_0 J_{r+2,1}}{\varepsilon_0 + P_0} \left(\Pi \dot{u}^\mu - \nabla^\mu \Pi + \Delta_\nu^\mu \partial_\lambda \pi^{\lambda\nu} \right) - \frac{1}{3} \nabla^\mu \left(m^2 \rho_{r-1} - \rho_{r+1} \right) \\
& - \Delta_\lambda^\mu \nabla_\nu \rho_{r-1}^{\lambda\nu} + r \rho_{r-1}^{\mu\nu} \dot{u}_\nu + (r-1) \rho_{r-2}^{\mu\nu\lambda} \sigma_{\lambda\nu} \ ,
\end{aligned} \tag{6.21}$$

$$\dot{\rho}_r^{\langle\mu\nu\rangle} - C_{r-1}^{\langle\mu\nu\rangle} = 2\alpha_r^{(2)}\sigma^{\mu\nu} - \frac{2}{7}\left[(2r+5)\rho_r^{\lambda\langle\mu} - 2m^2(r-1)\rho_{r-2}^{\lambda\langle\mu}\right]\sigma_\lambda^{\nu\rangle} + 2\rho_r^{\lambda\langle\mu}\omega_\lambda^{\nu\rangle}$$

$$+ \frac{2}{15}\left[(r+4)\rho_{r+2} - (2r+3)m^2\rho_r + (r-1)m^4\rho_{r-2}\right]\sigma^{\mu\nu}$$

$$+ \frac{2}{5}\nabla^{\langle\mu}\left(\rho_{r+1}^{\nu\rangle} - m^2\rho_{r-1}^{\nu\rangle}\right) - \frac{2}{5}\left[(r+5)\rho_{r+1}^{\langle\mu} - m^2r\rho_{r-1}^{\langle\mu}\right]\dot{u}^{\nu\rangle}$$

$$- \frac{1}{3}\left[(r+4)\rho_r^{\mu\nu} - m^2(r-1)\rho_{r-2}^{\mu\nu}\right]\theta$$

$$+ (r-1)\rho_{r-2}^{\mu\nu\lambda\rho}\sigma_{\lambda\rho} - \Delta_{\alpha\beta}^{\mu\nu}\nabla_\lambda\rho_{r-1}^{\alpha\beta\lambda} + r\rho_{r-1}^{\mu\nu\lambda}\dot{u}_\lambda \, , \tag{6.22}$$

where we introduced the generalized irreducible collision terms

$$C_r^{\langle\mu_1\cdots\mu_\ell\rangle} = \int dK\, E_k^r k^{\langle\mu_1}\cdots k^{\mu_\ell\rangle}C[f] \, . \tag{6.23}$$

We further used the definitions of the shear tensor $\sigma^{\mu\nu} \equiv \nabla^{\langle\mu}u^{\nu\rangle}$, the expansion scalar $\theta \equiv \nabla_\mu u^\mu$, and the vorticity tensor $\omega^{\mu\nu} \equiv (\nabla^\mu u^\nu - \nabla^\nu u^\mu)/2$ and we introduced $I^\mu \equiv \nabla^\mu\alpha_0$. All derivatives of α_0 and β_0 that appeared during the derivation of the above equations were replaced using the *exact* equations obtained from the conservation laws of particle number, energy, and momentum,

$$\dot{\alpha}_0 = \frac{1}{D_{20}}\left\{-J_{30}\left(n_0\theta + \partial_\mu n^\mu\right) + J_{20}\left[(\varepsilon_0 + P_0 + \Pi)\theta - \pi^{\mu\nu}\sigma_{\mu\nu}\right]\right\} \, , \tag{6.24}$$

$$\dot{\beta}_0 = \frac{1}{D_{20}}\left\{-J_{20}\left(n_0\theta + \partial_\mu n^\mu\right) + J_{10}\left[(\varepsilon_0 + P_0 + \Pi)\theta - \pi^{\mu\nu}\sigma_{\mu\nu}\right]\right\} \, , \tag{6.25}$$

$$\dot{u}^\mu = \frac{1}{\varepsilon_0 + P_0}\left(\nabla^\mu P_0 - \Pi\dot{u}^\mu + \nabla^\mu\Pi - \Delta_\alpha^\mu\partial_\beta\pi^{\alpha\beta}\right) \, , \tag{6.26}$$

cf. Eqs. (5.46), (5.58), and (5.59) in Chap. 5. The coefficients $\alpha_r^{(0)}$, $\alpha_r^{(1)}$, and $\alpha_r^{(2)}$ are functions of temperature and chemical potential and have the general form

$$\alpha_r^{(0)} = (1-r)I_{r1} - I_{r0} - \frac{1}{D_{20}}\left[G_{2r}(\varepsilon_0 + P_0) - G_{3r}n_0\right] \, , \tag{6.27}$$

$$\alpha_r^{(1)} = J_{r+1,1} - h_0^{-1}J_{r+2,1} \, , \tag{6.28}$$

$$\alpha_r^{(2)} = I_{r+2,1} + (r-1)I_{r+2,2} \, , \tag{6.29}$$

where we defined the thermodynamic functions

$$I_{nq}(\alpha_0, \beta_0) = \frac{1}{(2q+1)!!}\left\langle E_k^{n-2q}\left(-\Delta^{\alpha\beta}k_\alpha k_\beta\right)^q\right\rangle_0 \, , \quad J_{nq} = \left.\frac{\partial I_{nq}}{\partial\alpha_0}\right|_{\beta_0} \, , \tag{6.30}$$

$$G_{nm} = J_{n0}J_{m0} - J_{n-1,0}J_{m+1,0} \, , \quad D_{nq} = J_{n+1,q}J_{n-1,q} - J_{nq}^2 \, , \tag{6.31}$$

cf. also Eqs. (5.48), (5.49), and (5.55) in Chap. 5.

Using the matching conditions (6.15), the dissipative quantities appearing in the conservation laws can be (exactly) identified with the moments

$$\rho_0 = -\frac{3}{m^2}\,\Pi\,,\qquad \rho_0^\mu = n^\mu\,,\qquad \rho_0^{\mu\nu} = \pi^{\mu\nu}\,, \tag{6.32}$$

cf. Eqs. (5.26) and (5.28) in Chap. 5. We note that the derivation of the equations of motion for the irreducible moments is independent of the form of the expansion of the single-particle distribution introduced in the previous section.

6.3 Generalized Collision Term

In this section, we compute the collision integrals (6.23), expressing them in terms of the irreducible moments. This then formally closes the set of moment equations. We shall separate the collision term into two different components: one being linear in the non-equilibrium corrections $\phi_\mathbf{p}$ and the remaining being nonlinear in these functions. It is obvious that the first term will be linear in the irreducible moments, while the second one is nonlinear.

The first step is to express the collision term as a functional of $\phi_\mathbf{p}$, which was introduced in Eq. (6.1) as

$$\delta f_\mathbf{p} = f_\mathbf{p} - f_{0\mathbf{p}} = f_{0\mathbf{p}}\tilde{f}_{0\mathbf{p}}\phi_\mathbf{p}\,. \tag{6.33}$$

Using this definition, it is straightforward to demonstrate that

$$f_\mathbf{p} f_{\mathbf{p}'} = f_{0\mathbf{p}} f_{0\mathbf{p}'}\left(1 + \tilde{f}_{0\mathbf{p}'}\phi_{\mathbf{p}'} + \tilde{f}_{0\mathbf{p}}\phi_\mathbf{p} + \tilde{f}_{0\mathbf{p}}\tilde{f}_{0\mathbf{p}'}\phi_\mathbf{p}\phi_{\mathbf{p}'}\right)\,, \tag{6.34}$$

$$\tilde{f}_\mathbf{p}\tilde{f}_{\mathbf{p}'} = \tilde{f}_{0\mathbf{p}}\tilde{f}_{0\mathbf{p}'}\left(1 - af_{0\mathbf{p}'}\phi_{\mathbf{p}'} - af_{0\mathbf{p}}\phi_\mathbf{p} + a^2 f_{0\mathbf{p}} f_{0\mathbf{p}'}\phi_\mathbf{p}\phi_{\mathbf{p}'}\right)\,. \tag{6.35}$$

The terms linear in $\phi_\mathbf{p}$ will be collected into the linear collision term and computed in the next subsection, while the higher-order terms constitute the nonlinear collision term, respectively. We will elaborate on the calculation of the latter, but only in the classical (Boltzmann) limit, in the next-to-next subsection. An application of this scheme can be found in Ref. [23].

6.3.1 Computation of the Linear Collision Term

Substituting Eqs. (6.34) and (6.35) into Eq. (6.23), and keeping only terms that are linear in $\phi_\mathbf{k}$, one can derive the linearized collision term, $L[\phi]$, defined as $C[f] = L[\phi] + O(\phi^2)$. Explicitly, one obtains

$$L[\phi] = \frac{1}{\nu}\int dK' dP dP'\, W_{\mathbf{kk}'\to\mathbf{pp}'} f_{0\mathbf{k}} f_{0\mathbf{k}'}\tilde{f}_{0\mathbf{p}}\tilde{f}_{0\mathbf{p}'}\left(\phi_\mathbf{p} + \phi_{\mathbf{p}'} - \phi_\mathbf{k} - \phi_{\mathbf{k}'}\right)\,. \tag{6.36}$$

In order to derive this formula, we used that $C[f_0] = 0$ and the equality $f_{0\mathbf{p}} f_{0\mathbf{p}'} \tilde{f}_{0\mathbf{k}} \tilde{f}_{0\mathbf{k}'} = f_{0\mathbf{k}} f_{0\mathbf{k}'} \tilde{f}_{0\mathbf{p}} \tilde{f}_{0\mathbf{p}'}$.

Inserting Eq. (6.36) into the expression for the irreducible collision term (6.23), we obtain the linearized collision integral

$$
\begin{aligned}
L_{r-1}^{\langle \mu_1 \cdots \mu_\ell \rangle} &\equiv \int dK \, E_{\mathbf{k}}^{r-1} k^{\langle \mu_1} \cdots k^{\mu_\ell \rangle} L[\phi] \\
&= \frac{1}{\nu} \int dK \, dK' \, dP \, dP' \, W_{\mathbf{kk}' \to \mathbf{pp}'} f_{0\mathbf{k}} f_{0\mathbf{k}'} \tilde{f}_{0\mathbf{p}} \tilde{f}_{0\mathbf{p}'} \\
&\quad \times E_{\mathbf{k}}^{r-1} k^{\langle \mu_1} \cdots k^{\mu_\ell \rangle} \left(\phi_{\mathbf{p}} + \phi_{\mathbf{p}'} - \phi_{\mathbf{k}} - \phi_{\mathbf{k}'} \right) .
\end{aligned}
\tag{6.37}
$$

The next step is to substitute the moment expansion for $\phi_{\mathbf{k}}$, Eqs. (6.10) and (6.11), into Eq. (6.37), expressing it as a linear combination of the irreducible moments

$$
L_{r-1}^{\langle \mu_1 \cdots \mu_\ell \rangle} = - \sum_{m=0}^{\infty} \sum_{n=0}^{N_m} (\mathscr{A}_{rn})_{\nu_1 \cdots \nu_m}^{\mu_1 \cdots \mu_\ell} \, \rho_n^{\nu_1 \cdots \nu_m} ,
\tag{6.38}
$$

where we defined the tensor

$$
\begin{aligned}
(\mathscr{A}_{rn})_{\nu_1 \cdots \nu_m}^{\mu_1 \cdots \mu_\ell} &\equiv \frac{1}{\nu} \int dK \, dK' \, dP \, dP' \, W_{\mathbf{kk}' \to \mathbf{pp}'} f_{0\mathbf{k}} f_{0\mathbf{k}'} \tilde{f}_{0\mathbf{p}} \tilde{f}_{0\mathbf{p}'} E_{\mathbf{k}}^{r-1} k^{\langle \mu_1} \cdots k^{\mu_\ell \rangle} \\
&\quad \times \left(\mathcal{H}_{n\mathbf{k}}^{(m)} k_{\langle \nu_1} \cdots k_{\nu_m \rangle} + \mathcal{H}_{n\mathbf{k}'}^{(m)} k'_{\langle \nu_1} \cdots k'_{\nu_m \rangle} \right. \\
&\quad \left. - \mathcal{H}_{n\mathbf{p}}^{(m)} p_{\langle \nu_1} \cdots p_{\nu_m \rangle} - \mathcal{H}_{n\mathbf{p}'}^{(m)} p'_{\langle \nu_1} \cdots p'_{\nu_m \rangle} \right) .
\end{aligned}
\tag{6.39}
$$

Equation (6.38) looks like different tensor components of the irreducible moments contribute to a given tensor component of the collision term on the left-hand side. In fact, however, we will now show that the tensor components on both sides of this equation are the same, cf. Eq. (6.46).

The integral $(\mathscr{A}_{rn})_{\nu_1 \cdots \nu_m}^{\mu_1 \cdots \mu_\ell}$ is a tensor of rank $m + \ell$, which is symmetric under permutations of μ-type and ν-type indices, and which depends only on the equilibrium distribution function and the cross section. The former contains only one type of 4-vector, the fluid velocity u^μ. Therefore, $(\mathscr{A}_{rn})_{\nu_1 \cdots \nu_m}^{\mu_1 \cdots \mu_\ell}$ must be constructed from tensor structures made of u^μ and the metric tensor $g^{\mu\nu}$. Also, $(\mathscr{A}_{rn})_{\nu_1 \cdots \nu_m}^{\mu_1 \cdots \mu_\ell}$ was constructed to be orthogonal to u^μ and to satisfy the following property,

$$
\Delta_{\mu_1 \cdots \mu_\ell}^{\alpha_1 \cdots \alpha_\ell} \Delta_{\beta_1 \cdots \beta_m}^{\nu_1 \cdots \nu_m} (\mathscr{A}_{rn})_{\nu_1 \cdots \nu_m}^{\mu_1 \cdots \mu_\ell} = (\mathscr{A}_{rn})_{\beta_1 \cdots \beta_m}^{\alpha_1 \cdots \alpha_\ell} .
\tag{6.40}
$$

Since $(\mathscr{A}_{rn})_{\nu_1 \cdots \nu_m}^{\mu_1 \cdots \mu_\ell}$ is orthogonal to u^μ, it can only be constructed from combinations of projection operators, $\Delta^{\mu\nu}$. This already constrains $m + \ell$ to be an even number, since it is impossible to construct odd-ranked tensors solely from projection operators. This means that both ℓ and m are either even or odd. Therefore, the following type of terms could appear in $(\mathscr{A}_{rn})_{\nu_1 \cdots \nu_m}^{\mu_1 \cdots \mu_\ell}$:

(i) Terms where all μ-type indices pair up on projectors $\Delta^{\mu_i \mu_j}$ and all ν-type indices on projectors $\Delta_{\nu_p \nu_q}$, e.g.,

$$\Delta^{\mu_1 \mu_2} \cdots \Delta^{\mu_i \mu_j} \cdots \Delta^{\mu_{\ell-1} \mu_\ell} \Delta_{\nu_1 \nu_2} \cdots \Delta_{\nu_p \nu_q} \cdots \Delta_{\nu_{m-1} \nu_m} . \tag{6.41}$$

All possible permutations of the μ-type indices among themselves and ν-type indices among themselves are allowed.

(ii) Terms where at least one μ-type index pairs with a ν-type index on a projector, e.g.,

$$\Delta^{\mu_1}_{\nu_1} \Delta^{\mu_2 \mu_3} \cdots \Delta^{\mu_i \mu_j} \cdots \Delta^{\mu_{\ell-1} \mu_\ell} \Delta_{\nu_2 \nu_3} \cdots \Delta_{\nu_p \nu_q} \cdots \Delta_{\nu_{m-1} \nu_m} . \tag{6.42}$$

Again, all possible permutations of the μ-type and ν-type indices are allowed. If there is an odd number of projectors of the type $\Delta^{\mu_i}_{\nu_p}$, both ℓ and m must be odd. If there is an even number, both ℓ and m must be even, too. Without loss of generality, suppose that $\ell > m$. For $\ell + m$ to be even, ℓ must be $m + 2, m + 4, \ldots$. Then one could pair all ν-type indices with μ-type indices on projectors of the form $\Delta^{\mu_i}_{\nu_p}$, with some projectors left over which carry only μ-type indices, e.g., $\Delta^{\mu_j \mu_k}$.

(iii) If $\ell = m$, all μ-type indices could be paired up with ν-type indices on projectors of the form $\Delta^{\mu_i}_{\nu_p}$, with no left-over projectors like explained at the end of (ii),

$$\Delta^{\mu_1}_{\nu_1} \cdots \Delta^{\mu_\ell}_{\nu_\ell} . \tag{6.43}$$

Again, all permutations of the μ-type indices among themselves and ν-type indices among themselves are allowed.

Note that terms of the type (i) and (ii) by themselves do not satisfy the property (6.40). This happens because any term which contains at least one projector of the type $\Delta^{\mu_i \mu_j}$ or $\Delta_{\nu_p \nu_q}$ vanishes when contracted with $\Delta^{\alpha_1 \cdots \alpha_\ell}_{\mu_1 \cdots \mu_\ell} \Delta^{\nu_1 \cdots \nu_m}_{\beta_1 \cdots \beta_m}$. Therefore, $(\mathcal{A}_{rn})^{\mu_1 \cdots \mu_\ell}_{\nu_1 \cdots \nu_m}$ cannot be solely constructed from terms of type (i) and (ii), because otherwise it would vanish trivially, and the property (6.40) would not be satisfied. There must at least be one term of type (iii). However, this implies that $m = \ell$. This does not mean that terms of type (i) and (ii) do not appear; they do occur, but in such a way that Eq. (6.40) is satisfied. In summary, $(\mathcal{A}_{rn})^{\mu_1 \cdots \mu_\ell}_{\nu_1 \cdots \nu_m}$ has the form

$$(\mathcal{A}_{rn})^{\mu_1 \cdots \mu_\ell}_{\nu_1 \cdots \nu_m} = \delta_{\ell m} \left\{ \mathcal{A}^{(\ell)}_{rn} \Delta^{(\mu_1}_{(\nu_1} \cdots \Delta^{\mu_\ell)}_{\nu_\ell)} + [\text{terms of type (i) and (ii)}] \right\} , \tag{6.44}$$

where the parentheses denote the symmetrization of all Lorentz indices. Contracting Eq. (6.44) with $\Delta^{\alpha_1 \cdots \alpha_\ell}_{\mu_1 \cdots \mu_\ell} \Delta^{\nu_1 \cdots \nu_\ell}_{\beta_1 \cdots \beta_\ell}$ and using Eq. (6.40), we prove that

$$(\mathcal{A}_{rn})^{\alpha_1 \cdots \alpha_\ell}_{\beta_1 \cdots \beta_m} = \delta_{\ell m} \mathcal{A}^{(\ell)}_{rn} \Delta^{\alpha_1 \cdots \alpha_\ell}_{\beta_1 \cdots \beta_\ell} . \tag{6.45}$$

Finally, substituting Eq. (6.45) into Eq. (6.38), we derive

$$L_{r-1}^{\langle \mu_1 \cdots \mu_\ell \rangle} = -\sum_{n=0}^{\infty} \mathcal{A}_{rn}^{(\ell)} \rho_n^{\mu_1 \cdots \mu_\ell} . \tag{6.46}$$

This completes our goal to express the linear collision term as a linear combination of the irreducible moments. The coefficients $\mathcal{A}_{rn}^{(\ell)}$ can be obtained from the following projection of $(\mathcal{A}_{rn})_{\nu_1 \cdots \nu_\ell}^{\mu_1 \cdots \mu_\ell}$,

$$\mathcal{A}_{rn}^{(\ell)} = \frac{1}{\Delta_{\mu_1 \cdots \mu_\ell}^{\mu_1 \cdots \mu_\ell}} \Delta_{\mu_1 \cdots \mu_\ell}^{\nu_1 \cdots \nu_\ell} (\mathcal{A}_{rn})_{\nu_1 \cdots \nu_\ell}^{\mu_1 \cdots \mu_\ell}$$

$$= \frac{1}{\nu (2\ell + 1)} \int dK \, dK' dP \, dP' \, W_{\mathbf{kk}' \to \mathbf{pp}'} f_{0\mathbf{k}} f_{0\mathbf{k}'} \tilde{f}_{0\mathbf{p}} \tilde{f}_{0\mathbf{p}'} E_{\mathbf{k}}^{r-1} k^{\langle \mu_1} \cdots k^{\mu_\ell \rangle}$$

$$\times \left(\mathcal{H}_{\mathbf{k}n}^{(\ell)} k_{\langle \mu_1} \cdots k_{\mu_\ell \rangle} + \mathcal{H}_{\mathbf{k}'n}^{(\ell)} k'_{\langle \mu_1} \cdots k'_{\mu_\ell \rangle} \right.$$

$$\left. - \mathcal{H}_{\mathbf{p}n}^{(\ell)} p_{\langle \mu_1} \cdots p_{\mu_\ell \rangle} - \mathcal{H}_{\mathbf{p}'n}^{(\ell)} p'_{\langle \mu_1} \cdots p'_{\mu_\ell \rangle} \right) , \tag{6.47}$$

where we used that $\Delta_{\mu_1 \cdots \mu_\ell}^{\mu_1 \cdots \mu_\ell} = 2\ell + 1$. The coefficients $\mathcal{A}_{rn}^{(\ell)}$ are (invertible) matrices (the matrix structure is indicated by the row and column indices r and n) and contain all the information of the underlying microscopic theory. We explicitly compute some of them for the case of a classical gas of massless particles with constant cross section in Chap. 8.

We remark that, for $\ell = 0$ the terms with $n = 1, 2$, and for $\ell = 1$ the term with $n = 1$ are zero, because the moments ρ_1, ρ_2, and ρ_1^μ vanish due to the definition of the velocity field and the matching conditions, Eqs. (5.16) and (5.23). Therefore, in order to invert $\mathcal{A}_{rn}^{(\ell)}$, for $\ell = 0$ we have to exclude the second and third rows and columns, and for $\ell = 1$ the second row and column.

We note that similar matrices already appeared in the context of Chapman–Enskog theory and the method of moments, where they were referred to as $\mathcal{A}_{rn}^{\mathrm{s}}$, $\mathcal{A}_{rn}^{\mathrm{v}}$, and $\mathcal{A}_{rn}^{\mathrm{t}}$, cf. Eq. (5.93). Note that, if one replaces $\mathcal{H}_{\mathbf{k}n}^{(\ell)} \to E_{\mathbf{k}}^n$ in Eq. (6.47), both definitions actually agree (up to prefactors): $\mathcal{A}_{rn}^{\mathrm{s}} \to \mathcal{A}_{rn}^{(0)}$, $\mathcal{A}_{rn}^{\mathrm{v}} \to \beta_0 \mathcal{A}_{rn}^{(1)}$, and $\mathcal{A}_{rn}^{\mathrm{t}} \to \beta_0^2 \mathcal{A}_{rn}^{(2)}$. Therefore, the difference between Chapman–Enskog theory and the method of moments is entirely due to a different basis employed in the moment expansion in each method. In Chapman–Enskog theory the expansion basis employed was $1, E_{\mathbf{k}}, E_{\mathbf{k}}^2, \ldots$, while in the method of moments an orthogonal basis of polynomials was employed, $1, P_{\mathbf{k}1}, P_{\mathbf{k}2}, \ldots$. Since both bases are equivalent, the final result will be approximately the same in both methods, provided that one includes a sufficiently large number of basis elements.

6.3.2　Computation of the Nonlinear Collision Term

Now we turn to the computation of the nonlinear part of the collision integral, which can be performed with the same techniques employed in the previous subsection. Inspecting the previous formulas, Eqs. (6.23), (6.34), and (6.35), we observe that the collision integral is a quartic function of ϕ_k. For the sake of simplicity, we shall restrict ourselves to the case of classical particles, which respect Boltzmann statistics ($a = 0$). In this case, the dependence on ϕ_k becomes quadratic and the generalized collision integral can be formally written as

$$C_{r-1}^{\langle \mu_1 \cdots \mu_\ell \rangle} = L_{r-1}^{\langle \mu_1 \cdots \mu_\ell \rangle} + N_{r-1}^{\langle \mu_1 \cdots \mu_\ell \rangle} , \tag{6.48}$$

where $L_{r-1}^{\langle \mu_1 \cdots \mu_\ell \rangle}$ is the linear contribution discussed in the previous subsection and $N_{r-1}^{\langle \mu_1 \cdots \mu_\ell \rangle}$ is the quadratic contribution to the collision integral,

$$
\begin{aligned}
N_{r-1}^{\langle \mu_1 \cdots \mu_\ell \rangle} &\equiv \frac{1}{\nu} \int dK \, dK' \, dP \, dP' \, W_{kk' \to pp'} f_{0k} f_{0k'} \\
&\quad \times E_k^{r-1} k^{\langle \mu_1} \cdots k^{\mu_\ell \rangle} \left(\phi_p \phi_{p'} - \phi_k \phi_{k'} \right) \\
&= \frac{1}{\nu} \sum_{m,m'=0}^{\infty} \sum_{n=0}^{N_m} \sum_{n'=0}^{N_{m'}} \int dK \, dK' \, dP \, dP' \, W_{kk' \to pp'} f_{0k} f_{0k'} \\
&\quad \times E_k^{r-1} k^{\langle \mu_1} \cdots k^{\mu_\ell \rangle} \rho_n^{\alpha_1 \cdots \alpha_m} \rho_{n'}^{\beta_1 \cdots \beta_{m'}} \\
&\quad \times \left(\mathcal{H}_{pn}^{(m)} \mathcal{H}_{p'n'}^{(m')} P_{\langle \alpha_1} \cdots P_{\alpha_m \rangle} P'_{\langle \beta_1} \cdots P'_{\beta_{m'} \rangle} \right. \\
&\quad \left. - \mathcal{H}_{kn}^{(m)} \mathcal{H}_{k'n'}^{(m')} k_{\langle \alpha_1} \cdots k_{\alpha_m \rangle} k'_{\langle \beta_1} \cdots k'_{\beta_{m'} \rangle} \right) ,
\end{aligned}
\tag{6.49}
$$

where we have substituted Eq. (6.12) for the moment expansion of ϕ_p. Obviously, this nonlinear contribution is a linear combination of quadratic terms in the irreducible moments and can be written in the form

$$N_{r-1}^{\langle \mu_1 \cdots \mu_\ell \rangle} = \sum_{m'=0}^{\infty} \sum_{m=0}^{m'} \sum_{n=0}^{N_m} \sum_{n'=0}^{N_{m'}} \left(\mathcal{N}_{rnn'} \right)_{\alpha_1 \cdots \alpha_m \beta_1 \cdots \beta_{m'}}^{\mu_1 \cdots \mu_\ell} \rho_n^{\alpha_1 \cdots \alpha_m} \rho_{n'}^{\beta_1 \cdots \beta_{m'}} , \tag{6.50}$$

where we defined the tensor

$$
(\mathcal{N}_{rnn'})^{\mu_1\cdots\mu_\ell}_{\alpha_1\cdots\alpha_m\beta_1\cdots\beta_{m'}} = \frac{1}{\nu} \int dK\, dK'\, dP\, dP'\, W_{\mathbf{kk'}\to\mathbf{pp'}}\, f_{0\mathbf{k}}\, f_{0\mathbf{k'}}\, E_{\mathbf{k}}^{r-1} k^{\langle\mu_1}\cdots k^{\mu_\ell\rangle}
$$

$$
\times \Big[\mathcal{H}_{\mathbf{p}n}^{(m)}\, \mathcal{H}_{\mathbf{p}'n'}^{(m')}\, p_{\langle\alpha_1}\cdots p_{\alpha_m\rangle}\, p'_{\langle\beta_1}\cdots p'_{\beta_{m'}\rangle}
$$

$$
+ (1-\delta_{mm'})\, \mathcal{H}_{\mathbf{p}'n}^{(m)}\, \mathcal{H}_{\mathbf{p}n'}^{(m')}\, p'_{\langle\alpha_1}\cdots p'_{\alpha_m\rangle}\, p_{\langle\beta_1}\cdots p_{\beta_{m'}\rangle}
$$

$$
- \mathcal{H}_{\mathbf{k}n}^{(m)}\, \mathcal{H}_{\mathbf{k}'n'}^{(m')}\, k_{\langle\alpha_1}\cdots k_{\alpha_m\rangle}\, k'_{\langle\beta_1}\cdots k'_{\beta_{m'}\rangle}
$$

$$
- (1-\delta_{mm'})\, \mathcal{H}_{\mathbf{k}'n}^{(m)}\, \mathcal{H}_{\mathbf{k}n'}^{(m')}\, k'_{\langle\alpha_1}\cdots k'_{\alpha_m\rangle}\, k_{\langle\beta_1}\cdots k_{\beta_{m'}\rangle} \Big] .
$$

$$(6.51)$$

In comparison with Eq. (6.49), we have split the double sum $\sum_{m=0}^{\infty}\sum_{m'=0}^{\infty}$ into a double sum $\sum_{m'=0}^{\infty}\sum_{m=0}^{m'}$ and a double sum $\sum_{m=0}^{\infty}\sum_{m'=0}^{m}$, and subtracted the superfluous term $m = m'$ in the last sum with the help of a Kronecker delta. Then, we interchanged the summation indices $m \leftrightarrow m'$, $n \leftrightarrow n'$ in the second sum, together with the interchange of the sets $(\alpha_1, \ldots, \alpha_m) \leftrightarrow (\beta_1, \ldots, \beta_{m'})$. As for the linear collision term, the remaining challenge is to express this tensor as a linear combination of scalar terms, multiplied by quadratic terms in the irreducible moments.

The integral $(\mathcal{N}_{rnn'})^{\mu_1\cdots\mu_\ell}_{\alpha_1\cdots\alpha_m\beta_1\cdots\beta_{m'}}$ is a tensor of rank $m + m' + \ell$, which is symmetric under permutations of μ-type, α-type, and β-type indices, and which depends solely on equilibrium distribution functions and the cross section. The equilibrium distribution function contains only one 4-vector, i.e., the fluid 4-velocity u^μ. Therefore, $(\mathcal{N}_{rnn'})^{\mu_1\cdots\mu_\ell}_{\alpha_1\cdots\alpha_m\beta_1\cdots\beta_{m'}}$ must be constructed from tensor structures made of u^μ and the (inverse) metric tensor $g^{\mu\nu}$, or, equivalently, u^μ and $\Delta^{\mu\nu}$. Furthermore, $(\mathcal{N}_{rnn'})^{\mu_1\cdots\mu_\ell}_{\alpha_1\cdots\alpha_m\beta_1\cdots\beta_{m'}}$ must be orthogonal to u^μ, which implies that it can only be constructed from combinations of the rank-2 projection operators, $\Delta^{\mu\nu}$. As happened to $(\mathcal{A}_{rn})^{\alpha_1\cdots\alpha_\ell}_{\beta_1\cdots\beta_m}$, this constrains the rank of the tensor, $m + m' + \ell$, to be an even number. Finally, it must satisfy the following property:

$$
\Delta^{\mu_1\cdots\mu_\ell}_{\mu'_1\cdots\mu'_\ell} \Delta^{\alpha_1\cdots\alpha_m}_{\alpha'_1\cdots\alpha'_m} \Delta^{\beta_1\cdots\beta_{m'}}_{\beta'_1\cdots\beta'_{m'}} (\mathcal{N}_{rnn'})^{\mu_1\cdots\mu_\ell}_{\alpha_1\cdots\alpha_m\beta_1\cdots\beta_{m'}} = (\mathcal{N}_{rnn'})^{\mu'_1\cdots\mu'_\ell}_{\alpha'_1\cdots\alpha'_m\beta'_1\cdots\beta'_{m'}},
$$

$$(6.52)$$

Similar to the procedure employed to calculate $(\mathcal{A}_{rn})^{\mu_1\cdots\mu_\ell}_{\nu_1\cdots\nu_m}$, we collect all possible combinations of projection operators that can appear in $(\mathcal{N}_{rnn'})^{\mu_1\cdots\mu_\ell}_{\alpha_1\cdots\alpha_m\beta_1\cdots\beta_{m'}}$:

(i) Terms where all μ-type indices pair up on projectors, all α-type indices pair up on projectors, and all β-type indices pair up on projectors, e.g.,

$$
\Delta^{\mu_1\mu_2}\cdots\Delta^{\mu_{\ell-1}\mu_\ell}\Delta_{\alpha_1\alpha_2}\cdots\Delta_{\alpha_{m-1}\alpha_m}\Delta_{\beta_1\beta_2}\cdots\Delta_{\beta_{m'-1}\beta_{m'}} .
$$

$$(6.53)$$

All possible permutations of the μ-type, α-type, and β-type indices among themselves are allowed. In this case, ℓ, m, and m' must all be even.

(ii) Terms where at least one μ-type index pairs with an α-type index on a projector, or one μ-type index pairs with a β-type index on a projector, or one α-type index pairs with a β-type index on a projector, e.g.,

$$\Delta_{\alpha_1}^{\mu_1} \Delta^{\mu_2 \mu_3} \cdots \Delta^{\mu_{\ell-1}\mu_\ell} \Delta_{\alpha_2 \alpha_3} \cdots \Delta_{\alpha_{m-1}\alpha_m} \Delta_{\beta_1 \beta_2} \cdots \Delta_{\beta_{m'-1}\beta_{m'}} , \qquad (6.54)$$

$$\Delta_{\beta_1}^{\mu_1} \Delta^{\mu_2 \mu_3} \cdots \Delta^{\mu_{\ell-1}\mu_\ell} \Delta_{\alpha_1 \alpha_2} \cdots \Delta_{\alpha_{m-1}\alpha_m} \Delta_{\beta_2 \beta_3} \cdots \Delta_{\beta_{m'-1}\beta_{m'}} , \qquad (6.55)$$

$$\Delta_{\alpha_1 \beta_1} \Delta^{\mu_1 \mu_2} \cdots \Delta^{\mu_{\ell-1}\mu_\ell} \Delta_{\alpha_2 \alpha_3} \cdots \Delta_{\alpha_{m-1}\alpha_m} \Delta_{\beta_2 \beta_3} \cdots \Delta_{\beta_{m'-1}\beta_{m'}} . \qquad (6.56)$$

Again, all possible permutations of the μ-type, α-type, and β-type indices are allowed.

(iii) Terms where *each* μ-type, α-type, and β-type index pairs up with an index of another type. To guarantee that the μ-type indices have sufficiently many partners among the other two types of indices, one must have $\ell \le m + m'$. Similarly, in order for the α-type indices to pair up in this way, we have to require $m \le \ell + m'$. Finally, for the β-type indices, we need the condition $m' \le \ell + m$. In this case, only projectors of the type $\Delta_{\alpha_j}^{\mu_i}$, $\Delta_{\beta_j}^{\mu_i}$ or $\Delta_{\alpha_i \beta_j}$ exist, with no left-over projectors containing indices of the same type. Such terms have the form

$$\Delta_{\alpha_p}^{\mu_i} \Delta_{\beta_q}^{\mu_j} \Delta_{\alpha_r \beta_s} \cdots . \qquad (6.57)$$

Again, all permutations of the μ-type, α-type, and β-type indices among themselves are allowed.

It is important to emphasize that terms of the type (i) and (ii) by themselves do not satisfy the property (6.52), since any term which contains at least one projector of the type $\Delta^{\mu_i \mu_j}$, $\Delta_{\alpha_p \alpha_q}$, or $\Delta_{\beta_r \beta_s}$ vanishes when contracted with $\Delta_{\mu_1 \cdots \mu_\ell}^{\mu_1' \cdots \mu_\ell'} \Delta_{\alpha_1' \cdots \alpha_m'}^{\alpha_1 \cdots \alpha_m} \Delta_{\beta_1' \cdots \beta_{m'}'}^{\beta_1 \cdots \beta_{m'}}$. Therefore, $(\mathcal{N}_{rnn'})_{\alpha_1 \cdots \alpha_m \beta_1 \cdots \beta_{m'}}^{\mu_1 \cdots \mu_\ell}$ cannot be solely constructed from terms of type (i) and (ii) and there must be at least one term of type (iii).

Apparently terms of type (iii) are of special importance in this derivation and it is convenient to further discuss some of their properties. The inequalities that constrain terms of type (iii), i.e., $\ell \le m + m'$, $m \le \ell + m'$, $m' \le \ell + m$, can be solved and lead to

$$\ell = q + r , \quad m = p + r , \quad m' = p + q ,$$

with $p, q, r = 0, 1, 2, \ldots$. Since the index ℓ is always fixed in the summations appearing in Eq. (6.50), one can re-express the above equations as

$$m = p - q + \ell , \quad m' = p + q , \quad q \le \ell . \qquad (6.58)$$

For the purpose of deriving fluid dynamics, it is sufficient to consider the cases $\ell = 0$, $\ell = 1$, and $\ell = 2$:

6.3.2.1 $\ell = 0$

If $\ell = 0$, the equalities (6.58) imply that $m' = m = 0, 1, \ldots$ and, consequently, one must have

$$(\mathcal{N}_{rnn'})_{\alpha_1 \cdots \alpha_m \beta_1 \cdots \beta_{m'}} = \delta_{mm'} C_{(0)} \Delta_{(\alpha_1 \beta_1} \cdots \Delta_{\alpha_m \beta_m)} + [\text{terms of type (i) and (ii)}]. \tag{6.59}$$

Contracting Eq. (6.59) with $\Delta^{\alpha_1 \cdots \alpha_m}_{\alpha'_1 \cdots \alpha'_m} \Delta^{\beta_1 \cdots \beta_{m'}}_{\beta'_1 \cdots \beta'_{m'}}$ and using Eq. (6.52), we prove that

$$(\mathcal{N}_{rnn'})_{\alpha_1 \cdots \alpha_m \beta_1 \cdots \beta_{m'}} = \delta_{mm'} C_{(0)} \Delta_{\alpha_1 \cdots \alpha_m \beta_1 \cdots \beta_m}, \tag{6.60}$$

where $C_{(0)}$ is the trace of $(\mathcal{N}_{rnn'})_{\alpha_1 \cdots \alpha_m \beta_1 \cdots \beta_{m'}}$,

$$
\begin{aligned}
C_{(0)} &\equiv \left[\Delta^{\alpha_1 \cdots \alpha_m}_{\alpha_1 \cdots \alpha_m} \right]^{-1} \Delta^{\alpha_1 \cdots \alpha_m \beta_1 \cdots \beta_m} (\mathcal{N}_{rnn'})_{\alpha_1 \cdots \alpha_m \beta_1 \cdots \beta_m} \\
&= \frac{1}{(2m+1)\nu} \int dK \, dK' \, dP \, dP' \, W_{\mathbf{kk'} \to \mathbf{pp'}} f_{0\mathbf{k}} f_{0\mathbf{k'}} E^{r-1}_{\mathbf{k}} \\
&\qquad \times \left(\mathcal{H}^{(m)}_{\mathbf{p}n} \mathcal{H}^{(m)}_{\mathbf{p}'n'} p^{\langle \mu_1} \cdots p^{\mu_m \rangle} p'_{\langle \mu_1} \cdots p'_{\mu_m \rangle} \right. \\
&\qquad \left. - \mathcal{H}^{(m)}_{\mathbf{k}n} \mathcal{H}^{(m)}_{\mathbf{k}'n'} k^{\langle \mu_1} \cdots k^{\mu_m \rangle} k'_{\langle \mu_1} \cdots k'_{\mu_m \rangle} \right).
\end{aligned}
\tag{6.61}
$$

Therefore, the scalar nonlinear collision integral from Eq. (6.50) is given by

$$
\begin{aligned}
N_{r-1} &\equiv \sum_{m'=0}^{\infty} \sum_{m=0}^{m'} \sum_{n=0}^{N_m} \sum_{n'=0}^{N_{m'}} (\mathcal{N}_{rnn'})_{\alpha_1 \cdots \alpha_m \beta_1 \cdots \beta_{m'}} \rho_n^{\alpha_1 \cdots \alpha_m} \rho_{n'}^{\beta_1 \cdots \beta_{m'}} \\
&= \sum_{n=0}^{N_0} \sum_{n'=0}^{N_0} C^{0(0,0)}_{rnn'} \rho_n \rho_{n'} + \sum_{m=1}^{\infty} \sum_{n=0}^{N_m} \sum_{n'=0}^{N_m} C^{0(m,m)}_{rnn'} \rho_n^{\alpha_1 \cdots \alpha_m} \rho_{n', \alpha_1 \cdots \alpha_m}, \tag{6.62}
\end{aligned}
$$

where $C^{0(m,m)}_{rnn'}$ is the special case $\ell = 0$ of a more general coefficient,

$$
\begin{aligned}
C^{\ell(m, m+\ell)}_{rnn'} &= \frac{1}{(2m+2\ell+1)\nu} \int dK \, dK' \, dP \, dP' \, W_{\mathbf{kk'} \to \mathbf{pp'}} f_{0\mathbf{k}} f_{0\mathbf{k'}} E^{r-1}_{\mathbf{k}} k^{\langle \mu_1} \cdots k^{\mu_\ell \rangle} \\
&\quad \times \left[\mathcal{H}^{(m)}_{\mathbf{p}n} \mathcal{H}^{(m+\ell)}_{\mathbf{p}'n'} p^{\langle \nu_1} \cdots p^{\nu_m \rangle} p'_{\langle \mu_1} \cdots p'_{\mu_\ell} p'_{\nu_1} \cdots p'_{\nu_m \rangle} \right. \\
&\quad + (1 - \delta_{m, m+\ell}) \mathcal{H}^{(m)}_{\mathbf{p}'n} \mathcal{H}^{(m+\ell)}_{\mathbf{p}n'} p'^{\langle \nu_1} \cdots p'^{\nu_m \rangle} p_{\langle \mu_1} \cdots p_{\mu_\ell} p_{\nu_1} \cdots p_{\nu_m \rangle} \\
&\quad - \mathcal{H}^{(m)}_{\mathbf{k}n} \mathcal{H}^{(m+\ell)}_{\mathbf{k}'n'} k^{\langle \nu_1} \cdots k^{\nu_m \rangle} k'_{\langle \mu_1} \cdots k'_{\mu_\ell} k'_{\nu_1} \cdots k'_{\nu_m \rangle} \\
&\quad \left. - (1 - \delta_{m, m+\ell}) \mathcal{H}^{(m)}_{\mathbf{k}'n} \mathcal{H}^{(m+\ell)}_{\mathbf{k}n'} k'^{\langle \nu_1} \cdots k'^{\nu_m \rangle} k_{\langle \mu_1} \cdots k_{\mu_\ell} k_{\nu_1} \cdots k_{\nu_m \rangle} \right]. \tag{6.63}
\end{aligned}
$$

6.3.2.2 $\ell = 1$

For $\ell = 1$, Eq. (6.58) implies that $m' = m + 1$ (the other possibility is $m = m' + 1$, but the range of the sums in Eq. (6.50) excludes this case) and, consequently,

$$
(\mathcal{N}_{rnn'})^{\mu}_{\alpha_1 \cdots \alpha_m \beta_1 \cdots \beta_{m'}} = \delta_{m+1,m'} C_{(1)} \Delta^{\mu}_{(\beta_1} \Delta_{\beta_2 \alpha_1} \cdots \Delta_{\beta_{m+1} \alpha_m)}
$$
$$
+ \text{[terms of type (i) and (ii)] .} \tag{6.64}
$$

All permutations of the α-indices and β-indices among themselves are allowed, while permutations of the α-indices with the β-indices are forbidden. Contracting Eq. (6.64) with $\Delta^{\mu'}_{\mu} \Delta^{\alpha_1 \cdots \alpha_m}_{\alpha'_1 \cdots \alpha'_m} \Delta^{\beta_1 \cdots \beta_{m+1}}_{\beta'_1 \cdots \beta'_{m+1}}$ and using Eq. (6.52), we prove that

$$
(\mathcal{N}_{rnn'})^{\mu}_{\alpha_1 \cdots \alpha_m \beta_1 \cdots \beta_{m+1}} = C_{(1)} \Delta^{\mu}_{\alpha_1 \cdots \alpha_m \beta_1 \cdots \beta_{m+1}} . \tag{6.65}
$$

The coefficient $C_{(1)}$ is obtained from the trace of $(\mathcal{N}_{rnn'})^{\mu \alpha_1 \cdots \alpha_m}_{\beta_1 \cdots \beta_{m+1}}$, i.e.,

$$
C_{(1)} \equiv \left[\Delta^{\mu \alpha_1 \cdots \alpha_m}_{\beta_1 \cdots \beta_{m+1}} \Delta^{\beta_1 \cdots \beta_{m+1}}_{\mu \alpha_1 \cdots \alpha_m} \right]^{-1} \Delta^{\alpha_1 \cdots \alpha_m \beta_1 \cdots \beta_{m+1}}_{\mu} (\mathcal{N}_{rnn'})^{\mu}_{\alpha_1 \cdots \alpha_m \beta_1 \cdots \beta_{m+1}}
$$
$$
= \frac{1}{[2(m+1)+1]\nu} \int dK dK' dP dP' \, W_{\mathbf{kk'} \to \mathbf{pp'}} f_{0\mathbf{k}} f_{0\mathbf{k'}} E^{r-1}_{\mathbf{k}} k_{\mu}
$$
$$
\times \left(\mathcal{H}^{(m)}_{\mathbf{p}n} \mathcal{H}^{(m+1)}_{\mathbf{p}'n'} \, p_{\langle \alpha_1} \cdots p_{\alpha_m \rangle} \, p'^{\langle \mu} p'^{\alpha_1} \cdots p'^{\alpha_m \rangle} \right.
$$
$$
+ \mathcal{H}^{(m)}_{\mathbf{p}'n} \mathcal{H}^{(m+1)}_{\mathbf{p}n'} \, p'_{\langle \alpha_1} \cdots p'_{\alpha_m \rangle} \, p^{\langle \mu} p^{\alpha_1} \cdots p^{\alpha_m \rangle}
$$
$$
- \mathcal{H}^{(m)}_{\mathbf{k}n} \mathcal{H}^{(m+1)}_{\mathbf{k}'n'} \, k_{\langle \alpha_1} \cdots k_{\alpha_m \rangle} \, k'^{\langle \mu} k'^{\alpha_1} \cdots k'^{\alpha_m \rangle}
$$
$$
\left. - \mathcal{H}^{(m)}_{\mathbf{k}'n} \mathcal{H}^{(m+1)}_{\mathbf{k}n'} \, k'_{\langle \alpha_1} \cdots k'_{\alpha_m \rangle} \, k^{\langle \mu} k^{\alpha_1} \cdots k^{\alpha_m \rangle} \right) . \tag{6.66}
$$

Therefore, the nonlinear collision term for $\ell = 1$ becomes

$$
N^{\mu}_{r-1} \equiv \sum_{m'=0}^{\infty} \sum_{m=0}^{m'} \sum_{n=0}^{N_m} \sum_{n'=0}^{N_{m'}} (\mathcal{N}_{rnn'})^{\mu}_{\alpha_1 \cdots \alpha_m \beta_1 \cdots \beta_{m'}} \rho^{\alpha_1 \cdots \alpha_m}_n \rho^{\beta_1 \cdots \beta_{m'}}_{n'}
$$
$$
= \sum_{n=0}^{N_0} \sum_{n'=0}^{N_1} C^{1(0,1)}_{rnn'} \rho_n \rho^{\mu}_{n'} + \sum_{m=1}^{\infty} \sum_{n=0}^{N_m} \sum_{n'=0}^{N_{m+1}} C^{1(m,m+1)}_{rnn'} \rho^{\alpha_1 \cdots \alpha_m}_n \rho^{\mu}_{n',\alpha_1 \cdots \alpha_m} , \tag{6.67}
$$

where the coefficient $C^{1(m,m+1)}_{rnn'}$ is the $\ell = 1$ case of the general coefficient $C^{\ell(m,m+\ell)}_{rnn'}$ introduced in Eq. (6.63).

6.3.2.3 $\ell = 2$

For terms with $\ell = 2$, two solutions (with $m' \geq m$) are possible: $m' = m = 1, \ldots,$
and $m' = m + 2 = 2, 3, \ldots$. Therefore, two different type-(iii) tensors can be constructed, leading to

$$
\begin{aligned}
(\mathcal{N}_{rnn'})^{\mu\nu}_{\alpha_1 \cdots \alpha_m \beta_1 \cdots \beta_{m'}} &= \delta_{m+2,m'} C_{(2)} \Delta^{\mu}_{(\beta_1} \Delta^{\nu}_{\beta_2} \Delta_{\beta_3 \alpha_1} \cdots \Delta_{\beta_{m+2}\alpha_m)} \\
&+ \delta_{mm'} \mathcal{D}_{(2)} \Delta^{(\mu}_{(\beta_1} \Delta^{\nu)}_{\alpha_1} \Delta_{\beta_2 \alpha_2} \cdots \Delta_{\beta_m \alpha_m)} \\
&+ [\text{terms of type (i) and (ii)}] .
\end{aligned}
\tag{6.68}
$$

All permutations of the α-indices and β-indices among themselves are allowed, while permutations of the α-indices with the β-indices are forbidden. Contracting Eq. (6.68) with $\Delta^{\mu'\nu'}_{\mu\nu} \Delta^{\alpha_1 \cdots \alpha_m}_{\alpha'_1 \cdots \alpha'_m} \Delta^{\beta_1 \cdots \beta_{m'}}_{\beta'_1 \cdots \beta'_{m'}}$ and using Eq. (6.52), we prove that

$$
\begin{aligned}
(\mathcal{N}_{rnn'})^{\mu\nu}_{\alpha_1 \cdots \alpha_m \beta_1 \cdots \beta_{m'}} &= \delta_{m+2,m'} C_{(2)} \Delta^{\mu\nu}_{\alpha_1 \cdots \alpha_m \beta_1 \cdots \beta_{m+2}} \\
&+ \delta_{mm'} \mathcal{D}_{(2)} \Delta^{\mu\nu}_{\lambda_1 \sigma} \Delta^{\sigma}_{\lambda_2 \cdots \lambda_m \alpha_1 \cdots \alpha_m} \Delta^{\lambda_1 \cdots \lambda_m}_{\beta_1 \cdots \beta_m} ,
\end{aligned}
\tag{6.69}
$$

with the coefficients $C_{(2)}$ and $\mathcal{D}_{(2)}$ obtained from the corresponding traces of $(\mathcal{N}_{rnn'})^{\mu\nu}_{\alpha_1 \cdots \alpha_m \beta_1 \cdots \beta_{m'}}$ when $m' = m + 2$ and $m' = m$, respectively. That is, the coefficient $C_{(2)}$ is given by

$$
\begin{aligned}
C_{(2)} &\equiv \left[\Delta^{\mu\nu\alpha_1 \cdots \alpha_m}_{\beta_1 \cdots \beta_{m+2}} \Delta^{\beta_1 \cdots \beta_{m+2}}_{\mu\nu\alpha_1 \cdots \alpha_m} \right]^{-1} \Delta^{\alpha_1 \cdots \alpha_m \beta_1 \cdots \beta_{m+2}}_{\mu\nu} (\mathcal{N}_{rnn'})^{\mu\nu}_{\alpha_1 \cdots \alpha_m \beta_1 \cdots \beta_{m+2}} \\
&= \frac{1}{[2(m+\ell)+1]\nu} \Delta^{\mu\nu\alpha_1 \cdots \alpha_m \beta_1 \cdots \beta_{m+2}} \\
&\quad \times \int dK dK' dP dP' \, W_{\mathbf{kk'} \to \mathbf{pp'}} f_{0\mathbf{k}} f_{0\mathbf{k'}} \, E^{r-1}_{\mathbf{k}} k_{\langle \mu} k_{\nu \rangle} \\
&\quad\quad \times \Big(\mathcal{H}^{(m)}_{\mathbf{p}n} \mathcal{H}^{(m+2)}_{\mathbf{p'}n'} \, p_{\langle \alpha_1} \cdots p_{\alpha_m \rangle} \, p'^{\langle \mu} p'^{\nu} p'^{\alpha_1} \cdots p'^{\alpha_m \rangle} \\
&\quad\quad + \mathcal{H}^{(m)}_{\mathbf{p'}n} \mathcal{H}^{(m+2)}_{\mathbf{p}n'} \, p'_{\langle \alpha_1} \cdots p'_{\alpha_m \rangle} \, p^{\langle \mu} p^{\nu} p^{\alpha_1} \cdots p^{\alpha_m \rangle} \\
&\quad\quad - \mathcal{H}^{(m)}_{\mathbf{k}n} \mathcal{H}^{(m+2)}_{\mathbf{k'}n'} \, k_{\langle \alpha_1} \cdots k_{\alpha_m \rangle} \, k'^{\langle \mu} k'^{\nu} k'^{\alpha_1} \cdots k'^{\alpha_m \rangle} \\
&\quad\quad - \mathcal{H}^{(m)}_{\mathbf{k'}n} \mathcal{H}^{(m+2)}_{\mathbf{k}n'} \, k'_{\langle \alpha_1} \cdots k'_{\alpha_m \rangle} \, k^{\langle \mu} k^{\nu} k^{\alpha_1} \cdots k^{\alpha_m \rangle} \Big) ,
\end{aligned}
\tag{6.70}
$$

while $\mathcal{D}_{(2)}$ is

$$
\begin{aligned}
\mathcal{D}_{(2)} &\equiv \left[d^{(m)} \right]^{-1} \Delta^{\lambda_1 \sigma}_{\mu\nu} \Delta^{\lambda_2 \cdots \lambda_m \alpha_1 \cdots \alpha_m}_{\sigma} \Delta^{\beta_1 \cdots \beta_m}_{\lambda_1 \cdots \lambda_m} (\mathcal{N}_{rnn'})^{\mu\nu}_{\alpha_1 \cdots \alpha_m \beta_1 \cdots \beta_m} \\
&= \left[d^{(m)} \right]^{-1} \frac{1}{\nu} \int dK dK' dP dP' \, W_{\mathbf{kk'} \to \mathbf{pp'}} f_{0\mathbf{k}} f_{0\mathbf{k'}} \, E^{r-1}_{\mathbf{k}} k^{\langle \lambda_1} k_{\sigma \rangle} \\
&\quad \times \Big(\mathcal{H}^{(m)}_{\mathbf{p}n} \mathcal{H}^{(m)}_{\mathbf{p'}n'} \, p^{\langle \sigma} p^{\lambda_2} \cdots p^{\lambda_m} p'_{\langle \lambda_1} \cdots p'_{\lambda_m \rangle} \\
&\quad\quad - \mathcal{H}^{(m)}_{\mathbf{k}n} \mathcal{H}^{(m)}_{\mathbf{k'}n'} \, k^{\langle \sigma} k^{\lambda_2} \cdots k^{\lambda_m} k'_{\langle \lambda_1} \cdots k'_{\lambda_m \rangle} \Big) ,
\end{aligned}
\tag{6.71}
$$

where we defined $d^{(m)} = \Delta^{\rho_1 \tau}_{\lambda_1 \sigma} \Delta^{\rho_2 \cdots \rho_m \alpha_1 \cdots \alpha_m}_{\tau} \Delta^{\sigma}_{\lambda_2 \cdots \lambda_m \alpha_1 \cdots \alpha_m} \Delta^{\lambda_1 \cdots \lambda_m}_{\rho_1 \cdots \rho_m}$. The coefficients $C_{(2)}$ and $\mathcal{D}_{(2)}$ can be identified with the coefficients $C^{2(m,m+2)}_{rnn'}$ and $\mathcal{D}^{2(mm)}_{rnn'}$, respectively, where

$$
\mathcal{D}^{2(mm)}_{rnn'} = \frac{1}{d^{(m)} \nu} \int dK dK' dP dP' \, W_{\mathbf{kk'} \to \mathbf{pp'}} f_{0\mathbf{k}} f_{0\mathbf{k'}} \, E^{r-1}_{\mathbf{k}} k^{\langle \mu} k^{\nu \rangle}
$$

$$
\times \left(\mathcal{H}^{(m)}_{\mathbf{p}n} \mathcal{H}^{(m)}_{\mathbf{p}'n'} \, p_{\langle \mu} p^{\beta_2} \cdots p^{\beta_m \rangle} \, p'_{\langle \nu} p'_{\beta_2} \cdots p'_{\beta_m \rangle} \right.
$$

$$
\left. - \mathcal{H}^{(m)}_{\mathbf{k}n} \mathcal{H}^{(m)}_{\mathbf{k}'n'} \, k_{\langle \mu} k^{\beta_2} \cdots k^{\beta_m \rangle} \, k'_{\langle \nu} k'_{\beta_2} \cdots k'_{\beta_m \rangle} \right) . \quad (6.72)
$$

Thus, the nonlinear collision term for $\ell = 2$ becomes

$$
N^{\mu \nu}_{r-1} = \sum_{m=0}^{\infty} \sum_{n=0}^{N_m} \sum_{n'=0}^{N_{m'}} \delta_{m+2,m'} C^{2(m,m+2)}_{rnn'} \rho^{\alpha_1 \cdots \alpha_m}_{n} \rho^{\mu \nu}_{n' \alpha_1 \cdots \alpha_m}
$$

$$
+ \sum_{m=1}^{\infty} \sum_{n=0}^{N_m} \sum_{n'=0}^{N_{m'}} \delta_{mm'} \mathcal{D}^{2(m,m)}_{rnn'} \rho^{\langle \mu}_{n, \lambda_2 \cdots \lambda_m} \rho^{\nu \rangle \lambda_2 \cdots \lambda_m}_{n'} . \quad (6.73)
$$

Obtaining a general expression for the trace $d^{(m)}$ for an arbitrary value of m can be rather complicated. We restrict ourselves to the cases $m = 1, 2$, which are needed for the purposes of deriving hydrodynamics. It follows that, for $m = 1$,

$$
d^{(1)} \equiv \Delta^{\rho_1 \tau}_{\lambda_1 \sigma} \Delta^{\alpha_1}_{\tau} \Delta^{\sigma}_{\alpha_1} \Delta^{\lambda_1}_{\rho_1} = \Delta^{\lambda_1 \sigma}_{\lambda_1 \sigma} = 5 , \quad (6.74)
$$

while for $m = 2$ one obtains

$$
d^{(2)} \equiv \Delta^{\rho_1 \tau}_{\lambda_1 \sigma} \Delta^{\rho_2 \alpha_1 \alpha_2}_{\tau} \Delta^{\sigma}_{\lambda_2 \alpha_1 \alpha_2} \Delta^{\lambda_1 \lambda_2}_{\rho_1 \rho_2} = \frac{35}{12} .
$$

6.4   Summary

In this chapter, we have presented a relativistic generalization of the method of moments. We constructed an orthonormal basis in terms of irreducible tensors $k^{\langle \mu_1} \cdots k^{\mu_\ell \rangle}$ and polynomials in powers of $E_{\mathbf{k}}$, which allowed us to expand the deviation $\delta f_{\mathbf{k}}$ of the single-particle distribution function $f_{\mathbf{k}}$ from a reference state (which we took to be the distribution function $f_{0\mathbf{k}}$ in local thermodynamical equilibrium) in terms of the irreducible moments $\rho^{\mu_1 \cdots \mu_\ell}_n$ of the deviations of the distribution function from equilibrium. We then proceeded to derive exact equations of motion for these moments. Finally, we also showed how to compute the moments $C^{\mu_1 \cdots \mu_\ell}_n$ of the collision term in the Boltzmann equation, which appear in the equations of motion for the irreducible moments. We separated the moments into a linear and a nonlinear part and expressed the former as linear combination of irreducible moments, while the latter is, in the classical limit, a linear combination of quadratic terms in the irreducible moments.

The method of moments will be studied in more detail in the next chapter, where we shall consider simplified scenarios where the complete set of equations of motion for the moments can be derived, solved, and their fluid-dynamical limit systematically studied. In Chap. 8, we then demonstrate how the fluid-dynamical equations can be derived from the method of moments constructed in this chapter.

6.5 Appendix 1: Irreducible Projection Operators

In this appendix, we present the irreducible projection operators necessary to derive the irreducible moments of $\delta f_{\mathbf{k}}$. We start by recalling the definition of the irreducible projection operators [6,20,23],

$$
\Delta^{\mu_1 \cdots \mu_n}_{\nu_1 \cdots \nu_n} = \sum_{q=0}^{[n/2]} C(n, q) \frac{1}{\mathcal{N}_{nq}}
$$
$$
\times \sum_{\mathcal{P}^n_\mu \mathcal{P}^n_\nu} \Delta^{\mu_1 \mu_2} \cdots \Delta^{\mu_{2q-1} \mu_{2q}} \Delta_{\nu_1 \nu_2} \cdots \Delta_{\nu_{2q-1} \nu_{2q}} \Delta^{\mu_{2q+1}}_{\nu_{2q+1}} \cdots \Delta^{\mu_n}_{\nu_n} .
$$
(6.75)

Here, $[n/2]$ denotes the largest integer less than or equal to $n/2$, the coefficients $C(n, q)$ are defined as

$$
C(n, q) = (-1)^q \frac{(n!)^2}{(2n)!} \frac{(2n - 2q)!}{q!(n - q)!(n - 2q)!} ,
$$
(6.76)

and the second sum in Eq. (6.75) runs over all *distinct* permutations $\mathcal{P}^n_\mu \mathcal{P}^n_\nu$ of μ- and ν-type indices. The coefficient in front of this sum is just the inverse of the total number of these distinct permutations,

$$
\mathcal{N}_{nq} \equiv \frac{1}{(n - 2q)!} \left(\frac{n!}{2^q q!} \right)^2 .
$$
(6.77)

This number can be explained as follows: $(n!)^2$ is the number of *all* permutations of μ- and ν-type indices. In order to obtain the number of *distinct* permutations, one has to divide this by the number $(2^q)^2$ of permutations of μ- and ν-type indices on the same Δ-projectors (where only projectors with only μ- and only ν-type indices are considered), and by the number $(q!)^2$ of trivial reorderings of the sequence of these projectors. Finally, one also has to divide by the number $(n - 2q)!$ of trivial reorderings of the sequence of projectors with mixed indices.

The projectors (6.75) are symmetric under exchange of μ- and ν-type indices,

$$
\Delta^{\mu_1 \cdots \mu_n}_{\nu_1 \cdots \nu_n} = \Delta^{(\mu_1 \cdots \mu_n)}_{(\nu_1 \cdots \nu_n)} ,
$$
(6.78)

and traceless with respect to contraction of either μ- or ν-type indices,

$$\Delta^{\mu_1\cdots\mu_n}_{\nu_1\cdots\nu_n} g_{\mu_i\mu_j} = \Delta^{\mu_1\cdots\mu_n}_{\nu_1\cdots\nu_n} g^{\nu_i\nu_j} = 0 \quad \text{for any } i, j \, . \tag{6.79}$$

Moreover, upon complete contraction,

$$\Delta^{\mu_1\cdots\mu_n}_{\mu_1\cdots\mu_n} \equiv \Delta^{\mu_1\cdots\mu_n\nu_1\cdots\nu_n} g_{\mu_1\nu_1} \cdots g_{\mu_n\nu_n} = 2n + 1 \, , \tag{6.80}$$

cf. Eq. (23) in Chap. VI.2 of Ref. [6]. Note that the relation (6.80) means that the projection of an arbitrary tensor of rank n with respect to Eq. (6.75), i.e., $A^{\nu_1\cdots\nu_n} \Delta^{\mu_1\cdots\mu_n}_{\nu_1\cdots\nu_n}$ has $2n + 1$ independent tensor components.

In order to prove this, we note that an arbitrary tensor $A_d^{\mu_1\cdots\mu_n}$ of rank n in d-dimensional space-time has d^n independent components, because each of the n indices can assume d distinct values. Now consider a rank-n tensor which is completely symmetric with respect to the interchange of indices. This tensor can be constructed from the arbitrary tensor $A_d^{\mu_1\cdots\mu_n}$ via symmetrization,

$$A_d^{(\mu_1\cdots\mu_n)} = \frac{1}{n!} \sum_{\mathcal{P}_\mu} A_d^{\mu_1\cdots\mu_n} \, , \tag{6.81}$$

where the sum over $\mathcal{P}_\mu$ runs over all $n!$ permutations of the μ-type indices. The number of independent tensor components of such a symmetric tensor is given by the number of combinations with repetition to draw n elements from a set of d elements,

$$N_{dn}\left(A_d^{(\mu_1\cdots\mu_n)}\right) = \frac{(n + d - 1)!}{n!\,(d - 1)!} \, . \tag{6.82}$$

Let us now demand in addition that this tensor is traceless,

$$0 = A_d^{(\mu_1\cdots\mu_n)} g_{\mu_{n-1}\mu_n} \equiv A_d^{(\mu_1\cdots\mu_{n-2})} \, , \tag{6.83}$$

where the right-hand side defines a new symmetric tensor of rank $n - 2$. According to Eq. (6.82), this tensor has

$$N_{dn}\left(A_d^{(\mu_1\cdots\mu_{n-2})}\right) = \frac{(n + d - 3)!}{(n - 2)!\,(d - 1)!} \tag{6.84}$$

independent components. This is also the number of constraints by which the number of independent components of the original symmetric tensor $A_d^{(\mu_1\cdots\mu_n)}$ is reduced, if we demand that it is traceless in addition to being symmetric. Thus, the number of independent components of a symmetric traceless tensor is

$$\begin{aligned} N_{dn}\left(A_{d,\text{tr-less}}^{(\mu_1\cdots\mu_n)}\right) &= N_{dn}\left(A_d^{(\mu_1\cdots\mu_n)}\right) - N_{dn}\left(A_d^{(\mu_1\cdots\mu_{n-2})}\right) \\ &= \frac{(n + d - 1)!}{n!\,(d - 1)!} - \frac{(n + d - 3)!}{(n - 2)!\,(d - 1)!} \\ &= \frac{(n + d - 3)!}{n!\,(d - 2)!}(2n + d - 2) \, . \end{aligned} \tag{6.85}$$

Let us now require in addition that such a symmetric traceless tensor is orthogonal to a given 4-vector u^μ,

$$0 = A_{d,\text{tr}-\text{less}}^{(\mu_1 \cdots \mu_n)} u_{\mu_n} \equiv A_{d,\text{tr}-\text{less}}^{(\mu_1 \cdots \mu_{n-1})}. \tag{6.86}$$

The right-hand side defines a new symmetric traceless tensor of rank $n-1$ which, according to Eq. (6.85), has

$$N_{dn}\left(A_{d,\text{tr}-\text{less}}^{(\mu_1 \cdots \mu_{n-1})}\right) = \frac{(n+d-4)!}{(n-1)!\,(d-2)!}(2n+d-4) \tag{6.87}$$

independent components. This number reduces the number of independent components of the original symmetric traceless tensor, if we demand in addition that it is orthogonal to u^μ; thus the latter has

$$\begin{aligned}
N_{dn}\left(A_{d,\text{tr}-\text{less}\perp}^{(\mu_1 \cdots \mu_n)}\right) &= N_{dn}\left(A_{d,\text{tr}-\text{less}}^{(\mu_1 \cdots \mu_n)}\right) - N_{dn}\left(A_{d,\text{tr}-\text{less}}^{(\mu_1 \cdots \mu_{n-1})}\right)\\
&= \frac{(n+d-3)!}{n!\,(d-2)!}(2n+d-2) - \frac{(n+d-4)!}{(n-1)!\,(d-2)!}(2n+d-4)\\
&= \frac{(n+d-4)!}{n!\,(d-3)!}(2n+d-3)
\end{aligned} \tag{6.88}$$

independent components. Comparing this equation to Eq. (6.85), we realize that the orthogonality constraint (6.86) has effectively reduced the number of dimensions by one unit, $d \to d-1$. Subsequently demanding orthogonality to another 4-vector l^μ, as in Chap. 9, would reduce the number of dimensions by another unit, etc.

Now taking $d = 4$, Eq. (6.88) tells us that any symmetric traceless tensor of rank n, which is orthogonal to u^μ, has $N_{4n}(A_{4,\text{tr}-\text{less}\perp}^{(\mu_1 \cdots \mu_n)}) = 2n+1$ independent components. If this tensor is in addition orthogonal to another 4-vector l^μ, then Eq. (6.88) applies replacing $d = 4$ by $d = 3$, and we obtain $N_{3n}(A_{3,\text{tr}-\text{less}\perp}^{(\mu_1 \cdots \mu_n)}) = 2$ independent components. This result is independent of the tensor rank n.

6.6 Appendix 2: Thermodynamic Integrals and Properties

In this appendix, we compute the thermodynamic integrals $I_{i+n,q}$ in Eqs. (6.30). They are obtained by suitable projections of the tensors

$$I_i^{\mu_1 \cdots \mu_n} = \left\langle E_{\mathbf{k}}^i\, k^{\mu_1} \cdots k^{\mu_n} \right\rangle_0, \tag{6.89}$$

where the angular brackets denote the average over momentum space defined in Eq. (5.24). The subscript i on this quantity reflects the power of $E_{\mathbf{k}}$ in the definition of the moment. Due to the fact that the equilibrium distribution function depends only on the quantities α_0, β_0, and the flow velocity u^μ, the equilibrium moments can be expanded in terms of u^μ and the projector $\Delta^{\mu\nu}$ as

$$I_i^{\mu_1 \cdots \mu_n} = \sum_{q=0}^{[n/2]} (-1)^q\, b_{nq}\, I_{i+n,q}\, \Delta^{(\mu_1\mu_2} \cdots \Delta^{\mu_{2q-1}\mu_{2q}} u^{\mu_{2q+1}} \cdots u^{\mu_n)}, \tag{6.90}$$

where n, q are natural numbers while the sum runs over $0 \leq q \leq [n/2]$. Here, $[n/2]$ denotes the largest integer which is less than or equal to $n/2$.

The coefficient b_{nq} in Eq. (6.90) is defined as the number of distinct terms in the symmetrized tensor product

$$\Delta^{(\mu_1\mu_2} \cdots \Delta^{\mu_{2q-1}\mu_{2q}} u^{\mu_{2q+1}} \cdots u^{\mu_n)}$$

$$\equiv \frac{1}{b_{nq}} \sum_{\mathcal{P}^n_\mu} \Delta^{\mu_1\mu_2} \cdots \Delta^{\mu_{2q-1}\mu_{2q}} u^{\mu_{2q+1}} \cdots u^{\mu_n} , \tag{6.91}$$

where the sum runs over all distinct permutations of the n indices $\mu_1, \ldots, \mu_n$. The total number of permutations of n indices is $n!$. There are q projection operators $\Delta^{\mu_i\mu_j}$ and $n - 2q$ factors of u^{μ_k}. Permutations of the order of the $\Delta^{\mu_i\mu_j}$ and of the u^{μ_k} among themselves do not lead to distinct terms, so we need to divide the total number $n!$ by $q!(n - 2q)!$. Finally, since $\Delta^{\mu_i\mu_j}$ is a symmetric projection operator, a permutation of its indices does not lead to a distinct term. Since there are q such projection operators, there are 2^q permutations that also do not lead to distinct terms. Hence, the total number of distinct terms in the symmetrized tensor product is

$$b_{nq} \equiv \frac{n!}{2^q q! (n - 2q)!} = \frac{n! (2q - 1)!!}{(2q)! (n - 2q)!} , \tag{6.92}$$

which is identical to Eq. (A2) of Ref. [4].

In order to obtain the thermodynamic integrals $I_{i+n,q}$ by projection of the tensors $I_i^{\mu_1\cdots\mu_n}$, it is advantageous to use the orthogonality relation [6]

$$\Delta^{(\mu_1\mu_2} \cdots \Delta^{\mu_{2q-1}\mu_{2q}} u^{\mu_{2q+1}} \cdots u^{\mu_n)} \Delta_{(\mu_1\mu_2} \cdots \Delta_{\mu_{2q'-1}\mu_{2q'}} u_{\mu_{2q'+1}} \cdots u_{\mu_n)}$$

$$= \frac{(2q + 1)!!}{b_{nq}} \delta_{qq'} . \tag{6.93}$$

Let us prove this relation. First, it is clear that if $q \neq q'$ there are terms where a u^{μ_i} gets contracted with a $\Delta_{\mu_i\mu_j}$, which gives zero. The existence of the Kronecker delta is thus easily explained and we only need to prove Eq. (6.93) for $q = q'$. Second, as the same set of indices is symmetrized on both tensor products, it actually suffices to keep the set of indices fixed on one tensor, say in the order $\mu_1, \ldots, \mu_{2q}, \mu_{2q+1}, \ldots, \mu_n$, and symmetrize only the one on the other,

$$\Delta^{(\mu_1\mu_2} \cdots \Delta^{\mu_{2q-1}\mu_{2q}} u^{\mu_{2q+1}} \cdots u^{\mu_n)} \Delta_{(\mu_1\mu_2} \cdots \Delta_{\mu_{2q-1}\mu_{2q}} u_{\mu_{2q+1}} \cdots u_{\mu_n)}$$

$$= \Delta^{\mu_1\mu_2} \cdots \Delta^{\mu_{2q-1}\mu_{2q}} u^{\mu_{2q+1}} \cdots u^{\mu_n} \frac{1}{b_{nq}} \sum_{\mathcal{P}^n_\mu} \Delta_{\mu_1\mu_2} \cdots \Delta_{\mu_{2q-1}\mu_{2q}} u_{\mu_{2q+1}} \cdots u_{\mu_n} , \tag{6.94}$$

where we used Eq. (6.91). Among the terms in the sum over all distinct permutations, only those survive where the indices on the u's are $\mu_{2q+1}, \ldots, \mu_n$, just as in the term in front of the sum. (Otherwise, a u_{μ_i} will be contracted with a $\Delta^{\mu_i\mu_j}$, which gives zero.) Permutations among these indices do not lead to distinct terms. Using $u^\mu u_\mu = 1$, we thus obtain

$$\Delta^{(\mu_1\mu_2} \cdots \Delta^{\mu_{2q-1}\mu_{2q}} u^{\mu_{2q+1}} \cdots u^{\mu_n)} \Delta_{(\mu_1\mu_2} \cdots \Delta_{\mu_{2q-1}\mu_{2q}} u_{\mu_{2q+1}} \cdots u_{\mu_n)}$$

$$= \frac{1}{b_{nq}} \Delta^{\mu_1\mu_2} \cdots \Delta^{\mu_{2q-1}\mu_{2q}} \sum_{\mathcal{P}_\mu^{2q}} \Delta_{\mu_1\mu_2} \cdots \Delta_{\mu_{2q-1}\mu_{2q}} , \qquad (6.95)$$

where the sum now runs only over the distinct permutations of $2q$ indices $\mu_1, \ldots, \mu_{2q}$ on the Δ projectors. There are in total $(2q)!/(2^q q!) \equiv (2q-1)!!$ distinct terms, so that we obtain

$$\Delta^{(\mu_1\mu_2} \cdots \Delta^{\mu_{2q-1}\mu_{2q}} u^{\mu_{2q+1}} \cdots u^{\mu_n)} \Delta_{(\mu_1\mu_2} \cdots \Delta_{\mu_{2q-1}\mu_{2q}} u_{\mu_{2q+1}} \cdots u_{\mu_n)}$$

$$= \frac{(2q-1)!!}{b_{nq}} \Delta^{\mu_1\mu_2} \cdots \Delta^{\mu_{2q-1}\mu_{2q}} \Delta_{(\mu_1\mu_2} \cdots \Delta_{\mu_{2q-1}\mu_{2q})} . \qquad (6.96)$$

The proof of Eq. (6.93) is completed by proving that [6]

$$\Delta^{\mu_1\mu_2} \cdots \Delta^{\mu_{2q-1}\mu_{2q}} \Delta_{(\mu_1\mu_2} \cdots \Delta_{\mu_{2q-1}\mu_{2q})} = 2q + 1 . \qquad (6.97)$$

This is done by complete induction. Since $\Delta^{\mu_1\mu_2} \Delta_{\mu_1\mu_2} = \Delta^{\mu_1}_{\mu_1} = 3$, Eq. (6.97) obviously holds for $q = 1$. Now suppose it holds for q. Then, we have to show that it also holds for $q + 1$. In this case, using the definition of the symmetrized tensor,

$$\Delta^{\mu_1\mu_2} \cdots \Delta^{\mu_{2q+1}\mu_{2q+2}} \Delta_{(\mu_1\mu_2} \cdots \Delta_{\mu_{2q+1}\mu_{2q+2})}$$

$$= \frac{2^{q+1}(q+1)!}{(2q+2)!} \Delta^{\mu_1\mu_2} \cdots \Delta^{\mu_{2q+1}\mu_{2q+2}} \sum_{\mathcal{P}_\mu^{2q+2}} \Delta_{\mu_1\mu_2} \cdots \Delta_{\mu_{2q+1}\mu_{2q+2}} . \qquad (6.98)$$

Consider the contraction of $\Delta^{\mu_{2q+1}\mu_{2q+2}}$ with the sum over distinct permutations of $2q + 2$ indices $\mu_1, \ldots, \mu_{2q+2}$. There is one term in the sum where both indices are on the same Δ projector. This term is $\sim \Delta^{\mu_{2q+1}\mu_{2q+2}} \Delta_{\mu_{2q+1}\mu_{2q+2}} \equiv 3$. Then, there are $2q$ terms where the indices μ_{2q+1} and μ_{2q+2} are on different projectors, say $\Delta_{\mu_{2q+1}\mu_j} \Delta_{\mu_i\mu_{2q+2}}$. Contracting with $\Delta^{\mu_{2q+1}\mu_{2q+2}}$ gives a term $\sim \Delta_{\mu_i\mu_j}$, where both indices are from the set $\mu_1, \ldots, \mu_{2q}$. Putting this together and using Eq. (6.97) gives

$$\Delta^{\mu_1\mu_2} \cdots \Delta^{\mu_{2q+1}\mu_{2q+2}} \Delta_{(\mu_1\mu_2} \cdots \Delta_{\mu_{2q+1}\mu_{2q+2})}$$

$$= \frac{2(q+1)}{(2q+2)(2q+1)} \frac{2^q q!}{(2q)!} \Delta^{\mu_1\mu_2} \cdots \Delta^{\mu_{2q-1}\mu_{2q}} (2q+3) \sum_{\mathcal{P}_\mu^{2q}} \Delta_{\mu_1\mu_2} \cdots \Delta_{\mu_{2q-1}\mu_{2q}}$$

$$= \frac{2q+3}{2q+1} \Delta^{\mu_1\mu_2} \cdots \Delta^{\mu_{2q-1}\mu_{2q}} \Delta_{(\mu_1\mu_2} \cdots \Delta_{\mu_{2q-1}\mu_{2q})} \equiv 2q + 3 , \quad \text{q.e.d.} .$$

$$\qquad (6.99)$$

With the orthogonality relation (6.93), we now easily find by projecting Eq. (6.90) that

$$
\begin{aligned}
I_{i+n,q} &\equiv \frac{(-1)^q}{(2q+1)!!} \, I_i^{\mu_1 \cdots \mu_n} \, \Delta_{(\mu_1 \mu_2} \cdots \Delta_{\mu_{2q-1}\mu_{2q}} u_{\mu_{2q+1}} \cdots u_{\mu_n)} \\
&= \frac{(-1)^q}{(2q+1)!!} \int dK \, E_{\mathbf{k}u}^{i+n-2q} \left(\Delta^{\mu\nu} k_\mu k_\nu \right)^q f_{0\mathbf{k}} \,,
\end{aligned} \tag{6.100}
$$

where we used the definition (6.89) of the tensor $I_i^{\mu_1 \cdots \mu_n}$. With the definition of the thermodynamic average $\langle \cdots \rangle_0$, the second line yields Eq. (6.30).

Other useful relations are obtained from contracting two indices of the tensors (6.89) with a Δ-projector,

$$
I_i^{\mu_1 \cdots \mu_n} \Delta_{\mu_{n-1}\mu_n} = m^2 \, I_i^{\mu_1 \cdots \mu_{n-2}} - I_{i+2}^{\mu_1 \cdots \mu_{n-2}} \,. \tag{6.101}
$$

Comparison of Eq. (6.89) for $n = 0$ and Eq. (6.100) for $n = q = 0$ yields the identity

$$
I_{i,0} = I_i \,, \tag{6.102}
$$

and comparison of Eq. (6.89) for $n = 2$ and Eq. (6.100) for $n = 2, q = 1$

$$
I_{i+2,1} = -\frac{1}{3} I_i^{\mu\nu} \Delta_{\mu\nu} = -\frac{1}{3} \left(m^2 I_i - I_{i+2} \right) = -\frac{1}{3} \left(m^2 I_{i,0} - I_{i+2,0} \right) \,. \tag{6.103}
$$

The thermodynamic integrals (6.30) obey useful recursion relations, which are given here. Replacing $\left(\Delta^{\alpha\beta} k_\alpha k_\beta \right)^{q+1} = \left(\Delta^{\alpha\beta} k_\alpha k_\beta \right)^q \left(m^2 - E_{\mathbf{k}}^2 \right)$ in Eq. (6.30) we obtain for $0 \le q \le n/2$,

$$
I_{n+2,q} = m^2 \, I_{nq} + (2q + 3) \, I_{n+2,q+1} \,, \tag{6.104}
$$

cf. Eq. (5.54). For $n = q = 0$ this reads

$$
I_{20} = m^2 \, I_{00} + 3 I_{21} \,, \tag{6.105}
$$

which is consistent with Eq. (6.103) for $i = 0$. In the massless limit, this leads to the familiar relation $\varepsilon_0 = 3 P_0$.

6.7   Appendix 3: Orthogonality of the Irreducible Tensors

In this appendix, we derive the orthogonality condition (6.7) for the irreducible tensors (6.4). The derivation utilizes the relation

$$
k^{\langle \mu_1} \cdots k^{\mu_\ell \rangle} k_{\langle \mu_1} \cdots k_{\mu_\ell \rangle} = \frac{\ell!}{(2\ell - 1)!!} \left(\Delta^{\alpha\beta} k_\alpha k_\beta \right)^\ell \,. \tag{6.106}
$$

This relation is proved as follows. We first note that

$$k^{\langle \mu_1} \cdots k^{\mu_\ell \rangle} k_{\langle \mu_1} \cdots k_{\mu_\ell \rangle} = \Delta^{\mu_1 \cdots \mu_\ell}_{\beta_1 \cdots \beta_\ell} \Delta^{\alpha_1 \cdots \alpha_\ell}_{\mu_1 \cdots \mu_\ell} k_{\alpha_1} \cdots k_{\alpha_\ell} k^{\beta_1} \cdots k^{\beta_\ell}$$
$$= \Delta^{\alpha_1 \cdots \alpha_\ell}_{\beta_1 \cdots \beta_\ell} k_{\alpha_1} \cdots k_{\alpha_\ell} k^{\beta_1} \cdots k^{\beta_\ell} . \tag{6.107}$$

Now we insert the explicit form (6.75) of the projection operator and note that the contraction of all indices with the momenta reduces the second sum (including the prefactor $1/\mathcal{N}_{nq}$) to just a factor of $(\Delta^{\alpha\beta} k_\alpha k_\beta)^\ell$,

$$k^{\langle \mu_1} \cdots k^{\mu_\ell \rangle} k_{\langle \mu_1} \cdots k_{\mu_\ell \rangle} = \sum_{q=0}^{[\ell/2]} C(\ell, q) \left(\Delta^{\alpha\beta} k_\alpha k_\beta \right)^\ell . \tag{6.108}$$

The Legendre polynomials $P_\ell(z)$ have the representation [24]

$$P_\ell(z) = \frac{1}{2^\ell} \sum_{q=0}^{[\ell/2]} (-1)^q \frac{(2\ell - 2q)!}{q!(\ell - q)!(\ell - 2q)!} z^{\ell-2q} \equiv \frac{1}{2^\ell} \frac{(2\ell)!}{(\ell!)^2} \sum_{q=0}^{[\ell/2]} C(\ell, q) z^{\ell-2q} . \tag{6.109}$$

Since for all ℓ

$$1 \equiv P_\ell(1) = \frac{1}{2^\ell} \frac{(2\ell)!}{(\ell!)^2} \sum_{q=0}^{[\ell/2]} C(\ell, q) , \tag{6.110}$$

we derive the identity

$$\sum_{q=0}^{[\ell/2]} C(\ell, q) = \frac{2^\ell (\ell!)^2}{(2\ell)!} = \frac{\ell! \, 2^{\ell-1} (\ell - 1)!}{(2\ell - 1)!} \equiv \frac{\ell!}{(2\ell - 1)!!} , \tag{6.111}$$

where we have used the definition of the double factorial for odd numbers. Inserting this into Eq. (6.108) proves Eq. (6.106).

The orthogonality condition (6.7) is obtained from an integral of the type

$$M^{\langle \mu_1 \cdots \mu_\ell \rangle}_{\langle \nu_1 \cdots \nu_n \rangle} = \int dK \, F(E_{\mathbf{k}}) \, k^{\langle \mu_1} \cdots k^{\mu_\ell \rangle} k_{\langle \nu_1} \cdots k_{\nu_n \rangle} , \tag{6.112}$$

which is a tensor of rank $(\ell + n)$ that is (separately) symmetric under the permutation of μ-type and ν-type indices. In Appendix A of Ref. [20], it is proven that tensors of this type must obey the relation

$$M^{\langle \mu_1 \cdots \mu_\ell \rangle}_{\langle \nu_1 \cdots \nu_n \rangle} = \delta_{\ell n} \, M \, \Delta^{\mu_1 \cdots \mu_\ell}_{\nu_1 \cdots \nu_n} , \tag{6.113}$$

where M is an invariant scalar that can be computed by completely contracting the indices of $M^{\langle \mu_1 \cdots \mu_\ell \rangle}_{\langle \nu_1 \cdots \nu_n \rangle}$,

$$\mathcal{M} \equiv \frac{1}{\Delta^{\mu_1 \cdots \mu_\ell}_{\mu_1 \cdots \mu_\ell}} \int dK \, \mathrm{F}(E_{\mathbf{k}}) \, k^{\langle \mu_1} \cdots k^{\mu_\ell \rangle} k_{\langle \mu_1} \cdots k_{\mu_\ell \rangle}$$

$$= \frac{\ell!}{(2\ell+1)!!} \int dK \, \mathrm{F}(E_{\mathbf{k}}) \left(\Delta^{\alpha\beta} k_\alpha k_\beta \right)^\ell , \qquad (6.114)$$

where we have used Eqs. (6.80) and (6.106). This proves Eq. (6.7).

6.8   Appendix 4: Orthogonal Polynomials

The orthogonal polynomials $P^{(\ell)}_{\mathbf{k}n}$ are defined by orthogonality condition (6.8) with the measure $\omega^{(\ell)}$ given by Eq. (6.9).

The polynomials $P^{(\ell)}_{\mathbf{k}m}$ are generated by the set $1, E_{\mathbf{k}}, E^2_{\mathbf{k}}, \ldots$. We follow Ref. [25] and construct this orthogonal set via the Gram–Schmidt orthogonalization method and using the measure (6.9). Without loss of generality, the first polynomial of each series is always set to one, i.e., $P^{(\ell)}_{\mathbf{k}0} = a^{(\ell)}_{00} = 1$. Then, the weights $W^{(\ell)}$ from Eq. (6.9) can be determined enforcing that the normalization condition (6.8) is satisfied for $m = n = 0$,

$$\int dK \, \omega^{(\ell)} = 1 . \qquad (6.115)$$

This will lead to the following expressions for the weights,

$$W^{(\ell)} = \frac{(-1)^\ell}{\tilde{J}_{2\ell,\ell}} , \qquad (6.116)$$

where it was convenient to define the thermodynamic integrals $\tilde{J}_{nq}$

$$\tilde{J}_{nq} \left(\alpha_0, \beta_0 \right) = \int dK \, E^{n-2q}_{\mathbf{k}} \left(-\Delta_{\alpha\beta} k^\alpha k^\beta \right)^q f_{0\mathbf{k}} \tilde{f}_{0\mathbf{k}} \mathcal{G}_{\mathbf{k}} . \qquad (6.117)$$

With this knowledge, the remaining coefficients $a^{(\ell)}_{nm}$ can be extracted from Eq. (6.8). For a given value of n, the normalization/orthogonality conditions (6.8) lead to $n + 1$ equations that can be used to calculate $a^{(\ell)}_{nm}$ for $m = 0, \ldots, n$. Inverting such coefficients from Eq. (6.8) is a cumbersome task, but can be achieved with the proper numerical resources. As an example, let us discuss the case of $n = 1$, for an arbitrary value of ℓ. In this case, Eqs. (6.5) and (6.8) lead to

$$a^{(\ell)}_{10} \int dK \, \omega^{(\ell)} + a^{(\ell)}_{11} \int dK \, \omega^{(\ell)} E_{\mathbf{k}} = 0 , \qquad (6.118)$$

$$a^{(\ell)}_{10} \int dK \, \omega^{(\ell)} E_{\mathbf{k}} + a^{(\ell)}_{11} \int dK \, \omega^{(\ell)} E^2_{\mathbf{k}} = \frac{1}{a^{(\ell)}_{11}} . \qquad (6.119)$$

The solution of Eqs. (6.118) and (6.119) is

$$\frac{a_{10}^{(0)}}{a_{11}^{(0)}} = -\frac{\tilde{J}_{10}}{\tilde{J}_{00}}, \qquad (6.120)$$

$$\left[a_{11}^{(0)}\right]^2 = \frac{\tilde{J}_{00}^2}{\tilde{J}_{20}\tilde{J}_{00} - \tilde{J}_{10}^2}. \qquad (6.121)$$

As it turns out, it will always be sufficient to know, for a given n, the ratios, $a_{nm}^{(\ell)}/a_{nn}^{(\ell)}$, and the square of $a_{nn}^{(\ell)}$, $\left[a_{nn}^{(\ell)}\right]^2$, since only these quantities appear in the moment expansion of $\phi_{\mathbf{k}}$. Extracting the actual signs of the coefficients is irrelevant. General formulas for $a_{nm}^{(\ell)}/a_{nn}^{(\ell)}$ and $\left[a_{nn}^{(\ell)}\right]^2$ were derived in Ref. [20].

References

1. Chapman, S., Cowling, T.G.: The Mathematical Theory of Non-Uniform Gases, 3rd edn. Cambridge University Press, New York (1974)
2. Israel, W.: Ann. Phys. (N.Y.) **100**, 310 (1976)
3. Stewart, J.M.: Proc. Roy. Soc. A **357**, 59 (1977)
4. Israel, W., Stewart, J.M.: Ann. Phys. (N.Y.) **118**, 341 (1979)
5. Hiscock, W., Lindblom, L.: Phys. Rev. D **31**, 725 (1985)
6. de Groot, S.R., van Leeuwen, W.A., van Weert, Ch.G.: Relativistic Kinetic Theory—Principles and Applications. North-Holland (1980)
7. Hiscock,W., Lindblom, L.: Ann. Phys. (N.Y.), **151**, 466 (1983)
8. Hiscock, W., Lindblom, L.: Phys. Rev. D **35**, 3723 (1987)
9. Olson, T.S.: Annals Phys. **199**, 18 (1990)
10. Denicol, G.S., Kodama, T., Koide, T., Mota, P.: J. Phys. G **35** (2008)
11. Pu, S., Koide, T., Rischke, D.H.: Phys. Rev. D **81** (2010)
12. Grad, H.: Commun. Pure Appl. Math. **2**, 331 (1949)
13. Grad, H.: Comm. Pure Appl. Math. **2**, 325 (1949)
14. Chernikov, N.A.: Phys. Lett. **5**, 115 (1963)
15. Chernikov, N.A.: Acta Phys. Pol. **27**, 465 (1965)
16. Vignon, B.: Ann. Inst. H. Poincare **10**, 31 (1969)
17. Marle, C.: Ann. Inst. H. Poincare **10**, 127 (1969)
18. Kranys, M.: Phys. Lett. A **33**, 77 (1970)
19. Kranys, M.: Nuovo Cim. B **8**, 417 (1972)
20. Denicol, G.S., Niemi, H., Molnar, E., Rischke, D.H.: Phys. Rev. D **85** (2012)
21. Anderson, J.L.: J. Math. Phys. **15**, 1116 (1974)
22. Landau, L.D., Lifshitz, E.M.: Fluid Mechanics. Pergamon, New York (1959)
23. Molnar, E., Niemi, H., Denicol, G.S., Rischke, D.H.: Phys. Rev. D **89**(7), 074010 (2014). https://doi.org/10.1103/PhysRevD.89.074010, arXiv:1308.0785 [nucl-th]
24. Eq. (8.911.1) In: Gradshteyn, I.S., Ryzhik, I.M. (eds.) Table of Integrals, Series and Products, 4th edn. Academic (1980)
25. Cercignani, C., Kremer, G.M.: The Relativistic Boltzmann Equation: Theory and Applications. Birkhauser, Basel (2002)

Method of Moments: Convergence Properties

The method of moments has been widely employed to study the fluid-dynamical limit of the relativistic Boltzmann equation, with particular interest in understanding the hydrodynamic behavior of the quark-gluon plasma produced in ultrarelativistic heavy-ion collisions. As discussed in the previous chapter, this approach converts the Boltzmann equation (an integro-differential equation) into an infinite set of coupled partial differential equations for the moments of the single-particle distribution function. In particular, this method is expected to be very effective if the system in question can be described using a small number of degrees of freedom (or moments) and the system of moment equations can be truncated at a low order. This is expected to be the case in the fluid-dynamical regime, where the system can be just described in terms of the 14 fields contained in the conserved currents (the energy-momentum tensor, $T^{\mu\nu}$, and net-charge or particle 4-current, N^{μ}).

In the relativistic regime, the method of moments becomes essential since the truncation of the moment equations naturally leads to dynamical equations that are causal, at least in the linear regime. On the other hand, the Chapman-Enskog approximation [1], discussed in Sect. 5.2 of Chap. 5, leads to dissipative equations that are acausal and unstable [2] (see Chap. 2)—a problem that cannot be solved by adding higher-order corrections. The Chapman-Enskog expansion was considered a convenient method since it relies on an expansion in powers of a small dimensionless parameter, the Knudsen number—the ratio between the typical microscopic and macroscopic length scales of the system being described. In this context, the hydrodynamical regime is obtained by a low-order truncation of this series and was expected to be valid when the Knudsen number is sufficiently small. Nowadays, the situation is perceived to be more subtle, since, as discussed in Chap. 3, there are many evidences that the gradient expansion has a zero radius of convergence and, consequently, the domain of validity of truncations of such a series cannot be trivially defined.

© Springer Nature Switzerland AG 2021
G. S. Denicol and D. H. Rischke, *Microscopic Foundations of Relativistic Fluid Dynamics*,
Lecture Notes in Physics 990, https://doi.org/10.1007/978-3-030-82077-0_7

On the other hand, the method of moments is not constructed in terms of an expansion in a small parameter. It can be better described as an expansion in terms of degrees of freedom: truncating the moment equations at a higher order implies that one includes more moments and increases the number of dynamical variables describing the system. It is expected that increasing the number of moments or degrees of freedom will improve the description of the system and that, if the number of moments is sufficiently large, the full microscopic solution will be recovered. However, such convergence of the method of moments has not yet been properly investigated in relativistic systems, since it is extremely challenging to derive and solve the moment equations to arbitrarily high order.

The goal of this chapter is to investigate the convergence of the method of moments using the highly symmetric flow configurations described in Chap. 3: the Bjorken- and Gubser-flow scenarios. We shall demonstrate that these simplified, yet non-trivial, scenarios allow us to systematically derive the equations of motion for *all* the moments of the single-particle distribution function—something that is simply not possible when considering arbitrary flow profiles. This will enable us to study the convergence of the moment expansion and obtain an initial grasp on its domain of validity. This chapter is organized as follows: In Sect. 7.1, we discuss the relativistic Boltzmann equation in the relaxation-time approximation. Section 7.2 applies the method of moments to ultrarelativistic gases undergoing Bjorken flow and systematically investigates the convergence of the method of moments. In Sect. 7.3, the same discussion is performed for ultrarelativistic gases undergoing Gubser flow. We then end this chapter with a brief summary of the results.

7.1   Boltzmann Equation and Fluid-Dynamical Variables

For the purpose of this chapter, our starting point is the relativistic Boltzmann equation in *curved* space [3,4],

$$k^\mu \partial_\mu f_{\mathbf{k}} + \Gamma^\lambda_{\mu i} k_\lambda k^\mu \frac{\partial f_{\mathbf{k}}}{\partial k_i} = C[f] , \tag{7.1}$$

where $k^\mu = (k^0, \mathbf{k})^T$ is the 4-momentum of the particles, $\Gamma^\lambda_{\mu\nu}$ is the Christoffel symbol, and, as usual, we employ the notation $f_{\mathbf{k}} \equiv f(x^\mu, \mathbf{k})$. Here, we shall restrict our system to be a gas of *massless* and *classical* particles, for which the 4-momentum satisfies the condition, $g_{\mu\nu}k^\mu k^\nu = 0$, with $g_{\mu\nu}$ being the metric tensor.

For the sake of simplicity, we will always simplify the collision term using the relaxation-time approximation (RTA) [5,6]. In the relativistic regime, this approximation was constructed by Anderson and Witting [6] and assumes the following form for the collision operator

$$C[f] = -E_{\mathbf{k}} \frac{f_{\mathbf{k}} - f_{0\mathbf{k}}}{\tau_R(T)} , \tag{7.2}$$

where τ_R is the relaxation time, $E_{\mathbf{k}} = u_\mu k^\mu$, and $f_{0\mathbf{k}} = \exp(-\beta_0 E_{\mathbf{k}})$ is the classical local-equilibrium distribution function, with $\beta_0 = 1/T$ being the inverse temperature, and u^μ the local 4-velocity of the system. Note that the relaxation time is chosen to be a function of the temperature, $\tau_R(T)$. Unless stated otherwise, we shall assume that

$$\tau_R = \frac{c}{T}, \tag{7.3}$$

which corresponds to the relaxation time of a conformal system, with c being a constant. As we will see below, within the RTA and in the massless limit, the shear viscosity, η, is related to the relaxation time as [7]

$$\tau_R = \frac{5\eta}{T s_0} \implies \frac{\eta}{s_0} = \frac{c}{5},$$

where s_0 is the entropy density. As expected for a conformal system, the ratio of the shear viscosity-to-entropy density is a constant, here determined by the parameter c. Here, we will always fix the parameter c by choosing the value of η/s_0.

In the RTA, the temperature and velocity are new fields introduced into the Boltzmann equation and, in this sense, they must be properly defined before one can solve the equation. Here, we define T using the Landau matching condition [8], (see Sect. 1.3.1 of Chap. 1 for a detailed discussion on matching conditions) in order to guarantee energy-momentum conservation

$$\varepsilon = \int dK \, E_{\mathbf{k}}^2 f_{\mathbf{k}} \equiv \int dK \, E_{\mathbf{k}}^2 f_{0\mathbf{k}}, \tag{7.4}$$

where, similar to the notation used in the previous chapters

$$dK \equiv d_{\text{dof}} \frac{d^3\mathbf{k}}{(2\pi)^3 k^0 \sqrt{-g}}$$

is the Lorentz-invariant momentum-space volume, with $d^3\mathbf{k}$ built from the (covariant) spatial components of k_μ, d_{dof} is the number of internal degrees of freedom, and g the determinant of the metric. The above condition, when applied to a gas of massless, classical particles, fixes the temperature of the system as function of the energy density via the formula

$$\varepsilon = \frac{3}{\pi^2} d_{\text{dof}} T^4. \tag{7.5}$$

We determine the velocity following Landau's prescription [8], where it is defined as the eigenvector of the energy-momentum tensor, $T^{\mu\nu}$, which has the energy density as its eigenvalue, that is, $T^{\mu\nu} u_\mu = \varepsilon u^\nu$ (see Sect. 1.3.3 of Chap. 1 for more details).

Now that the temperature and 4-velocity are defined, we must determine their equations of motion. Such equations are obtained using the continuity equation satisfied by the energy-momentum tensor

$$D_\mu T^{\mu\nu} = 0 \,, \tag{7.6}$$

where D_μ is a covariant derivative (see Sect. 3.1 of Chap. 3 for more details). In kinetic theory, the energy-momentum tensor is expressed as the second reducible moment of the single-particle distribution function

$$T^{\mu\nu} = \int dK \, k^\mu k^\nu f_{\mathbf{k}} \,. \tag{7.7}$$

In order to define the fluid-dynamical variables of interest, $T^{\mu\nu}$ is decomposed in terms of the fluid 4-velocity as explained in Sect. 5.1 of Chap. 5,

$$T^{\mu\nu} = \varepsilon u^\mu u^\nu - P_0 \Delta^{\mu\nu} + \pi^{\mu\nu} \,. \tag{7.8}$$

We use the same notation employed in the previous chapters: we introduced the projection operator $\Delta^{\mu\nu} = g^{\mu\nu} - u^\mu u^\nu$, the isotropic pressure $P_0 = -\Delta_{\mu\nu} T^{\mu\nu}/3$ (the bulk-viscous pressure $\Pi \equiv 0$ for massless particles), and the shear-stress tensor $\pi^{\mu\nu} = \Delta^{\mu\nu}_{\alpha\beta} T^{\alpha\beta}$. For the definition of the latter quantity, we employed the symmetric and traceless projection operator $\Delta^{\mu\nu}_{\alpha\beta} = (\Delta^\mu_\alpha \Delta^\nu_\beta + \Delta^\mu_\beta \Delta^\nu_\alpha)/2 - \Delta^{\mu\nu} \Delta_{\alpha\beta}/3$. Since the system is made of massless particles, the energy-momentum tensor is traceless, $T^\mu_\mu = 0$, and the isotropic pressure is related to the energy density via $\varepsilon = 3P_0$.

Finally, the equations of motion are obtained by combining the conservation laws with the Boltzmann equation in the RTA,

$$u^\mu \partial_\mu \varepsilon + \frac{4}{3}\varepsilon D_\mu u^\mu - \pi^{\mu\nu} D_\mu u_\nu = 0 \,, \tag{7.9}$$

$$\frac{4}{3}\varepsilon u^\mu D_\mu u^\lambda - \frac{1}{3}\Delta^{\lambda\mu}\partial_\mu \varepsilon + \Delta^\lambda_\nu D_\mu \pi^{\mu\nu} = 0 \,, \tag{7.10}$$

$$k^\mu \partial_\mu f + \Gamma^\lambda_{\mu i} k_\lambda k^\mu \frac{\partial f_{\mathbf{k}}}{\partial k_i} + u_\mu k^\mu \frac{f_{\mathbf{k}} - f_{0\mathbf{k}}}{\tau_R} = 0 \,, \tag{7.11}$$

where we used the fact that the covariant derivative of a scalar quantity is identical to the partial derivative, cf. Eq. (3.1). The covariant derivatives of the velocity and shear-stress tensor are expressed in terms of the Christoffel symbols in the following manner, cf. Eqs. (3.2) and (3.9) in Sect. 3.1 of Chap. 3,

$$D_\mu u^\lambda = \partial_\mu u^\lambda + \Gamma^\lambda_{\mu\nu} u^\nu \,,$$

$$D_\mu \pi^{\mu\nu} = \frac{1}{\sqrt{-g}} \partial_\mu \left(\sqrt{-g}\, \pi^{\mu\nu}\right) + \Gamma^\nu_{\mu\lambda} \pi^{\mu\lambda} \,. \tag{7.12}$$

We shall refer to the set (7.9)–(7.11) of coupled equations as the Anderson-Witting-Boltzmann (AWB) equations. It is important to emphasize that these equations of motion are not linear, since both $f_{0\mathbf{k}}$ and τ_R have a temperature dependence and the shear-stress tensor is an integral of the distribution function. Thus, the AWB equations are actually a set of nonlinear integro-differential equations that couple the energy density, velocity field, and single-particle distribution function.

Solving the AWB equations is certainly a non-trivial task. Nevertheless, under some simplifying assumptions, exact and analytic solutions can be found. For example, exact solutions in 0+1 dimensions, assuming Bjorken flow [9], and in 1+1 dimensions, assuming Gubser flow [10, 11], have already been found. For details, the reader is referred to Refs. [12, 13] for solutions under the Bjorken scaling hypothesis and Refs. [14, 15] for solutions assuming Gubser flow. In this chapter, we show how the method of moments can be used to solve the AWB equations and use such solutions to discuss its level of accuracy.

7.2 Bjorken Flow

First we consider a *homogeneous* system of massless particles expanding in a space-time described by the metric [9],

$$ds^2 = g_{\mu\nu}dx^\mu dx^\nu = d\tau^2 - \left(dx^2 + dy^2 + \tau^2 d\eta_s^2\right) , \tag{7.13}$$

as discussed in Sect. 3.2 of Chap. 3. We use the same notation employed in Chap. 3, in which τ and η_s are the hyperbolic coordinates, defined via Eq. (3.19). The nonzero Christoffel symbols related to the metric tensor in Eq. (7.13) are given in Eq. (3.22), while the determinant of this metric is $\sqrt{-g} = \tau$. Since the system is homogeneous, all fields will depend *solely* on the time variable τ. As already explained in Chap. 3, in Cartesian coordinates this system corresponds to a gas that is expanding in the (longitudinal) z-direction.

The assumption that the system is homogeneous in hyperbolic coordinates implies that a trivial velocity field, $u^\mu = (1, 0, 0, 0)^T$, solves Eq. (7.10) in this coordinate system. Nevertheless, as already discussed in Chap. 3, the system is not really static and the metric (7.13) will induce a nonzero expansion rate and shear tensor

$$\theta \equiv \frac{1}{\sqrt{-g}}\partial_\mu(\sqrt{-g}\, u^\mu) = \frac{1}{\tau} , \tag{7.14}$$

$$\sigma_{\mu\nu} = \Delta_{\mu\nu}^{\alpha\beta} D_\alpha u_\beta = \text{diag}\left(0, \frac{1}{3\tau}, \frac{1}{3\tau}, -\frac{2}{3}\tau\right) , \tag{7.15}$$

cf. Eqs. (3.29) and (3.30). As already explained, the spatial gradients of *any* scalar function will always be zero and any 4-vector that is orthogonal to u^μ must always be zero, which prohibits the existence of any diffusion 4-current. Finally, in this geometry, the vorticity tensor is also always zero, $\omega^{\mu\nu} = \Delta_{\mu\alpha}\Delta_{\nu\beta}\left(D^\alpha u^\beta - D^\beta u^\alpha\right)/2 = 0$.

We have already derived the equations of motion for the energy density in Bjorken flow in Chap. 3

$$u_\nu D_\mu T^{\mu\nu} = \frac{d\varepsilon}{d\tau} + \frac{4}{3\tau}\varepsilon - \frac{\bar{\pi}}{\tau} = 0 , \tag{7.16}$$

where we employed the following parametrization of the shear-stress tensor, $\pi^{\mu\nu} =$ diag $\left(0, \bar{\pi}/2, \bar{\pi}/2, -\bar{\pi}/\tau^2\right)$, cf. Eq. (3.33). For a dilute gas, the field $\bar{\pi}$ can be expressed as the following integral of the single-particle distribution function, $f_{\mathbf{k}}$

$$\bar{\pi} = \pi_{\eta_s}^{\eta_s} = -\int dK\, k_0^2 \left[\left(\frac{k_{\eta_s}}{k_0 \tau}\right)^2 - \frac{1}{3}\right] f_{\mathbf{k}}\,, \tag{7.17}$$

with the on-shell condition relating the energy of the particle to its momentum components,

$$k_0 = \sqrt{k_x^2 + k_y^2 + \frac{k_{\eta_s}^2}{\tau^2}}\,. \tag{7.18}$$

Finally, all terms in the AWB equations involving Christoffel symbols vanish (in fact, in Eq. (7.11) there is a cancellation of the term $\sim \Gamma_{\eta_s \eta_s}^\tau$ with that $\sim \Gamma_{\tau \eta_s}^{\eta_s}$) and, thus, Eq. (7.11) can be cast into the very simple form

$$\partial_\tau f_{\mathbf{k}} = -\frac{f_{\mathbf{k}} - f_{0\mathbf{k}}}{\tau_R}\,. \tag{7.19}$$

7.2.1 Method of Moments

As already mentioned, the AWB equations (7.9)–(7.11) are a set of nonlinear integro-differential equations and are rather challenging to solve in this form. Our strategy is to use the method of moments and convert these integro-differential equations into an infinite set of ordinary differential equations that, once truncated, can be solved using simple numerical methods. This procedure will be discussed in the following.

The first step is to expand the single-particle momentum distribution using a complete set of functions [16, 17]. In Chap. 6, we expanded the *non-equilibrium correction* to the single-particle distribution function using a complete basis composed of irreducible tensors, $k^{\langle \mu_1} \cdots k^{\mu_\ell \rangle}$, $\ell = 0, \ldots, \infty$, and orthogonal polynomials, $P_{\mathbf{k}n}^{(\ell)}$, $n = 0, \ldots, \infty$. For a system made of *classical* particles, the moment expansion can be written as

$$\phi_{\mathbf{k}} \equiv \frac{f_{\mathbf{k}} - f_{0\mathbf{k}}}{f_{0\mathbf{k}}} = \sum_{n=0}^{\infty} \sum_{\ell=0}^{\infty} P_{\mathbf{k}n}^{(\ell)} c_n^{\mu_1 \cdots \mu_\ell} k_{\langle \mu_1} \cdots k_{\mu_\ell \rangle}\,. \tag{7.20}$$

Above, we have already set the arbitrary function, $\mathcal{G}_{\mathbf{k}}$, introduced in Eq. (6.1) of Sect. 6.1, in Chap. 6, to one, $\mathcal{G}_{\mathbf{k}} = 1$. The tensors $c_n^{\mu_1 \cdots \mu_\ell}$, which play the role of the expansion coefficients, are found using the orthogonality relations satisfied by the basis elements, derived in Appendices 6.7 and 6.8 of Chap. 6. The irreducible tensors, $c_n^{\mu_1 \cdots \mu_\ell}$, are then given by the following integral of $\delta f_{\mathbf{k}} = f_{\mathbf{k}} - f_{0\mathbf{k}}$

$$c_n^{\mu_1 \cdots \mu_\ell} = \frac{W^{(\ell)}}{\ell!} \int dK\, P_{\mathbf{k}n}^{(\ell)} k^{\langle \mu_1} \cdots k^{\mu_\ell \rangle} \delta f_{\mathbf{k}}(\tau)\,, \tag{7.21}$$

cf. Eq. (6.11) in Sect. 6.1, where, for massless, classical particles, the normalization function $W^{(\ell)}$ is given by $W^{(\ell)} = (-1)^\ell / I_{2\ell,\ell}$, with the thermodynamic integrals

$$
\begin{aligned}
I_{2\ell,\ell} &= \frac{1}{(2\ell+1)!!} \int dK \left(-\Delta_{\alpha\beta} k^\alpha k^\beta\right)^\ell f_{0\mathbf{k}} \\
&\xrightarrow[m \to 0]{} \frac{1}{(2\ell+1)!!} \int dK \, E_{\mathbf{k}}^{2\ell} f_{0\mathbf{k}} \equiv \frac{I_{2\ell,0}}{(2\ell+1)!!} ,
\end{aligned}
\tag{7.22}
$$

as defined in Eq. (5.48).

We now discuss how the moment expansion can be simplified due to the stringent symmetries imposed in the Bjorken-flow scenario. For instance, isotropy in the transverse plane allows to simplify the momentum dependence of $f_{\mathbf{k}}$: it can only depend on $k_\perp \equiv \sqrt{k_x^2 + k_y^2}$, and not on the azimuthal angle. Due to the on-shell condition (7.18), $k_\perp$ can be expressed in terms of the two independent variables k_{η_s} and k_0. Regarding the space-time dependence of $f_{\mathbf{k}}$, the assumptions of boost invariance in the z-direction and homogeneity and isotropy in the spatial transverse (x, y)-plane imply that $f_{\mathbf{k}}$ can only depend on the time variable τ. Altogether, $f_{\mathbf{k}}$ is then defined on a three-dimensional hypersurface of phase space, $f_{\mathbf{k}}(\tau) = f(\tau, k_{\eta_s}, k_0)$. It is convenient to use a covariant notation, introducing the normalized, space-like 4-vector, z^μ, constructed to be orthogonal to the 4-velocity u^μ, as

$$
z_\mu z^\mu = -1 , \quad z_\mu u^\mu = 0 .
$$

For Bjorken flow, we have in hyperbolic coordinates $u_\mu = (1, 0, 0, 0)$, thus we are free to choose $z_\mu \equiv (0, 0, 0, -\tau)$. Therefore, we may write the single-particle distribution function in the highly symmetric Bjorken-flow configuration as

$$
f_{\mathbf{k}}(\tau) \equiv f(\tau, z_\mu k^\mu, u_\mu k^\mu) .
\tag{7.23}
$$

Given the symmetries of Bjorken flow, we see from Eq. (7.21) that the tensors $c_n^{\mu_1 \cdots \mu_\ell}$ must be constructed from combinations of u^μ, z^μ, and the inverse metric tensor $g^{\mu\nu}$. Naturally, only combinations that form irreducible tensors of rank ℓ are allowed. This poses a significant constraint on the form of such tensors

$$
c_n^{\mu_1 \cdots \mu_\ell} \equiv \hat{c}_{n,\ell}^\delta (\tau) \, z^{\langle \mu_1} \cdots z^{\mu_\ell \rangle} ,
\tag{7.24}
$$

where the τ-dependent scalar coefficients $\hat{c}_{n,\ell}^\delta (\tau)$ are obtained from the following projection

$$
\hat{c}_{n,\ell}^\delta (\tau) = (-1)^\ell \frac{(2\ell-1)!!}{\ell!} z_{\langle \mu_1} \cdots z_{\mu_\ell \rangle} c_n^{\mu_1 \cdots \mu_\ell} .
\tag{7.25}
$$

Here we used that

$$
z_{\langle \mu_1} \cdots z_{\mu_\ell \rangle} z^{\langle \mu_1} \cdots z^{\mu_\ell \rangle} = (-1)^\ell \frac{\ell!}{(2\ell-1)!!} ,
\tag{7.26}
$$

which can be proven analogously to Eq. (6.106) in Appendix 6.7 (note that $\Delta^{\alpha\beta} z_\alpha z_\beta = -1$ on account of the definition of z_μ). Therefore, the moment expansion (7.20) becomes

$$\phi_{\mathbf{k}} = \sum_{n=0}^{\infty} \sum_{\ell=0}^{\infty} P_{\mathbf{k}n}^{(\ell)} \hat{c}_{n,\ell}^{\delta}(\tau) \, z^{\langle\mu_1} \cdots z^{\mu_\ell\rangle} k_{\langle\mu_1} \cdots k_{\mu_\ell\rangle} \, . \tag{7.27}$$

Using the properties of the projection operators $\Delta_{\nu_1\cdots\nu_\ell}^{\mu_1\cdots\mu_\ell}$ listed in Eq. (6.75) in Appendix 6.5, it is straightforward to demonstrate that

$$z^{\langle\mu_1} \cdots z^{\mu_\ell\rangle} k_{\langle\mu_1} \cdots k_{\mu_\ell\rangle} = \sum_{q=0}^{[\ell/2]} C(\ell, q) \left(\Delta_{\alpha\beta} k^\alpha k^\beta\right)^q \left(\Delta_{\alpha\beta} z^\alpha z^\beta\right)^q \left(\Delta_{\alpha\beta} k^\alpha z^\beta\right)^{\ell-2q}$$

$$= E_{\mathbf{k}}^\ell \sum_{q=0}^{[\ell/2]} C(\ell, q) \cos^{\ell-2q}\Theta \, , \tag{7.28}$$

where we introduced the scalar angle Θ as

$$\cos\Theta \equiv \frac{z_\mu k^\mu}{u_\mu k^\mu} \equiv \frac{k_{\eta_s}}{\tau k_0} \, . \tag{7.29}$$

The coefficient $C(\ell, q)$ was given in Eq. (6.76). Using Ref. [18], as well as the identity $(2\ell)! = 2^\ell \ell! (2\ell - 1)!!$, we can immediately relate the contracted tensors $z^{\langle\mu_1} \cdots z^{\mu_\ell\rangle} k_{\langle\mu_1} \cdots k_{\mu_\ell\rangle}$ to the Legendre polynomials

$$z^{\langle\mu_1} \cdots z^{\mu_\ell\rangle} k_{\langle\mu_1} \cdots k_{\mu_\ell\rangle} = \frac{\ell!}{(2\ell - 1)!!} E_{\mathbf{k}}^\ell P_\ell(\cos\Theta) \, , \tag{7.30}$$

where $P_\ell(\cos\Theta)$ is the Legendre polynomial of ℓ-th rank.

Furthermore, in the massless limit, the orthogonal polynomials $P_{\mathbf{k}n}^{(\ell)}$, constructed in Appendix 6.8 of Chap. 6, can be identified with the associated Laguerre polynomials. This can be seen directly from their orthogonality relation (6.8) which for massless, classical particles takes the form

$$\int_0^{\infty} dx \, x^{2\ell+1} e^{-x} \, P_{\mathbf{k}n}^{(\ell)} P_{\mathbf{k}m}^{(\ell)} = (2\ell + 1)! \, \delta_{nm} \, ,$$

and where we have to replace $E_{\mathbf{k}} \equiv Tx$ in the definition of $P_{\mathbf{k}n}^{(\ell)}$. If we compare this to the orthogonality condition satisfied by the associated Laguerre polynomials $L_n^{(\alpha)}(x)$ [19]

$$\int_0^{\infty} dx \, x^\alpha e^{-x} \, L_n^{(\alpha)}(x) L_m^{(\alpha)}(x) = \frac{(n + \alpha)!}{n!} \delta_{nm} \, ,$$

it is possible (but *only* in the massless limit) to relate $P_{\mathbf{k}n}^{(\ell)}$ with $L_n^{(2\ell+1)}(x)$ in the following way

$$\sqrt{\frac{(n+2\ell+1)!}{n!\,(2\ell+1)!}}\,P_{\mathbf{k}n}^{(\ell)} = L_n^{(2\ell+1)}(x)\,. \tag{7.31}$$

Thus, for massless particles the moment expansion (7.27) in the highly symmetric Bjorken-flow configuration can be re-expressed as an expansion in terms of associated Laguerre polynomials and Legendre polynomials

$$\phi_{\mathbf{k}} = \sum_{n=0}^{\infty}\sum_{\ell=0}^{\infty}\hat{c}_{n,\ell}(\tau)\,E_{\mathbf{k}}^{\ell}L_n^{(2\ell+1)}(\beta_0 E_{\mathbf{k}})\,P_{\ell}(\cos\Theta)\,, \tag{7.32}$$

where we defined the new expansion coefficients $\hat{c}_{n,\ell}(\tau)$ as

$$\hat{c}_{n,\ell}(\tau) \equiv \frac{\ell!}{(2\ell-1)!!}\sqrt{\frac{n!\,(2\ell+1)!}{(n+2\ell+1)!}}\,\hat{c}_{n,\ell}^{\delta}(\tau)$$

$$= \frac{n!\,(2\ell+1)!}{(n+2\ell+1)!}\frac{1}{(2\ell-1)!!\,I_{2\ell,\ell}}\int dK\,E_{\mathbf{k}}^{\ell}L_n^{(2\ell+1)}(\beta_0 E_{\mathbf{k}})\,P_{\ell}(\cos\Theta)\,\delta f_{\mathbf{k}}\,.$$

To prove the second line, insert Eq. (7.32) into $\delta f_{\mathbf{k}} = f_{0\mathbf{k}}\phi_{\mathbf{k}}$ and use the orthogonality relations for the associated Laguerre polynomials and Legendre polynomials, as well as the definition (7.22) of the function $I_{2\ell,\ell}$.

It is possible to further simplify the moment expansion, expressing it as an expansion for the full distribution function $f_{\mathbf{k}}$, and not just for the non-equilibrium correction $\delta f_{\mathbf{k}} = f_{0\mathbf{k}}\phi_{\mathbf{k}}$. To this end, we note that the equilibrium version of the expansion coefficients

$$\hat{c}_{n,\ell}^{\text{eq}}(\tau) \equiv \frac{n!\,(2\ell+1)!}{(n+2\ell+1)!}\frac{1}{(2\ell-1)!!I_{2\ell,\ell}}$$

$$\times \int dK\,E_{\mathbf{k}}^{\ell}L_n^{(2\ell+1)}(\beta_0 E_{\mathbf{k}})\,P_{\ell}(\cos\Theta)\,f_{0\mathbf{k}} = 0\,,$$

vanishes for all values of n and ℓ, except $n = \ell = 0$. (For the proof, note that $\int_{-1}^{1} dx\,P_{\ell}(x) = 2\delta_{\ell,0}$ and $\int_0^{\infty} dx\,e^{-x}x^{\alpha}L_n^{(\alpha)} = \delta_{n,0}$, cf. Ref. [20]). That is

$$\hat{c}_{n,\ell}^{\text{eq}}(\tau) = 0\,,\quad \forall \ell \neq 0\quad \text{and}\quad \hat{c}_{n,0}^{\text{eq}}(\tau) = 0\,,\quad \forall n \neq 0\,.$$

We also note that $\hat{c}_{0,0}^{\text{eq}}(\tau) = 1$. With these properties, we can prove that the $\hat{c}_{n,\ell}^{\text{eq}}(\tau)$ satisfy the sum rule

$$1 = \sum_{n=0}^{\infty}\sum_{\ell=0}^{\infty}\hat{c}_{n,\ell}^{\text{eq}}(\tau)\,E_{\mathbf{k}}^{\ell}L_n^{(2\ell+1)}(\beta_0 E_{\mathbf{k}})\,P_{\ell}(\cos\Theta)\,. \tag{7.33}$$

Therefore, using $f_{\mathbf{k}} = f_{0\mathbf{k}}(1 + \phi_{\mathbf{k}})$ and substituting the expansions (7.32) and (7.33), it is possible to write the full distribution function $f_{\mathbf{k}}$ in the form of a moment expansion

$$f_{\mathbf{k}} = f_{0\mathbf{k}} \sum_{n=0}^{\infty} \sum_{\ell=0}^{\infty} c_{n,\ell}(\tau) E_{\mathbf{k}}^{\ell} L_n^{(2\ell+1)}(\beta_0 E_{\mathbf{k}}) P_\ell(\cos \Theta) \,, \qquad (7.34)$$

where the expansion coefficients, $c_{n,\ell}(\tau) \equiv \hat{c}_{n,\ell}(\tau) + \hat{c}_{n,\ell}^{\mathrm{eq}}(\tau)$, are now moments of the *full* single-particle distribution function

$$c_{n,\ell}(\tau) = \frac{n!\,(2\ell+1)!}{(n+2\ell+1)!} \frac{1}{(2\ell-1)!!\,I_{2\ell,\ell}} \int dK\, E_{\mathbf{k}}^{\ell} L_n^{(2\ell+1)}(\beta_0 E_{\mathbf{k}}) P_\ell(\cos \Theta)\, f_{\mathbf{k}} \,.$$

As we shall see next, it is extremely useful to recast the problem in terms of an expansion in moments of the full single-particle distribution function. Nevertheless, this is just a matter of convenience. The calculations performed in the remainder of this chapter could have also been performed using moments of $\delta f_{\mathbf{k}}$. We note that, due to the reflection symmetry around the η_s-axis in Bjorken flow, we have $f(\tau, k_{\eta_s}, k_0) = f(\tau, -k_{\eta_s}, k_0)$, and only integrals with an even power of $\cos \Theta = k_{\eta_s}/(k_0 \tau)$ will be nonzero. Hence, only Legendre polynomials of even order will contribute to the moment expansion.

For the sake of convenience, one can make use of a different, but equivalent set of moments of the distribution function

$$\rho_{n,m} \equiv \int dK\, k_0^{n+1} P_{2m}(\cos \Theta)\, f_{\mathbf{k}} \,, \qquad (7.35)$$

replacing the associated Laguerre polynomials with argument $\beta_0 E_{\mathbf{k}} = \beta_0 k_0$ by simple powers of k_0. These moments are equivalent to the irreducible moments defined in the previous chapter and in Refs. [16,17] (the moments defined in the previous chapter, $\rho_n^{\mu_1 \cdots \mu_{2m}}$ are related to $\rho_{n+2m,m}$ via $z_{\langle \mu_1} \cdots z_{\mu_{2m} \rangle} \rho_n^{\mu_1 \cdots \mu_{2m}} \sim \rho_{n+2m,m}$, cf. Eq. (7.30)). Naturally, these two choices of moments are related by the definition of the associated Laguerre polynomials

$$L_n^{(2\ell+1)}(\beta_0 E_{\mathbf{k}}) = \sum_{m=0}^{n} (-1)^m \binom{n+2\ell+1}{n-m} \frac{(\beta_0 E_{\mathbf{k}})^m}{m!} \,,$$

which leads to the relation

$$\begin{aligned}
c_{n,2\ell}(\tau) = {} & \frac{n!\,(4\ell+1)!}{(n+4\ell+1)!} \frac{1}{(4\ell-1)!!\,I_{4\ell,2\ell}} \\
& \times \sum_{m=0}^{n} \frac{(-1)^m \beta_0^m}{m!} \binom{n+4\ell+1}{n-m} \rho_{2\ell+m-1,\ell} \,.
\end{aligned} \qquad (7.36)$$

Therefore, once the moments $\rho_{n,\ell}$ are known, it is possible to determine the expansion coefficients $c_{n,2\ell}(\tau)$.

The energy density is identified as the moment $\rho_{1,0}$ and, consequently, the matching conditions impose the following constraint on this moment

$$\rho_{1,0} = \rho_{1,0}^{eq} \equiv \int dK \, k_0^2 f_{0\mathbf{k}} \; . \tag{7.37}$$

We note that, because $\int_{-1}^{1} dx \, P_m(x) = 2\delta_{m,0}$, for any $m > 0$ the equilibrium value of any moment is zero, $\rho_{n,m}^{eq} = 0$. On the other hand, for $m = 0$, the equilibrium value of the moments can be trivially expressed in terms of powers of temperature

$$\rho_{n,0}^{eq} = \int dK \, k_0^{n+1} f_{0\mathbf{k}} = d_{\text{dof}} \frac{(n+2)!}{2\pi^2} T^{n+3} \; . \tag{7.38}$$

7.2.2   Moment Equations

The equation of motion for an arbitrary moment, $\rho_{n,m}$, is obtained by multiplying Eq. (7.19) with $k_0^{n+1} P_{2m}(\cos\Theta)$ and integrating over dK—a procedure analogous to the one applied in Sect. 6.2 of Chap. 6,

$$\int dK \, k_0^{n+1} P_{2m}(\cos\Theta) \, \partial_\tau f_{\mathbf{k}} = \frac{\rho_{n,m}^{eq}}{\tau_R} - \frac{\rho_{n,m}}{\tau_R} \; .$$

We note that the measure dK, the energy k_0, and the scalar angle Θ all contain an implicit time dependence, due to the metric, which must be taken into account when further evaluating the above equation. After some straightforward calculation, one derives the following dynamical equation for $\rho_{n,m}$

$$\partial_\tau \rho_{n,m} + \frac{\rho_{n,m}}{\tau_R} + \frac{\rho_{n,m}}{\tau} = \frac{\rho_{n,m}^{eq}}{\tau_R} - 2m \frac{(2m-1)(n+2m+1)}{(4m+1)(4m-1)} \frac{\rho_{n,m-1}}{\tau}$$
$$- \frac{2m(2m+1) + n(8m^2 + 4m - 1)}{(4m-1)(4m+3)} \frac{\rho_{n,m}}{\tau}$$
$$- (n-2m) \frac{(2m+1)(2m+2)}{(4m+1)(4m+3)} \frac{\rho_{n,m+1}}{\tau} \; . \tag{7.39}$$

In order to obtain this equation, we made use of the following properties of the Legendre polynomials

$$(n+1) P_{n+1}(x) = (2n+1) x P_n(x) - n P_{n-1}(x) \; , \tag{7.40}$$
$$(x^2 - 1) \frac{d}{dx} P_n(x) = nx P_n(x) - n P_{n-1}(x) \; . \tag{7.41}$$

Here, due to the symmetries of the system, it is rather straightforward to derive the equations of motion for all moments (in the previous chapter, we only did this for the irreducible moments of rank 0, 1, and 2).

As already demonstrated in the previous chapter and in Refs. [16, 17], we obtain a hierarchy of equations of motion, where in order to solve for a moment $\rho_{n,m}$, we always need to know the next moment $\rho_{n,m+1}$ and the previous moment $\rho_{n,m-1}$ (here, the index m corresponds to the tensor rank of the irreducible moment, implying that the dynamics of a lower-rank tensor always couples to the dynamics of a higher-rank tensor). Thus, in order to determine the time evolution of the energy density, $\varepsilon = \rho_{1,0}$, we need to know all moments $\rho_{1,m}$, with m ranging from 1 to infinity, $m = 1, 2, \ldots$. Using Eq. (7.17), we see that the shear-stress tensor itself is related to one of these moments, namely

$$\rho_{1,1} = \frac{3}{2} \int dK \, k_0^2 \left(\cos^2 \Theta - \frac{1}{3} \right) f_{\mathbf{k}} = -\frac{3}{2} \bar{\pi} , \qquad (7.42)$$

where we used $P_2(x) \equiv (3x^2 - 1)/2$.

Finally, it is important to emphasize that moments with different values of n do not couple to each other. This is actually not a general feature of the moment equations, but a consequence of the RTA—moments of different n are in principle coupled by the moment equations when using a general collision term, see Eq. (6.46) in Chap. 6. Thus, in the RTA, if we want to solely describe the moments that appear or couple to the energy-momentum tensor, we only need to worry about one single value of n, i.e., $n = 1$. This is also a strength of the method of moments: we do not need to solve the full Boltzmann equation in order to understand some of its moments. In the following, since we only solve for the moments $\rho_{1,m}$, we will always omit the index that is held fixed

$$\rho_{1,m} \equiv \rho_m . \qquad (7.43)$$

Note that taking $m = 0$ in Eq. (7.39), we obtain the usual equation of motion for the energy density

$$\partial_\tau \varepsilon + \frac{4}{3} \frac{\varepsilon}{\tau} = -\frac{2}{3} \frac{\rho_1}{\tau} = \frac{\bar{\pi}}{\tau} , \qquad (7.44)$$

while taking $m = 1$, we obtain an equation of motion for the shear-stress tensor

$$\partial_\tau \bar{\pi} + \frac{\bar{\pi}}{\tau_R} = \frac{16}{45} \frac{\varepsilon}{\tau} - \frac{38}{21} \frac{\bar{\pi}}{\tau} - \frac{8}{35} \frac{\rho_2}{\tau} . \qquad (7.45)$$

The only way to effectively solve this set of equations is to truncate the moment expansion of $f_{\mathbf{k}}$ and describe our system using a finite number of moments. Naturally, the goal is to perform this task using the smallest possible number of moments. The expectation is that, if the mean free path of the particles is sufficiently small, it may be possible to describe the energy-momentum tensor using a sufficiently small number of moments, saving computational time, and also obtaining a more clear picture of the degrees of freedom required to describe the system in question.

Here, we define $2M$ as the rank of the last moment included in the moment expansion (7.34), i.e.

$$f_{\mathbf{k}} = f_{0\mathbf{k}} \sum_{n=0}^{\infty} \sum_{\ell=0}^{M} c_{n,2\ell}(\tau)\, k_0^{2\ell}\, L_n^{(4\ell+1)}(\beta_0 k_0)\, P_{2\ell}(\cos\Theta)\,, \qquad (7.46)$$

where we are already only effectively summing over *even* values of ℓ. Once this truncation is imposed, we can show that the moment $\rho_{n,M+1}$, which appears in the equation of motion for $\rho_{n,M}$, must be zero. The proof consists of inserting Eq. (7.46) into Eq. (7.35)

$$\begin{aligned}
\rho_{n,M+1} &= \sum_{n=0}^{\infty} \sum_{\ell=0}^{M} c_{n,2\ell}(\tau) \int dK\, k_0^{2\ell+n+1} L_n^{(4\ell+1)}(\beta_0 k_0)\, P_{2M+2}(\cos\Theta)\, P_{2\ell}(\cos\Theta)\, f_{0\mathbf{k}} \\
&= \sum_{n=0}^{\infty} \sum_{\ell=0}^{M} c_{n,2\ell}(\tau) \int_0^{\infty} \frac{dk}{(2\pi)^2} k^{2\ell+n+2} L_n^{(4\ell+1)}(\beta_0 k)\, f_{0\mathbf{k}} \int_{-1}^{1} dx\, P_{2M+2}(x)\, P_{2\ell}(x) \\
&= 0\,,
\end{aligned} \qquad (7.47)$$

because of the orthogonality of the Legendre polynomials (note that the sum runs only up to M, but $M+1$ would be required for a non-trivial result of the x-integral). This holds for all values of n—not only for $n=1$, which is the value assumed at this point.

Following this scheme, the lowest possible truncation is $M=0$, i.e., one only includes the very first moment $\rho_0 = \varepsilon$. This leads to the equations of ideal fluid dynamics (see Chap. 3), i.e., Eq. (7.44) becomes

$$\partial_\tau \varepsilon + \frac{4}{3}\frac{\varepsilon}{\tau} = 0\,, \qquad (7.48)$$

while Eq. (7.45) becomes a trivial identity. The next possible truncation is to include only the first two moments, ρ_0 and ρ_1, which will actually lead to the equations of dissipative fluid dynamics, cf. Eq. (3.38) (with $\Pi = 0$, as applies in the ultrarelativistic limit)

$$\partial_\tau \bar{\pi} + \frac{\bar{\pi}}{\tau_R} = \frac{16}{45}\frac{\varepsilon}{\tau} - \frac{38}{21}\frac{\bar{\pi}}{\tau}\,, \qquad (7.49)$$

where the relaxation time τ_R is identified as the shear relaxation time, $\tau_\pi \equiv \tau_R$, and the shear viscosity is given by $\eta = 4\varepsilon\tau_R/15 = \tau_R\,(\varepsilon + P_0)\,/5$. If the transport coefficient $\delta_{\pi\pi}/\tau_R = 4/3$ (as expected for a fluid composed of massless particles), the transport coefficient $\tau_{\pi\pi}/\tau_R$ is then identified to be $10/7$.

Any higher-order truncation, $M > 1$, goes beyond the traditional fluid-dynamical description, but is also not equivalent to a fully microscopic description, which is only recovered as the number of moments included is taken to infinity. In the following, we investigate how many moments are required in order to obtain a satisfactory description of the energy-momentum tensor (and other moments that couple to it), for viscosities ranging from $\eta/s_0 = 0.1$–10.

We note that, when $2m = n$, the term proportional to $\rho_{n,m+1}$ in Eq. (7.39) vanishes. In this case, the equation of motion becomes closed and can be solved without relying on any truncation. An example is $n = 2$. In this case, the equations for $m = 0, 1$ become

$$\partial_\tau \rho_{2,0} + \frac{\rho_{2,0} - \rho_{2,0}^{eq}}{\tau_R} = -\frac{5}{3} \frac{\rho_{2,0}}{\tau} - \frac{4}{3} \frac{\rho_{2,1}}{\tau},$$

$$\partial_\tau \left(\frac{\rho_{2,1}}{\rho_{2,0}^{eq}} \right) + \frac{\rho_{2,1}}{\rho_{2,0}^{eq}} \frac{\partial_\tau \rho_{2,0}^{eq}}{\rho_{2,0}^{eq}} + \frac{\rho_{2,1}}{\rho_{2,0}^{eq}\tau_R} = -\frac{2}{3\tau} \frac{\rho_{2,0}}{\rho_{2,0}^{eq}} - \frac{7}{3\tau} \frac{\rho_{2,1}}{\rho_{2,0}^{eq}}. \qquad (7.50)$$

Here we used the fact that $\rho_{2,1}^{eq} = 0$. However, the closure of the set of moment equations only happens in the RTA. A general collision term would lead to a coupling of other moments (with different n, m).

7.2.3   Results

Figures 7.1, 7.2, and 7.3 show the moments $\bar{\pi}$, ρ_2, and ρ_6, respectively, normalized to the thermodynamic pressure $P_0 = \varepsilon/3$, as functions of τ/τ_R. The initial time in all calculations is $\tau_0 = 0.1$ fm, the initial temperature is $T_0 = 1$ GeV, and the system is initially in local equilibrium ($\rho_m^{eq} = 0, m > 0$). The different panels in each figure show simulations with a different value of shear viscosity, while the different curves in each plot correspond to a different truncation of the moment expansion.

In all calculations shown, the solution converges after a sufficiently large number of moments is included. However, the number of moments required to obtain a convergent solution depends strongly on the chosen value of η/s_0 and on the moment one is attempting to describe. Smaller viscosities clearly lead to a faster convergence of the solution. As can be seen in Fig. 7.1, when $\eta/s_0 = 0.1$ (left panel), the solution of $\bar{\pi}/P$ including three moments has already converged and the solution with two moments (dissipative fluid dynamics) is already a very good approximation of the

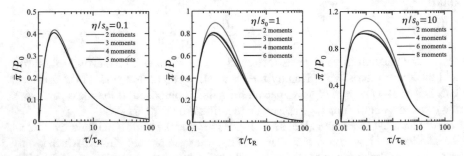

Fig. 7.1 The shear-stress tensor normalized to the thermodynamic pressure, $\bar{\pi}/P_0$, as a function of τ/τ_R for several values of the shear viscosity-to-entropy density ratio: $\eta/s_0 = 0.1$ (left panel), $\eta/s_0 = 1$ (middle panel), and $\eta/s_0 = 10$ (right panel)

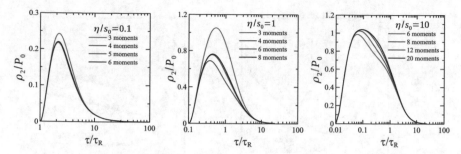

Fig. 7.2 The moment ρ_2 normalized by to thermodynamic pressure as a function of τ/τ_R for several values of the shear viscosity-to-entropy density ratio: $\eta/s_0 = 0.1$ (left panel), $\eta/s_0 = 1$ (middle panel), and $\eta/s_0 = 10$ (right panel)

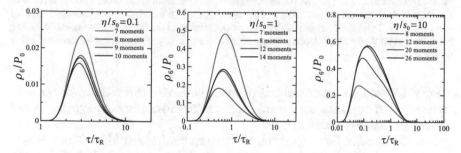

Fig. 7.3 The moment ρ_6 normalized to the thermodynamic pressure as a function of τ/τ_R for several values of the shear viscosity-to-entropy density ratio: $\eta/s_0 = 0.1$ (left panel), $\eta/s_0 = 1$ (middle panel), and $\eta/s_0 = 10$ (right panel)

final answer. When $\eta/s_0 = 1$ (middle panel) one needs at least four moments to obtain a convergent solution, even though one already obtains a reasonable approximation with three moments. Finally, when $\eta/s_0 = 10$ (right panel) one needs at least eight moments to obtain a convergent solution, with a reasonable approximation being obtained already with four moments.

Furthermore, when describing the moments ρ_2 and ρ_6, the convergence of the method of moments is visibly slower, as can be seen in Figs. 7.2 and 7.3. First of all, independent of the value of the shear viscosity, one needs at least three moments in order to describe ρ_2, while one needs at least seven moments in order to describe ρ_6. For $\eta/s_0 = 0.1$ (left panel), ρ_2 requires only four moments to converge, while for $\eta/s_0 = 1$ (middle panel), it requires at least six moments, and for $\eta/s_0 = 10$ (right panel), ρ_2 requires more than twelve moments to converge. The moment ρ_6 requires at least ten moments to converge for $\eta/s_0 = 0.1$, at least 14 moments for $\eta/s_0 = 1$, and more than 20 moments to converge when $\eta/s_0 = 10$. Clearly, the method of moments can become very time-consuming when attempting to describe moments with a large index, but can be very useful when describing moments with a small index or when the viscosity is small. We note that simulations with even a 1000 moments can be easily performed on a laptop computer, so it is still possible to

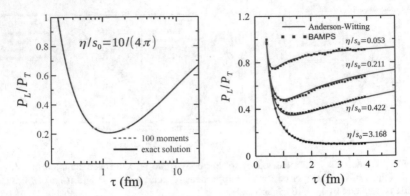

Fig. 7.4 *Left*: The ratio P_L/P_T as a function of τ for $\eta/s_0 = 10/(4\pi)$. The solid line is the exact solution calculated in Ref. [12], while the red dashed line is the result calculated with the method of moments including 100 moments. *Right*: The ratio P_L/P_T as a function of τ for several values of shear viscosity. The solid lines are the solutions calculated with the method of moments including 100 moments while the black dots are the solutions of the full Boltzmann equation calculated using BAMPS, Ref. [21]

reasonably describe a large number of moments with this method. However, we never needed to include such a large number of moments to obtain convergent solutions for the cases studied in this chapter.

In the left panel of Fig. 7.4, we compare the result calculated with 100 moments to the exact solution for the shear-stress tensor calculated numerically in Refs. [12,13]. The quantity plotted is the ratio of the longitudinal pressure over the transverse pressure

$$\frac{P_L}{P_T} = \frac{P_0 - \bar{\pi}}{P_0 + \bar{\pi}/2}, \tag{7.51}$$

as a function of τ. Our calculations were made with the same initial condition and transport coefficients as the calculation performed in Ref. [12,13], with $\eta/s_0 = 10/4\pi$, an initial time of $\tau_0 = 0.25$ fm, and an initial temperature of $T_0 = 0.6$ GeV. We see that both calculations are in perfect agreement—a similar agreement for this quantity could have been obtained including a smaller number of moments.

We also compare our convergent solution with solutions of the full Boltzmann equation in Bjorken scaling from Ref. [21] calculated numerically using the BAMPS algorithm [22]. The cross sections in these simulations have no momentum dependence, but were chosen to have a temperature dependence in such a way that η/s_0 is constant. We perform calculations using the method of moments with the same viscosity coefficient, initial time, temperature, and shear-stress tensor as in Ref. [21]. Namely, the initial time, in this case, is $\tau_0 = 0.4$ fm, the initial temperature is $T_0 = 0.5$ GeV, and the system is initially in local equilibrium. The comparison is shown in the right panel of Fig. 7.4, for four different $\eta/s_0 = 0.053$, 0.211, 0.422, and 3.168. We observe that both calculations are in very good agreement. This need not be the case, since our simulation was performed assuming the RTA, while the calculations performed with BAMPS take into account the full collision term for binary collisions. Therefore, the agreement found for such a wide range of shear viscosity values indicates that, at least when aiming to describe the shear-stress tensor, the RTA is not such a bad assumption. A similar result was also found in the context of anisotropic hydrodynamics in Ref. [23].

7.3 Gubser Flow

We now consider a homogeneous system of massless particles expanding according to the metric proposed by Gubser [10, 11], cf. Eq. (3.102),

$$
\frac{ds^2}{\tau^2} = g_{\mu\nu}dx^\mu dx^\nu = d\rho^2 - \left(\cosh^2\rho\, d\theta^2 + \cosh^2\rho\, \sin^2\theta\, d\phi^2 + d\eta_s^2\right) \, , \quad (7.52)
$$

where we introduced the Gubser coordinates

$$
\sinh\rho \equiv -\frac{1 - (q\tau)^2 + (qr)^2}{2q\tau} \, , \qquad \tan\theta \equiv \frac{2qr}{1 + (q\tau)^2 - (qr)^2} \, , \quad (7.53)
$$

with $r = \sqrt{x^2 + y^2}$ and $\phi = \arctan(y/x)$ being polar coordinates in the transverse (x, y)-plane, cf. Eq. (3.101). The measure ds^2 is rescaled by a factor τ^2, cf. Eq. (3.110), which means that all quantities in this coordinate system are dimensionless. The parameter q appearing in the definitions of ρ and θ is arbitrary and will not play any role as long as we remain in the Gubser coordinate system. It will only matter when we transform back to hyperbolic coordinates. When doing so, one also has to restore the dimension of all fields; see, e.g., Refs. [10, 11].

The nonzero Christoffel symbols in this case are given in Eq. (3.117). We list them here again, in the following dispensing with indicating Weyl-rescaled quantities by a hat: $\Gamma^\rho_{\theta\theta} = \cosh\rho \sinh\rho$, $\Gamma^\rho_{\phi\phi} = \sin^2\theta \cosh\rho \sinh\rho$, $\Gamma^\theta_{\rho\theta} = \Gamma^\phi_{\rho\phi} = \tanh\rho$, $\Gamma^\theta_{\phi\phi} = -\sin\theta\cos\theta$, $\Gamma^\phi_{\theta\phi} = \cot\theta$, and the determinant of this metric is $\sqrt{-g} = \sin\theta \cosh^2\rho$, cf. Eq. (3.118). Note that, since the system is homogeneous, all fields will depend solely on the time-like variable ρ, without displaying any dependence on θ, ϕ, and η_s. Also, note that ρ can assume negative values, ranging from $-\infty$ to $+\infty$. If we transform back to hyperbolic coordinates, this homogeneous system maps to a radially symmetric expanding system. This is more general than the case considered in the previous section, where no radial expansion was considered.

Similar to the Bjorken-flow case, the assumption that the system is homogeneous leads to a trivial velocity field, $u^\mu = (1, 0, 0, 0)^T$. However, the system has a finite expansion rate and shear tensor now given by Eqs. (3.119) and (3.120), respectively, i.e.

$$
\theta = 2\tanh\rho \, ,
$$

$$
\sigma_{\mu\nu} = \mathrm{diag}\left(0, -\frac{1}{3}\cosh\rho\sinh\rho, -\frac{1}{3}\sin^2\theta\cosh\rho\sinh\rho, \frac{2}{3}\tanh\rho\right) \, .
$$

As happened for Bjorken flow, heat flow or diffusion does not exist and the vorticity tensor is also zero. Finally, the shear-stress tensor is diagonal, having the following general form, cf. Eq. (3.122)

$$\pi_\nu^\mu = \text{diag}\left(0, \frac{\bar{\pi}}{2}, \frac{\bar{\pi}}{2}, -\bar{\pi}\right), \tag{7.54}$$

and thus, the continuity equation for the energy can be written as in Eq. (3.123), i.e.

$$u_\nu D_\mu T^{\mu\nu} = \frac{d\varepsilon}{d\rho} + \frac{8}{3}\varepsilon \tanh \rho - \bar{\pi} \tanh \rho = 0, \tag{7.55}$$

where we used that $\pi^{\mu\nu}\sigma_{\mu\nu} = \bar{\pi} \tanh \rho$.

Considering the Boltzmann equation (7.11), we note that the ϕ- and η_s-derivatives vanish by symmetry, while all terms with nonzero Christoffel symbols cancel against each other, except the one $\sim \Gamma_{\phi\theta}^\phi$

$$\left(k^\rho \partial_\rho + k^\theta \partial_\theta\right) f_\mathbf{k} + \Gamma_{\phi\theta}^\phi k_\phi k^\phi \frac{\partial f_\mathbf{k}}{\partial k_\theta} = -k^\rho \frac{f_\mathbf{k} - f_{0\mathbf{k}}}{\tau_R}. \tag{7.56}$$

Now consider the fact that all particles are on-shell, i.e., their energy k_ρ is given in terms of the other components as

$$k_\rho = \sqrt{\frac{k_\theta^2}{\cosh^2\rho} + \frac{k_\phi^2}{\cosh^2\rho \sin^2\theta} + k_{\eta_s}^2}. \tag{7.57}$$

For fixed k_ρ, k_ϕ, and k_{η_s}, this means that k_θ and θ are not independent variables. In fact, rewriting $\partial_\theta \equiv (\partial k_\theta / \partial\theta)\, \partial/\partial k_\theta$ and using Eq. (7.57), we observe that the term $\sim \partial_\theta$ and that with the remaining Christoffel symbol just cancel each other, leading to a Boltzmann equation of a similarly simple form as in the case of Bjorken flow

$$\partial_\rho f_\mathbf{k} = -\frac{1}{\tau_R}\left(f_\mathbf{k} - f_{0\mathbf{k}}\right), \tag{7.58}$$

with $f_{0\mathbf{k}} = \exp\left(-\beta_0 k_\rho\right)$.

7.3.1 Method of Moments

As before, the moment expansion will correspond to an expansion of the distribution function using Laguerre and Legendre polynomials, as shown in Eq. (7.34), but now, because of the Weyl rescaling of the metric, the angle Θ appearing in the Legendre polynomials is defined by the relation

$$\cos \Theta \equiv \frac{k_{\eta_s}}{k_\rho}, \tag{7.59}$$

i.e., without the additional factor τ in the denominator. The distribution function can then be expressed in terms of moments that are very similar to those used in the previous section for Bjorken flow, i.e.

$$\rho_{n,m} = \int dK \, k_\rho^{n+1} P_{2m}(\cos \Theta) \, f_{\mathbf{k}} \, . \tag{7.60}$$

Once more, since the distribution function is invariant under reflections around $\eta_s = 0$, only the moments with an even power in k_{η_s} contribute to the moment expansion (and hence the index $2m$ of the Legendre polynomials in Eq. (7.60)). We note that, as before, the moment $\rho_{1,0}$ is identified as the energy density and, due to the matching conditions, is constructed to be equal to its equilibrium value, $\rho_{1,0} \equiv \rho_{1,0}^{\text{eq}}$. The shear-stress tensor is related to the moment $\rho_{1,1}$ in the following way

$$\bar{\pi} \equiv -\pi_{\eta_s}^{\eta_s} = \frac{2}{3}\rho_{1,1} \, .$$

Finally, in equilibrium, all moments with $m > 0$ vanish, $\rho_{n,m>0}^{\text{eq}} = 0$.

As before, the equation of motion for each moment $\rho_{n,m}$ is obtained by multiplying Eq. (7.11) by $k_\rho^{n+1} P_{2m}(\cos \Theta)$ and integrating over

$$dK \equiv d_{\text{dof}} \frac{dk_\theta \, dk_\phi \, dk_{\eta_s}}{(2\pi)^3 k_\rho \sin \theta \cosh^2 \rho} \, .$$

After some manipulations using Eqs. (7.40) and (7.41), one arrives at

$$
\begin{aligned}
\partial_\rho \rho_{n,m} + \frac{\rho_{n,m}}{\tau_R} &= \frac{\rho_{n,m}^{\text{eq}}}{\tau_R} + 2m \frac{(2m-1)(n+2m+1)}{(4m+1)(4m-1)} \rho_{n,m-1} \tanh \rho \\
&+ \left[\frac{4m^2(n+2m+1)}{(4m+1)(4m-1)} + \frac{(n-2m)(2m+1)^2}{(4m+1)(4m+3)} - (n+2) \right] \rho_{n,m} \tanh \rho \\
&+ (n-2m) \frac{(2m+1)(2m+2)}{(4m+1)(4m+3)} \rho_{n,m+1} \tanh \rho \, .
\end{aligned}
\tag{7.61}
$$

This equation of motion is very similar to Eq. (7.39) derived in the previous section, with only minor changes. We see that moments with different values of n do not couple to each other (a feature of the RTA) and can be solved independently. On the other hand, moments with a given value of m also couple to a higher-order moment of rank $m + 1$ and to a lower-order moment of rank $m - 1$. This leads to an infinite set of coupled moment equations, which must be truncated at a given value of m. From now on, we follow the same strategy used when assuming Bjorken flow and fix $n = 1$, keeping all the moments that couple to the energy-momentum tensor. As before, we omit the index that is held fixed and write the moments as $\rho_{1,m} \equiv \rho_m$.

Then, taking $m = 0$ leads to the equation of motion for the energy density, cf. Eq. (3.123),

$$\partial_\rho \varepsilon + \frac{8}{3}\varepsilon \tanh \rho = \frac{2}{3}\rho_1 \tanh \rho = \bar{\pi} \tanh \rho \, . \tag{7.62}$$

Taking $m = 1$, leads to an equation of motion for $\bar{\pi}$

$$\partial_\rho \bar{\pi} + \frac{\bar{\pi}}{\tau_R} = \frac{16}{45} \varepsilon \tanh \rho - \frac{46}{21} \bar{\pi} \tanh \rho - \frac{8}{35} \rho_2 \tanh \rho \,. \tag{7.63}$$

And, taking $m = 2$, we obtain the equation

$$\partial_\rho \rho_2 + \frac{\rho_2}{\tau_R} = \frac{12}{7} \bar{\pi} \tanh \rho - \frac{172}{77} \rho_2 \tanh \rho - \frac{10}{11} \rho_3 \tanh \rho \,. \tag{7.64}$$

In the last equation, we see that shear-stress tensor acts as a Navier–Stokes term for the moment ρ_2. In a similar manner, ρ_2 will act as a Navier–Stokes term for ρ_3, and so on.

The goal is now to truncate the moment equations and investigate the solutions of the moments that couple to the energy-momentum tensor. We aim to describe the system using the smallest possible number of moments, ρ_m, $m = 0, \ldots, M$—with M being the last moment included. Then, the moment ρ_{M+1}, which appears in the equation of motion for ρ_M, is simply set to zero (this is a direct consequence of the truncation of the moment expansion (7.11)). Following this scheme, the lowest possible truncation is to only include the very first moment $\rho_0 = \varepsilon$, which will lead to the equations of ideal fluid dynamics. The next possible truncation is to include the first two moments, ρ_0 and ρ_1, leading to the equations of dissipative fluid dynamics, as derived from the Boltzmann equation using the 14-moment approximation [7, 16] (see also Refs. [15, 24], which give the fluid-dynamical equations assuming Gubser flow). In the following, we investigate how many moments are required in order to obtain a satisfactory description of the energy-momentum tensor (and other moments that couple to it), for viscosities ranging from $\eta/s_0 = 0.1$–10.

For this purpose, we must choose an initial condition for the moments. As was already mentioned, the Gubser-flow profile corresponds to a system that is expanding radially in hyperbolic coordinates. We fix the temperature at the center of the fireball $(r = 0)$, at a time of $\tau = 0.1$ fm, to be 1 GeV. We also impose that the system size is approximately 10 fm, at the initial time. These are typical scales encountered in fluid-dynamical simulations of the quark-gluon plasma created in ultrarelativistic heavy-ion collisions. Such conditions fix the parameter q, introduced in Eq. (7.53), to be $q = 0.4$ fm^{-1}. Finally, we assume that the shear-stress tensor and all higher-order moments vanish in the center of the fireball at the initial time, i.e., that the center of the system is in equilibrium at $\tau = 0.1$ fm. All this information corresponds to the following initial state in Gubser coordinates

$$\varepsilon \, (\rho = -3) = 0.2 \,, \quad \bar{\pi} \, (\rho = -3) = 0 \,.$$

In the left panel of Fig. 7.5, we show calculations performed using $\eta/s_0 = 0.1$, including two moments (fluid dynamics), ten moments, 20 moments, and 100 moments. We see a clear sign of convergence of the series: in the range $\rho = -6, \ldots, 6$, the solution including ten moments has already converged and the solution with two moments can be considered a good approximation, with disagreements

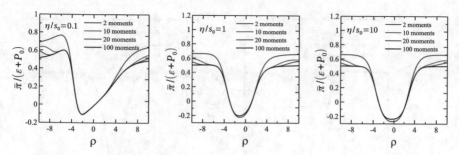

Fig. 7.5 The shear-stress tensor normalized by the enthalpy density, $\bar{\pi}/(\varepsilon + P_0)$, as a function of ρ for several values of the shear viscosity-to-entropy density ratio: $\eta/s_0 = 0.1$ (left panel), $\eta/s_0 = 1$ (middle panel), and $\eta/s_0 = 10$ (right panel)

Fig. 7.6 The shear-stress tensor normalized by the enthalpy density, $\bar{\pi}/(\varepsilon + P_0)$, as a function of ρ for $\eta/s_0 = 10$. Results including 100 (red dashed line) and 200 (black solid line) moments are shown

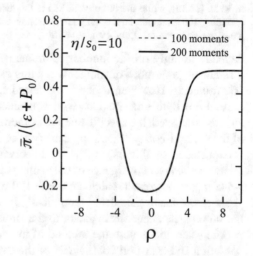

mostly at large values of $|\rho|$. We note that values of ρ ranging from $-6, \ldots, 6$ already describe a wide range of values in radius and time in hyperbolic space, e.g., at a time $\tau = 10$ fm, values of radius $r < 20$ fm are contained in values of ρ between $-2.5 < \rho < 1.5$.

We also show calculations using $\eta/s_0 = 1$ (middle panel) and $\eta/s_0 = 10$ (right panel). We see that these calculations also show signs of convergence, with larger values of viscosity converging slower (i.e., they require a larger number of moments). As the shear viscosity increases, the solution for $\bar{\pi}/(\varepsilon + P_0)$ gradually becomes more symmetric around $\rho = 0$, as also happens in Israel–Stewart theory; see Ref. [24]. In Fig. 7.6, we confirm the convergence of the method of moments for large values of $|\rho|$ showing that simulations with 100 and 200 moments, for $\eta/s_0 = 10$, provide basically the same result for $\bar{\pi}/(\varepsilon + P_0)$.

In particular, it is important to mention that $\bar{\pi}/(\varepsilon + P_0)$ tends to a constant as ρ is taken to infinity (or minus infinity). This feature is qualitatively captured by the method of moments, but we should note that the asymptotic value of $\bar{\pi}/(\varepsilon + P_0)$

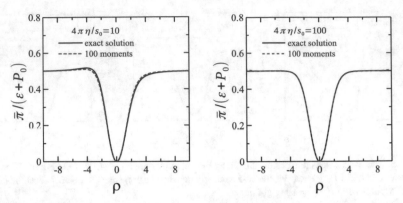

Fig. 7.7 The ratio $\bar{\pi}/(\varepsilon + P_0)$ as a function of ρ. The solid line is the exact solution calculated in Refs. [14,15], while the red dashed line is the result calculated with the method of moments including 100 moments. In the left panel, we show the result for $4\pi\eta/s_0 = 10$, while in the right panel, we show the result for $4\pi\eta/s_0 = 100$

depends strongly on the number of moments included in the calculation. It takes a rather large number of moments for this asymptotic value to converge, roughly 100 moments. However, if one is interested solely in the description of the system around $\rho = 0$, this can be done with a considerably smaller number of moments (as already discussed, the specific number of moments required will depend on the value of the shear viscosity-to-entropy density ratio chosen). A hydrodynamic description is expected at small values of $|\rho|$, and hence the hydrodynamic limit appears to be related to the fast convergence of the method of moments. In the left panel of Fig. 7.7, we compare our result, calculated with 100 moments, to the exact solution for the shear-stress tensor calculated numerically in Refs. [14,15]. The quantity plotted is the ratio of the shear-stress tensor to the enthalpy density, $\varepsilon + P_0$, as a function of ρ. The calculations with the method of moments were made with the same initial condition and transport coefficients of the calculation performed in Refs. [14,15], with shear viscosities $\eta/s_0 = 10/(4\pi)$ and $100/(4\pi)$, and for a system that is in equilibrium with an energy density $\varepsilon = 1$ at $\rho = 0$. We see that both calculations are in good agreement, with small differences around $\rho = -4.4$ when $\eta/s_0 = 10/4\pi$. Such a small difference is not due to the lack of convergence of the method of moments, since we already showed in Fig. 7.6 that 100 moments are more than enough to obtain a convergent solution for $\bar{\pi}/(\varepsilon + P_0)$. Such a difference between both results disappears when setting $\eta/s_0 = 100/(4\pi)$, as can be seen in the right panel of Fig. 7.7.

For the sake of completeness, we show in Figs. 7.8 and 7.9 the convergence of the moments ρ_2 and ρ_6, for $\eta/s_0 = 10$. The left panels of both plots show that convergence is achieved for $|\rho| \lesssim 4$, with a relatively small number of moments of order ~ 10. It is clear that the higher-order moments require the inclusion of a larger number of moments in order to obtain a convergent solution, as was also found in the case of Bjorken flow. The right panels of Figs. 7.8 and 7.9 show that convergence is much slower when trying to describe ρ_2 and ρ_6 for larger values of $|\rho|$, in which case absolute convergence can require more than 100 moments.

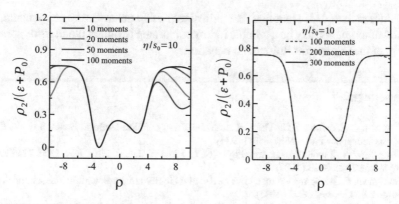

Fig. 7.8 The ratio $\rho_2/(\varepsilon + P_0)$ as a function of ρ for $\eta/s_0 = 10$

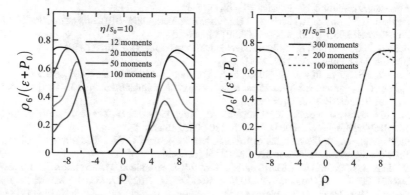

Fig. 7.9 The ratio $\rho_6/(\varepsilon + P_0)$ as a function of ρ for $\eta/s_0 = 10$

7.4 Summary

In Chap. 6, we presented a relativistic generalization of the method of moments. In this chapter, we implemented this method considering ultrarelativistic gases undergoing highly symmetric flow configurations: Bjorken and Gubser flow. We further simplified the relativistic Boltzmann equation by imposing the relaxation-time approximation (RTA). In these cases, we were able to derive the complete equations of motion for arbitrary moments of the single-particle distribution function (something that was not possible for arbitrary flow configurations) and could investigate how to systematically truncate them. We showed that, even though the method of moments is not an expansion in terms of a small dimensionless parameter, it still appears to converge to exact solutions of the Boltzmann equation, if a sufficiently large number of moments are included. Therefore, in contrast to the Chapman-Enskog expansion, the method of moments may be a reliable intermediate step to obtain the fluid-dynamical regime of the relativistic Boltzmann equation.

References

1. Chapman, S., Cowling, T.G.: The Mathematical Theory of Non-Uniform Gases, 3rd edn. Cambridge University Press, New York (1974)
2. Hiscock, W.A., Lindblom, L.: Ann. Phys. (N.Y.) **151** 466 (1983); Phys. Rev. D **31** 725 (1985); Phys. Rev. D **35** 3723 (1987)
3. Cercignani, C., Medeiros Kremer, G.: The Relativistic Boltzmann Equation: Theory and Applications. Birkhauser, Basel (2002)
4. de Groot, S.R., van Leewen, W.A., van Weert, Ch.G.: Relativistic Kinetic Theory: Principles and Applications. North-Holland (1980)
5. Bhatnagar, P.L., Gross, E.P., Krook, M.: Phys. Rev. **94**, 511 (1954)
6. Anderson, J., Witting, H.: Physica **74**, 466 (1974)
7. Denicol, G.S., Koide, T., Rischke, D.H.: Phys. Rev. Lett. **105** (2010). https://doi.org/10.1103/PhysRevLett.105.162501, arXiv:1004.5013 [nucl-th]
8. Landau, L.D., Lifshitz, E.M.: Fluid Mechanics. Pergamon, New York (1959)
9. Bjorken, J.D.: Phys. Rev. D **27**, 140 (1983)
10. Gubser, S.S.: Phys. Rev. D **82** (2010). https://doi.org/10.1103/PhysRevD.82.085027
11. Gubser, S.S., Yarom, A.: Nucl. Phys. B **846**, 469 (2011). https://doi.org/10.1016/j.nuclphysb.2011.01.012, arXiv:1012.1314 [hep-th]
12. Florkowski, W., Ryblewski, R., Strickland, M.: Nucl. Phys. A **916**, 249 (2013). https://doi.org/10.1016/j.nuclphysa.2013.08.004, arXiv:1304.0665 [nucl-th]
13. Florkowski, W., Ryblewski, R., Strickland, M.: Phys. Rev. C **88**, (2013). https://doi.org/10.1103/PhysRevC.88.024903. arXiv:1305.7234 [nucl-th]
14. Denicol, G.S., Heinz, U.W., Martinez, M., Noronha, J., Strickland, M.: Phys. Rev. Lett. **113**(20), 202301 (2014). https://doi.org/10.1103/PhysRevLett.113.202301, arXiv:1408.5646 [hep-ph]
15. Denicol, G.S., Heinz, U.W., Martinez, M., Noronha, J., Strickland, M.: Phys. Rev. D **90**(12), 125026 (2014). https://doi.org/10.1103/PhysRevD.90.125026, arXiv:1408.7048 [hep-ph]
16. Denicol, G.S., Niemi, H., Molnar, E., Rischke, D.H.: Phys. Rev. D **85**, 114047 (2012). Erratum: [Phys. Rev. D **91**(3), 039902 (2015)]. https://doi.org/10.1103/PhysRevD.85.114047, https://doi.org/10.1103/PhysRevD.91.039902, arXiv:1202.4551 [nucl-th]
17. Denicol, G.S.: J. Phys. G **41**(12) (2014). https://doi.org/10.1088/0954-3899/41/12/124004
18. Eq. (8.911.1) In: Gradshteyn, I.S., Ryzhik, I.M. (eds.) Table of Integrals, Series and Products, 4th edn. Academic (1980)
19. Eq. (7.414.3) In: Gradshteyn, I.S., Ryzhik, I.M. (eds.) Table of Integrals, Series and Products, 4th edn. Academic (1980)
20. Eq. (7.414.8) In: Gradshteyn, I.S., Ryzhik, I.M. (eds.) Table of Integrals, Series and Products, 4th edn. Academic (1980)
21. El, A., Xu, Z., Greiner, C.: Phys. Rev. C **81** (2010). https://doi.org/10.1103/PhysRevC.81.041901, arXiv:0907.4500 [hep-ph]
22. Xu, Z., Greiner, C.: Phys. Rev. C **71**, 064901 (2005); Phys. Rev. C **76**, 024911 (2007)
23. Almaalol, D., Strickland, M.: arXiv:1801.10173 [hep-ph]
24. Marrochio, H., Noronha, J., Denicol, G.S., Luzum, M., Jeon, S., Gale, C.: Phys. Rev. C **91**(1) (2015)

Fluid Dynamics from the Method of Moments

8

The two widespread methods to derive relativistic fluid dynamics from the Boltzmann equation, the Chapman-Enskog expansion [1] and the method of moments as proposed by Israel and Stewart [2–4], both described in Chap. 5, have flaws. In the relativistic case, the Chapman-Enskog expansion leads to equations of fluid dynamics that are acausal and intrinsically unstable [5–7] and, consequently, should not be applied to derive the equations of relativistic fluid dynamics from kinetic theory.

On the other hand, the method of moments [2–4,8] is in principle free of such problems and leads to transient fluid-dynamical equations that are causal and stable [9,10]. The method of moments was first developed by Grad [11] for non-relativistic systems. In Grad's original work, the single-particle distribution function is expanded around its local-equilibrium value in terms of a complete set of Hermite polynomials [12]. This expansion is truncated and the distribution function is finally expressed in terms of 13 fluid-dynamical variables: the velocity field, the temperature, the chemical potential, the heat-conduction current, and the shear-stress tensor. In this case, the heat-conduction current and shear-stress tensor become independent dynamical variables, which satisfy partial differential equations that describe their relaxation toward their respective Navier–Stokes values. We emphasize that, in Grad's theory, the gradient expansion appears only as an asymptotic (long-time limit) solution [13–17], as shown in Chap. 5. This confirms the idea originally proposed by Grad that the series generated by the Chapman-Enskog procedure is asymptotic rather than convergent [18].

Nevertheless, Grad's method has still one major drawback: unlike the Chapman-Enskog expansion, it lacks a small parameter, such as the Knudsen number, in which one can do power-counting, and thus systematically improve the approximation [19]. This deficiency, together with the bad performance of Grad's method in comparison to microscopic calculations [20], have led researchers to abandon this approach for some time. However, recently a lot of effort has been made to reformulate the method

© Springer Nature Switzerland AG 2021

G. S. Denicol and D. H. Rischke, *Microscopic Foundations of Relativistic Fluid Dynamics*, Lecture Notes in Physics 990, https://doi.org/10.1007/978-3-030-82077-0_8

of moments into a more reliable tool to describe non-equilibrium phenomena for large Knudsen numbers [20]. For instance, in Refs. [21–23], Grad's equations were regularized to have a wider domain of validity in Knudsen number and were then shown to be in good agreement with microscopic calculations. Such approaches, however, were only formulated for non-relativistic systems.

The generalization of Grad's method of moments to relativistic systems has been pursued by several authors [24–29]. The most widely employed approach is due to Israel and Stewart; see Sect. 5.3 in Chap. 5. However, Israel's and Stewart's theory shares the same problems as Grad's original approach: it lacks a parameter in which one can do systematic power counting of corrections to the local-equilibrium distribution function.

It was recently confirmed that, at least for some special problems, the Israel–Stewart equations are not in good agreement with the numerical solution of the Boltzmann equation [30–32]. Attempts to improve Israel's and Stewart's theory were initially made in Refs. [33–36], but Israel's and Stewart's 14-moment approximation was still used. In this chapter, we discuss the more complete derivation of fluid dynamics proposed in Ref. [37], which does not rely on the 14-moment approximation.

We present a general derivation of relativistic fluid dynamics from the Boltzmann equation [37] employing the generalized method of moments presented in Chap. 6. The main point of this approach, in contrast with the traditional 14-moment approximation, is that it does not close the fluid-dynamical equations of motion by an *ad hoc* truncation of the moment expansion. Instead, the reduction of the degrees of freedom is performed by identifying the microscopic time scales of the Boltzmann equation and considering only the slowest ones. In addition, the equations of motion for the dissipative quantities are truncated according to a systematic power-counting scheme in Knudsen and inverse Reynolds number. We conclude that the equations of motion can be closed in terms of only 14 dynamical variables, as long as we only keep terms of first order in either Knudsen or inverse Reynolds number, or the product of both. We show that, even though the equations of motion are closed in terms of these 14 fields, the transport coefficients carry information about all the moments of the distribution function.

This chapter is organized as follows. In Sect. 8.1, we introduce our power-counting scheme in terms of Knudsen and inverse Reynolds numbers. Then, by diagonalizing the linear part of the set of moment equations derived in Chap. 6, we demonstrate how to identify the slowest microscopic time scale of the Boltzmann equation for each dissipative current. We shall derive dynamical equations for the slowest modes, but approximate faster modes by their asymptotic solution for long times. This will then lead, in Sect. 8.3, to the complete set of fluid-dynamical equations which contains *all terms up to second order in Knudsen and inverse Reynolds numbers* R_i^{-1}, i.e., $O(\mathrm{Kn}^2, R_i^{-1}R_j^{-1}, \mathrm{Kn}\,R_i^{-1})$, where the indices $i, j = \Pi, n, \pi$ refer to the particular way of defining the inverse Reynolds numbers (see below). In Sect. 8.4, we first demonstrate the validity of our approach by restricting the calculation to the 14-moment approximation and recovering the results derived in Chap. 5 (and of Ref. [33]) for the transport coefficients of an ultrarelativistic, classical gas with constant cross section. We then show how to successively improve the expression for the transport coefficients by extending the number of moments to $14 + 9 \times n$. We explicitly study the cases $n = 1, 2$ and 3. In Sect. 8.6, we derive hyperbolic equations of motion for transient fluid dynamics that are valid up to second order in Knudsen number. In Sect. 8.7, we compare the solutions of the derived fluid-dynamical equations to numerical solutions of the Boltzmann equation for certain shock-wave scenarios. We analyze the domain of validity of the derived equations, depending on the choice of the cross section. We end this chapter with a brief discussion and summary of the results obtained.

8.1 Power Counting

In Chap. 6, the single-particle distribution function $f_{\mathbf{k}}$ is expanded in terms of an orthogonal basis in 4-momentum space. The expansion basis contains two basic ingredients. The first are the *irreducible* tensors, 1, $k^{\langle \mu \rangle}$, $k^{\langle \mu_1} k^{\mu_2 \rangle}$, ..., $k^{\langle \mu_1} \cdots k^{\mu_\ell \rangle}$, ..., which form a *complete and orthogonal* set, analogous to the spherical harmonics [8,37]. We used the notation $A^{\langle \mu_1 \cdots \mu_\ell \rangle} \equiv \Delta^{\mu_1 \cdots \mu_\ell}_{\nu_1 \cdots \nu_\ell} A^{\nu_1 \cdots \nu_\ell}$, with $\Delta^{\mu_1 \cdots \mu_m}_{\nu_1 \cdots \nu_m}$ being symmetrized and, except for $m = 1$, traceless projectors onto the subspaces orthogonal to u^μ, as defined in Appendix 6.5 of Chap. 6. In the local rest frame of the fluid, the irreducible tensors are formed from the spatial components of the 4-momentum k^μ. Note that the expansion of $f_{\mathbf{k}}$ in Israel–Stewart theory is not in terms of the irreducible tensors $k^{\langle \mu_1} \cdots k^{\mu_\ell \rangle}$, but in terms of the tensors $k^{\mu_1} \cdots k^{\mu_\ell}$, which are complete but neither irreducible nor orthogonal.

The second ingredient are orthogonal polynomials in the energy in the local rest frame, $E_{\mathbf{k}} = u^\mu k_\mu$, $P^{(\ell)}_{n\mathbf{k}} = \sum_{r=0}^{n} a^{(\ell)}_{nr} E^r_{\mathbf{k}}$, constructed in Appendix 6.8 of Chap. 6. Then, $f_{\mathbf{k}}$ is expanded as

$$f_{\mathbf{k}} = f_{0\mathbf{k}} + f_{0\mathbf{k}} \tilde{f}_{0\mathbf{k}} \sum_{\ell=0}^{\infty} \sum_{n=0}^{N_\ell} \mathcal{H}^{(\ell)}_{n\mathbf{k}} \, \rho^{\mu_1 \cdots \mu_\ell}_n \, k_{\langle \mu_1} \cdots k_{\mu_\ell \rangle} \,, \tag{8.1}$$

where $f_{0\mathbf{k}} = \left[\exp \left(\beta_0 E_{\mathbf{k}} - \alpha_0 \right) + a \right]^{-1}$ is the local-equilibrium distribution function, with $\alpha_0 = \mu / T$ being the ratio of chemical potential over temperature and $\beta_0 = 1/T$ the inverse temperature, while $a = \pm 1$ for fermions/bosons and $a = 0$ for classical Boltzmann particles. We further introduced polynomials of order N_ℓ in energy, $\mathcal{H}^{(\ell)}_{n\mathbf{k}} \equiv \left(W^{(\ell)} / \ell! \right) \sum_{m=n}^{N_\ell} a^{(\ell)}_{mn} P^{(\ell)}_{m\mathbf{k}}$, with a normalization constant $W^{(\ell)}$, and the irreducible moments of $\delta f_{\mathbf{k}} = f_{\mathbf{k}} - f_{0\mathbf{k}}$

$$\rho^{\mu_1 \cdots \mu_\ell}_r \equiv \int dK \, E^r_{\mathbf{k}} \, k^{\langle \mu_1} \cdots k^{\mu_\ell \rangle} \delta f_{\mathbf{k}} \,. \tag{8.2}$$

The values of α_0 and β_0 are defined by the matching conditions for particle-number density and energy density, $n \equiv n_0 = \langle E_{\mathbf{k}} \rangle_0$, $\varepsilon \equiv \varepsilon_0 = \langle E^2_{\mathbf{k}} \rangle_0$, where $\langle \cdots \rangle_0 \equiv \int dK \, (\cdots) \, f_{0\mathbf{k}}$. The matching conditions and the definition of u^μ according to the Landau-frame choice imply that the following irreducible moments vanish: $\rho_1 = \rho_2 = \rho^\mu_1 = 0$.

The equations of motion for ρ_r, ρ_r^μ, and $\rho_r^{\mu\nu}$ together with their respective transport coefficients were derived in Chap. 6, cf. Eqs. (6.20)–(6.22). For the sake of convenience, we again quote these equations of motion

$$\dot{\rho}_r - C_{r-1} = \alpha_r^{(0)}\theta - \frac{G_{2r}}{D_{20}}\Pi\theta + \frac{G_{2r}}{D_{20}}\pi^{\mu\nu}\sigma_{\mu\nu} + \frac{G_{3r}}{D_{20}}\partial_\mu n^\mu + (r-1)\rho_{r-2}^{\mu\nu}\sigma_{\mu\nu}$$
$$+ r\rho_{r-1}^\mu\dot{u}_\mu - \nabla_\mu\rho_{r-1}^\mu - \frac{1}{3}\left[(r+2)\rho_r - (r-1)m^2\rho_{r-2}\right]\theta\,, \qquad (8.3)$$

$$\dot{\rho}_r^{\langle\mu\rangle} - C_{r-1}^{\langle\mu\rangle} = \alpha_r^{(1)}I^\mu + \rho_r^\nu\omega_\nu^\mu + \frac{1}{3}\left[(r-1)m^2\rho_{r-2}^\mu - (r+3)\rho_r^\mu\right]\theta$$
$$+ \frac{1}{5}\left[(2r-2)m^2\rho_{r-2}^\nu - (2r+3)\rho_r^\nu\right]\sigma_\nu^\mu + \frac{1}{3}\left[m^2r\rho_{r-1} - (r+3)\rho_{r+1}\right]\dot{u}^\mu$$
$$+ \frac{\beta_0 J_{r+2,1}}{\varepsilon_0 + P_0}\left(\Pi\dot{u}^\mu - \nabla^\mu\Pi + \Delta_\nu^\mu\partial_\lambda\pi^{\lambda\nu}\right) - \frac{1}{3}\nabla^\mu\left(m^2\rho_{r-1} - \rho_{r+1}\right)$$
$$- \Delta_\lambda^\mu\nabla_\nu\rho_{r-1}^{\lambda\nu} + r\rho_{r-1}^{\mu\nu}\dot{u}_\nu + (r-1)\rho_{r-2}^{\mu\nu\lambda}\sigma_{\lambda\nu}\,, \qquad (8.4)$$

$$\dot{\rho}_r^{\langle\mu\nu\rangle} - C_{r-1}^{\langle\mu\nu\rangle} = 2\alpha_r^{(2)}\sigma^{\mu\nu} - \frac{2}{7}\left[(2r+5)\rho_r^{\lambda\langle\mu} - 2m^2(r-1)\rho_{r-2}^{\lambda\langle\mu}\right]\sigma_\lambda^{\nu\rangle} + 2\rho_r^{\lambda\langle\mu}\omega_\lambda^{\nu\rangle}$$
$$+ \frac{2}{15}\left[(r+4)\rho_{r+2} - (2r+3)m^2\rho_r + (r-1)m^4\rho_{r-2}\right]\sigma^{\mu\nu}$$
$$+ \frac{2}{5}\nabla^{\langle\mu}\left(\rho_{r+1}^{\nu\rangle} - m^2\rho_{r-1}^{\nu\rangle}\right) - \frac{2}{5}\left[(r+5)\rho_{r+1}^{\langle\mu} - m^2r\rho_{r-1}^{\langle\mu}\right]\dot{u}^{\nu\rangle}$$
$$- \frac{1}{3}\left[(r+4)\rho_r^{\mu\nu} - m^2(r-1)\rho_{r-2}^{\mu\nu}\right]\theta$$
$$+ (r-1)\rho_{r-2}^{\mu\nu\lambda\rho}\sigma_{\lambda\rho} - \Delta_{\alpha\beta}^{\mu\nu}\nabla_\lambda\rho_{r-1}^{\alpha\beta\lambda} + r\rho_{r-1}^{\mu\nu\lambda}\dot{u}_\lambda\,, \qquad (8.5)$$

where $C_r^{\langle\mu_1\cdots\mu_\ell\rangle}$ are the irreducible moments of the collision term

$$C_r^{\langle\mu_1\cdots\mu_\ell\rangle} \equiv \int dK\, E_{\mathbf{k}}^r k^{\langle\mu_1}\cdots k^{\mu_\ell\rangle}C[f]\,. \qquad (8.6)$$

We also employ the usual notation, where $\sigma^{\mu\nu} \equiv \nabla^{\langle\mu}u^{\nu\rangle}$ is the shear tensor, $\theta \equiv \nabla_\mu u^\mu$ the expansion scalar, and $\omega^{\mu\nu} \equiv (\nabla^\mu u^\nu - \nabla^\nu u^\mu)/2$ the vorticity tensor. The thermodynamic functions $\alpha_r^{(0)}$, $\alpha_r^{(1)}$, $\alpha_r^{(2)}$, G_{nq}, and D_{nq} were introduced and defined in Eqs. (6.27), (6.28), (6.29), and (6.31). Finally, in deriving these equations, Landau matching conditions [38] were employed.

In Chap. 6, we further expressed the irreducible moments of the collision term, $C_{r-1}^{\langle\mu_1\cdots\mu_\ell\rangle}$, in terms of the irreducible moments of $\delta f_{\mathbf{k}}$

$$C_{r-1}^{\langle\mu_1\cdots\mu_\ell\rangle} = -\sum_{n=0}^{N_\ell}\mathcal{A}_{rn}^{(\ell)}\rho_n^{\mu_1\cdots\mu_\ell} + (\text{terms nonlinear in }\delta f)\,, \qquad (8.7)$$

where the scalars $\mathcal{A}_{rn}^{(\ell)}$ emerged from the linear part of the collision term, cf. Eq. (6.47) in Sect. 6.3.1

$$\mathscr{A}_{rn}^{(\ell)} = \frac{1}{\nu(2\ell+1)} \int dK dK' dP dP' W_{\mathbf{kk'} \to \mathbf{pp'}} f_{0\mathbf{k}} f_{0\mathbf{k'}} \tilde{f}_{0\mathbf{p}} \tilde{f}_{0\mathbf{p'}} E_{\mathbf{k}}^{r-1} k^{\langle \nu_1} \cdots k^{\nu_\ell \rangle}$$

$$\times \left(\mathcal{H}_{\mathbf{k}n}^{(\ell)} k_{\langle \nu_1} \cdots k_{\nu_\ell \rangle} + \mathcal{H}_{\mathbf{k'}n}^{(\ell)} k'_{\langle \nu_1} \cdots k'_{\nu_\ell \rangle} - \mathcal{H}_{\mathbf{p}n}^{(\ell)} p_{\langle \nu_1} \cdots p_{\nu_\ell \rangle} - \mathcal{H}_{\mathbf{p'}n}^{(\ell)} p'_{\langle \nu_1} \cdots p'_{\nu_\ell \rangle} \right),$$

$$(8.8)$$

while (terms nonlinear in δf) correspond to terms that are nonlinear in the irreducible moments of $\delta f_{\mathbf{k}}$, cf. Sect. 6.3.2.

It was found that there is an infinite number of coupled equations (labeled by the index r) for these moments, and the equations for the moments up to rank two contain moments of rank higher than two. In general, one would have to solve this infinite set of coupled equations in order to determine the time evolution of the system. However, in the fluid-dynamical limit, it is expected that the macroscopic dynamics of a given system simplifies, and therefore, it can be described by the conserved currents N^μ and $T^{\mu\nu}$ alone. From the kinetic point of view, it is usually believed that the validity of the fluid-dynamical limit can be quantified by the Knudsen number

$$\mathrm{Kn} = \frac{\lambda}{L}, \tag{8.9}$$

where λ and L are typical microscopic and macroscopic length or time scales of the system, respectively, as already defined in Chap. 1. The relevant macroscopic scales are usually estimated from the gradients of fluid-dynamical quantities, while the microscopic scales are of the order of the mean free path or time between collisions. It is generally assumed that when there is a clear separation of the microscopic and macroscopic scales, i.e., when $\mathrm{Kn} \ll 1$, the microscopic details can be safely integrated out and the dynamics of the system can be described using only a few macroscopic fields.

Furthermore, we also expect fluid dynamics to be valid near local thermal equilibrium, i.e., when $\delta f_{\mathbf{k}} \ll f_{0\mathbf{k}}$. We can quantify the deviation from equilibrium in terms of the macroscopic variables by defining a set of ratios of dissipative quantities to the equilibrium pressure or density. These can be understood as generalizations of the inverse Reynolds number and will be denoted as

$$\mathrm{R}_\Pi^{-1} \equiv \frac{|\Pi|}{P_0}, \quad \mathrm{R}_n^{-1} \equiv \frac{|n^\mu|}{n_0}, \quad \mathrm{R}_\pi^{-1} \equiv \frac{|\pi^{\mu\nu}|}{P_0}. \tag{8.10}$$

Since the non-equilibrium moments are integrals of $\delta f_{\mathbf{k}}$, while the equilibrium pressure and particle-number density are integrals over the equilibrium distribution function $f_{0\mathbf{k}}$, these ratios may be used to quantify how significantly the system deviates from equilibrium.

With this in mind, it is clear that these two measures, the Knudsen number and the inverse Reynolds number, should be used to quantify the proximity of the system to the fluid-dynamical limit. In general, these two measures are independent of each other, e.g., a system can be initialized in such a way that the Knudsen number is large, but the inverse Reynolds number is small or vice versa. When deriving transient fluid dynamics, one should not *a priori* assume that $\mathrm{Kn} \sim \mathrm{R}_i^{-1}$: while the inverse Reynolds and Knudsen numbers are certainly related, their relation is in principle dynamical and is precisely what we aim to find. Only for asymptotically long times, when the solutions of the dynamical equations approach their Navier-Stokes values, one typically has $\text{Kn} \sim \text{R}_i^{-1}$, as will be discussed in more detail below.

In the traditional 14-moment approximation introduced by Israel and Stewart [4], the fluid-dynamical limit is implemented by a truncation of the expansion of the distribution function, which corresponds neither to a truncation in Knudsen nor in inverse Reynolds number. In this sense, the domain of validity of the equations of motion obtained via the traditional 14-moment approximation is not clear, because it is not possible to determine the order of the terms that were neglected. In order to obtain a closed set of macroscopic equations with a clear domain of validity in both Kn and R_i^{-1}, another truncation procedure is necessary. Its derivation is the main purpose of this chapter.

8.2   Resummed Transient Relativistic Fluid Dynamics

The exact equations of motion (8.3)–(8.5) contain infinitely many degrees of freedom, given by the irreducible moments of the distribution function, and also infinitely many microscopic time scales, related to the coefficients $\mathcal{A}_{rn}^{(\ell)}$. As was argued in Chap. 4 (see also Ref. [13]), the slowest microscopic time scale should dominate the dynamics at long times, i.e., in the fluid-dynamical limit. In order to extract the relevant relaxation time scales, we have to determine the normal modes of Eqs. (8.3)–(8.5), i.e., we diagonalize the part which is linear in the irreducible moments $\rho_r^{\mu_1 \cdots \mu_\ell}$. These are the linear terms on the left-hand sides arising from Eq. (8.7) and the first terms on the right-hand sides. The nonlinear terms from Eq. (8.7) as well as the remaining terms on the right-hand sides of Eqs. (8.3)–(8.5), which are products of the moments and gradients of the primary fluid-dynamical variables α_0, β_0, and u^μ, or which are gradients of the moments, are not considered in the diagonalization procedure. Identifying and separating the microscopic time scales of the Boltzmann equation is also the basic step for obtaining general relations between the irreducible moments and the dissipative currents and, as we shall see, closing the equations of motion in terms of N^μ and $T^{\mu\nu}$.

To this end, we shall introduce the matrix $\Omega^{(\ell)}$, which diagonalizes $\mathcal{A}^{(\ell)}$

$$\left(\Omega^{-1} \right)^{(\ell)} \mathcal{A}^{(\ell)} \Omega^{(\ell)} = \text{diag} \left(\chi_0^{(\ell)}, \ldots, \chi_j^{(\ell)}, \ldots \right) , \tag{8.11}$$

where $\chi_j^{(\ell)}$ are the eigenvalues of $\mathcal{A}^{(\ell)}$. We further define the tensors $X_i^{\mu_1 \cdots \mu_\ell}$ as

$$X_i^{\mu_1 \cdots \mu_\ell} \equiv \sum_{j=0}^{N_\ell} \left(\Omega^{-1} \right)_{ij}^{(\ell)} \rho_j^{\mu_1 \cdots \mu_\ell} . \tag{8.12}$$

These are the eigenmodes of the linearized Boltzmann equation. Multiplying Eq. (8.7) with $\left(\Omega^{-1} \right)^{(\ell)}$ from the left and using Eqs. (8.11) and (8.12), we obtain

$$\sum_{j=0}^{N_\ell} \left(\Omega^{-1}\right)_{ij}^{(\ell)} C_{j-1}^{\langle\mu_1\cdots\mu_\ell\rangle} = -\chi_i^{(\ell)} X_i^{\mu_1\cdots\mu_\ell} + \text{(terms nonlinear in } \delta f) , \qquad (8.13)$$

where we do not sum over the index i on the right-hand side of the equation. Then we multiply Eqs. (8.3)–(8.5) with $\left(\Omega^{-1}\right)_{ir}^{(\ell)}$ and sum over r. Using Eq. (8.13), we obtain the equations of motion for the new variables $X_i^{\mu_1\cdots\mu_\ell}$,

$$\dot{X}_i + \chi_i^{(0)} X_i = \beta_i^{(0)}\theta \quad + \quad \text{(higher-order terms)} ,$$
$$\dot{X}_i^{\langle\mu\rangle} + \chi_i^{(1)} X_i^\mu = \beta_i^{(1)} I^\mu \quad + \quad \text{(higher-order terms)} ,$$
$$\dot{X}_i^{\langle\mu\nu\rangle} + \chi_i^{(2)} X_i^{\mu\nu} = \beta_i^{(2)}\sigma^{\mu\nu} \quad + \quad \text{(higher-order terms)} , \qquad (8.14)$$

where we introduced the transport coefficients

$$\beta_i^{(0)} = \sum_{j=0,\neq 1,2}^{N_0} \left(\Omega^{-1}\right)_{ij}^{(0)} \alpha_j^{(0)} ,$$

$$\beta_i^{(1)} = \sum_{j=0,\neq 1}^{N_1} \left(\Omega^{-1}\right)_{ij}^{(1)} \alpha_j^{(1)} , \qquad (8.15)$$

$$\beta_i^{(2)} = 2\sum_{j=0}^{N_2} \left(\Omega^{-1}\right)_{ij}^{(2)} \alpha_j^{(2)} .$$

With "higher-order terms" in Eqs. (8.14) we refer to the terms nonlinear in δf from Eq. (8.13) as well as to the higher-order terms on the right-hand sides of Eqs. (8.3)–(8.5). As expected, the equations of motion for the tensors $X_i^{\mu_1\cdots\mu_\ell}$ decouple in the linear regime, displaying a relaxation-type behavior. Without loss of generality, we order the tensors $X_r^{\mu_1\cdots\mu_\ell}$ according to increasing $\chi_r^{(\ell)}$, i.e., $\chi_r^{(\ell)} < \chi_{r+1}^{(\ell)} \; \forall \ell$.

By diagonalizing Eqs. (8.3)–(8.5), we were able to identify the microscopic time scales of the Boltzmann equation given by the inverse of the coefficients $\chi_r^{(\ell)}$. It is clear that, if the nonlinear terms in Eqs. (8.14) are sufficiently small, each tensor $X_r^{\mu_1\cdots\mu_\ell}$ relaxes exponentially and independently to its respective asymptotic value, given by the first term on the right-hand sides of Eqs. (8.14) (divided by the corresponding $\chi_r^{(\ell)}$), on a time scale $\sim 1/\chi_r^{(\ell)}$. We will refer to these asymptotic solutions as *Navier–Stokes values*. By neglecting all relaxation scales, i.e., taking the limit $\chi_r^{(\ell)} \to \infty$ with the ratio $\beta_r^{(\ell)}/\chi_r^{(\ell)}$ held fixed, all irreducible moments $\rho_r^{\mu_1\cdots\mu_\ell}$ become proportional to gradients of α_0, β_0, and u^μ, and we obtain a Chapman-Enskog-type solution, which at first order in the Knudsen number results in the relativistic Navier–Stokes equations of fluid dynamics. As already discussed, these equations provide solutions which are unstable and acausal, hence they cannot provide a proper description of relativistic fluids.

The solution for this problem was already mentioned in the previous chapters of this book. To obtain causal and stable equations, one must take into account the characteristic time scales over which the bulk-viscous pressure, the particle-diffusion current, and the shear-stress tensor relax toward their asymptotic Navier–Stokes values. As already shown in Chap. 4 (see also Ref. [13]), in the fluid-dynamical limit, these are given by the slowest microscopic time scales of the underlying microscopic theory.

In practice, this is implemented by assuming that only the slowest modes with rank 2 and smaller, X_0, X_0^μ, and $X_0^{\mu\nu}$, remain in the transient regime and satisfy the partial differential equations (8.14)

$$
\begin{aligned}
\dot{X}_0 + \chi_0^{(0)} X_0 &= \beta_0^{(0)} \theta \quad + \text{ (higher-order terms) ,} \\
\dot{X}_0^{\langle\mu\rangle} + \chi_0^{(1)} X_0^\mu &= \beta_0^{(1)} I^\mu \quad + \text{ (higher-order terms) ,} \\
\dot{X}_0^{\langle\mu\nu\rangle} + \chi_0^{(2)} X_0^{\mu\nu} &= \beta_0^{(2)} \sigma^{\mu\nu} \quad + \text{ (higher-order terms) ,}
\end{aligned} \qquad (8.16)
$$

while the modes described by faster relaxation scales, i.e., X_r, X_r^μ, and $X_r^{\mu\nu}$, for any $r > 0$, will be approximated by their asymptotic solutions

$$
X_r \simeq \frac{\beta_r^{(0)}}{\chi_r^{(0)}} \theta \quad + \text{ (higher-order terms) ,}
$$

$$
X_r^\mu \simeq \frac{\beta_r^{(1)}}{\chi_r^{(1)}} I^\mu \quad + \text{ (higher-order terms) ,}
$$

$$
X_r^{\mu\nu} \simeq \frac{\beta_r^{(2)}}{\chi_r^{(2)}} \sigma^{\mu\nu} \quad + \text{ (higher-order terms) .} \qquad (8.17)
$$

While this approximation is similar to the Chapman-Enskog expansion, Eqs. (8.16) go beyond the Chapman-Enskog expansion since they include the transient dynamics of the slowest eigenmodes.

Note that, for $r \geq 1$, X_r, X_r^μ, and $X_r^{\mu\nu}$ are of first order in Knudsen number, $O(\text{Kn})$. The reason is that the gradient terms θ, I^μ, and $\sigma^{\mu\nu}$ are proportional to L^{-1}, while $1/\chi_r^{(\ell)}$ is proportional to λ. The other coefficients $\beta_r^{(\ell)}$ are simply functions of the thermodynamic variables α_0, β_0.

Furthermore, in order to obtain the traditional equations of fluid dynamics given in terms of the conserved currents, it was already mentioned that there should not appear any tensor $X_r^{\mu\nu\lambda\cdots}$ of rank higher than 2. Neglecting such tensors is justified because they have asymptotic solutions which are at least $O(\text{Kn}^2, \text{Kn}\,R_i^{-1})$, i.e., beyond the order we consider here. This can be seen by noting that it is impossible to construct irreducible tensors of rank larger than 2 in terms of single powers of space-like gradients of temperature, chemical potential, or velocity, i.e., it is impossible to construct irreducible tensors of rank larger than 2 that are only of first order in Knudsen number.

Equations (8.17) enable us to approximate, up to a given order in Knudsen number, the irreducible moments that do not appear in the conserved currents in terms of those that do occur, namely, the particle-diffusion current, the bulk-viscous pressure, and the shear-stress tensor. Using relations (8.17), it is possible to prove that, for all $r, n \geq 1$,

$$X_n = \frac{\chi_r^{(0)}}{\chi_n^{(0)}} \frac{\beta_n^{(0)}}{\beta_r^{(0)}} X_r + \text{(higher-order terms)},$$ (8.18)

$$X_n^\mu = \frac{\chi_r^{(1)}}{\chi_n^{(1)}} \frac{\beta_n^{(1)}}{\beta_r^{(1)}} X_r^\mu + \text{(higher-order terms)},$$ (8.19)

$$X_n^{\mu\nu} = \frac{\chi_r^{(2)}}{\chi_n^{(2)}} \frac{\beta_n^{(2)}}{\beta_r^{(2)}} X_r^{\mu\nu} + \text{(higher-order terms)}.$$ (8.20)

Then, choosing, e.g., $r = 3$ for the scalar modes, $r = 2$ for the vector modes, and $r = 1$ for the tensor modes in the above relations, we can write

$$X_n = \frac{\chi_3^{(0)}}{\chi_n^{(0)}} \frac{\beta_n^{(0)}}{\beta_3^{(0)}} X_3 + \text{(higher-order terms)},$$ (8.21)

$$X_n^\mu = \frac{\chi_2^{(1)}}{\chi_n^{(1)}} \frac{\beta_n^{(1)}}{\beta_2^{(1)}} X_2^\mu + \text{(higher-order terms)},$$ (8.22)

$$X_n^{\mu\nu} = \frac{\chi_1^{(2)}}{\chi_n^{(2)}} \frac{\beta_n^{(2)}}{\beta_1^{(2)}} X_1^{\mu\nu} + \text{(higher-order terms)}.$$ (8.23)

Next, we invert Eq. (8.12)

$$\rho_r^{\mu_1 \cdots \mu_\ell} = \sum_{j=0}^{N_\ell} \Omega_{rn}^{(\ell)} X_n^{\mu_1 \cdots \mu_\ell},$$ (8.24)

and, using Eqs. (8.17), we obtain

$$\rho_r \simeq \Omega_{r0}^{(0)} X_0 + \left(\frac{\chi_3^{(0)}}{\beta_3^{(0)}} \sum_{n=3}^{N_0} \Omega_{rn}^{(0)} \frac{\beta_n^{(0)}}{\chi_n^{(0)}} \right) X_3 = \Omega_{r0}^{(0)} X_0 + O(\text{Kn}),$$

$$\rho_r^\mu \simeq \Omega_{r0}^{(1)} X_0^\mu + \left(\frac{\chi_2^{(1)}}{\beta_2^{(1)}} \sum_{n=2}^{N_1} \Omega_{rn}^{(1)} \frac{\beta_n^{(1)}}{\chi_n^{(1)}} \right) X_2^\mu = \Omega_{r0}^{(1)} X_0^\mu + O(\text{Kn}),$$

$$\rho_r^{\mu\nu} \simeq \Omega_{r0}^{(2)} X_0^{\mu\nu} + \left(\frac{\chi_1^{(2)}}{\beta_1^{(2)}} \sum_{n=1}^{N_2} \Omega_{rn}^{(2)} \frac{\beta_n^{(2)}}{\chi_n^{(2)}} \right) X_1^{\mu\nu} = \Omega_{r0}^{(2)} X_0^{\mu\nu} + O(\text{Kn}).$$ (8.25)

Here, we indicated that the contributions from the modes X_3, X_2^μ, and $X_1^{\mu\nu}$ are of order $O(\text{Kn})$, cf. Eq. (8.17).

The dissipative quantities appearing in the conservation laws can be exactly identified with the irreducible moments ρ_0, ρ_0^μ, and $\rho_0^{\mu\nu}$ in the following way

$$\rho_0 = -\frac{3}{m^2} \Pi, \quad \rho_0^\mu = n^\mu, \quad \rho_0^{\mu\nu} = \pi^{\mu\nu}.$$ (8.26)

Thus, taking $i = 0$ in Eqs. (8.25) and, without loss of generality, setting $\Omega_{00}^{(\ell)} = 1$, we obtain

$$-\frac{3}{m^2}\Pi \simeq X_0 + \left(\frac{\chi_3^{(0)}}{\beta_3^{(0)}} \sum_{n=3}^{N_0} \Omega_{0n}^{(0)} \frac{\beta_n^{(0)}}{\chi_n^{(0)}}\right) X_3 = X_0 + O(\mathrm{Kn}),$$

$$n^\mu \simeq X_0^\mu + \left(\frac{\chi_2^{(1)}}{\beta_2^{(1)}} \sum_{n=2}^{N_1} \Omega_{0n}^{(1)} \frac{\beta_n^{(1)}}{\chi_n^{(1)}}\right) X_2^\mu = X_0^\mu + O(\mathrm{Kn}),$$

$$\pi^{\mu\nu} \simeq X_0^{\mu\nu} + \left(\frac{\chi_1^{(2)}}{\beta_1^{(2)}} \sum_{n=1}^{N_2} \Omega_{0n}^{(2)} \frac{\beta_n^{(2)}}{\chi_n^{(2)}}\right) X_1^{\mu\nu} = X_0^{\mu\nu} + O(\mathrm{Kn}). \qquad (8.27)$$

Substituting Eqs. (8.27) into Eqs. (8.25) and using Eq. (8.15), we obtain

$$\frac{m^2}{3}\rho_r \simeq -\Omega_{r0}^{(0)}\Pi + \frac{\chi_3^{(0)}}{\beta_3^{(0)}}\left(\zeta_r - \Omega_{r0}^{(0)}\zeta_0\right)X_3 = -\Omega_{r0}^{(0)}\Pi + O(\mathrm{Kn}),$$

$$\rho_r^\mu \simeq \Omega_{r0}^{(1)}n^\mu + \frac{\chi_2^{(1)}}{\beta_2^{(1)}}\left(\varkappa_r - \Omega_{r0}^{(1)}\varkappa_0\right)X_2^\mu = \Omega_{r0}^{(1)}n^\mu + O(\mathrm{Kn}),$$

$$\rho_r^{\mu\nu} \simeq \Omega_{r0}^{(2)}\pi^{\mu\nu} + 2\frac{\chi_1^{(2)}}{\beta_1^{(2)}}\left(\eta_r - \Omega_{r0}^{(2)}\eta_0\right)X_1^{\mu\nu} = \Omega_{r0}^{(2)}\pi^{\mu\nu} + O(\mathrm{Kn}),$$

$$\rho_i^{\mu\nu\lambda\cdots} \simeq O(\mathrm{Kn}^2, \mathrm{Kn}\,R_i^{-1}), \qquad (8.28)$$

where we defined the transport coefficients

$$\zeta_r \equiv \frac{m^2}{3}\sum_{n=0,\neq 1,2}^{N_0} \tau_{rn}^{(0)}\alpha_n^{(0)}, \quad \varkappa_r \equiv \sum_{n=0,\neq 1}^{N_1} \tau_{rn}^{(1)}\alpha_n^{(1)}, \quad \eta_r \equiv \sum_{n=0}^{N_2} \tau_{rn}^{(2)}\alpha_n^{(2)}, \quad (8.29)$$

with

$$\tau_{rn}^{(\ell)} \equiv \sum_{m=0}^{N_\ell} \Omega_{rm}^{(\ell)}\frac{1}{\chi_m^{(\ell)}}\left(\Omega^{-1}\right)_{mn}^{(\ell)}. \qquad (8.30)$$

These coefficients define the inverse of $\mathcal{A}^{(\ell)}$, $\tau^{(\ell)} \equiv \left(\mathcal{A}^{-1}\right)^{(\ell)}$, cf. Eq. (8.11). Note that, for $\ell = 0$, one excludes the $m = 1, 2$ terms in the sum, and for $\ell = 1$ the $m = 1$ term. In order to obtain the fourth equation (8.28), we further used that $X_r^{\mu_1\cdots\mu_\ell} \sim O(\mathrm{Kn}^2, \mathrm{Kn}\,R_i^{-1})$ for $\ell \geq 3$. In the next subsection, we shall identify the coefficients ζ_0, $\varkappa_0$, and η_0 as the bulk-viscosity, particle-diffusion, and shear-viscosity coefficient, respectively.

So far, we have proved that, by taking into account only the slowest relaxation time scales, all irreducible moments of the deviation of the single-particle distribution function from the equilibrium one can be related, *up to first order in Knudsen number*, $O(\mathrm{Kn})$, to the dissipative currents, Π, n^μ, and $\pi^{\mu\nu}$. This demonstrates that in this limit, it is possible to reduce the number of dynamical variables in Eqs. (8.3)–(8.5) to quantities appearing in the conserved currents. This will be explicitly shown in the next section. We note that, so far, we also explicitly kept the terms $\sim X_3$, X_2^μ, $X_1^{\mu\nu}$, for the purpose of extending traditional fluid dynamics, which only considers the 14 quantities appearing in the conservation equations as dynamical variables, to a fluid-dynamical theory with a larger number of dynamical variables; see Sect. 8.6.

We remark that similar relations between the irreducible moments and the dissipative currents can also be obtained with the 14-moment approximation, but with a different set of proportionality coefficients [39]. However, in the traditional 14-moment approximation, such relations are obtained by explicitly truncating the moment expansion and, as a result, they are not of a definite order in powers of Knudsen number. This is the reason why the 14-moment approximation does not give rise to equations of motion with a definite domain of validity in Knudsen and inverse Reynolds numbers.

Note, however, that the relations (8.28) are only valid for the moments $\rho_r^{\mu_1\cdots\mu_\ell}$ with positive r. This is not a problem since similar relations can also be obtained for the irreducible moments with negative r. The moment expansion developed in Chap. 6, Sect. 6.1, was constructed in terms of a complete basis, and therefore, any moment that does not appear in this expansion must be linearly related to those that do appear. This means that, using this moment expansion, it is possible to express the moments with negative r in terms of the ones with positive r. This was already done in Chap. 6, Sect. 6.1, cf. Eq. (6.16). Here, we re-express this formula using a different notation

$$\rho_{-r}^{\nu_1\cdots\nu_\ell} = \sum_{n=0}^{N_\ell} \mathcal{F}_{rn}^{(\ell)} \rho_n^{\nu_1\cdots\nu_\ell} , \qquad (8.31)$$

where we defined the following thermodynamic integral

$$\mathcal{F}_{rn}^{(\ell)} = \frac{\ell!}{(2\ell+1)!!} \int dK \, f_{0\mathbf{k}} \tilde{f}_{0\mathbf{k}} E_{\mathbf{k}}^{-r} \mathcal{H}_{\mathbf{k}n}^{(\ell)} \left(\Delta^{\alpha\beta} k_\alpha k_\beta\right)^\ell . \qquad (8.32)$$

Therefore, Eq. (8.28) reads for negative values of r

$$\frac{m^2}{3}\rho_{-r} = -\gamma_r^{(0)}\Pi + \left[\frac{\chi_3^{(0)}}{\beta_3^{(0)}} \sum_{n=0,\neq 1,2}^{N_0} \mathcal{F}_{rn}^{(0)}\left(\zeta_n - \Omega_{n0}^{(0)}\zeta_0\right)\right] X_3$$

$$= -\gamma_r^{(0)}\Pi + O(\mathrm{Kn}) ,$$

$$\rho_{-r}^\mu = \gamma_r^{(1)}n^\mu + \left[\frac{\chi_2^{(1)}}{\beta_2^{(1)}} \sum_{n=0,\neq 1}^{N_1} \mathcal{F}_{rn}^{(1)}\left(\varkappa_n - \Omega_{n0}^{(1)}\varkappa_0\right)\right] X_2^\mu$$

$$= \gamma_r^{(1)}n^\mu + O(\mathrm{Kn}) , \qquad (8.33)$$

$$\rho_{-r}^{\mu\nu} = \gamma_r^{(2)}\pi^{\mu\nu} + 2\left[\frac{\chi_1^{(2)}}{\beta_1^{(2)}} \sum_{n=0}^{N_2} \mathcal{F}_{rn}^{(2)}\left(\eta_n - \Omega_{n0}^{(2)}\eta_0\right)\right] X_1^{\mu\nu}$$

$$= \gamma_r^{(2)}\pi^{\mu\nu} + O(\mathrm{Kn}) ,$$

$$\rho_{-r}^{\mu\nu\cdots} = O(\mathrm{Kn}^2, \mathrm{Kn}\,\mathrm{R}_i^{-1}) ,$$

where we introduced the coefficients

$$
\gamma_r^{(0)} = \sum_{n=0,\neq 1,2}^{N_0} \mathcal{F}_{rn}^{(0)} \Omega_{n0}^{(0)} , \quad \gamma_r^{(1)} = \sum_{n=0,\neq 1}^{N_1} \mathcal{F}_{rn}^{(1)} \Omega_{n0}^{(1)} , \quad \gamma_r^{(2)} = \sum_{n=0}^{N_2} \mathcal{F}_{rn}^{(2)} \Omega_{n0}^{(2)} .
$$

(8.34)

8.3 Resummed Transient Relativistic Fluid Dynamics: 14 Dynamical Variables

Now we are ready to close Eqs. (8.3)–(8.5) in terms of the 14 quantities appearing in N^μ and $T^{\mu\nu}$, i.e., the five primary fluid-dynamical quantities α_0, β_0, and u^μ, as well as the nine dissipative currents Π, n^μ, and $\pi^{\mu\nu}$, and to derive the fluid-dynamical equations of motion for these 14 dynamical variables.

For this purpose, it is convenient to use the inverse of $\mathcal{A}^{(\ell)}$, $\tau^{(\ell)} = \left(\mathcal{A}^{-1}\right)^{(\ell)}$, cf. Eq. (8.30), which by definition satisfies $\tau^{(\ell)}\mathcal{A}^{(\ell)} = \mathbf{1}$. Hence, it is straightforward to rewrite Eq. (8.7) as

$$
\sum_{j=0}^{N_\ell} \tau_{ij}^{(\ell)} C_{j-1}^{\langle \mu_1 \cdots \mu_\ell \rangle} = -\rho_i^{\mu_1 \cdots \mu_\ell} + (\text{terms nonlinear in } \delta f) .
$$

(8.35)

Then we multiply Eqs. (8.3)–(8.5) by $\tau_{nr}^{(\ell)}$, sum over r, and substitute Eq. (8.35). Next, we use Eqs. (8.28) and (8.33) to replace all irreducible moments $\rho_i^{\mu_1 \cdots \mu_\ell}$ appearing in the equations by the fluid-dynamical variables. Additionally, all covariant time derivatives of α_0, β_0, and u^μ are replaced by spatial gradients of fluid-dynamical variables using the conservation laws in the form shown in Eqs. (6.24), (6.25), and (6.26). The resulting equations of motion can be written in the following form

$$
\tau_\Pi \dot{\Pi} + \Pi = -\zeta\theta + \mathcal{J} + \mathcal{K} + \mathcal{R} ,
$$

(8.36)

$$
\tau_n \dot{n}^{\langle \mu \rangle} + n^\mu = \varkappa I^\mu + \mathcal{J}^\mu + \mathcal{K}^\mu + \mathcal{R}^\mu ,
$$

(8.37)

$$
\tau_\pi \dot{\pi}^{\langle \mu\nu \rangle} + \pi^{\mu\nu} = 2\eta\sigma^{\mu\nu} + \mathcal{J}^{\mu\nu} + \mathcal{K}^{\mu\nu} + \mathcal{R}^{\mu\nu} .
$$

(8.38)

We remark that in order to derive these equations of motion, we substituted X_3, X_2^μ, and $X_1^{\mu\nu}$ in Eqs. (8.28) and (8.33) by their Navier–Stokes values, i.e.

$$
X_3 = \frac{\beta_3^{(0)}}{\chi_3^{(0)}}\theta , \quad X_2^\mu = \frac{\beta_2^{(1)}}{\chi_2^{(1)}}I^\mu , \quad X_1^{\mu\nu} = \frac{\beta_1^{(2)}}{\chi_1^{(2)}}\sigma^{\mu\nu} ,
$$

(8.39)

and used Eq. (8.11) in the following form

$$
\sum_{j=0}^{N_\ell} \tau_{ij}^{(\ell)} \Omega_{jm}^{(\ell)} = \Omega_{im}^{(\ell)} \frac{1}{\chi_m^{(\ell)}} .
$$

(8.40)

We further made use of the following relations

$$
\dot{\theta} = -\sigma^{\mu\nu}\sigma_{\mu\nu} + \omega^{\mu\nu}\omega_{\mu\nu} - \frac{1}{3}\theta^2 + \frac{1}{\varepsilon_0 + P_0}\nabla_\mu F^\mu - \frac{2(\varepsilon_0 + P_0) + \beta_0 J_{30}}{(\varepsilon_0 + P_0)^3}F_\mu F^\mu
$$
$$
- \frac{(\varepsilon_0 + P_0)J_{20} - n_0 J_{30}}{(\varepsilon_0 + P_0)^3}F_\mu I^\mu + O\left(\mathrm{Kn}^2 \mathrm{R}_i^{-1}\right), \tag{8.41}
$$

$$
\dot{I}^{\langle\mu\rangle} = \frac{(\varepsilon_0 + P_0)J_{20} - n_0 J_{30}}{D_{20}}\nabla^\mu \theta - \left\{\frac{\partial}{\partial\beta_0}\left[\beta_0\frac{(\varepsilon_0 + P_0)J_{20} - n_0 J_{30}}{D_{20}}\right]\right\}\frac{1}{\varepsilon_0 + P_0}\theta F^\mu
$$
$$
+ \left\{\left[\left(\frac{\partial}{\partial\alpha_0} + h_0^{-1}\frac{\partial}{\partial\beta_0}\right)\frac{(\varepsilon_0 + P_0)J_{20} - n_0 J_{30}}{D_{20}}\right] - \frac{1}{3}\right\}\theta I^\mu
$$
$$
- \sigma^{\mu\alpha}I_\alpha - \omega^{\mu\alpha}I_\alpha - \frac{1}{3}\theta I^\mu + O\left(\mathrm{Kn}^2 \mathrm{R}_i^{-1}\right), \tag{8.42}
$$

$$
\dot{\sigma}^{\langle\mu\nu\rangle} = -\sigma_\lambda^{\langle\mu}\sigma^{\nu\rangle\lambda} + 2\sigma_\lambda^{\langle\mu}\omega^{\nu\rangle\lambda} - \omega_\lambda^{\langle\mu}\omega^{\nu\rangle\lambda} - \frac{2}{3}\sigma^{\mu\nu}\theta + \frac{1}{\varepsilon_0 + P_0}\nabla^{\langle\mu}F^{\nu\rangle}
$$
$$
- \frac{2(\varepsilon_0 + P_0) + \beta_0 J_{30}}{(\varepsilon_0 + P_0)^3}F^{\langle\mu}F^{\nu\rangle} - \frac{(\varepsilon_0 + P_0)J_{20} - n_0 J_{30}}{(\varepsilon_0 + P_0)^3}F^{\langle\mu}I^{\nu\rangle} + O\left(\mathrm{Kn}^2 \mathrm{R}_i^{-1}\right), \tag{8.43}
$$

where we defined $F^\mu \equiv \nabla^\mu P_0$.

In the above equations of motion, all nonlinear terms and couplings to other currents were collected in the tensors $\mathcal{J}, \mathcal{K}, \mathcal{R}, \mathcal{J}^\mu, \mathcal{K}^\mu, \mathcal{R}^\mu, \mathcal{J}^{\mu\nu}, \mathcal{K}^{\mu\nu}$, and $\mathcal{R}^{\mu\nu}$. The tensors $\mathcal{J}, \mathcal{J}^\mu$, and $\mathcal{J}^{\mu\nu}$ contain all terms of first order in Knudsen and inverse Reynolds numbers

$$
\mathcal{J} = -\ell_{\Pi n}\nabla_\mu n^\mu - \tau_{\Pi n}n^\mu F_\mu - \delta_{\Pi\Pi}\Pi\theta - \lambda_{\Pi n}n^\mu I_\mu + \lambda_{\Pi\pi}\pi^{\mu\nu}\sigma_{\mu\nu},
$$
$$
\mathcal{J}^\mu = -\tau_n n_\nu\omega^{\nu\mu} - \delta_{nn}n^\mu\theta - \ell_{n\Pi}\nabla^\mu\Pi + \ell_{n\pi}\Delta^{\mu\nu}\nabla_\lambda\pi_\nu^\lambda + \tau_{n\Pi}\Pi F^\mu
$$
$$
- \tau_{n\pi}\pi^{\mu\nu}F_\nu - \lambda_{nn}n_\nu\sigma^{\mu\nu} + \lambda_{n\Pi}\Pi I^\mu - \lambda_{n\pi}\pi^{\mu\nu}I_\nu,
$$
$$
\mathcal{J}^{\mu\nu} = 2\tau_\pi\pi_\lambda^{\langle\mu}\omega^{\nu\rangle\lambda} - \delta_{\pi\pi}\pi^{\mu\nu}\theta - \tau_{\pi\pi}\pi^{\lambda\langle\mu}\sigma_\lambda^{\nu\rangle} + \lambda_{\pi\Pi}\Pi\sigma^{\mu\nu}
$$
$$
- \tau_{\pi n}n^{\langle\mu}F^{\nu\rangle} + \ell_{\pi n}\nabla^{\langle\mu}n^{\nu\rangle} + \lambda_{\pi n}n^{\langle\mu}I^{\nu\rangle}. \tag{8.44}
$$

The tensors $\mathcal{K}, \mathcal{K}^\mu$, and $\mathcal{K}^{\mu\nu}$ contain all terms of second order in Knudsen number

$$
\mathcal{K} = \tilde{\zeta}_1\omega_{\mu\nu}\omega^{\mu\nu} + \tilde{\zeta}_2\sigma_{\mu\nu}\sigma^{\mu\nu} + \tilde{\zeta}_3\theta^2 + \tilde{\zeta}_4 I_\mu I^\mu
$$
$$
+ \tilde{\zeta}_5 F_\mu F^\mu + \tilde{\zeta}_6 I_\mu F^\mu + \tilde{\zeta}_7\nabla_\mu I^\mu + \tilde{\zeta}_8\nabla_\mu F^\mu,
$$
$$
\mathcal{K}^\mu = \tilde{\varkappa}_1\sigma^{\mu\nu}I_\nu + \tilde{\varkappa}_2\sigma^{\mu\nu}F_\nu + \tilde{\varkappa}_3 I^\mu\theta + \tilde{\varkappa}_4 F^\mu\theta
$$
$$
+ \tilde{\varkappa}_5\omega^{\mu\nu}I_\nu + \tilde{\varkappa}_6\Delta_\lambda^\mu\partial_\nu\sigma^{\lambda\nu} + \tilde{\varkappa}_7\nabla^\mu\theta,
$$
$$
\mathcal{K}^{\mu\nu} = \tilde{\eta}_1\omega_\lambda^{\langle\mu}\omega^{\nu\rangle\lambda} + \tilde{\eta}_2\theta\sigma^{\mu\nu} + \tilde{\eta}_3\sigma^{\lambda\langle\mu}\sigma_\lambda^{\nu\rangle} + \tilde{\eta}_4\sigma_\lambda^{\langle\mu}\omega^{\nu\rangle\lambda}
$$
$$
+ \tilde{\eta}_5 I^{\langle\mu}I^{\nu\rangle} + \tilde{\eta}_6 F^{\langle\mu}F^{\nu\rangle} + \tilde{\eta}_7 I^{\langle\mu}F^{\nu\rangle} + \tilde{\eta}_8\nabla^{\langle\mu}I^{\nu\rangle} + \tilde{\eta}_9\nabla^{\langle\mu}F^{\nu\rangle}. \tag{8.45}
$$

The tensors $\mathcal{R}, \mathcal{R}^\mu$, and $\mathcal{R}^{\mu\nu}$ contain all terms of second order in inverse Reynolds number

$$\mathcal{R} = \varphi_1 \Pi^2 + \varphi_2 n_\mu n^\mu + \varphi_3 \pi_{\mu\nu} \pi^{\mu\nu} ,$$
$$\mathcal{R}^\mu = \varphi_4 n_\nu \pi^{\mu\nu} + \varphi_5 \Pi n^\mu ,$$
$$\mathcal{R}^{\mu\nu} = \varphi_6 \Pi \pi^{\mu\nu} + \varphi_7 \pi^{\lambda\langle\mu} \pi_\lambda^{\nu\rangle} + \varphi_8 n^{\langle\mu} n^{\nu\rangle} . \tag{8.46}$$

In Eqs. (8.36)–(8.38), terms of order $O(\text{Kn}^3)$, $O(R_i^{-1} R_j^{-1} R_k^{-1})$, $O(\text{Kn}^2 R_i^{-1})$, and $O(\text{Kn}\, R_i^{-1} R_j^{-1})$ were omitted.

Note that the equations of motion are closed in terms of 14 dynamical variables, even without making use of the 14-moment approximation. This means that the reduction of degrees of freedom was not obtained by a direct truncation of the moment expansion, but by a separation of the microscopic time scales and the power-counting scheme itself. The information about all other moments are actually included in the transport coefficients, as will be shown later. If we neglect the terms of second order in Knudsen and inverse Reynolds number, Eqs. (8.45) and (8.46), respectively, we recover the equations of motion that are of the same form as those derived via the 14-moment approximation [33]. However, even in this case, the coefficients in Eq. (8.44) and the relaxation times in Eqs. (8.36)–(8.38) are not the same as those calculated from the 14-moment approximation of Israel and Stewart.

The resulting equations of motion (8.36)–(8.38) contain 57 transport coefficients. In particular, the viscosity coefficients and relaxation times of the dissipative currents are found to be

$$\zeta \equiv \zeta_0 = \frac{m^2}{3} \sum_{r=0,\neq 1,2}^{N_0} \tau_{0r}^{(0)} \alpha_r^{(0)} ,$$

$$\varkappa \equiv \varkappa_0 = \sum_{r=0,\neq 1}^{N_1} \tau_{0r}^{(1)} \alpha_r^{(1)} , \tag{8.47}$$

$$\eta \equiv \eta_0 = \sum_{r=0}^{N_2} \tau_{0r}^{(2)} \alpha_r^{(2)} ,$$

$$\tau_\Pi \equiv \frac{1}{\chi_0^{(0)}} , \quad \tau_n \equiv \frac{1}{\chi_0^{(1)}} , \quad \tau_\pi \equiv \frac{1}{\chi_0^{(2)}} ,$$

cf. Eqs. (8.16) and (8.29). Note that, in general, these transport coefficients depend not only on one moment of the distribution function, but also on all moments of the corresponding rank ℓ, i.e., the contributions of higher moments of the distribution function are resummed as contributions to the microscopic formulas for the transport coefficients. For this reason, we shall refer to this formalism as Resummed Transient Relativistic Fluid Dynamics (RTRFD).

As in Chapman-Enskog theory, the viscosity coefficients can only be obtained by inverting $\mathcal{A}^{(\ell)}$. As a matter of fact, the microscopic formulas for ζ, $\varkappa$, and η obtained above are equivalent to those obtained from the Chapman-Enskog expansion; see Chap. 5, Sect. 5.2. However, to obtain the transient dynamics of the fluid, characterized by the relaxation times, it is also necessary to find the eigenvalues and eigenvectors of $\mathcal{A}^{(\ell)}$.

In practice, the moment expansion of the single-particle distribution function is always truncated and the matrices $\mathcal{A}^{(\ell)}$, $\Omega^{(\ell)}$, and $\tau^{(\ell)}$ will actually be finite. The truncation of this expansion was already introduced as an upper limit, N_ℓ, in the corresponding summations. In principle, one should only truncate the moment expansion when the values of all relevant transport coefficients have converged. Note that different transport coefficients may require a different number of moments to converge.

8.4   Transport Coefficients

In this section, we compute the transport coefficients for several cases. First, we considered the lowest possible truncation scheme for the moment expansion, with $N_0 = 2$, $N_1 = 1$, and $N_2 = 0$. In this case, the distribution function is expanded in terms of 14 moments and is actually equivalent to the one obtained via Israel's and Stewart's 14-moment Ansatz. Second, we consider the next simplest case and take $N_0 = 3$, $N_1 = 2$, and $N_2 = 1$. Then, the distribution function is characterized by 23 moments, and consequently we shall refer to this case as 23-moment approximation. Finally, we include 32 and 41 moments and verify the convergence of the transport coefficients.

We also compute the transport coefficients of the terms appearing in $\mathcal{J}$, $\mathcal{J}^\mu$, and $\mathcal{J}^{\mu\nu}$, the explicit expressions of which are given in Appendix 8.9. Note, however, that we are using a linear approximation to the collision term. Nonlinear contributions could, in principle, also enter the transport coefficients in the equations of motion (8.36)–(8.38), but will not be calculated here. For this reason, we also do not compute any coefficient of the terms of order $O(R_i^{-1} R_j^{-1})$, i.e., those entering $\mathcal{R}$, $\mathcal{R}^\mu$, and $\mathcal{R}^{\mu\nu}$, cf. Eq. (8.46), since all of them originate exclusively from nonlinear contributions to the collision term. An investigation of these terms can be found in Ref. [40]. Appendix I of that reference also contains explicit expressions for the transport coefficients contained in $\mathcal{K}$, $\mathcal{K}^\mu$, and $\mathcal{K}^{\mu\nu}$, cf. Eq. (8.45). One should note that they vanish in the 14-moment approximation ($N_0 = 2$, $N_1 = 1$, $N_2 = 0$).

8.4.1   14-Moment Approximation

The 14-moment approximation is recovered by truncating all summations at $N_0 = 2$, $N_1 = 1$, and $N_2 = 0$. For this specific truncation, $\mathcal{A}^{(\ell)}$ is just a number (because for $\mathcal{A}^{(0)}$, we have to exclude the second and third rows and columns and for $\mathcal{A}^{(1)}$, the second row and column), and thus

$$\tau^{(\ell)} = \frac{1}{\mathcal{A}^{(\ell)}}, \quad \Omega^{(\ell)} = 1, \quad \chi^{(\ell)} = \mathcal{A}^{(\ell)}.$$

Then, the equations of motion and transport coefficients reduce to those derived in Refs. [33,41] and reproduced in Chap. 5, Sect. 5.3.

Table 8.1 The transport coefficients for the equation of motion of the particle-diffusion current, calculated for a classical gas with constant cross section in the ultrarelativistic limit and in the 14-moment approximation.

$\varkappa$	$\tau_n[\lambda]$	$\delta_{nn}[\tau_n]$	$\lambda_{nn}[\tau_n]$	$\lambda_{n\pi}[\tau_n]$	$\ell_{n\pi}[\tau_n]$	$\tau_{n\pi}[\tau_n]$
$3/(16\sigma)$	$9/4$	1	$3/5$	$\beta_0/20$	$\beta_0/20$	$\beta_0/(80P_0)$

Table 8.2 The transport coefficients for the equation of motion of the shear-stress tensor, calculated for a classical gas with constant cross section in the ultrarelativistic limit and in the 14-moment approximation

η	$\tau_\pi[\lambda]$	$\tau_{\pi\pi}[\tau_\pi]$	$\lambda_{\pi n}[\tau_\pi]$	$\delta_{\pi\pi}[\tau_\pi]$	$\ell_{\pi n}[\tau_\pi]$	$\tau_{\pi n}[\tau_\pi]$
$4/(3\sigma\beta_0)$	$5/3$	$10/7$	0	$4/3$	0	0

For a classical gas of hard spheres with total cross section σ, in the massless limit, the integrals $\mathcal{A}^{(1)} = \mathcal{A}^{(1)}_{00}$ and $\mathcal{A}^{(2)} = \mathcal{A}^{(2)}_{00}$ can be computed and have the following simple form

$$\mathcal{A}^{(1)} = \frac{4}{9\lambda} , \tag{8.48}$$

$$\mathcal{A}^{(2)} = \frac{3}{5\lambda} , \tag{8.49}$$

where we defined the mean free path $\lambda \equiv 1/(n_0\sigma)$. The details of this calculation are shown in Appendix 8.10. The coefficients in the ultrarelativistic limit, $m\beta_0 \to 0$, can then be calculated analytically. The coefficients of order $O(\text{Kn } R_i^{-1})$ are collected for the shear viscosity and particle diffusion in Tables 8.1 and 8.2, respectively. Note that, in this limit, the bulk-viscous pressure vanishes, and thus we do not need to compute $\mathcal{A}^{(0)}$ or any coefficient related to bulk-viscous pressure.

8.4.2 23-Moment Approximation and Beyond

In order to better understand our result (8.47), we compute the first correction to the expressions in Tables 8.1 and 8.2. For this purpose, we consider $N_0 = 3$, $N_1 = 2$, and $N_2 = 1$. Then, $\mathcal{A}^{(\ell)}$, $\Omega^{(\ell)}$, and $\tau^{(\ell)}$ are 2×2 matrices that can be computed from the collision integral (6.47). We obtain the elements of $\mathcal{A}^{(1,2)}$, their inverses $\tau^{(1,2)}$, and $\Omega^{(1,2)}$ as

$$\mathcal{A}^{(1)} = \frac{1}{3\lambda} \begin{pmatrix} 2 & \beta_0^2/30 \\ -4/\beta_0^2 & 1 \end{pmatrix} , \quad \mathcal{A}^{(2)} = \frac{1}{\lambda} \begin{pmatrix} 9/10 & -\beta_0/20 \\ 4/(3\beta_0) & 1/3 \end{pmatrix} , \tag{8.50}$$

$$\tau^{(1)} = \frac{3}{8}\lambda \begin{pmatrix} 15/4 & -\beta_0^2/8 \\ 15/\beta_0^2 & 15/2 \end{pmatrix} , \quad \tau^{(2)} = \frac{1}{11}\lambda \begin{pmatrix} 10 & 3\beta_0/2 \\ -40/\beta_0 & 27 \end{pmatrix} , \tag{8.51}$$

$$\Omega^{(1)} = \begin{pmatrix} 1 & 1 \\ -\left(15 + \sqrt{105}\right)/\beta_0^2 & \left(-15 + \sqrt{105}\right)/\beta_0^2 \end{pmatrix} , \quad \Omega^{(2)} = \begin{pmatrix} 1 & 1 \\ 8/\beta_0 & 10/(3\beta_0) \end{pmatrix} , \tag{8.52}$$

see Appendix 8.10 for details. For all matrices with $\ell = 1$, the second row and column have been removed. The eigenvalues of $\mathcal{A}^{(1)}$ and $\mathcal{A}^{(2)}$ are

$$\chi_0^{(1)} = \frac{1}{2\lambda}\left(1 - \sqrt{\frac{7}{135}}\right) , \quad \chi_2^{(1)} = \frac{1}{2\lambda}\left(1 + \sqrt{\frac{7}{135}}\right) , \tag{8.53}$$

$$\chi_0^{(2)} = \frac{1}{2\lambda} , \quad \chi_1^{(2)} = \frac{11}{15\lambda} . \tag{8.54}$$

Note that the next largest eigenvalue following $\chi_0^{(1)}$ is $\chi_2^{(1)}$, not $\chi_1^{(1)}$ (following our convention to erase all rows and columns with index 1 in the matrices for $\ell = 1$).

Using Eq. (8.47), we calculate the corrected values for the particle-number diffusion coefficient and diffusion relaxation time and for the shear viscosity and shear relaxation time

$$\varkappa = \frac{21}{128}n_0\lambda \simeq 0.164\, n_0\lambda , \tag{8.55}$$

$$\tau_n = \frac{90}{45 - \sqrt{105}}\lambda \simeq 2.5897\,\lambda , \tag{8.56}$$

$$\eta = \frac{14}{11}P_0\lambda \simeq 1.2727\, P_0\lambda , \tag{8.57}$$

$$\tau_\pi = 2\lambda , \tag{8.58}$$

where we used that, in the massless and classical limit

$$\alpha_0^{(1)} = \frac{1}{12}n_0 , \quad \alpha_2^{(1)} = -\frac{1}{\beta_0}P_0 , \tag{8.59}$$

$$\alpha_0^{(2)} = \frac{4}{5}P_0 , \quad \alpha_1^{(2)} = \frac{4}{\beta_0}P_0 . \tag{8.60}$$

As before, the coefficients in the ultrarelativistic limit, $m\beta_0 \to 0$, can then be calculated analytically. The coefficients of order $O(\mathrm{Kn}\,\mathrm{R}_i^{-1})$ are collected for the equations of motion of particle-diffusion current and shear-stress tensor in Tables 8.3 and 8.4, respectively.

In order to obtain these expressions, we used the results from Appendix 8.11 and that, in the massless and classical limit, $D_{20} = 3P_0^2$. Note that most of the transport coefficients are corrected by the inclusion of more moments in the computation. The coefficients related to the shear-stress tensor are less affected by the additional

Table 8.3 The transport coefficients for the equation of motion of the particle-diffusion current, calculated for a classical gas with constant cross section in the ultrarelativistic limit and in the 23-moment approximation

$\varkappa$	$\tau_n[\lambda]$	$\delta_{nn}[\tau_n]$	$\lambda_{nn}[\tau_n]$	$\lambda_{n\pi}[\tau_n]$	$\ell_{n\pi}[\tau_n]$	$\tau_{n\pi}[\tau_n]$
$21/(128\sigma)$	2.59	1.00	0.96	$0.054\beta_0$	$0.118\beta_0$	$0.0295\beta_0/P_0$

Table 8.4 The transport coefficients for the equation of motion of the shear-stress tensor, calculated for a classical gas with constant cross section in the ultrarelativistic limit and in the 23-moment approximation

η	$\tau_\pi[\lambda]$	$\tau_{\pi\pi}[\tau_\pi]$	$\lambda_{\pi n}[\tau_\pi]$	$\delta_{\pi\pi}[\tau_\pi]$	$\ell_{\pi n}[\tau_\pi]$	$\tau_{\pi n}[\tau_\pi]$
$14/(11\sigma\beta_0)$	2	134/77	$0.344/\beta_0$	4/3	$-0.689/\beta_0$	$-0.689/n_0$

Table 8.5 The transport coefficients for the equation of motion of the particle-diffusion current, calculated for a classical gas with constant cross section in the ultrarelativistic limit, in the 14, 23, 32, and 41-moment approximations

Number of moments	$\varkappa$	$\tau_n[\lambda]$	$\delta_{nn}[\tau_n]$	$\lambda_{nn}[\tau_n]$	$\lambda_{n\pi}[\tau_n]$	$\ell_{n\pi}[\tau_n]$	$\tau_{n\pi}[\tau_n]$
14	$3/(16\sigma)$	9/4	1	3/5	$\beta_0/20$	$\beta_0/20$	$\beta_0/(80P_0)$
23	$21/(128\sigma)$	2.59	1.0	0.96	$0.054\beta_0$	$0.118\beta_0$	$0.0295\beta_0/P_0$
32	$0.1605/\sigma$	2.57	1.0	0.93	$0.052\beta_0$	$0.119\beta_0$	$0.0297\beta_0/P_0$
41	$0.1596/\sigma$	2.57	1.0	0.92	$0.052\beta_0$	$0.119\beta_0$	$0.0297\beta_0/P_0$

Table 8.6 The transport coefficients for the equation of motion of the shear-stress tensor, calculated for a classical gas with constant cross section in the ultrarelativistic limit, in the 14, 23, 32, and 41-moment approximations

Number of moments	η	$\tau_\pi[\lambda]$	$\tau_{\pi\pi}[\tau_\pi]$	$\lambda_{\pi n}[\tau_\pi]$	$\delta_{\pi\pi}[\tau_\pi]$	$\ell_{\pi n}[\tau_\pi]$	$\tau_{\pi n}[\tau_\pi]$
14	$4/(3\sigma\beta_0)$	5/3	10/7	0	4/3	0	0
23	$14/(11\sigma\beta_0)$	2	134/77	$0.344/\beta_0$	4/3	$-0.689/\beta_0$	$-0.689/n_0$
32	$1.268/(\sigma\beta_0)$	2	1.69	$0.254/\beta_0$	4/3	$-0.687/\beta_0$	$-0.687/n_0$
41	$1.267/(\sigma\beta_0)$	2	1.69	$0.244/\beta_0$	4/3	$-0.685/\beta_0$	$-0.685/n_0$

moments when compared to the particle-diffusion coefficients. This might explain the poor agreement between Israel–Stewart theory and numerical solutions of the Boltzmann equation in Ref. [32] regarding heat flow and fugacity.

We further checked the convergence of this approach by taking 32 and 41 moments. In this case, the matrices $\mathcal{A}^{(1,2)}$, $\tau^{(1,2)}$, and $\Omega^{(1,2)}$ were computed numerically. There is a clear tendency of convergence as we increase the number of moments. For the particular case of classical particles with constant cross section, 32 moments seem sufficient; see Tables 8.5 and 8.6 for the results.

8.5   Discussion: Navier–Stokes Limit and Causality

Note that one of the main features of transient theories of fluid dynamics is the relaxation of the dissipative currents toward their Navier–Stokes values, with time scales given by the transport coefficients τ_Π, τ_n, and τ_π. From the Boltzmann equation, Navier–Stokes theory is obtained by means of the Chapman-Enskog expansion, cf. Chap. 5, which describes an asymptotic solution of the single-particle distribution function. It is already clear from the previous section that the equations of motion derived in this chapter approach Navier–Stokes-type solutions at asymptotically long times, in which the dissipative currents are solely expressed in terms of gradients of fluid-dynamical variables.

It is interesting to investigate, however, if our equations approach the correct Navier–Stokes theory, i.e., if the viscosity coefficients obtained via our method are equivalent to the ones obtained via Chapman-Enskog theory. It should be noted that this is neither the case for Grad's theory nor for Israel's and Stewart's theory [2–4, 8, 33]. The transport coefficients computed within these theories only coincide, if we use the minimal truncation scheme in Chapman-Enskog theory, as explained in Chap. 5. We remark that, after taking into account further corrections to the shear-viscosity coefficient, see Eq. (8.57) and Table 8.6, our results approach the solution obtained using Chapman-Enskog theory, $\eta_{NS} = 1.2654/(\beta_0 \sigma)$ [8]. In principle, there is no reason for the method of moments to attain a different Navier–Stokes limit than Chapman-Enskog theory. As a matter of fact, if the same basis of irreducible tensors $k^{\langle \mu_1} \dots k^{\mu_\ell \rangle}$ and polynomials $P_{kn}^{(\ell)}$ is used in both calculations, they both yield the same result, even order by order.

It is important to mention that the terms $\mathcal{K}$, $\mathcal{K}^\mu$, and $\mathcal{K}^{\mu\nu}$, which are of second order in Knudsen number, lead to several problems. The terms which contain second-order spatial derivatives of u^μ, α_0, and P_0, e.g., $\nabla_\mu I^\mu$, $\nabla_\mu F^\mu$, $\nabla^{\langle \mu} I^{\nu \rangle}$, $\nabla^{\langle \mu} F^{\nu \rangle}$, $\Delta_\alpha^\mu \partial_\nu \sigma^{\alpha \nu}$, and $\nabla^\mu \theta$, are especially problematic since they change the boundary conditions of the equations. In relativistic systems these derivatives, even though they are made of space-like vectors, also contain time derivatives and thus require initial values. This means that, by including them, one would have to specify not only the initial spatial distribution of the fluid-dynamical variables but also the spatial distribution of their time derivatives. In practice, this implies that we would be increasing the number of fluid-dynamical degrees of freedom.

There is an even more serious problem. By including terms of order higher than one in the Knudsen number, the transport equations become parabolic. In a relativistic theory, this comes with disastrous consequences, since the solutions are acausal and consequently unstable [5–7]. For this reason, we do not compute the transport coefficients for these higher-order terms in this chapter.

If one wants to include terms of higher order in Knudsen number, it is mandatory to include also second-order co-moving time derivatives of the dissipative quantities. Or, equivalently, one could promote the moments ρ_3, ρ_2^μ, $\rho_1^{\mu\nu}$, or further ones, to dynamical variables. We will show how to do this in the next section.

8.6 Resummed Transient Relativistic Fluid Dynamics: 23 Dynamical Variables

As already discussed, the terms of higher order in Knudsen number render the equations of motion parabolic, *despite* the existence of a relaxation time. Therefore, describing the fluid up to a higher order in Knudsen number using Eqs. (8.36)–(8.38) is problematic since, in order to do so, one would have to solve parabolic equations in a covariant setup. In this section, we show how to solve this problem and derive transient fluid-dynamical equations of motion that are hyperbolic even up to second order in the Knudsen number.

The parabolic and, thus, acausal nature of the equations of motion (8.36)–(8.38) can be understood as follows: The main assumption of RTRFD is to approximate the quickly varying eigenmodes of the Boltzmann equation by their asymptotic (i.e., Navier–Stokes) values. This approximation happened in Eq. (8.17), while the substitution of the eigenmodes X_3, X_2^μ, and $X_1^{\mu\nu}$ by their Navier–Stokes values occurred in Eq. (8.39). It was this last step that rendered Eqs. (8.36)–(8.38) parabolic, since in this substitution, it is implicitly assumed that these eigenmodes relax instantaneously to their corresponding Navier–Stokes values, consequently leading to acausal behavior.

In order to obtain hyperbolic equations of motion which do not simply neglect terms of order $O(\mathrm{Kn}^2)$, it is necessary to refrain from the substitution (8.39). This can be simply done by keeping X_3, X_2^μ, and $X_1^{\mu\nu}$ in Eqs. (8.28) as independent dynamical variables instead of replacing them by their Navier–Stokes values.

In this case, it is convenient to replace the modes X_3, X_2^μ, and $X_1^{\mu\nu}$ in Eqs. (8.28) and (8.33) by irreducible moments of the distribution function. In principle, any of the irreducible moments ρ_r, ρ_r^μ, and $\rho_r^{\mu\nu}$ can be used to replace these modes as independent dynamical variables. Without loss of generality, we choose ρ_3, ρ_2^μ, and $\rho_1^{\mu\nu}$. After Π ($\ell = 0$), n^μ ($\ell = 1$), and $\pi^{\mu\nu}$ ($\ell = 2$), these are the irreducible moments with the lowest power of $E_\mathbf{k}$ under the integral (8.2) which appear in the expansion of the single-particle distribution function (8.1). We can relate X_3, X_2^μ, and $X_1^{\mu\nu}$ to $\Pi, n^\mu, \pi^{\mu\nu}, \rho_3, \rho_2^\mu$, and $\rho_1^{\mu\nu}$, by taking in Eq. (8.28) $r = 3$ for the scalar irreducible moment, $r = 2$ for the first-rank irreducible moment, and $r = 1$ for the second-rank irreducible moment, obtaining

$$
\frac{\chi_3^{(0)}}{\beta_3^{(0)}} \left(\zeta_3 - \Omega_{30}^{(0)} \zeta_0 \right) X_3 = \frac{m^2}{3} \rho_3 + \Omega_{30}^{(0)} \Pi \; ,
$$

$$
\frac{\chi_2^{(1)}}{\beta_2^{(1)}} \left(\varkappa_2 - \Omega_{20}^{(1)} \varkappa_0 \right) X_2^\mu = \rho_2^\mu - \Omega_{20}^{(1)} n^\mu \; ,
$$

$$
2\frac{\chi_1^{(2)}}{\beta_1^{(2)}} \left(\eta_1 - \Omega_{10}^{(2)} \eta_0 \right) X_1^{\mu\nu} = \rho_1^{\mu\nu} - \Omega_{10}^{(2)} \pi^{\mu\nu} \; . \tag{8.61}
$$

Then, we substitute the relations (8.61) into Eqs. (8.28) and (8.33), effectively removing the dependence of the irreducible moments on the eigenmodes X_3, X_2^μ, and $X_1^{\mu\nu}$. We obtain the following new relations for the irreducible moments with positive r

$$
\frac{m^2}{3} \rho_r \simeq \lambda_{r0}^{(0)} \Pi + \lambda_{r3}^{(0)} \rho_3 + O(\mathrm{Kn}^2, \, \mathrm{Kn}\,\mathrm{R}_i^{-1}, \, \mathrm{R}_i^{-2}) \; ,
$$

$$
\rho_r^\mu = \lambda_{r0}^{(1)} n^\mu + \lambda_{r2}^{(1)} \rho_2^\mu + O(\mathrm{Kn}^2, \, \mathrm{Kn}\,\mathrm{R}_i^{-1}, \, \mathrm{R}_i^{-2}) \; ,
$$

$$
\rho_r^{\mu\nu} = \lambda_{r0}^{(2)} \pi^{\mu\nu} + \lambda_{r1}^{(2)} \rho_1^{\mu\nu} + O(\mathrm{Kn}^2, \, \mathrm{Kn}\,\mathrm{R}_i^{-1}, \, \mathrm{R}_i^{-2}) \; , \tag{8.62}
$$

and for the irreducible moments with negative r

$$\frac{m^2}{3}\rho_{-r} = \left(\sum_{n=0,\neq 1,2}^{N_0} \mathcal{F}_{rn}^{(0)}\lambda_{n0}^{(0)}\right)\Pi + \left(\sum_{n=0,\neq 1,2}^{N_0} \mathcal{F}_{rn}^{(0)}\lambda_{n3}^{(0)}\right)\rho_3 + O(\mathrm{Kn}^2,\ \mathrm{Kn}\,\mathrm{R}_i^{-1},\ \mathrm{R}_i^{-2})\,,$$

$$\rho_{-r}^{\mu} = \left(\sum_{n=0,\neq 1}^{N_1} \mathcal{F}_{rn}^{(1)}\lambda_{n0}^{(1)}\right)n^{\mu} + \left(\sum_{n=0,\neq 1}^{N_1} \mathcal{F}_{rn}^{(1)}\lambda_{n2}^{(1)}\right)\rho_2^{\mu} + O(\mathrm{Kn}^2,\ \mathrm{Kn}\,\mathrm{R}_i^{-1},\ \mathrm{R}_i^{-2})\,,$$

$$\rho_{-r}^{\mu\nu} = \left(\sum_{n=0}^{N_2} \mathcal{F}_{rn}^{(2)}\lambda_{n0}^{(2)}\right)\pi^{\mu\nu} + \left(\sum_{n=0}^{N_2} \mathcal{F}_{rn}^{(2)}\lambda_{n1}^{(2)}\right)\rho_1^{\mu\nu} + O(\mathrm{Kn}^2,\ \mathrm{Kn}\,\mathrm{R}_i^{-1},\ \mathrm{R}_i^{-2})\,. \quad (8.63)$$

The thermodynamic functions $\mathcal{F}_{rn}^{(\ell)}$ were defined in Eq. (8.32) and we introduced the auxiliary functions

$$\lambda_{r0}^{(0)} = \frac{\Omega_{30}^{(0)}\zeta_r - \Omega_{r0}^{(0)}\zeta_3}{\zeta_3 - \Omega_{30}^{(0)}\zeta_0}\,, \qquad \lambda_{r3}^{(0)} = \frac{m^2}{3}\frac{\zeta_r - \Omega_{r0}^{(0)}\zeta_0}{\zeta_3 - \Omega_{30}^{(0)}\zeta_0}\,, \qquad (8.64)$$

$$\lambda_{r0}^{(1)} = \frac{\Omega_{20}^{(1)}\varkappa_r - \Omega_{r0}^{(1)}\varkappa_2}{\Omega_{20}^{(1)}\varkappa_0 - \varkappa_2}\,, \qquad \lambda_{r2}^{(1)} = \frac{\varkappa_0\Omega_{r0}^{(1)} - \varkappa_r}{\varkappa_0\Omega_{20}^{(1)} - \varkappa_2}\,, \qquad (8.65)$$

$$\lambda_{r0}^{(2)} = \frac{\Omega_{10}^{(2)}\eta_r - \Omega_{r0}^{(2)}\eta_1}{\Omega_{10}^{(2)}\eta_0 - \eta_1}\,, \qquad \lambda_{r1}^{(2)} = \frac{\Omega_{r0}^{(2)}\eta_0 - \eta_r}{\Omega_{10}^{(2)}\eta_0 - \eta_1}\,. \qquad (8.66)$$

Using the relations (8.62), (8.63), it is possible to close Eqs. (8.3)–(8.5) in terms of $\Pi, n^{\mu}, \pi^{\mu\nu}, \rho_3, \rho_2^{\mu}$, and $\rho_1^{\mu\nu}$. The resulting equations of motion will be hyperbolic up to a higher order in Knudsen number when compared with Eqs. (8.36)–(8.38).

The equations of motion are obtained as explained in Sect. 8.3. We multiply Eqs. (8.3)–(8.5) by $\tau_{nr}^{(\ell)}$, substitute Eq. (8.35), and sum over r. The only difference is that now we use Eqs. (8.62) and (8.63) to replace all irreducible moments $\rho_i^{\mu_1\cdots\mu_\ell}$ appearing in the equations by $\Pi, n^{\mu}, \pi^{\mu\nu}, \rho_3, \rho_2^{\mu}$, and $\rho_1^{\mu\nu}$. The resulting equations of motion can be written as

$$\hat{\tau}_{\Pi}\dot{\vec{\Pi}} + \vec{\Pi} = -\vec{\zeta}\theta - \hat{\delta}_{\Pi\Pi}\vec{\Pi}\theta - \hat{\ell}_{\Pi n}\nabla_{\mu}\vec{n}^{\mu} - \hat{\tau}_{\Pi n}\vec{n}_{\mu}F^{\mu}$$
$$\qquad - \hat{\lambda}_{\Pi n}\vec{n}_{\mu}I^{\mu} + \hat{\lambda}_{\Pi\pi}\vec{\pi}^{\mu\nu}\sigma_{\mu\nu}\,,$$

$$\hat{\tau}_n\dot{\vec{n}}^{\langle\mu\rangle} + \vec{n}^{\mu} = \vec{\varkappa}I^{\mu} - \hat{\tau}_n\vec{n}_{\nu}\omega^{\nu\mu} - \hat{\delta}_{nn}\vec{n}^{\mu}\theta + \hat{\ell}_{n\pi}\Delta^{\mu\nu}\partial_{\lambda}\vec{\pi}_{\nu}^{\lambda} - \hat{\ell}_{n\Pi}\nabla^{\mu}\vec{\Pi}$$
$$\qquad - \hat{\tau}_{n\pi}\vec{\pi}^{\mu\nu}F_{\nu} + \hat{\tau}_{n\Pi}\vec{\Pi}F^{\mu} - \hat{\lambda}_{nn}\vec{n}_{\nu}\sigma^{\mu\nu} - \hat{\lambda}_{n\pi}\vec{\pi}^{\mu\nu}I_{\nu} + \hat{\lambda}_{n\Pi}\vec{\Pi}I^{\mu}\,,$$

$$\hat{\tau}_{\pi}\dot{\vec{\pi}}^{\langle\mu\nu\rangle} + \vec{\pi}^{\mu\nu} = 2\vec{\eta}\sigma^{\mu\nu} + 2\hat{\tau}_{\pi}\vec{\pi}_{\lambda}^{\langle\mu}\omega^{\nu\rangle\lambda} - \hat{\delta}_{\pi\pi}\vec{\pi}^{\mu\nu}\theta - \hat{\tau}_{\pi\pi}\vec{\pi}^{\lambda\langle\mu}\sigma_{\lambda}^{\nu\rangle}$$
$$\qquad - \hat{\tau}_{\pi n}\vec{n}^{\langle\mu}F^{\nu\rangle} + \hat{\ell}_{\pi n}\nabla^{\langle\mu}\vec{n}^{\nu\rangle} + \hat{\lambda}_{\pi n}\vec{n}^{\langle\mu}I^{\nu\rangle} + \hat{\lambda}_{\pi\Pi}\vec{\Pi}\sigma^{\mu\nu}\,,$$
$$\qquad\qquad\qquad\qquad\qquad\qquad\qquad\qquad\qquad\qquad\qquad\qquad (8.67)$$

where we defined the vectors

$$\vec{\Pi} \equiv \begin{pmatrix} \Pi \\ \rho_3 \end{pmatrix}\,, \qquad \vec{n}^{\mu} \equiv \begin{pmatrix} n^{\mu} \\ \rho_2^{\mu} \end{pmatrix}\,, \qquad \vec{\pi}^{\mu\nu} \equiv \begin{pmatrix} \pi^{\mu\nu} \\ \rho_1^{\mu\nu} \end{pmatrix}\,. \qquad (8.68)$$

In order to obtain the above equations, all covariant time derivatives of α_0, β_0, and u^μ were replaced by spatial gradients of fluid-dynamical variables using the conservation laws in the form shown in Eqs. (6.24), (6.25), and (6.26).

In this approximation, RTRFD becomes a theory with 23 dynamical variables, while in the previous approximation, i.e., Eqs. (8.36)–(8.38), there were only 14 dynamical variables. These equations of motion are hyperbolic and neglect terms of $O(R_i^{-1}Kn^2, Kn^3, R_i^{-2})$, in contrast to Eqs. (8.36)–(8.38) that become hyperbolic by neglecting terms of $O(Kn^2)$. Above, $\hat{\tau}_\Pi$, $\hat{\tau}_n$, $\hat{\tau}_\pi$, $\hat{\ell}_{\Pi n}$, $\hat{\ell}_{n\Pi}$, $\hat{\ell}_{n\pi}$, $\hat{\ell}_{\pi n}$, $\hat{\delta}_{\Pi\Pi}$, $\hat{\delta}_{nn}$, $\hat{\delta}_{\pi\pi}$, $\hat{\tau}_{\Pi n}$, $\hat{\tau}_{n\pi}$, $\hat{\tau}_{n\Pi}$, $\hat{\tau}_{\pi\pi}$, $\hat{\tau}_{\pi n}$, $\hat{\lambda}_{\Pi n}$, $\hat{\lambda}_{\Pi\pi}$, $\hat{\lambda}_{nn}$, $\hat{\lambda}_{n\Pi}$, $\hat{\lambda}_{n\pi}$, $\hat{\lambda}_{\pi n}$, and $\hat{\lambda}_{\pi\Pi}$ are 2×2 matrices, while $\vec{\zeta}$, $\vec{\varkappa}$, and $\vec{\eta}$ are two-component vectors.

As happened before, even though closed in terms of 23 moments, the transport coefficients will depend on all the moments of the distribution function. The microscopic formulas for these transport coefficients were computed for a gas of massless particles and are shown in Appendix 8.12. For a gas of classical particles with a constant cross section σ, the values for the diffusion and viscosity coefficients, $\vec{\varkappa}$ and $\vec{\eta}$, and for the relaxation-time matrices, $\hat{\tau}_n$ and $\hat{\tau}_\pi$, are

$$\frac{\vec{\varkappa}}{\lambda n_0} = \begin{pmatrix} 0.1596 \\ -2.3616/\beta_0^2 \end{pmatrix}, \quad \frac{\vec{\eta}}{\lambda P_0} = \begin{pmatrix} 1.268 \\ 6.929/\beta_0 \end{pmatrix},$$

$$\frac{\hat{\tau}_n}{\lambda} = \begin{pmatrix} 1.295 & -0.053\beta_0^2 \\ 5.18/\beta_0^2 & 2.787 \end{pmatrix}, \quad \frac{\hat{\tau}_\pi}{\lambda} = \begin{pmatrix} 0.912 & 0.136\beta_0 \\ -3.647/\beta_0 & 2.456 \end{pmatrix}. \tag{8.69}$$

The transport coefficients of the nonlinear terms in the equation of motion for $\vec{n}^\mu$ are

$$\frac{\hat{\delta}_{nn}}{\lambda} = \begin{pmatrix} 1.295 & -0.0883\beta_0^2 \\ 5.18/\beta_0^2 & 4.645 \end{pmatrix}, \quad \frac{\hat{\lambda}_{nn}}{\lambda} = \begin{pmatrix} 0.524 & -0.0341\beta_0^2 \\ 2.096/\beta_0^2 & 2.863 \end{pmatrix},$$

$$\frac{\hat{\lambda}_{n\pi}}{\lambda} = \begin{pmatrix} 0.1677\beta_0 & -0.0288\beta_0^2 \\ 0.6708/\beta_0 & -0.1147 \end{pmatrix}, \quad \frac{\hat{\tau}_{n\pi}}{\lambda} = \frac{1}{4P_0}\begin{pmatrix} 0 & 0.0973\beta_0^2 \\ 0 & -2.6106 \end{pmatrix},$$

$$\frac{\hat{\ell}_{n\pi}}{\lambda} = \begin{pmatrix} -0.4723\beta_0 & 0.0973\beta_0^2 \\ 13.111/\beta_0 & -2.611 \end{pmatrix}, \tag{8.70}$$

while those in the equation of motion for $\vec{\pi}^{\mu\nu}$ are

$$\frac{\hat{\delta}_{\pi\pi}}{\lambda} = -\frac{4}{3}\begin{pmatrix} 0.912 & 0.17\beta_0 \\ -3.647/\beta_0 & 3.0698 \end{pmatrix}, \quad \frac{\hat{\tau}_{\pi\pi}}{\lambda} = \begin{pmatrix} 1.5688 & 0.2261\beta_0 \\ -6.2751/\beta_0 & 5.0956 \end{pmatrix}, \tag{8.71}$$

$$\frac{\hat{\tau}_{\pi n}}{\lambda} = \frac{1}{P_0}\begin{pmatrix} 0.2228/\beta_0 & 0.0714\beta_0 \\ 0.8913/\beta_0^2 & 1.5144 \end{pmatrix}, \quad \frac{\hat{\ell}_{\pi n}}{\lambda} = \begin{pmatrix} 0.2228/\beta_0 & 0.0476\beta_0 \\ 0.8913/\beta_0^2 & 1.0096 \end{pmatrix},$$

$$\frac{\hat{\lambda}_{\pi n}}{\lambda} = \begin{pmatrix} 0.1186/\beta_0 & 0.0084\beta_0 \\ -0.4744/\beta_0^2 & -0.0338 \end{pmatrix}. \tag{8.72}$$

In the massless limit, the scalar moments become less important (Π is exactly zero and ρ_3 is small) and, for this reason, we did not compute the microscopic formulas or the transport coefficients associated with these moments. These transport coefficients were computed including a total of 41 moments, as was done for the transport coefficients of the terms $\mathcal{J}^\mu$ and $\mathcal{J}^{\mu\nu}$ in the last section.

As already mentioned, this approach increases the domain of validity of the equations of motion without making them parabolic. However, this does not solve the intrinsic problem of the coarse-graining procedure: if one attempts to go to an even higher order in Kn, the equations become once more parabolic. This can be solved in a similar fashion, by again increasing the number of dynamical variables describing the system. However, a complete and causal description of the system can only be obtained by the microscopic theory itself, i.e., by solving the Boltzmann equation or, equivalently, considering an infinite number of moments.

8.7 Comparisons with Microscopic Theory

In the last sections, a systematic derivation of transient relativistic fluid dynamics from the Boltzmann equation was introduced. The main difference between Israel–Stewart theory and RTRFD is that the latter does not truncate the moment expansion of the single-particle distribution function. Instead, dynamical equations for all its moments are considered and solved by separating the slowest microscopic time scale from the faster ones. Then, the resulting fluid-dynamical equations are truncated according to a systematic power-counting scheme using the inverse Reynolds number $R^{-1} \sim |n^\mu|/n \sim |\pi^{\mu\nu}|/P_0$ and the Knudsen number $Kn = \lambda/L$, with λ being the mean free path and L a characteristic macroscopic distance scale, e.g., $L^{-1} \sim \partial_\mu u^\mu$. The values of the transport coefficients of RTRFD are obtained by re-summing the contributions from all moments of the single-particle distribution function, similar to what happens in the Chapman-Enskog expansion [1].

In this section, we reproduce the findings of Ref. [42] and we test the validity of RTRFD by comparing it to solutions of the relativistic Boltzmann equation, as previously done in Refs. [30–32,43]. We then demonstrate that this method is able to handle problems with strong initial gradients in pressure or particle-number density. This resolves the differences between the solution of Israel–Stewart theory and of the Boltzmann equation observed in Ref. [43]. We conclude that these differences were caused by the uncontrolled truncation procedure of the expansion of the single-particle distribution function in terms of Lorentz tensors in 4-momentum as employed in Israel–Stewart theory.

8.7.1 Stationary Shock Solutions

We consider a (3+1)-dimensional gas of classical massless particles with a constant cross section σ. For the sake of simplicity, the distribution of particles is assumed to be homogeneous in the (y, z)-plane, being allowed to vary only in the longitudinal x-direction. This effectively leads to a (1+1)-dimensional problem. We consider two different types of initial conditions:

In **case I**, the system is initialized with a homogeneous fugacity distribution, $\lambda \equiv e^{\alpha_0} \equiv 1$ (here, λ denotes the fugacity, not the mean free path), but with an inhomogeneous pressure profile in the longitudinal direction, i.e., the system is in chemical, but not mechanical equilibrium. The pressure profile is constructed by smoothly connecting two temperature states, $T(-\infty) = 0.4\,\mathrm{GeV}$ (the temperature at $x \to -\infty$) and $T(+\infty) = 0.25\,\mathrm{GeV}$ (the temperature at $x \to \infty$) using a Woods-Saxon parametrization with a thickness parameter $D = 0.3$ fm. In this scheme, by taking the limit $D \to 0$, we obtain the pressure profile of the Riemann problem.

In **case II**, the pressure is assumed to be homogeneous, i.e., the system is initially in mechanical equilibrium, with a value of $P_0 = d_{\mathrm{dof}} T^4(-\infty)/\pi^2$ (the degeneracy factor is taken to be $d_{\mathrm{dof}} = 16$, as appropriate for a gas of gluons, the gauge bosons of the theory of the strong interaction, quantum chromodynamics, and $T(-\infty) = 0.4\,\mathrm{GeV}$). On the other hand, the system is not in chemical equilibrium and the fugacity distribution is obtained by smoothly connecting two fugacity states, $\lambda(-\infty) = 1$ (the fugacity at $x \to -\infty$) and $\lambda(+\infty) = 0.2$ (the fugacity at $x \to \infty$) using a Woods-Saxon parametrization with a thickness parameter $D = 0.3$ fm.

In both cases, matter is initialized in *local* thermodynamic equilibrium, i.e., with all dissipative currents and eigenmodes of the Boltzmann equation set to zero, and at rest, i.e., with velocity $u^\mu = (1, 0, 0, 0)^T$. These initial conditions are shown in Fig. 8.1.

In both cases, we consider two exemplary values for the cross section, $\sigma = 2$ and 8 mb, and consider the solutions after the system has evolved for 6 fm in time. We compare the solution of the Boltzmann equation with that of traditional Israel–Stewart theory (including terms omitted in the original work [2,4] but quoted in Ref. [34]), as well as with RTRFD at various levels of approximation. Equations (8.36)–(8.38) contain 13 moments as independent dynamical variables (14, if we include bulk-viscous pressure). The calculation of the transport coefficients in these equations can be done with increasing accuracy, as more irreducible moments are considered in the moment expansion. The lowest possible accuracy is reached if no more than the original 13 (14, in the case of non-vanishing bulk-viscous pressure) irreducible moments are considered for the calculation of the transport coefficients. At the next level, we include one more set of irreducible moments of tensor-rank one and two

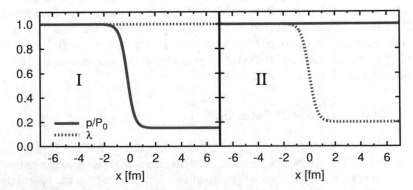

Fig. 8.1 Initial conditions for cases I and II. Figure taken from Ref. [42]

(and one more scalar moment in the case of non-vanishing bulk-viscous pressure), which leads to a total of 21 (23, in the case of non-vanishing bulk-viscous pressure) irreducible moments. In this way, the number of irreducible moments entering the transport coefficients increases by 8 (9) at each successive level of approximation. For the purpose of this comparison, we found that going to the third level of iteration, i.e., considering $13 + 8 \times 3 = 37$ moments ($14 + 9 \times 3 = 41$ in the case of non-vanishing bulk-viscous pressure) is sufficient to reach the desired accuracy in the values of the transport coefficients. In the following, we shall compare RTRFD with 13 dynamical degrees of freedom and with the transport coefficients computed with 13 and with 37 moments. We shall term these variants of RTRFD "13/13" and "13/37", respectively. In addition, we also solve Eqs. (8.67). These contain 21 dynamical degrees of freedom. We compute the corresponding transport coefficients using 37 moments. We shall refer to this variant of RTRFD as "21/37". In the following figures, the numerical solutions of the Boltzmann equation is always displayed by open dots, the results of Israel–Stewart theory by black dash-dotted lines, the solution of RTRFD "13/13" by green dashed lines, that of RTRFD "13/37" by blue dotted lines, and that of RTRFD "21/37" by solid red curves.

In Fig. 8.2, we show the fugacity (top) and thermodynamic pressure (bottom) and in Fig. 8.3 the heat flow $q^\mu \equiv -(\varepsilon + P_0)n^\mu/n$ (top) and shear-stress tensor (bottom) for case I. The Boltzmann equation and the fluid-dynamical theories were solved for $\sigma = 2$ mb (shown in the left panels of each figure) and for $\sigma = 8$ mb

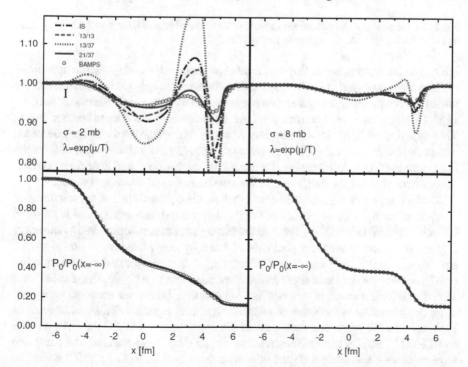

Fig. 8.2 Fugacity and pressure profiles at $t = 6$ fm for case I, for $\sigma = 2$ mb (left panels) and $\sigma = 8$ mb (right panels). Figure taken from Ref. [42]

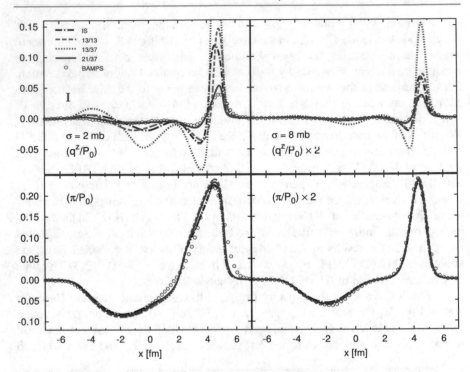

Fig. 8.3 Heat-flow and shear-stress tensor profiles at $t = 6$ fm for case I, for $\sigma = 2$ mb (left panels) and $\sigma = 8$ mb (right panels). Figure taken from Ref. [42]

(shown in the right panels). For $\sigma = 8$ mb, the thermodynamic pressure and shear-stress tensor computed in all fluid-dynamical theories are in good agreement with the numerical solutions of the Boltzmann equation. As we decrease the cross section, we expect the agreement between macroscopic and microscopic theory to become worse. This explains why, for $\sigma = 2$ mb, the pressure and shear-stress tensor computed within fluid-dynamical theories deviate more strongly from those computed via the microscopic theory. Nevertheless, compared to the fugacity and heat-flow profiles, the agreement is not too bad, even for the smaller value of the cross section.

The initial pressure gradient in case I drives, via conservation of momentum, the creation of large velocity gradients. On the other hand, the gradient of fugacity is initially zero and turns out to remain small throughout the evolution. In this situation, higher-order terms involving gradients of velocity and of the shear-stress tensor, e.g., $\widetilde{\varkappa}_6 \Delta^\mu_\lambda \partial_\nu \sigma^{\lambda\nu} \subset \mathcal{K}^\mu$ and $\ell_{n\pi} \Delta^{\mu\nu} \nabla_\lambda \pi^\lambda_\nu \subset \mathcal{J}^\mu$ in the particle-diffusion equation (8.37), become of the same order as the respective (first-order) Navier–Stokes term $\varkappa I^\mu$. Therefore, if terms of this type are not properly taken into account, we expect large deviations from the solution of the Boltzmann equation. This can be seen in Figs. 8.2 and 8.3 when comparing Israel–Stewart theory, RTRFD "13/13", as well as RTRFD "13/37" with the Boltzmann result. In all of these variants, the parabolic term $\sim \widetilde{\varkappa}_6$ is either absent (Israel-Stewart theory and RTRFD "13/13") or has to be dismissed (RTRFD "13/37") for reasons of causality. In addition, Israel–Stewart theory and RTRFD "13/13" do not have the correct value for $\ell_{n\pi}$, because we did not include a sufficiently large number of irreducible moments in its computation. Although RTRFD "13/37" features (within the desired accuracy) the correct value for this transport coefficient (as well as for $\tilde{\varkappa}_6$), it does even worse in describing the fugacity and heat-flow profiles than the previous two theories. This is because the term $\sim \tilde{\varkappa}_6$ could not be taken into account for reasons of causality, although it is of the same order of magnitude as the term $\sim \ell_{n\pi}$. These problems of fluid-dynamical theories with only 13 dynamical variables are resolved by RTRFD "21/37", which is the only fluid-dynamical theory considered here that contains *all* contributions of second order in the Knudsen number in a hyperbolic fashion.

In Fig. 8.4, we show the fugacity (top) and thermodynamic pressure (bottom) and in Fig. 8.5 the heat-flow (top) and shear-stress tensor profiles (bottom) for case II. As before, the Boltzmann equation and the fluid-dynamical theories considered were solved for $\sigma = 2$ mb (shown in the left panels) and for $\sigma = 8$ mb (shown in the right panels). Again, we expect and see a better agreement between fluid dynamics and the Boltzmann equation for the larger value of the cross section. While the fugacity profiles are in good agreement with the solution of the Boltzmann equation for all fluid-dynamical theories and both values of the cross section, the heat flow is not well described in Israel–Stewart theory and in RTRFD "13/13": Israel–Stewart theory predicts values for the heat flow which are smaller in magnitude than the Boltzmann equation, while RTRFD "13/13" predicts larger values, even for $\sigma = 8$ mb. On the

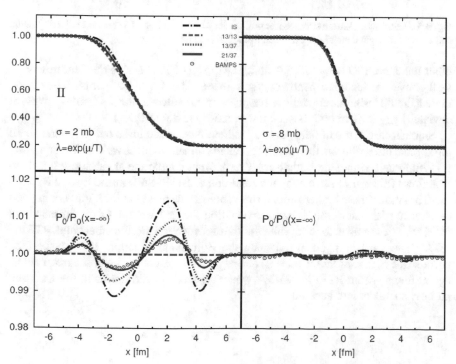

Fig. 8.4 Fugacity and thermodynamic pressure profiles at $t = 6$ fm for case II, for $\sigma = 2$ mb (left panels) and $\sigma = 8$ mb (right panels). Figure taken from Ref. [42]

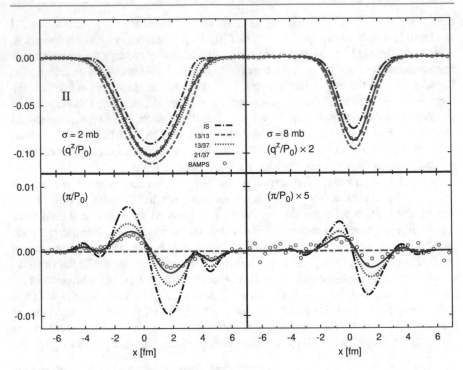

Fig. 8.5 Heat-flow and shear-stress tensor profiles at $t = 6$ fm for case II, for $\sigma = 2$ mb (left panels) and $\sigma = 8$ mb (right panels). Figure taken from Ref. [42]

other hand, both RTRFD "13/37" and RTRFD "21/37" describe the heat flow very well or even perfectly, respectively, for both values of the cross section. The reason is that the diffusion coefficient $\varkappa$ has the correct value in these theories (while it deviates by $\sim 30\%$ in both Israel-Stewart theory and RTRFD "13/13").

Since in case II the initial pressure gradient is zero and turns out to remain small throughout the evolution, the velocity gradients remain small as well. In this situation, it is important to include higher-order terms that couple the shear-stress tensor to heat flow. This is the reason why the solutions of Israel–Stewart theory and RTRFD "13/13" (where these higher-order terms vanish in the massless limit) are not in good agreement with that of the Boltzmann equation for the thermodynamic pressure and the shear-stress tensor, for both values of the cross section. On the other hand, RTRFD "13/37" does a better job in matching the Boltzmann equation. It is not perfect, because the higher-order terms $\sim \widetilde{\eta}_5$ and $\sim \widetilde{\eta}_8$ were dropped. The best agreement is, again, found within RTRFD "21/37" where all second-order terms in the Knudsen number are taken into account.

8.8 Summary

In this chapter, we have presented a general and consistent derivation of relativistic fluid dynamics from the Boltzmann equation using the method of moments. The main difference of our approach, termed Resummed Transient Relativistic Fluid Dynamics (RTRFG), to Israel–Stewart theory is that we did not close the fluid-dynamical equations of motion by truncating the expansion of the distribution function. Instead, we kept all terms in the moment expansion and truncated the exact equations of motion according to a power-counting scheme in Knudsen and inverse Reynolds number. Contrary to many calculations, we did not assume that the inverse Reynolds and Knudsen numbers are of the same order. As a matter of fact, in order to obtain relaxation-type equations, we had to explicitly include the slowest microscopic time scales, which are shown to be the characteristic times over which dissipative currents relax toward their asymptotic Navier–Stokes solutions. Thus, Navier–Stokes theory, or the Chapman-Enskog expansion is already included in our formulation as an asymptotic limit of the dynamical equations.

We concluded that the equations of motion can be closed in terms of only 14 dynamical variables, as long as we only keep terms of second order in Knudsen and/or inverse Reynolds number. Even though the equations of motion are closed in terms of these 14 fields, the transport coefficients carry information about all moments of the distribution function (all different relaxation scales of the irreducible moments). The bulk-viscosity, particle-diffusion, and shear-viscosity coefficients agree with the values obtained via Chapman-Enskog theory. We then showed how to use this formalism to derive equations of motion that are hyperbolic and, at the same time, include terms up to second order in the Knudsen number.

Finally, we compared the derived equations of motion of RTRFD at various levels of approximation with numerical solutions of the Boltzmann equation for two different types of shock solutions (labeled cases I and II). The initial conditions for cases I and II were chosen in such a way that considerably different spatial profiles are generated throughout the fluid-dynamical evolution. In case I, the pressure gradient is initially large, which gives rise to large velocity gradients in the later stages of the evolution. This means that, in case I, the shear-stress tensor is mainly generated by its corresponding Navier–Stokes term, i.e., by gradients of velocity. On the other hand, the fugacity gradient is initially zero in case I, and remains relatively small throughout the evolution of the fluid. Therefore, in case I the heat flow is not mainly created by its Navier–Stokes term, i.e., by the gradient of fugacity, but by the coupling term to the shear tensor and shear-stress tensor, i.e., the terms $\Delta^{\mu\nu}\nabla_\lambda \pi_\nu^\lambda$ and $\Delta^{\mu\nu}\nabla_\lambda \sigma_\nu^\lambda$ in Eq. (8.37). Therefore, in this case, the higher-order terms in Knudsen number must be included and one really needs to solve the hyperbolic equations derived in this chapter to obtain a good agreement. The fact that Israel–Stewart theory is always deviating from the microscopic theory when it concerns heat flow means that it does not predict correctly the terms of order one and two in Knudsen number. On the other hand, in RTRFD "13/37" and RTRFD "21/37" all transport coefficients are computed with a sufficiently large number of irreducible moments. This guarantees that all terms of the desired order are included and is the reason for the better agreement of these fluid-dynamical theories with the microscopic theory. The reason why RTRFD "13/37" fails in certain situations is that important terms have to be neglected in order to preserve hyperbolicity and causality.

In case II, the fugacity gradient is initially large while the pressure gradient is zero. This means that the heat flow originates mainly from its Navier–Stokes term, while the shear-tress tensor originates mainly from its coupling to heat flow, i.e., the terms $\nabla^{<\mu} n^{\nu>}$, $\nabla^{<\mu} I^{\nu>}$, $I^{<\mu} I^{\nu>}$, and $n^{<\mu} I^{\nu>}$ in Eq. (8.38). The fact that the heat flow calculated from Israel–Stewart theory deviates from the solution given by the microscopic theory even in this case is evidence that the Navier–Stokes term of this theory does not contain the correct transport coefficient. The coupling of the shear-stress tensor with the heat flow in Israel–Stewart theory is also not correctly taken into account.

In conclusion, the resummation of irreducible moments for the computation of the transport coefficients is essential to obtain a good agreement with the microscopic theory. It provides not only the correct values for the shear-viscosity and heat-conduction coefficients, but also for the transport coefficients that couple the respective dissipative currents. Moreover, in situations where higher-order terms are important, one has to make sure to include them in a hyperbolic way, and not simply drop relevant contributions because they are parabolic. These two factors resolve differences between the solution of Israel–Stewart theory and of the Boltzmann equation observed in Ref. [43].

As expected, and explicitly demonstrated in this chapter, the agreement between solutions of RTRFD and the Boltzmann equation depends on the value of the cross section σ. For the cases considered here, we obtain a good agreement for $\sigma = 8$ mb, while for $\sigma = 2$ mb we start to notice small deviations. In order to improve the agreement for smaller values of the cross section, we would have to include more moments of the Boltzmann equation to describe the state of the system, i.e., such moments would have to contribute not only to the values of the transport coefficients, but also as independent dynamical variables. This is consistent with the discussion developed in Chap. 7, where it was shown that the method of moments converges faster when the viscosity is smaller, or, equivalently, the cross section is larger.

8.9   Appendix 1: Transport Coefficients in Eq. (8.44)

In this appendix, we list all transport coefficients appearing in Eq. (8.44). The transport coefficients in the equation for the bulk-viscous pressure are

$$\ell_{\Pi n} = -\frac{m^2}{3} \left(\gamma_1^{(1)} \tau_{00}^{(0)} - \sum_{r=0,\neq 1,2}^{N_0} \tau_{0r}^{(0)} \frac{G_{3r}}{D_{20}} + \sum_{r=0}^{N_0-3} \tau_{0,r+3}^{(0)} \Omega_{r+2,0}^{(1)} \right) , \tag{8.73}$$

$$\tau_{\Pi n} = \frac{m^2}{3\,(\varepsilon_0 + P_0)} \left\{ \tau_{00}^{(0)} \frac{\partial \gamma_1^{(1)}}{\partial \ln \beta_0} - \sum_{r=0,\neq 1,2}^{N_0} \tau_{0r}^{(0)} \frac{G_{3r}}{D_{20}} \right.$$

$$+ \sum_{r=0}^{N_0-3} \tau_{0,r+3}^{(0)} \left[\frac{\partial \Omega_{r+2,0}^{(1)}}{\partial \ln \beta_0} + (r+3)\, \Omega_{r+2,0}^{(1)} \right] \Bigg\} , \tag{8.74}$$

$$\delta_{\Pi\Pi} = \frac{1}{3}\left(2 + m^2 \gamma_2^{(0)}\right) \tau_{00}^{(0)} - \frac{m^2}{3} \sum_{r=0,\neq 1,2}^{N_0} \tau_{0r}^{(0)} \frac{G_{2r}}{D_{20}} + \frac{1}{3} \sum_{r=0}^{N_0-3} (r+5)\, \tau_{0,r+3}^{(0)} \Omega_{r+3,0}^{(0)}$$

$$+ \sum_{r=3}^{N_0} \tau_{0r}^{(0)} \left[\frac{(\varepsilon_0 + P_0) J_{20} - n_0 J_{30}}{D_{20}} \frac{\partial \Omega_{r0}^{(0)}}{\partial \alpha_0} + \frac{(\varepsilon_0 + P_0) J_{10} - n_0 J_{20}}{D_{20}} \frac{\partial \Omega_{r0}^{(0)}}{\partial \beta_0} \right]$$

$$- \frac{m^2}{3} \sum_{r=0}^{N_0-5} (r+4)\, \tau_{0,r+5}^{(0)} \Omega_{r+3,0}^{(0)} , \tag{8.75}$$

$$\lambda_{\Pi n} = -\frac{m^2}{3} \left[\left(\frac{\partial \gamma_1^{(1)}}{\partial \alpha_0} + \frac{1}{h_0} \frac{\partial \gamma_1^{(1)}}{\partial \beta_0} \right) \tau_{00}^{(0)} + \sum_{r=0}^{N_0-3} \tau_{0,r+3}^{(0)} \left(\frac{\partial \Omega_{r+2,0}^{(1)}}{\partial \alpha_0} + \frac{1}{h_0} \frac{\partial \Omega_{r+2,0}^{(1)}}{\partial \beta_0} \right) \right] , \tag{8.76}$$

$$\lambda_{\Pi\pi} = \frac{m^2}{3} \left[\gamma_2^{(2)} \tau_{00}^{(0)} - \sum_{r=0,\neq 1,2}^{N_0} \tau_{0r}^{(0)} \frac{G_{2r}}{D_{20}} - \sum_{r=0}^{N_0-3} (r+2)\, \tau_{0,r+3}^{(0)} \Omega_{r+1,0}^{(2)} \right] , \tag{8.77}$$

where $h_0 = (\varepsilon_0 + P_0)/n_0$ is the enthalpy per particle.

The transport coefficients in the equation for the particle-diffusion current are

$$\delta_{nn} = \left(1 + \frac{1}{3} m^2 \gamma_2^{(1)}\right) \tau_{00}^{(1)} - \frac{1}{3} m^2 \sum_{r=0}^{N_1-2} (r+1)\, \tau_{0,r+2}^{(1)} \Omega_{r0}^{(1)} + \frac{1}{3} \sum_{r=2}^{N_1} (r+3)\, \tau_{0r}^{(1)} \Omega_{r0}^{(1)}$$

$$+ \sum_{r=2}^{N_1} \tau_{0r}^{(1)} \left[\frac{(\varepsilon_0 + P_0) J_{20} - n_0 J_{30}}{D_{20}} \frac{\partial \Omega_{r0}^{(1)}}{\partial \alpha_0} + \frac{(\varepsilon_0 + P_0) J_{10} - n_0 J_{20}}{D_{20}} \frac{\partial \Omega_{r0}^{(1)}}{\partial \beta_0} \right] , \tag{8.78}$$

$$\ell_{n\Pi} = \left(\frac{1}{h_0} - \gamma_1^{(0)} \right) \tau_{00}^{(1)} + \sum_{r=0}^{N_1-2} \tau_{0,r+2}^{(1)} \frac{\beta_0 J_{r+4,1}}{\varepsilon_0 + P_0} + \frac{1}{m^2} \sum_{r=0}^{N_1-2} \tau_{0,r+2}^{(1)} \Omega_{r+3,0}^{(0)}$$

$$- \sum_{r=0}^{N_1-4} \tau_{0,r+4}^{(1)} \Omega_{r+3,0}^{(0)} , \tag{8.79}$$

$$\tau_{n\Pi} = \frac{1}{\varepsilon_0 + P_0} \left(\left(\frac{1}{h_0} - \frac{\partial \gamma_1^{(0)}}{\partial \ln \beta_0} \right) \tau_{00}^{(1)} - \sum_{r=0}^{N_1-4} \tau_{0,r+4}^{(1)} \left[(r+4)\, \Omega_{r+3,0}^{(0)} + \frac{\partial \Omega_{r+3,0}^{(0)}}{\partial \ln \beta_0} \right] \right.$$

$$+ \left. \sum_{r=0}^{N_1-2} \tau_{0,r+2}^{(1)} \left\{ \frac{\beta_0 J_{r+4,1}}{\varepsilon_0 + P_0} + \frac{1}{m^2} \left[(r+5)\, \Omega_{r+3,0}^{(0)} + \frac{\partial \Omega_{r+3,0}^{(0)}}{\partial \ln \beta_0} \right] \right\} \right) , \tag{8.80}$$

$$\ell_{n\pi} = \left(\frac{1}{h_0} - \gamma_1^{(2)} \right) \tau_{00}^{(1)} + \sum_{r=0}^{N_1-2} \tau_{0,r+2}^{(1)} \left(\frac{\beta_0 J_{r+4,1}}{\varepsilon_0 + P_0} - \Omega_{r+1,0}^{(2)} \right) , \tag{8.81}$$

$$\tau_{n\pi} = \frac{1}{\varepsilon_0 + P_0} \left\{ \left(\frac{1}{h_0} - \frac{\partial \gamma_1^{(2)}}{\partial \ln \beta_0} \right) \tau_{00}^{(1)} \right.$$

$$+ \left. \sum_{r=0}^{N_1-2} \tau_{0,r+2}^{(1)} \left[\frac{\beta_0 J_{r+4,1}}{\varepsilon_0 + P_0} - \frac{\partial \Omega_{r+1,0}^{(2)}}{\partial \ln \beta_0} - (r+2) \Omega_{r+1,0}^{(2)} \right] \right\} , \tag{8.82}$$

$$\lambda_{nn} = \frac{1}{5}\left(3 + 2m^2\gamma_2^{(1)}\right)\tau_{00}^{(1)} - \frac{2}{5}m^2 \sum_{r=0,r\neq1}^{N_1-2}(r+1)\,\tau_{0,r+2}^{(1)}\Omega_{r0}^{(1)} + \frac{1}{5}\sum_{r=2}^{N_1}(2r+3)\,\tau_{0r}^{(1)}\,\Omega_{r0}^{(1)}\,,$$

$$\tag{8.83}$$

$$\lambda_{n\Pi} = \tau_{00}^{(1)}\left(\frac{1}{h_0}\frac{\partial\gamma_1^{(0)}}{\partial\beta_0} + \frac{\partial\gamma_1^{(0)}}{\partial\alpha_0}\right) - \frac{1}{m^2}\sum_{r=0}^{N_1-2}\tau_{0,r+2}^{(1)}\left(\frac{1}{h_0}\frac{\partial\Omega_{r+3,0}^{(0)}}{\partial\beta_0} + \frac{\partial\Omega_{r+3,0}^{(0)}}{\partial\alpha_0}\right)$$

$$+ \sum_{r=0}^{N_1-4}\tau_{0,r+4}^{(1)}\left(\frac{1}{h_0}\frac{\partial\Omega_{r+3,0}^{(0)}}{\partial\beta_0} + \frac{\partial\Omega_{r+3,0}^{(0)}}{\partial\alpha_0}\right)\,,$$

$$\tag{8.84}$$

$$\lambda_{n\pi} = \left(\frac{1}{h_0}\frac{\partial\gamma_1^{(2)}}{\partial\beta_0} + \frac{\partial\gamma_1^{(2)}}{\partial\alpha_0}\right)\tau_{00}^{(1)} + \sum_{r=0}^{N_1-2}\tau_{0,r+2}^{(1)}\left(\frac{1}{h_0}\frac{\partial\Omega_{r+1,0}^{(2)}}{\partial\beta_0} + \frac{\partial\Omega_{r+1,0}^{(2)}}{\partial\alpha_0}\right)\,. \tag{8.85}$$

The transport coefficients in the equation for the shear-stress tensor are

$$\delta_{\pi\pi} = \frac{1}{3}m^2\gamma_2^{(2)}\tau_{00}^{(2)} + \frac{1}{3}\sum_{r=0}^{N_2}(r+4)\,\tau_{0r}^{(2)}\Omega_{r0}^{(2)} - \frac{1}{3}m^2\sum_{r=0}^{N_2-2}(r+1)\,\tau_{0,r+2}^{(2)}\Omega_{r0}^{(2)}$$

$$+ \sum_{r=0}^{N_2}\tau_{0r}^{(2)}\left[\frac{(\varepsilon_0+P_0)J_{10}-n_0J_{20}}{D_{20}}\frac{\partial\Omega_{r0}^{(2)}}{\partial\beta_0} + \frac{(\varepsilon_0+P_0)J_{20}-n_0J_{30}}{D_{20}}\frac{\partial\Omega_{r0}^{(2)}}{\partial\alpha_0}\right]\,, \tag{8.86}$$

$$\tau_{\pi\pi} = \frac{2}{7}\sum_{r=0}^{N_2}(2r+5)\,\tau_{0r}^{(2)}\Omega_{r0}^{(2)} + \frac{4}{7}m^2\gamma_2^{(2)}\tau_{00}^{(2)} - \frac{4}{7}m^2\sum_{r=0}^{N_2-2}(r+1)\,\tau_{0,r+2}^{(2)}\Omega_{r0}^{(2)}\,, \tag{8.87}$$

$$\lambda_{\pi\Pi} = \frac{6}{5}\tau_{00}^{(2)} + \frac{2}{5}m^2\gamma_2^{(0)}\tau_{00}^{(2)} - \frac{2}{5m^2}\sum_{r=0}^{N_2-1}(r+5)\,\tau_{0,r+1}^{(2)}\Omega_{r+3,0}^{(0)}$$

$$+ \frac{2}{5}\sum_{r=3}^{N_2}(2r+3)\,\tau_{0r}^{(2)}\Omega_{r0}^{(0)} - \frac{2}{5}m^2\sum_{r=0,\neq1,2}^{N_2-2}(r+1)\,\tau_{0,r+2}^{(2)}\Omega_{r0}^{(0)}\,, \tag{8.88}$$

$$\tau_{\pi n} = \frac{2}{5(\varepsilon_0+P_0)}\left\{-m^2\tau_{00}^{(2)}\frac{\partial\gamma_1^{(1)}}{\partial\ln\beta_0} - m^2\sum_{r=0}^{N_2-3}\tau_{0,r+3}^{(2)}\frac{\partial\Omega_{r+2,0}^{(1)}}{\partial\ln\beta_0}\right.$$

$$\left. -m^2\sum_{r=0,\neq1}^{N_2-1}(r+1)\,\tau_{0,r+1}^{(2)}\Omega_{r0}^{(1)} + \sum_{r=0}^{N_2-1}\tau_{0,r+1}^{(2)}\left[(r+6)\,\Omega_{r+2,0}^{(1)} + \frac{\partial\Omega_{r+2,0}^{(1)}}{\partial\ln\beta_0}\right]\right\}\,,$$

$$\tag{8.89}$$

$$\ell_{\pi n} = -\frac{2}{5}m^2\gamma_1^{(1)}\tau_{00}^{(2)} + \frac{2}{5}\sum_{r=0}^{N_2-1}\tau_{0,r+1}^{(2)}\Omega_{r+2,0}^{(1)} - \frac{2}{5}m^2\sum_{r=0,\neq1}^{N_2-1}\tau_{0,r+1}^{(2)}\Omega_{r0}^{(1)}\,, \tag{8.90}$$

$$\lambda_{\pi n} = -\frac{2}{5}m^2\tau_{00}^{(2)}\left(\frac{1}{h_0}\frac{\partial\gamma_1^{(1)}}{\partial\beta_0} + \frac{\partial\gamma_1^{(1)}}{\partial\alpha_0}\right) + \frac{2}{5}\sum_{r=0}^{N_2-1}\tau_{0,r+1}^{(2)}\left(\frac{1}{h_0}\frac{\partial\Omega_{r+2,0}^{(1)}}{\partial\beta_0} + \frac{\partial\Omega_{r+2,0}^{(1)}}{\partial\alpha_0}\right)$$

$$- \frac{2}{5}m^2\sum_{r=0}^{N_2-3}\tau_{0,r+3}^{(2)}\left(\frac{1}{h_0}\frac{\partial\Omega_{r+2,0}^{(1)}}{\partial\beta_0} + \frac{\partial\Omega_{r+2,0}^{(1)}}{\partial\alpha_0}\right)\,. \tag{8.91}$$

8.10   Appendix 2: Calculation of the Collision Integrals

In this appendix, we calculate the collision integrals (6.47) for a classical gas, i.e., $\tilde{f}_{0k} = 1$, of hard spheres in the ultrarelativistic limit, $m\beta_0 \ll 1$. Then, Eq. (6.47) becomes

$$\mathcal{A}_{rn}^{(\ell)} = \frac{1}{\nu\,(2\ell+1)} \int dKdK'dPdP'\, W_{\mathbf{kk'}\to\mathbf{pp'}}\, f_{0\mathbf{k}} f_{0\mathbf{k'}} E_{\mathbf{k}}^{r-1} k^{\langle\nu_1}\cdots k^{\nu_\ell\rangle}$$
$$\times \left(\mathcal{H}_{\mathbf{k}n}^{(\ell)} k_{\langle\nu_1}\cdots k_{\nu_\ell\rangle} + \mathcal{H}_{\mathbf{k'}n}^{(\ell)} k'_{\langle\nu_1}\cdots k'_{\nu_\ell\rangle} - \mathcal{H}_{\mathbf{p}n}^{(\ell)} p_{\langle\nu_1}\cdots p_{\nu_\ell\rangle} - \mathcal{H}_{\mathbf{p'}n}^{(\ell)} p'_{\langle\nu_1}\cdots p'_{\nu_\ell\rangle}\right). \tag{8.92}$$

The functions $\mathcal{H}_{\mathbf{k}n}^{(\ell)}$ were defined in Eq. (6.13). The transition rate $W_{\mathbf{kk'}\to\mathbf{pp'}}$ is written in terms of the differential cross section $\sigma(s,\Theta)$ as

$$W_{\mathbf{kk'}\to\mathbf{pp'}} = s\sigma(s,\Theta_s)\,(2\pi)^6\,\delta^{(4)}\left(k^\mu + k'^\mu - p^\mu - p'^\mu\right). \tag{8.93}$$

The variables s and Θ_s are defined as

$$s = \left(k+k'\right)^2, \quad \cos\Theta_s = \frac{\left(k-k'\right)_\mu \left(p-p'\right)^\mu}{\left(k-k'\right)^2}. \tag{8.94}$$

We further define the total cross section as the integral

$$\sigma(s) = \frac{2\pi}{\nu} \int d\Theta_s\, \sin\Theta_s\, \sigma(s,\Theta_s). \tag{8.95}$$

In order to calculate $\mathcal{A}_{rn}^{(\ell)}$ it is convenient to first define the tensors

$$X_{\mu\nu\gamma_1\cdots\gamma_m}^n = \frac{1}{\nu} \int dKdK'dPdP'\, W_{\mathbf{kk'}\to\mathbf{pp'}}\, f_{0\mathbf{k}} f_{0\mathbf{k'}} E_{\mathbf{k}}^n k_\mu k_\nu$$
$$\times \left(k_{\gamma_1}\cdots k_{\gamma_m} + k'_{\gamma_1}\cdots k'_{\gamma_m} - p_{\gamma_1}\cdots p_{\gamma_m} - p'_{\gamma_1}\cdots p'_{\gamma_m}\right). \tag{8.96}$$

The collision integrals $\mathcal{A}_{rn}^{(\ell)}$ can always be expressed as linear combinations of contractions/projections of $X_{\mu\nu\gamma_1\cdots\gamma_m}^n$. For the purpose of this appendix, we only need $X_{\mu\nu\gamma_1\cdots\gamma_m}^n$ for $m = 2$ and 3. For now, we concentrate on calculating these integrals. We separate $X_{\mu\nu\gamma_1\cdots\gamma_m}^n$ as

$$X_{\mu\nu\gamma_1\cdots\gamma_m}^n = A_{\mu\nu\gamma_1\cdots\gamma_m}^n + B_{\mu\nu\gamma_1\cdots\gamma_m}^n, \tag{8.97}$$

with

$$
A^n_{\mu\nu\gamma_1\cdots\gamma_m} = \frac{1}{\nu} \int dKdK' dPdP' W_{\mathbf{kk'}\to\mathbf{pp'}} f_{0\mathbf{k}} f_{0\mathbf{k'}} E^n_{\mathbf{k}} k_\mu k_\nu \left(k_{\gamma_1}\cdots k_{\gamma_m} + k'_{\gamma_1}\cdots k'_{\gamma_m} \right) ,
$$

$$
B^n_{\mu\nu\gamma_1\cdots\gamma_m} = -\frac{1}{\nu} \int dKdK' dPdP' W_{\mathbf{kk'}\to\mathbf{pp'}} f_{0\mathbf{k}} f_{0\mathbf{k'}} E^n_{\mathbf{k}} k_\mu k_\nu \left(p_{\gamma_1}\cdots p_{\gamma_m} + p'_{\gamma_1}\cdots p'_{\gamma_m} \right) .
$$

$$(8.98)$$

The $dPdP'$ integration in the first tensor, $A^n_{\mu\nu\gamma_1\cdots\gamma_m}$, can be immediately performed and written in terms of the total cross section, $\sigma(s)$, as

$$
A^n_{\mu\nu\gamma_1\cdots\gamma_m} = \int dKdK' f_{0\mathbf{k}} f_{0\mathbf{k'}} E^n_{\mathbf{k}} k_\mu k_\nu \left(k_{\gamma_1}\cdots k_{\gamma_m} + k'_{\gamma_1}\cdots k'_{\gamma_m} \right) \frac{s}{2} \sigma(s) . \quad (8.99)
$$

The calculation of the second tensor, $B^n_{\mu\nu\gamma_1\cdots\gamma_m}$, is cumbersome. First, we write it in the general form

$$
B^n_{\mu\nu\gamma_1\cdots\gamma_m} = -\int dKdK' f_{0\mathbf{k}} f_{0\mathbf{k'}} E^n_{\mathbf{k}} k_\mu k_\nu \Theta_{\gamma_1\cdots\gamma_m} , \quad (8.100)
$$

where we introduced the tensor

$$
\Theta^{\gamma_1\cdots\gamma_m} = \frac{2}{\nu} \int dPdP' W_{\mathbf{kk'}\to\mathbf{pp'}} p^{\gamma_1}\cdots p^{\gamma_m} . \quad (8.101)
$$

The integral $\Theta^{\gamma_1\cdots\gamma_m}$ is an m-th rank tensor. For isotropic cross sections, this tensor can only depend on the normalized total momentum of the collision $\tilde{P}^\mu_T \equiv s^{-1/2}\left(k^\mu + k'^\mu\right) \equiv s^{-1/2}P^\mu_T$. Thus, the tensor structure of $\Theta^{\gamma_1\cdots\gamma_m}$ must be constructed from combinations of $\tilde{P}^\mu_T$ and the projection operator orthogonal to $\tilde{P}^\mu_T$, $\Delta^{\mu\nu}_P = g^{\mu\nu} - \tilde{P}^\mu_T \tilde{P}^\nu_T$. In general,

$$
\Theta^{\gamma_1\cdots\gamma_m} = \sum_{q=0}^{[m/2]} (-1)^q a_{mq} C_{mq} C^{\gamma_1\cdots\gamma_m}_q , \quad (8.102)
$$

where we defined

$$
a_{mq} = \frac{m!}{(m-2q)!2q!} (2q-1)!! ,
$$

$$
C^{\gamma_1\cdots\gamma_m}_q = \Delta^{(\gamma_1\gamma_2}_P \cdots \Delta^{\gamma_{2q-1}\gamma_{2q}}_P \tilde{P}^{\gamma_{2q+1}}_T \cdots \tilde{P}^{\gamma_m)}_T ,
$$

$$
C_{mq} = \frac{2}{\nu(2q+1)!!} \int dPdP' W_{\mathbf{kk'}\to\mathbf{pp'}} \left(\tilde{P}^\mu_T p_\mu \right)^{m-2q} \left(-\Delta^{\alpha\beta}_P p_\alpha p_\beta \right)^q .
$$

$$(8.103)$$

As usual, the parentheses () around the indices denote the symmetrization of the tensor. For example,

$$\Theta_{\mu\nu} = C_{20}\tilde{P}_{T\mu}\tilde{P}_{T\nu} - C_{21}\Delta_{P\mu\nu} \,,$$

$$\Theta_{\mu\nu\lambda} = C_{30}\tilde{P}_{T\mu}\tilde{P}_{T\nu}\tilde{P}_{T\lambda} - C_{31}\left(\Delta_{P\mu\nu}\tilde{P}_{T\lambda} + \Delta_{P\mu\lambda}\tilde{P}_{T\nu} + \Delta_{P\nu\lambda}\tilde{P}_{T\mu}\right) \,. \quad (8.104)$$

The integrals C_{nq} are scalars and can be computed in any frame. It is most convenient to calculate them in the center-of-momentum frame, where, $\tilde{P}_T^\mu = (1,0,0,0)^T$ and $\Delta_P^{\mu\nu} = \text{diag}(0,-1,-1,-1)$. Then, it is straightforward to prove that

$$C_{nq} = \frac{\sigma(s)}{2^n(2q+1)!!}s^{(n-2q+1)/2}\left(s-4m^2\right)^{(2q+1)/2} \xrightarrow[m\to 0]{} \frac{\sigma(s)}{2^n(2q+1)!!}s^{(n+2)/2} \,. \quad (8.105)$$

In the massless limit, the tensors $X_{\mu\nu\alpha\beta}^n$ and $X_{\mu\nu\alpha\beta\gamma}^n$ become

$$X_{\mu\nu\alpha\beta}^n = \frac{1}{2}\int dKdK'\, f_{0\mathbf{k}}f_{0\mathbf{k}'}E_{\mathbf{k}}^n k_\mu k_\nu\, s\sigma(s)\left(k_\alpha k_\beta + k_\alpha'k_\beta' - \frac{2}{3}P_{T\alpha}P_{T\beta} + \frac{s}{6}g_{\alpha\beta}\right) \,,$$

$$X_{\mu\nu\alpha\beta\gamma}^n = \frac{1}{2}\int dKdK'\, f_{0\mathbf{k}}f_{0\mathbf{k}'}E_{\mathbf{k}}^n k_\mu k_\nu\, s\sigma(s)\Big[k_\alpha k_\beta k_\gamma + k_\alpha'k_\beta'k_\gamma'$$
$$- \frac{1}{2}P_{T\alpha}P_{T\beta}P_{T\gamma} + \frac{s}{12}\left(g_{\alpha\beta}P_{T\gamma} + g_{\alpha\gamma}P_{T\beta} + g_{\beta\gamma}P_{T\alpha}\right)\Big] \,, \quad (8.106)$$

where we used that, in the massless limit, $s = 2k^\lambda k_\lambda'$.

8.10.1 Particle-Diffusion Current

For the collision integrals related to the particle-number diffusion current, we need the following two contractions

$$\Delta^{\mu\alpha}u^\nu u^\beta X_{\mu\nu\alpha\beta}^n = -\sigma\left(I_{10}I_{n+5,1} - 4I_{21}I_{n+4,1} - I_{31}I_{n+3,1}\right) \,,$$

$$\Delta^{\mu\alpha}u^\nu u^\beta u^\gamma X_{\mu\nu\alpha\beta\gamma}^n = -\frac{\sigma}{2}\left(3I_{10}I_{n+6,1} - 11I_{21}I_{n+5,1} - 5I_{31}I_{n+4,1} - 3I_{41}I_{n+3,1}\right) \,. \quad (8.107)$$

To obtain the above relations, we used the definitions (6.30) and Eq. (6.45). In the massless and classical limits, the integrals $I_{nq} = J_{nq}$ can be calculated analytically

$$I_{nq} = d_{\text{dof}}\frac{e^{\alpha_0}}{(2q+1)!!}\frac{1}{2\pi^2}\frac{(n+1)!}{\beta_0^{n+2}} = \frac{(n+1)!}{(2q+1)!!}\frac{P_0}{2\beta_0^{n-2}} \,. \quad (8.108)$$

Then,

$$\Delta^{\mu\alpha} u^{\nu} u^{\beta} X^{-2}_{\mu\nu\alpha\beta} = \frac{4}{3} n_0 \sigma \frac{P_0}{\beta_0} ,$$

$$\Delta^{\mu\alpha} u^{\nu} u^{\beta} X^{0}_{\mu\nu\alpha\beta} = -24 n_0 \sigma \frac{P_0}{\beta_0^3} ,$$

$$\Delta^{\mu\alpha} u^{\nu} u^{\beta} u^{\gamma} X^{-2}_{\mu\nu\alpha\beta\gamma} = 12 n_0 \sigma \frac{P_0}{\beta_0^2} ,$$

$$\Delta^{\mu\alpha} u^{\nu} u^{\beta} u^{\gamma} X^{0}_{\mu\nu\alpha\beta\gamma} = -280 n_0 \sigma \frac{P_0}{\beta_0^4} . \tag{8.109}$$

As a consistency check, we confirm that $\Delta^{\mu\alpha} u^{\nu} u^{\beta} X^{-1}_{\mu\nu\alpha\beta} = \Delta^{\mu\alpha} u^{\nu} u^{\beta} u^{\gamma} X^{-1}_{\mu\nu\alpha\beta\gamma} = 0$.

The components of $\mathcal{A}^{(1)}$ change according to the number of moments included. In the 14-moment approximation, using Eqs. (6.5) and (6.13), we obtain

$$\mathcal{A}^{(1)}_{00} = \frac{W^{(1)}}{3} a^{(1)}_{10} a^{(1)}_{11} \Delta^{\mu\alpha} u^{\nu} u^{\beta} X^{-2}_{\mu\nu\alpha\beta} = \frac{4}{9} n_0 \sigma . \tag{8.110}$$

In the 23-moment approximation, e.g., considering three polynomials in the expansion (8.1) for $\ell = 1$

$$\mathcal{A}^{(1)}_{r0} = \frac{W^{(1)}}{3} \left[\left(a^{(1)}_{10} a^{(1)}_{11} + a^{(1)}_{20} a^{(1)}_{21} \right) \Delta^{\mu\alpha} u^{\nu} u^{\beta} X^{r-2}_{\mu\nu\alpha\beta} + a^{(1)}_{20} a^{(1)}_{22} \Delta^{\mu\alpha} u^{\nu} u^{\beta} u^{\gamma} X^{r-2}_{\mu\nu\alpha\beta\gamma} \right] ,$$

$$\mathcal{A}^{(1)}_{r2} = \frac{W^{(1)}}{3} \left(a^{(1)}_{22} a^{(1)}_{21} \Delta^{\mu\alpha} u^{\nu} u^{\beta} X^{r-2}_{\mu\nu\alpha\beta} + a^{(1)}_{22} a^{(1)}_{22} \Delta^{\mu\alpha} u^{\nu} u^{\beta} u^{\gamma} X^{r-2}_{\mu\nu\alpha\beta\gamma} \right) . \tag{8.111}$$

Then, using the results from Appendix 6.8 for the coefficients $a^{(\ell)}_{nq}$ together with Eqs. (8.108) and (8.109), we obtain

$$\mathcal{A}^{(1)}_{00} = \frac{2}{3} n_0 \sigma , \quad \mathcal{A}^{(1)}_{02} = \frac{\beta_0^2}{90} n_0 \sigma ,$$

$$\mathcal{A}^{(1)}_{20} = -\frac{4}{3\beta_0^2} n_0 \sigma , \quad \mathcal{A}^{(1)}_{22} = \frac{1}{3} n_0 \sigma . \tag{8.112}$$

8.10.2 Shear-Stress Tensor

For the collision integrals related to the shear-stress tensor, we need the following two contractions

$$\Delta^{\mu\nu\alpha\beta} X^{n}_{\mu\nu\alpha\beta} = \frac{10}{3} \sigma \left(I_{10} I_{n+5,2} + 4 I_{21} I_{n+4,2} \right) ,$$

$$\Delta^{\mu\nu\alpha\beta} u^{\gamma} X^{n}_{\mu\nu\alpha\beta\gamma} = 5 \sigma \left(I_{10} I_{n+6,2} - I_{21} I_{n+5,2} + 2 I_{31} I_{n+4,2} \right) . \tag{8.113}$$

In order to obtain the above relations, we used the definitions (6.30) and Eq. (6.45). Using Eq. (8.108)

$$\Delta^{\mu\nu\alpha\beta} X^{-1}_{\mu\nu\alpha\beta} = 24\sigma \frac{P_0^2}{\beta_0} ,$$

$$\Delta^{\mu\nu\alpha\beta} X^0_{\mu\nu\alpha\beta} = \frac{400}{3}\sigma \frac{P_0^2}{\beta_0^2} ,$$

$$\Delta^{\mu\nu\alpha\beta} u^\gamma X^{-1}_{\mu\nu\alpha\beta\gamma} = 132\sigma \frac{P_0^2}{\beta_0^2} ,$$

$$\Delta^{\mu\nu\alpha\beta} u^\gamma X^0_{\mu\nu\alpha\beta\gamma} = 880\sigma \frac{P_0^2}{\beta_0^3} . \tag{8.114}$$

The components of $\mathcal{A}^{(2)}$ change according to the number of moments included. In the 14-moment approximation, using Eqs. (6.13) and (6.5), we obtain

$$\mathcal{A}^{(2)}_{00} = \frac{W^{(2)}}{10} \Delta^{\mu\nu\alpha\beta} X^{-1}_{\mu\nu\alpha\beta} = \frac{3}{5} n_0 \sigma , \tag{8.115}$$

where we used Eqs. (8.108) and (8.114), together with the results from Appendix 6.8.

In the 23-moment approximation, e.g., considering two polynomials in the expansion (8.1), for $\ell = 2$

$$\mathcal{A}^{(2)}_{r0} = \frac{W^{(2)}}{10} \left(1 + a^{(2)}_{10} a^{(2)}_{10} \right) \Delta^{\mu\nu\alpha\beta} X^{r-1}_{\mu\nu\alpha\beta} + \frac{W^{(2)}}{10} a^{(2)}_{10} a^{(2)}_{11} \Delta^{\mu\nu\alpha\beta} u^\gamma X^{r-1}_{\mu\nu\alpha\beta\gamma} ,$$

$$\mathcal{A}^{(2)}_{r1} = \frac{W^{(2)}}{10} a^{(2)}_{11} a^{(2)}_{10} \Delta^{\mu\nu\alpha\beta} X^{r-1}_{\mu\nu\alpha\beta} + \frac{W^{(2)}}{10} a^{(2)}_{11} a^{(2)}_{11} \Delta^{\mu\nu\alpha\beta} u^\gamma X^{r-1}_{\mu\nu\alpha\beta\gamma} . \tag{8.116}$$

Then, using once more the results from Appendix 6.8 and Eqs. (8.108) and (8.114), we obtain

$$\mathcal{A}^{(2)}_{00} = \frac{9}{10} n_0 \sigma , \quad \mathcal{A}^{(2)}_{01} = -\frac{1}{20} \beta_0 n_0 \sigma ,$$

$$\mathcal{A}^{(2)}_{10} = \frac{4}{3\beta_0} n_0 \sigma , \quad \mathcal{A}^{(2)}_{11} = \frac{1}{3} n_0 \sigma . \tag{8.117}$$

We did not calculate the coefficients related to the bulk-viscous pressure, since this quantity vanishes in the massless limit. Also, if the mass was taken to be finite, some of the steps taken in this appendix would not be possible.

8.11   Appendix 3: Calculation of $\gamma_1^{(2)}$

In this appendix, we compute the quantity $\gamma_1^{(2)}$ in the 14-moment approximation and the 23-moment approximation. Among all the $\gamma_i^{(\ell)}$ appearing in the transport coefficients listed in Appendix 8.9 this is the only one that survives in the ultrarelativistic limit, all the others are accompanied by factors of m^2 or couple to Π, which vanishes in this limit. The variable $\gamma_1^{(2)}$ was defined in the main text

$$\gamma_1^{(2)} = \sum_{n=0}^{N_2} \mathcal{F}_{rn}^{(2)} \Omega_{n0}^{(2)} . \tag{8.118}$$

The first step is to compute the thermodynamic integral

$$\mathcal{F}_{rn}^{(\ell)} = \frac{\ell!}{(2\ell+1)!!} \int dK \, f_{0\mathbf{k}} \tilde{f}_{0\mathbf{k}} E_{\mathbf{k}}^{-r} \mathcal{H}_{\mathbf{k}n}^{(\ell)} \left(\Delta^{\alpha\beta} k_\alpha k_\beta \right)^\ell . \tag{8.119}$$

8.11.1   14-Moment Approximation

In this case, $N_1 = 1$ and $N_2 = 0$, and

$$\gamma_1^{(2)} = \mathcal{F}_{10}^{(2)} . \tag{8.120}$$

Also, in the 14-moment approximation

$$\mathcal{H}_{\mathbf{k}0}^{(2)} \equiv \frac{W^{(2)}}{2!} a_{00}^{(2)} P_{\mathbf{k}0}^{(2)} = \frac{W^{(2)}}{2!} . \tag{8.121}$$

In the massless and classical limit

$$\mathcal{H}_{\mathbf{k}0}^{(2)} = \frac{\beta_0^2}{8P_0} , \tag{8.122}$$

and finally

$$\gamma_1^{(2)} = \frac{\beta_0^2}{4P_0} \frac{1}{5!!} \int dK \, f_{0\mathbf{k}} E_{\mathbf{k}}^{-1} \left(\Delta^{\alpha\beta} k_\alpha k_\beta \right)^2 = \frac{\beta_0}{5} . \tag{8.123}$$

8.11.2 23-Moment Approximation

In this case, $N_1 = 2$ and $N_2 = 1$, and

$$\gamma_1^{(2)} = \mathcal{F}_{10}^{(2)} + \Omega_{10}^{(2)} \mathcal{F}_{11}^{(2)} . \tag{8.124}$$

Also, in the 23-moment approximation

$$
\begin{aligned}
\mathcal{H}_{k0}^{(2)} &= \frac{W^{(2)}}{2!} \left(1 + a_{10}^{(2)} P_{k1}^{(2)} \right) = \frac{W^{(2)}}{2!} \left[1 + \left(a_{10}^{(2)} \right)^2 + a_{10}^{(2)} a_{11}^{(2)} E_k \right] , \\
\mathcal{H}_{k1}^{(2)} &= \frac{W^{(2)}}{2!} a_{11}^{(2)} P_{k1}^{(2)} = \frac{W_{(2)}}{2!} \left[a_{10}^{(2)} a_{11}^{(2)} + \left(a_{11}^{(2)} \right)^2 E_k \right] .
\end{aligned} \tag{8.125}
$$

We know that

$$W^{(2)} = \frac{\beta_0^2}{4P_0} , \quad \left(a_{11}^{(2)} \right)^2 = \frac{\beta_0^2}{6} , \quad \frac{a_{10}^{(2)}}{a_{11}^{(2)}} = -\frac{6}{\beta_0} . \tag{8.126}$$

Thus,

$$
\begin{aligned}
\mathcal{H}_{k0}^{(2)} &= \frac{\beta_0^2}{8P_0} \left(7 - \beta_0 E_k \right) , \\
\mathcal{H}_{k1}^{(2)} &= \frac{\beta_0^3}{8P_0} \left(-1 + \frac{1}{6} \beta_0 E_k \right) ,
\end{aligned} \tag{8.127}
$$

and

$$
\begin{aligned}
\mathcal{F}_{10}^{(2)} &= \frac{\beta_0^2}{4P_0} \frac{1}{5!!} \int dK \, f_{0k} E_k^{-1} \left(7 - \beta_0 E_k \right) \left(\Delta^{\alpha\beta} k_\alpha k_\beta \right)^2 = \frac{2}{5} \beta_0 , \\
\mathcal{F}_{11}^{(2)} &= \frac{\beta_0^3}{4P_0} \frac{1}{5!!} \int dK \, f_{0k} E_k^{-1} \left(-1 + \frac{1}{6} \beta_0 E_k \right) \left(\Delta^{\alpha\beta} k_\alpha k_\beta \right)^2 = -\frac{\beta_0^2}{30} .
\end{aligned} \tag{8.128}
$$

Substituting $\Omega^{(2)}$ from Eq. (8.50), we obtain

$$\gamma_1^{(2)} = \frac{2}{15} \beta_0 = 0.133 \beta_0 . \tag{8.129}$$

8.12   Appendix 4: Transport Coefficients in Sect. 8.6

In this appendix, we list all transport coefficients appearing in the extension of fluid dynamics discussed in Sect. 8.6. The microscopic formulas for the diffusion and viscosity coefficients, $\vec{\varkappa}$ and $\vec{\eta}$, and for the relaxation-time matrices, $\hat{\tau}_n$ and $\hat{\tau}_\pi$, are

$$
\vec{\varkappa} = \sum_{k=0,\neq 1}^{N_1} \alpha_k^{(1)} \begin{pmatrix} \tau_{0k}^{(1)} \\ \tau_{2k}^{(1)} \end{pmatrix} , \quad \vec{\eta} = \sum_{k=0}^{N_2} \alpha_k^{(2)} \begin{pmatrix} \tau_{0k}^{(2)} \\ \tau_{1k}^{(2)} \end{pmatrix} , \tag{8.130}
$$

$$
\hat{\tau}_n = \sum_{r=0,\neq 1}^{N_1} \begin{pmatrix} \tau_{0r}^{(1)}\lambda_{r0}^{(1)} & \tau_{0r}^{(1)}\lambda_{r2}^{(1)} \\ \tau_{2r}^{(1)}\lambda_{r0}^{(1)} & \tau_{2r}^{(1)}\lambda_{r2}^{(1)} \end{pmatrix} , \quad \hat{\tau}_\pi = \sum_{r=0}^{N_2} \begin{pmatrix} \tau_{0r}^{(2)}\lambda_{r0}^{(2)} & \tau_{0r}^{(2)}\lambda_{r1}^{(2)} \\ \tau_{1r}^{(2)}\lambda_{r0}^{(2)} & \tau_{1r}^{(2)}\lambda_{r1}^{(2)} \end{pmatrix} . \tag{8.131}
$$

The transport coefficients of the nonlinear terms in the equation of motion for $\vec{n}^{\mu}$ are

$$
\hat{\delta}_{nn} = \frac{1}{3} \sum_{r=0,\neq 1}^{N_1} \begin{pmatrix} 3\tau_{0r}^{(1)}\lambda_{r0}^{(1)} & 5\tau_{0r}^{(1)}\lambda_{r2}^{(1)} \\ 3\tau_{2r}^{(1)}\lambda_{r0}^{(1)} & 5\tau_{2r}^{(1)}\lambda_{r2}^{(1)} \end{pmatrix} , \tag{8.132}
$$

$$
\hat{\lambda}_{nn} = \frac{1}{5} \sum_{r=0,\neq 1}^{N_1} (2r+3) \begin{pmatrix} \tau_{0r}^{(1)}\lambda_{r0}^{(1)} & \tau_{0r}^{(1)}\lambda_{r2}^{(1)} \\ \tau_{2r}^{(1)}\lambda_{r0}^{(1)} & \tau_{2r}^{(1)}\lambda_{r2}^{(1)} \end{pmatrix} , \tag{8.133}
$$

$$
\hat{\lambda}_{n\pi} = \frac{1}{4} \left[\sum_{r=0}^{N_2} \begin{pmatrix} \tau_{00}^{(1)}\mathcal{F}_{1r}^{(2)}\lambda_{r0}^{(2)} & 2\tau_{00}^{(1)}\mathcal{F}_{1r}^{(2)}\lambda_{r1}^{(2)} \\ \tau_{20}^{(1)}\mathcal{F}_{1r}^{(2)}\lambda_{r0}^{(2)} & 2\tau_{20}^{(1)}\mathcal{F}_{1r}^{(2)}\lambda_{r1}^{(2)} \end{pmatrix} \right.
$$
$$
\left. + \sum_{r=2}^{N_1} \begin{pmatrix} (1-r)\,\tau_{0r}^{(1)}\lambda_{r-1,0}^{(2)} & (2-r)\,\tau_{0r}^{(1)}\lambda_{r-1,1}^{(2)} \\ (1-r)\,\tau_{2r}^{(1)}\lambda_{r-1,0}^{(2)} & (2-r)\,\tau_{2r}^{(1)}\lambda_{r-1,1}^{(2)} \end{pmatrix} \right] , \tag{8.134}
$$

$$
\hat{\tau}_{n\pi} = -4P_0 \left[\sum_{r=2}^{N_1} \begin{pmatrix} 0 & \tau_{0r}^{(1)}\lambda_{r-1,1}^{(2)} \\ 0 & \tau_{2r}^{(1)}\lambda_{r-1,1}^{(2)} \end{pmatrix} + \sum_{r=0}^{N_2} \begin{pmatrix} 0 & \tau_{00}^{(1)}\mathcal{F}_{1r}^{(2)}\lambda_{r1}^{(2)} \\ 0 & \tau_{20}^{(1)}\mathcal{F}_{1r}^{(2)}\lambda_{r1}^{(2)} \end{pmatrix} \right] , \tag{8.135}
$$

$$
\hat{\ell}_{n\pi} = -\sum_{r=0}^{N_2} \begin{pmatrix} \tau_{00}^{(1)}\mathcal{F}_{1r}^{(2)}\lambda_{r0}^{(2)} & \tau_{00}^{(1)}\mathcal{F}_{1r}^{(2)}\lambda_{r1}^{(2)} \\ \tau_{20}^{(1)}\mathcal{F}_{1r}^{(2)}\lambda_{r0}^{(2)} & \tau_{20}^{(1)}\mathcal{F}_{1r}^{(2)}\lambda_{r1}^{(2)} \end{pmatrix} + \frac{\beta_0}{4P_0} \sum_{r=0,\neq 1}^{N_1} \begin{pmatrix} \tau_{0r}^{(1)} I_{r+2,1} & 0 \\ \tau_{2r}^{(1)} I_{r+2,1} & 0 \end{pmatrix}
$$
$$
- \sum_{r=2}^{N_1} \begin{pmatrix} \tau_{0r}^{(1)}\lambda_{r-1,0}^{(2)} & \tau_{0r}^{(1)}\lambda_{r-1,1}^{(2)} \\ \tau_{2r}^{(1)}\lambda_{r-1,0}^{(2)} & \tau_{2r}^{(1)}\lambda_{r-1,1}^{(2)} \end{pmatrix} , \tag{8.136}
$$

while those in the equation of motion for $\vec{\pi}^{\,\mu\nu}$ are

$$\hat{\delta}_{\pi\pi} = \frac{1}{3}\sum_{r=0}^{N_2}\begin{pmatrix} 4\tau_{0r}^{(2)}\lambda_{r0}^{(2)} & 5\tau_{0r}^{(2)}\lambda_{r1}^{(2)} \\ 4\tau_{1r}^{(2)}\lambda_{r0}^{(2)} & 5\tau_{1r}^{(2)}\lambda_{r1}^{(2)} \end{pmatrix}, \tag{8.137}$$

$$\hat{\tau}_{\pi\pi} = \frac{2}{7}\sum_{r=0}^{N_2}(2r+5)\begin{pmatrix} \tau_{0r}^{(2)}\lambda_{r0}^{(2)} & \tau_{0r}^{(2)}\lambda_{r1}^{(2)} \\ \tau_{1r}^{(2)}\lambda_{r0}^{(2)} & \tau_{1r}^{(2)}\lambda_{r1}^{(2)} \end{pmatrix}, \tag{8.138}$$

$$\hat{\tau}_{\pi n} = \frac{1}{5P_0}\sum_{r=1}^{N_2}\begin{pmatrix} 2\tau_{0r}^{(2)}\lambda_{r+1,0}^{(1)} & 3\tau_{0r}^{(2)}\lambda_{r+1,2}^{(1)} \\ 2\tau_{1r}^{(2)}\lambda_{r+1,0}^{(1)} & 3\tau_{1r}^{(2)}\lambda_{r+1,2}^{(1)} \end{pmatrix}, \tag{8.139}$$

$$\hat{\ell}_{\pi n} = \frac{2}{5}\sum_{r=1}^{N_2}\begin{pmatrix} \tau_{0r}^{(2)}\lambda_{r+1,0}^{(1)} & \tau_{0r}^{(2)}\lambda_{r+1,2}^{(1)} \\ \tau_{1r}^{(2)}\lambda_{r+1,0}^{(1)} & \tau_{1r}^{(2)}\lambda_{r+1,2}^{(1)} \end{pmatrix}, \tag{8.140}$$

$$\hat{\lambda}_{\pi n} = -\frac{1}{10}\sum_{r=2}^{N_2}\begin{pmatrix} (1+r)\,\tau_{0r}^{(2)}\lambda_{r+1,0}^{(1)} & \tau_{0r}^{(2)}\,(r-1)\,\lambda_{r+1,2}^{(1)} \\ (1+r)\,\tau_{1r}^{(2)}\lambda_{r+1,0}^{(1)} & \tau_{1r}^{(2)}\,(r-1)\,\lambda_{r+1,2}^{(1)} \end{pmatrix}. \tag{8.141}$$

References

1. Chapman, S., Cowling, T.G.: The Mathematical Theory of Non-Uniform Gases. 3rd edn. Cambridge University Press, New York (1974)
2. Israel, W.: Ann. Phys. (N.Y.) **100**, 310 (1976)
3. Stewart, J.M.: Proc. Roy. Soc. A **357**, 59 (1977)
4. Israel, W., Stewart, J.M.: Ann. Phys. (N.Y.) **118**, 341 (1979)
5. Hiscock, W., Lindblom, L.: Phys. Rev. D **31**, 725 (1985)
6. Hiscock, W., Lindblom, L.: Ann. Phys. (N.Y.) **151**, 466 (1983)
7. Hiscock, W., Lindblom, L.: Phys. Rev. D **35**, 3723 (1987)
8. de Groot, S.R., van Leeuwen, W.A., van Weert, Ch.G.: Relativistic Kinetic Theory—Principles and Applications. North-Holland (1980)
9. Denicol, G.S., Kodama, T., Koide, T., Mota, P.: J. Phys. G **35**, 115102 (2008)
10. Pu, S., Koide, T., Rischke, D.H.: Phys. Rev. D **81**, 114039 (2010)
11. Grad, H.: Commun. Pure Appl. Math. **2**, 331 (1949)
12. Grad, H.: Comm. Pure Appl. Math. **2**, 325 (1949)
13. Denicol, G.S., Noronha, J., Niemi, H., Rischke, D.H.: Phys. Rev. D **83**, 074019 (2011)
14. Reinecke, S., Kremer, G.M.: Phys. Rev. A **42**, 815–820 (1990)
15. Reinecke, S., Kremer, G.M.: Continuum Mech. Thermodyn. **8**, 121–130 (1996)
16. Struchtrup, H.: IMA Vol. Math Appl. **735**. Springer (2004)
17. Karlin, I.V., Gorban, A.N.: Ann. Phys. (Leipzig) **11**, 783 (2002)
18. Grad, H.: Phys. Fluids **6**, 147 (1963)
19. Struchtrup, H., Taheri, P.: IMA J. Appl. Math. (published online 2011)
20. see for example M. Torrilhon, : Cont. Mech. Thermodyn. **21**, 341 (2009). and references therein
21. Struchtrup, H., Torrilhon, M.: Phys. Fluids **15/9**, 2668 (2003)
22. Torrilhon, M., Struchtrup, H.: J. Fluid Mech. **513**, 171 (2004)
23. Struchtrup, H., Torrilhon, M.: Phys. Rev. Lett. **99**, 014502 (2007)
24. Chernikov, N.A.: Phys. Lett. **5**, 115 (1963)
25. Chernikov, N.A.: Acta Phys. Pol. **27**, 465 (1965)
26. Vignon, B.: Ann. Inst. H. Poincare **10**, 31 (1969)
27. Marle, C.: Ann. Inst. H. Poincare **10**, 127 (1969)

28. Kranys, M.: Phys. Lett. A **33**, 77 (1970)
29. Kranys, M.: Nuovo Cim. B **8**, 417 (1972)
30. Huovinen, P., Molnar, D.: Phys. Rev. C **79**, 014906 (2009)
31. El, A., Xu, Z., Greiner, C.: Phys. Rev. C **81**, 041901 (2010)
32. Bouras, I., Molnar, E., Niemi, H., Xu, Z., El, A., Fochler, O., Greiner, C., Rischke, D.H.: Phys. Rev. C **82**, 024910 (2010). [arXiv:1006.0387 [hep-ph]]
33. Denicol, G.S., Koide, T., Rischke, D.H.: Phys. Rev. Lett. **105**, 162501 (2010)
34. Betz, B., Henkel, D., Rischke, D.H.: Prog. Part. Nucl. Phys. **62**, 556–561 (2009)
35. Betz, B., Henkel, D., Rischke, D.H.: J. Phys. G **36**, 064029 (2009)
36. Betz, B., Denicol, G.S., Koide, T., Molnar, E., Niemi, H., Rischke, D.H.: EPJ Web Conf. **13**, 07005 (2011)
37. Denicol, G.S., Niemi, H., Molnar, E., Rischke, D.H.: To be published in Phys. Rev. D. arXiv:1202.4551 [nucl-th]
38. Landau, L.D., Lifshitz, E.M.: Fluid Mechanics. Pergamon, New York (1959)
39. Denicol, G.S., Molnár, E., Niemi, H., Rischke, D.H.: Eur. Phys. J. A **48**, 170 (2012). https://doi.org/10.1140/epja/i2012-12170-x, arXiv:1206.1554 [nucl-th]
40. Molnar, E., Niemi, H., Denicol, G.S., Rischke, D.H.: Phys. Rev. D **89**(7), 074010 (2014). https://doi.org/10.1103/PhysRevD.89.074010, arXiv:1308.0785 [nucl-th]
41. Denicol, G.S., Huang, X.G., Koide, T., Rischke, D.H.: Phys. Lett. B **708**, 174–178 (2012). https://doi.org/10.1016/j.physletb.2012.01.018, arXiv:1003.0780 [hep-th]
42. Denicol, G.S., Niemi, H., Bouras, I., Molnar, E., Xu, Z., Rischke, D.H., Greiner, C.: Phys. Rev. D **89**(7), 074005 (2014). https://doi.org/10.1103/PhysRevD.89.074005, arXiv:1207.6811 [nucl-th]
43. Bouras, I., Molnar, E., Niemi, H., Xu, Z., El, A., Fochler, O., Greiner, C., Rischke, D.H.: Phys. Rev. Lett. **103**, 032301 (2009). https://doi.org/10.1103/PhysRevLett.103.032301, arXiv:0902.1927 [hep-ph]
44. Cercignani, C., Kremer, G.M.: The Relativistic Boltzmann Equation: Theory and Applications. Birkhauser, Basel (2002)

Method of Moments: Anisotropic Reference State

In Chap. 6, the single-particle distribution function $f_\mathbf{k}$ has been expanded around the distribution function $f_{0\mathbf{k}}$ in local thermodynamical equilibrium. The range of applicability of this expansion is (at least formally) restricted to situations where the deviation $\delta f_\mathbf{k} \equiv f_\mathbf{k} - f_{0\mathbf{k}}$ is small, $|\delta f_\mathbf{k}| \ll f_{0\mathbf{k}}$. The deviation $\delta f_\mathbf{k}$ is defined as a power series in the rest-frame energy $E_\mathbf{k}$, the irreducible tensors in momentum space, $k^{\langle \mu_1} \ldots k^{\mu_\ell \rangle}$, and the irreducible moments $\rho_{\mu_1 \ldots \mu_\ell}$ of $\delta f_\mathbf{k}$ which, through their equations of motion, are proportional to powers of gradients of fluid-dynamical variables. Thus, $\delta f_\mathbf{k}$ is formally given by an expansion in powers of the Knudsen number. There are, however, situations where gradients or the Knudsen number, and thus $\delta f_\mathbf{k}$, become so large that a power-series expansion is expected to break down. One of these situations is, e.g., the initial stage of ultrarelativistic heavy-ion collisions. In this case, the gradient of the fluid velocity in beam (z-) direction is of the order of the inverse lifetime of the system, $\partial_z v_z \sim 1/t$ [1], which can, in principle, become arbitrarily large as one approaches the moment of impact of the colliding nuclei (at $t = 0$). This large gradient is reflected in a single-particle distribution function which is highly anisotropic in z-direction in momentum space. In this case, one should modify the expansion, explicitly taking into account deviations from local thermodynamical equilibrium to all orders.

Generalizing the discussion of Chap. 6, one should take an anisotropic single-particle distribution function, called $\hat{f}_{0\mathbf{k}}$ in the following, as starting point for an expansion of the general single-particle distribution function, i.e.

$$f_\mathbf{k} \equiv \hat{f}_{0\mathbf{k}} + \delta \hat{f}_\mathbf{k} \,. \tag{9.1}$$

While this looks similar to Eq. (6.1), the rationale behind an expansion around $\hat{f}_{0\mathbf{k}}$ instead of around $f_{0\mathbf{k}}$ as in Eq. (6.1) is the following: in the case of a pronounced anisotropy, $\delta f_\mathbf{k}$ in Eq. (6.1) may be of similar magnitude (or even larger) than $f_{0\mathbf{k}}$, i.e., an expansion around the local-equilibrium distribution $f_{0\mathbf{k}}$ converges badly. However, taking a suitably chosen $\hat{f}_{0\mathbf{k}}$, we ensure that $|\delta \hat{f}_\mathbf{k}| \ll |\delta f_\mathbf{k}|$, so that the convergence properties of the series expansion are vastly improved.

© Springer Nature Switzerland AG 2021
G. S. Denicol and D. H. Rischke, *Microscopic Foundations of Relativistic Fluid Dynamics*,
Lecture Notes in Physics 990, https://doi.org/10.1007/978-3-030-82077-0_9

In the isotropic case, one uses a set of orthogonal polynomials in energy

$$E_{\mathbf{k}u} \equiv k^{\mu} u_{\mu} , \tag{9.2}$$

and irreducible tensors in momentum, $1, k^{\langle \mu \rangle}, k^{\langle \mu} k^{\nu \rangle}, \ldots$, in the expansion of $\delta f_{\mathbf{k}}$. (In this chapter, the energy variable will be denoted by $E_{\mathbf{k}u}$ instead of $E_{\mathbf{k}}$, in order to distinguish it from the momentum variable $E_{\mathbf{k}l}$; see Eq. (9.3) below.) However, in the case that the anisotropy singles out a certain direction in space, besides u^{μ} there is an additional space-like 4-vector, l^{μ}, which defines the direction of the anisotropy (in the case of heavy-ion collisions usually taken to be the z-direction) and, since it must be space-like, can be chosen to be orthogonal to the time-like fluid 4-velocity vector u^{μ}, $l^{\mu} u_{\mu} = 0$. In place of $\delta f_{\mathbf{k}}$ in Eq. (6.1), one now needs to expand $\delta \hat{f}_{\mathbf{k}}$ in Eq. (9.1). This expansion involves orthogonal polynomials in the *two* variables $E_{\mathbf{k}u}$ and the particle momentum in the direction of the anisotropy

$$E_{\mathbf{k}l} \equiv -k^{\mu} l_{\mu} , \tag{9.3}$$

as well as irreducible tensors which are orthogonal to *both u^{μ} and l^{μ}*. The derivation of the equations of motion for the irreducible moments of $\delta \hat{f}_{\mathbf{k}}$ is the main goal of the present chapter. In this way, we provide a starting point for a systematically improvable framework for anisotropic dissipative fluid dynamics.

This chapter is organized as follows. In Sect. 9.1, we introduce the tensor decomposition of fluid-dynamical variables with respect to the time-like fluid 4-velocity u^{μ} and the space-like 4-vector l^{μ} ($l^{\mu} l_{\mu} = -1$), which is usually chosen to point into the direction of the spatial anisotropy. In Sect. 9.2, we study the limiting case where the single-particle distribution function is a given function of l^{μ} and u^{μ} and, consequently, where only tensor structures proportional to u^{μ}, l^{μ}, their direct product, and the two-space projector orthogonal to both u^{μ} and l^{μ} [2–5],

$$\varXi^{\mu\nu} \equiv \Delta^{\mu\nu} + l^{\mu} l^{\nu} = g^{\mu\nu} - u^{\mu} u^{\nu} + l^{\mu} l^{\nu} , \tag{9.4}$$

appear in the moments of the single-particle distribution function. (A further possibility would be the rank-two tensor $\epsilon^{\mu\nu\alpha\beta} u_{\alpha} l_{\beta}$, but this is not considered in the following.) In Sect. 9.3, we present the general expansion of the single-particle distribution function $f_{\mathbf{k}}$ around the anisotropic state $\hat{f}_{0\mathbf{k}}$. In analogy to Refs. [6,7], this is done in terms of an orthogonal basis of irreducible tensors in momentum space. However, in contrast to Chap. 6, these tensors are not only orthogonal to u^{μ}, but also to l^{μ}. Then, in Sect. 9.4, taking moments of the Boltzmann equation we derive the equations of motion for the irreducible moments of the single-particle distribution function up to tensor-rank two. These equations are not yet closed and need to be truncated in order to derive the fluid-dynamical equations of motion in terms of conserved quantities, i.e., the particle 4-current N^{μ} and the energy-momentum tensor $T^{\mu\nu}$. In Sect. 9.5, we study the explicit form of the collision integral. Section 9.6 concludes this chapter with a summary. The material of this chapter is based on Ref. [8].

9.1 Fluid-Dynamical Variables

In this section, we introduce the tensor decomposition of the fluid-dynamical variables with respect to the time-like fluid 4-velocity u^μ and the space-like 4-vector l^μ.

As usual, the velocity of fluid-dynamical flow is specified in terms of the time-like 4-vector $u^\mu = \gamma\,(1, v^x, v^y, v^z)^T$, which is taken to be normalized, $u^\mu u_\mu = 1$. In order to specify the direction of a possible anisotropy in a given system, we define a space-like 4-vector l^μ

$$l^\mu l_\mu = -l^2 \,, \tag{9.5}$$

where $l^2 > 0$ characterizes the strength of the anisotropy. Furthermore, l^μ is taken to be orthogonal to the 4-flow velocity

$$u^\mu l_\mu = 0 \,. \tag{9.6}$$

In the context of heavy-ion collisions, this 4-vector would be chosen to point into the direction of the beam axis (usually the z-axis).

In general, a space-like 4-vector can be written in the form

$$l^\mu\,(t, \mathbf{x}) = l\,\gamma_l\,\left(1, \ell^x, \ell^y, \ell^z\right)^T \,, \tag{9.7}$$

where $\gamma_l \equiv (\ell^2 - 1)^{-1/2}$ follows from the normalization condition (9.5). If one is only interested in the direction of the anisotropy, and not its magnitude, l^μ can be normalized to one, i.e., $l \equiv 1$. The orthogonality of the normalized l^μ to the flow velocity (9.6) gives the constraint

$$u^\mu l_\mu \equiv \gamma\gamma_l(1 - \mathbf{v} \cdot \boldsymbol{\ell}) = 0 \,, \tag{9.8}$$

which may serve to express one component of l^μ by the others, provided the corresponding component of u^μ does not vanish. Thus, in general, a normalized l^μ has two independent components. For instance, for purely longitudinal flow $u^\mu = \gamma_z(1, 0, 0, v^z)^T$, with $\gamma_z \equiv \left(1 - v_z^2\right)^{-1/2}$, and one can determine ℓ^z from Eq. (9.8) as $\ell^z = 1/v^z$. Without loss of generality it is possible to set $\ell^x = \ell^y = 0$, such that l^μ is completely specified

$$l^\mu = \gamma_l\,\left(1, 0, 0, 1/v^z\right)^T \equiv \gamma_z\,\left(v^z, 0, 0, 1\right)^T \,. \tag{9.9}$$

One may use this form even if the flow is three-dimensional, since it still fulfills the requirements (9.5) (with $l = 1$) and (9.6) [9–11]. In the co-moving frame or local rest

(LR) frame of matter, $u^{\mu}_{LR} = (1, 0, 0, 0)^T$ (independent of the physical meaning of the 4-velocity), hence the direction of the anisotropy corresponds to the longitudinal or z-direction of the coordinate system, $l^{\mu}_{LR} = (0, 0, 0, 1)^T$.

An anisotropy in a system singles out another direction besides the direction of fluid flow. In our case, this is the direction characterized by l^{μ}. Thus, it is natural to generalize the projection operator $\Delta^{\mu\nu} = g^{\mu\nu} - u^{\mu}u^{\nu}$ to the projection operator (9.4) onto the two-dimensional subspace orthogonal to both u^{μ} and l^{μ} [2–5]. Note that $\varXi^{\mu\nu} = \varXi^{\nu\mu}$, $\varXi^{\mu\nu}u_{\nu} = \varXi^{\mu\nu}l_{\nu} = 0$, and $\varXi^{\mu}_{\mu} = 2$.

Let us briefly introduce our notational conventions. The projection of an arbitrary 4-vector A^{μ} orthogonal to both u^{μ} and l^{μ} will be denoted by

$$A^{\{\mu\}} = \varXi^{\mu\nu} A_{\nu} \, . \tag{9.10}$$

The corresponding projection of an arbitrary rank-two tensor is defined as

$$A^{\{\mu\nu\}} = \varXi^{\mu\nu}_{\alpha\beta} A^{\alpha\beta} \, , \tag{9.11}$$

where the corresponding symmetric, orthogonal, and traceless projection operator is

$$\varXi^{\mu\nu}_{\alpha\beta} = \frac{1}{2} \left(\varXi^{\mu}_{\alpha} \varXi^{\nu}_{\beta} + \varXi^{\mu}_{\beta} \varXi^{\nu}_{\alpha} \right) - \frac{1}{2} \varXi^{\mu\nu} \varXi_{\alpha\beta} \, . \tag{9.12}$$

The projection operators of arbitrary tensor rank are specified in Appendix 9.7.

The tensor decomposition of a 4-vector with respect to u^{μ}, l^{μ}, and $\varXi^{\mu\nu}$ reads

$$A^{\mu} = A^{\nu}u_{\nu} u^{\mu} - A^{\nu}l_{\nu} l^{\mu} + A^{\{\mu\}} \, . \tag{9.13}$$

Analogously, the tensor decomposition of the 4-gradient with respect to u^{μ}, l^{μ}, and $\varXi^{\mu\nu}$ reads as

$$\partial_{\mu} = u_{\mu}D + l_{\mu}D_l + \tilde{\nabla}_{\mu} \, , \tag{9.14}$$

where

$$D = u^{\nu}\partial_{\nu} \, , \tag{9.15}$$

$$D_l = -l^{\nu}\partial_{\nu} \, , \tag{9.16}$$

$$\tilde{\nabla}_{\mu} \equiv \varXi_{\mu\nu}\partial^{\nu} = \partial_{\{\mu\}} \, . \tag{9.17}$$

According to the specific choice of Eq. (9.9) in the LR frame $D_{l,LR} = \partial/\partial z$ corresponds to the derivative in the direction of the anisotropy, while the operator $\tilde{\nabla}_{LR\,\mu} = (0, \partial/\partial x, \partial/\partial y, 0)$ is the spatial gradient in the remaining transverse directions orthogonal to both u^{μ} and l^{μ}.

The tensor decomposition of a rank-two tensor with respect to u^μ, l^μ, and $\varXi^{\mu\nu}$ reads

$$
\begin{aligned}
A^{\mu\nu} &= A^{\alpha\beta}u_\alpha u_\beta\,u^\mu u^\nu + A^{\alpha\beta}l_\alpha l_\beta\,l^\mu l^\nu - A^{\alpha\beta}u_\alpha l_\beta\,u^\mu l^\nu - A^{\alpha\beta}u_\beta l_\alpha\,u^\nu l^\mu \\
&\quad + \varXi^{\mu\alpha}A_{\alpha\beta}u^\beta\,u^\nu + \varXi^{\nu\beta}A_{\alpha\beta}u^\alpha\,u^\mu - \varXi^{\mu\alpha}A_{\alpha\beta}l^\beta\,l^\nu - \varXi^{\nu\beta}A_{\alpha\beta}l^\alpha\,l^\mu \\
&\quad + \frac{1}{2}A^{\alpha\beta}\varXi_{\alpha\beta}\,\varXi^{\mu\nu} + \varXi^{\mu\nu}_{\alpha\beta}A^{\alpha\beta} + \varXi^\mu_\alpha \varXi^\nu_\beta A^{[\alpha\beta]}\ .
\end{aligned}
\tag{9.18}
$$

With this decomposition, we obtain

$$
\partial_\mu u_\nu = u_\mu Du_\nu + l_\mu D_l u_\nu + \frac{1}{2}\tilde{\theta}\,\varXi_{\mu\nu} - l_\beta l_\nu \tilde{\nabla}_\mu u^\beta + \tilde{\sigma}_{\mu\nu} + \tilde{\omega}_{\mu\nu}\ ,
\tag{9.19}
$$

where we have made use of the fact that u^μ and l^μ are normalized and orthogonal to each other. Note that $-l_\beta l_\nu \tilde{\nabla}_\mu u^\beta = -l_\beta l_{(\mu}\tilde{\nabla}_{\nu)}u^\beta + l_\beta l_{[\mu}\tilde{\nabla}_{\nu]}u^\beta$ can in principle also be further separated into a symmetric and an antisymmetric part.

In Eq. (9.19), we defined the following quantities in the subspace orthogonal to both u^μ and l^μ: the transverse expansion scalar

$$
\tilde{\theta} \equiv \tilde{\nabla}_\mu u^\mu\ ,
\tag{9.20}
$$

the transverse shear tensor

$$
\tilde{\sigma}^{\mu\nu} \equiv \partial^{\{\mu}u^{\nu\}} = \tilde{\nabla}^{(\mu}u^{\nu)} - \frac{1}{2}\tilde{\theta}\,\varXi^{\mu\nu} + l_\beta l^{(\mu}\tilde{\nabla}^{\nu)}u^\beta\ ,
\tag{9.21}
$$

and the transverse vorticity

$$
\tilde{\omega}^{\mu\nu} \equiv \varXi^{\mu\alpha}\varXi^{\nu\beta}\partial_{[\alpha}u_{\beta]} = \tilde{\nabla}^{[\mu}u^{\nu]} - l_\beta l^{[\mu}\tilde{\nabla}^{\nu]}u^\beta\ .
\tag{9.22}
$$

Similar to Eq. (9.19) for $\partial_\mu u_\nu$, we also need the decomposition of $\partial_\mu l_\nu$

$$
\partial_\mu l_\nu \equiv u_\mu Dl_\nu + l_\mu D_l l_\nu + \frac{1}{2}\tilde{\theta}_l\,\varXi_{\mu\nu} + u_\beta u_\nu \tilde{\nabla}_\mu l^\beta + \tilde{\sigma}_{l,\mu\nu} + \tilde{\omega}_{l,\mu\nu}\ ,
\tag{9.23}
$$

where we defined the quantities

$$
\tilde{\theta}_l \equiv \tilde{\nabla}_\mu l^\mu\ ,
\tag{9.24}
$$

$$
\tilde{\sigma}_l^{\mu\nu} \equiv \partial^{\{\mu}l^{\nu\}} = \tilde{\nabla}^{(\mu}l^{\nu)} - \frac{1}{2}\tilde{\theta}_l\,\varXi^{\mu\nu} - u_\beta u^{(\mu}\tilde{\nabla}^{\nu)}l^\beta\ ,
\tag{9.25}
$$

$$
\tilde{\omega}_l^{\mu\nu} \equiv \varXi^{\mu\alpha}\varXi^{\nu\beta}\partial_{[\alpha}l_{\beta]} = \tilde{\nabla}^{[\mu}l^{\nu]} + u_\beta u^{[\mu}\tilde{\nabla}^{\nu]}l^\beta\ .
\tag{9.26}
$$

One can now tensor-decompose k^μ using Eq. (9.13)

$$
k^\mu = E_{ku}u^\mu + E_{kl}l^\mu + k^{\{\mu\}}\ .
\tag{9.27}
$$

In the LR frame and with the choice (9.9) for l^μ, the quantities defined in Eq. (9.3) are $E_{\mathbf{k}l,LR} = k^z$, i.e., the component of three-momentum in l^μ-direction, and $k_{LR}^{\{\mu\}} = (0, k^x, k^y, 0)^T$, i.e., the components of three-momentum orthogonal to l^μ.

For on-shell particles

$$k^\mu k_\mu = E_{\mathbf{k}u}^2 - E_{\mathbf{k}l}^2 + k^{\{\mu\}} k_{\{\mu\}} = m^2 \, , \tag{9.28}$$

where m is the rest mass of the particle, while

$$k^{\{\mu\}} k_{\{\mu\}} = \varXi^{\alpha\beta} k_\alpha k_\beta \, . \tag{9.29}$$

From Eq. (9.18)

$$k^\mu k^\nu = E_{\mathbf{k}u}^2 u^\mu u^\nu + E_{\mathbf{k}l}^2 l^\mu l^\nu + 2 E_{\mathbf{k}u} E_{\mathbf{k}l} u^{(\mu} l^{\nu)} + 2 E_{\mathbf{k}u} k^{(\{\mu\}} u^{\nu)} + 2 E_{\mathbf{k}l} k^{(\{\mu\}} l^{\nu)}$$
$$+ \frac{1}{2} k^{\{\alpha\}} k_{\{\alpha\}} \varXi^{\mu\nu} + k^{\{\mu} k^{\nu\}} \, , \tag{9.30}$$

where we used $\varXi_{\alpha\beta}^{\mu\nu} k^\alpha k^\beta = k^{\{\mu} k^{\nu\}}$. Higher-rank tensors formed from dyadic products of k^μ can be decomposed in a similar manner.

The particle 4-current and the energy-momentum tensor can also be decomposed with respect to u^μ, the space-like 4-vector l^μ, and the projection tensor $\varXi^{\mu\nu}$. Using Eqs. (9.27) and (9.30) in the definitions

$$N^\mu \equiv \langle k^\mu \rangle \, , \quad T^{\mu\nu} \equiv \langle k^\mu k^\nu \rangle \, , \quad \langle \cdots \rangle \equiv \int dK \, (\cdots) \, f_{\mathbf{k}} \, , \tag{9.31}$$

cf. Eqs. (5.4), (5.5) with Eq. (5.6), as well as the fact that the energy-momentum tensor is symmetric, we obtain

$$N^\mu = n u^\mu + n_l l^\mu + V_\perp^\mu \, , \tag{9.32}$$

$$T^{\mu\nu} = \varepsilon u^\mu u^\nu + 2 M u^{(\mu} l^{\nu)} + P_l l^\mu l^\nu - P_\perp \varXi^{\mu\nu}$$
$$+ 2 W_{\perp u}^{(\mu} u^{\nu)} + 2 W_{\perp l}^{(\mu} l^{\nu)} + \pi_\perp^{\mu\nu} \, . \tag{9.33}$$

Here, the particle density n and the energy density ε read

$$n \equiv \langle E_{\mathbf{k}u} \rangle = N^\mu u_\mu \, , \tag{9.34}$$

$$\varepsilon \equiv \langle E_{\mathbf{k}u}^2 \rangle = T^{\mu\nu} u_\mu u_\nu \, . \tag{9.35}$$

The part of the particle-diffusion current in the l^μ-direction is denoted by

$$n_l \equiv \langle E_{\mathbf{k}l} \rangle = -N^\mu l_\mu \, , \tag{9.36}$$

while the particle-diffusion current orthogonal to both 4-vectors is denoted by

$$V_\perp^\mu \equiv \left\langle k^{\{\mu\}} \right\rangle = \Xi_\nu^\mu N^\nu . \tag{9.37}$$

The pressure in the transverse direction is denoted by

$$P_\perp \equiv -\frac{1}{2} \left\langle \Xi^{\mu\nu} k_\mu k_\nu \right\rangle = -\frac{1}{2} T^{\mu\nu} \Xi_{\mu\nu} , \tag{9.38}$$

while the pressure in the longitudinal direction is

$$P_l \equiv \left\langle E_{\mathbf{k}l}^2 \right\rangle = T^{\mu\nu} l_\mu l_\nu . \tag{9.39}$$

The projection of the energy-momentum tensor in both u^μ- and l^ν-direction is denoted by

$$M \equiv \left\langle E_{\mathbf{k}u} E_{\mathbf{k}l} \right\rangle = -T^{\mu\nu} u_\mu l_\nu . \tag{9.40}$$

The projections in either u^μ- or l^μ-direction and orthogonal to both directions are denoted by

$$W_{\perp u}^\mu \equiv \left\langle E_{\mathbf{k}u} k^{\{\mu\}} \right\rangle = \Xi_\alpha^\mu T^{\alpha\beta} u_\beta , \tag{9.41}$$

$$W_{\perp l}^\mu \equiv \left\langle E_{\mathbf{k}l} k^{\{\mu\}} \right\rangle = -\Xi_\alpha^\mu T^{\alpha\beta} l_\beta , \tag{9.42}$$

$$\pi_\perp^{\mu\nu} \equiv \left\langle k^{\{\mu} k^{\nu\}} \right\rangle = \Xi_{\alpha\beta}^{\mu\nu} T^{\alpha\beta} . \tag{9.43}$$

From these definitions, it is evident that $V_\perp^\mu u_\mu = V_\perp^\mu l_\mu = 0$ as well as $W_{\perp u}^\mu u_\mu = W_{\perp u}^\mu l_\mu = 0$ and $W_{\perp l}^\mu u_\mu = W_{\perp l}^\mu l_\mu = 0$. The transverse shear-stress tensor $\pi_\perp^{\mu\nu}$ is the part of the energy-momentum tensor that is symmetric, $\pi_\perp^{\mu\nu} = \pi_\perp^{\nu\mu}$, traceless, $\pi_\perp^{\mu\nu} g_{\mu\nu} = 0$, and orthogonal to both preferred 4-vectors, $\pi_\perp^{\mu\nu} u_\mu = \pi_\perp^{\mu\nu} l_\mu = 0$. Note that the various subscripts u (for projection onto the direction of u^μ), l (for projection onto the direction of l^μ) and $\perp$ (for projection onto the direction "perpendicular" to both u^μ and l^μ) serve as reminders of the directions that the various quantities are projected onto.

The isotropic pressure P is related to the longitudinal and transverse pressure components (9.39) and (9.38) via Eq. (9.4)

$$P \equiv -\frac{1}{3} \Delta^{\mu\nu} T_{\mu\nu} = \frac{1}{3} \left(l^\mu l^\nu - \Xi^{\mu\nu} \right) T_{\mu\nu} = \frac{1}{3} \left(P_l + 2P_\perp \right) . \tag{9.44}$$

This result is independent on how far off the system is from local thermodynamical equilibrium. In case that $P = P_\perp = P_l$, the pressure is isotropic, but the system may not be in local thermodynamical equilibrium, because the bulk-viscous pressure

$$\Pi_{\text{iso}} \equiv P - P_0 , \tag{9.45}$$

may be non-zero. Here, P_0 is the pressure in local thermodynamical equilibrium.

Furthermore, not only the isotropic pressure separates into two parts, but also the particle-diffusion current V^μ and the energy-diffusion current W^μ are split according to the direction defined by l^μ and the direction perpendicular to it

$$V^\mu \equiv \Delta^{\mu\nu} N_\nu = - \left(l^\mu l^\nu - \Xi^{\mu\nu} \right) N_\nu = n_l\, l^\mu + V_\perp^\mu \,, \qquad (9.46)$$

$$W^\mu \equiv \Delta^{\mu\nu} T_{\nu\lambda} u^\lambda = - \left(l^\mu l^\nu - \Xi^{\mu\nu} \right) T_{\nu\lambda} u^\lambda = M\, l^\mu + W_{\perp u}^\mu \,. \qquad (9.47)$$

Finally, using the projection operator (9.12) and the definition (9.4), we can show that

$$\pi^{\mu\nu} \equiv \Delta_{\alpha\beta}^{\mu\nu} T^{\alpha\beta} = \pi_\perp^{\mu\nu} + 2\, W_{\perp l}^{(\mu} l^{\nu)} + \frac{1}{3} \left(P_l - P_\perp \right) \left(2 l^\mu l^\nu + \Xi^{\mu\nu} \right) \,. \qquad (9.48)$$

Equations (9.44)–(9.48) relate the fluid-dynamical quantities decomposed with respect to u^μ and $\Delta^{\mu\nu}$ to those decomposed with respect to u^μ, l^μ, and $\Xi^{\mu\nu}$. Note that, in terms of independent degrees of freedom, these two decompositions are completely equivalent. In general, N^μ has four, while $T^{\mu\nu}$ has ten independent components. The decomposition with respect to (a given 4-vector) u^μ and the projector $\Delta^{\mu\nu}$ also contains 14 independent dynamical variables. These are the three scalars n, ε, and P, the two vectors V^μ and W^μ, each with three independent components, while the shear-stress tensor $\pi^{\mu\nu}$ has five independent components. On the other hand, the decomposition with respect to (given) u^μ, l^μ, and $\Xi^{\mu\nu}$ has the six scalars n, ε, n_l, M, P_l, and $P_\perp$, the three vectors $V_\perp^\mu$, $W_{\perp u}^\mu$, and $W_{\perp l}^\mu$, with two independent components each, whereas the shear-stress tensor in the transverse direction, $\pi_\perp^{\mu\nu}$, possesses only two independent components.

The choice of the fluid 4-velocity u^μ is not unique. The two most popular choices to fix the local rest frame of the fluid are the Eckart frame [12], which follows the flow of particles, and the Landau frame [13], which follows the flow of energy. Consequently, in the Eckart frame, there is no diffusion of particles (or charges) relative to u^μ, so that

$$V^\mu = 0\,, \quad V_\perp^\mu = 0\,, \quad n_l = 0\,, \qquad (9.49)$$

where we used Eq. (9.46). In the Landau frame, the energy-diffusion current vanishes

$$W^\mu = 0\,, \quad W_{\perp u}^\mu = 0\,, \quad M = 0\,, \qquad (9.50)$$

where we used Eq. (9.47). In both cases, three of the 14 independent fluid-dynamical variables are replaced by the three independent components of u^μ, so the total number of independent variables is still 14.

It is also possible to assign a physical meaning to the (so far fixed) 4-vector l^μ. This vector would then become a dynamical variable with two independent components. A clever choice of frame could then be used to eliminate two of the 14 independent fluid-dynamical variables and replace them with the two independent components of l^μ. This is, however, not what is commonly done. In the initial stage of a heavy-ion collision, the single-particle distribution function is highly anisotropic in the beam (z-) direction. Therefore, l^μ is usually taken to be

$$l^\mu = \gamma_z (v^z, 0, 0, 1)^T , \qquad (9.51)$$

see Refs. [9–11] and Eq. (9.9). Here, v^z is the z-component of the fluid three-velocity $\mathbf{v}$, and $\gamma_z = (1 - v_z^2)^{-1/2}$. One can easily convince oneself that $l^\mu l_\mu = -1$ and $l^\mu u_\mu = 0$. Since v^z is uniquely determined by u^μ, this choice of l^μ does not represent a new dynamical quantity, and l^μ is completely fixed once u^μ is known.

9.2 Anisotropic State

In this section, we discuss the case where the single-particle distribution function has a given anisotropic shape in momentum space. The local-equilibrium distribution function is

$$f_{0\mathbf{k}}(\alpha_0, \beta_0 E_{\mathbf{k}u}) \equiv [\exp(-\alpha_0 + \beta_0 E_{\mathbf{k}u}) + a]^{-1} , \qquad (9.52)$$

where $\beta_0 = 1/T, \alpha_0 = \beta_0 \mu, a = \pm 1, 0$ for fermions/bosons, or Boltzmann particles, respectively.

We will denote the anisotropic single-particle distribution function as $\hat{f}_{0\mathbf{k}} = \hat{f}_{0\mathbf{k}}\left(\hat{\alpha}, \hat{\beta}_u E_{\mathbf{k}u}, \hat{\beta}_l E_{\mathbf{k}l}\right)$. At this point, the functional dependence on $\hat{\beta}_l E_{\mathbf{k}l}$ does not need to be specified. All we need to know is that this combination of variables parametrizes the momentum anisotropy. The single-particle distribution function is now a function of the scalars $\hat{\alpha}$, $\hat{\beta}_u$, and $\hat{\beta}_l$ as well as of two distinct 4-vectors, the flow velocity u^μ and the vector l^μ parametrizing the direction of the anisotropy. All these quantities are functions of x^μ. Such distribution functions are common in plasma physics where the presence of magnetic fields introduces a momentum anisotropy and so the particle momenta parallel and perpendicular to the magnetic field are different; as in the case of the bi-Maxwellian, the drifting Maxwellian, or the loss-cone distribution functions [14]. Analogously, $\hat{\beta}_l$ can be thought of as an additional parameter characterizing the temperature difference between the directions parallel and perpendicular to the z-axis.

We will also assume that

$$\lim_{\hat{\beta}_l \to 0} \hat{f}_{0\mathbf{k}}\left(\hat{\alpha}, \hat{\beta}_u E_{\mathbf{k}u}, \hat{\beta}_l E_{\mathbf{k}l}\right) = f_{0\mathbf{k}}\left(\hat{\alpha}, \hat{\beta}_u E_{\mathbf{k}u}\right) , \qquad (9.53)$$

i.e., that in the limit of vanishing anisotropy the single-particle distribution function assumes the local-equilibrium form (9.52). The assumption (9.53) has no impact on our formulation of anisotropic dissipative fluid dynamics, it is merely physically natural. We furthermore demand that

$$\left(\frac{\partial \hat{f}_{0\mathbf{k}}}{\partial \hat{\alpha}}\right)_{\hat{\beta}_u, \hat{\beta}_l} = \hat{f}_{0\mathbf{k}}\left(1 - a \hat{f}_{0\mathbf{k}}\right) . \qquad (9.54)$$

This further constraint on the form of $\hat{f}_{0\mathbf{k}}$ is naturally respected in the limit of vanishing anisotropy, cf. Eq. (9.52). There is no real physical reason that we should require it also for $\hat{\beta}_l \neq 0$, but it simplifies the following calculations. For instance, the spheroidal distribution function proposed in Ref. [15] and used in Refs. [9–11, 16–20],

$$
\hat{f}_{0\mathbf{k}} \equiv \left[\exp\left(-\hat{\alpha} + \hat{\beta}_u \sqrt{ E_{\mathbf{k}u}^2 + \frac{\hat{\beta}_l^2}{\hat{\beta}_u^2} E_{\mathbf{k}l}^2 } \right) + a \right]^{-1} , \tag{9.55}
$$

satisfies the above constraints and can be used for explicit calculations.

In analogy to Eq. (6.89), we now introduce a set of generalized moments of $\hat{f}_{0\mathbf{k}}$ of tensor-rank n

$$
\hat{\mathcal{I}}_{ij}^{\mu_1\cdots\mu_n} \equiv \left\langle E_{\mathbf{k}u}^i \, E_{\mathbf{k}l}^j \, k^{\mu_1}\cdots k^{\mu_n} \right\rangle_{\hat{0}} , \tag{9.56}
$$

where, similar to Eqs. (5.6) and (5.24),

$$
\langle \cdots \rangle_{\hat{0}} = \int dK \, (\cdots) \, \hat{f}_{0\mathbf{k}} , \tag{9.57}
$$

and the subscripts i and j denote the powers of $E_{\mathbf{k}u}$ and $E_{\mathbf{k}l}$, respectively. These generalized moments can be expanded in terms of the two 4-vectors u^μ, l^μ, and the tensor $\Xi^{\mu\nu}$

$$
\hat{\mathcal{I}}_{ij}^{\mu_1\cdots\mu_n} = \sum_{q=0}^{[n/2]} \sum_{r=0}^{n-2q} (-1)^q \, b_{nrq} \, \hat{\mathcal{I}}_{i+j+n,\,j+r,\,q}
$$
$$
\times \, \Xi^{(\mu_1\mu_2}\cdots\Xi^{\mu_{2q-1}\mu_{2q}} l^{\mu_{2q+1}}\cdots l^{\mu_{2q+r}} u^{\mu_{2q+r+1}}\cdots u^{\mu_n)} , \tag{9.58}
$$

where n, r, and q are natural numbers, r counts the number of 4-vectors l^μ, and q the number of Ξ projectors in the expansion. The symmetrized tensor products $\Xi^{(\mu_1\mu_2}\cdots\Xi^{\mu_{2q-1}\mu_{2q}} l^{\mu_{2q+1}}\cdots l^{\mu_{2q+r}} u^{\mu_{2q+r+1}}\cdots u^{\mu_n)}$ are discussed in Appendix 9.8. The symmetrization yields

$$
b_{nrq} \equiv \frac{n!}{2^q q! \, r! \, (n-r-2q)!} \tag{9.59}
$$

distinct terms, see Appendix 9.8. Finally, the generalized thermodynamic integrals $\hat{\mathcal{I}}_{nrq}$ are defined as

$$
\hat{\mathcal{I}}_{nrq}\left(\hat{\alpha}, \hat{\beta}_u, \hat{\beta}_l \right) = \frac{(-1)^q}{(2q)!!} \left\langle E_{\mathbf{k}u}^{n-r-2q} \, E_{\mathbf{k}l}^r \left(\Xi^{\mu\nu} k_\mu k_\nu \right)^q \right\rangle_{\hat{0}} , \tag{9.60}
$$

where the double factorial of an even number is $(2q)!! \equiv 2^q q!$. The corresponding generalized auxiliary thermodynamic integrals are defined with the help of Eq. (9.54) and similar to Eq. (6.30)

$$\hat{J}_{nrq} \equiv \left(\frac{\partial \hat{I}_{nrq}}{\partial \hat{\alpha}} \right)_{\beta_u, \beta_l} = \frac{(-1)^q}{(2q)!!} \int dK \, E_{\mathbf{k}u}^{n-r-2q} \, E_{\mathbf{k}l}^r \left(\Xi^{\mu\nu} k_\mu k_\nu \right)^q \hat{f}_{0\mathbf{k}} \left(1 - a \hat{f}_{0\mathbf{k}} \right) .$$

(9.61)

As in Eq. (6.101), one can easily show that

$$\hat{I}_{i,j}^{\mu_1 \cdots \mu_n} = m^2 \, \hat{I}_{i,j}^{\mu_1 \cdots \mu_{n-2}} - \hat{I}_{i+2,j}^{\mu_1 \cdots \mu_{n-2}} + \hat{I}_{i,j+2}^{\mu_1 \cdots \mu_{n-2}} .$$

(9.62)

As in Eq. (6.102), comparison of Eqs. (9.56) and (9.60) in the case $q = 0$ leads to

$$\hat{I}_{i+j,j,0} \equiv \hat{I}_{ij} ,$$

(9.63)

while for $q = 1$, using the explicit form of $\Xi^{\mu\nu}$ and the on-shell condition $k^\mu k_\mu = m^2$, comparison of Eqs. (9.56) and (9.60) yields

$$\hat{I}_{i+j+2,j,1} \equiv -\frac{1}{2} \hat{I}_{ij}^{\mu\nu} \Xi_{\mu\nu} = -\frac{1}{2} \left(m^2 \, \hat{I}_{ij} - \hat{I}_{i+2,j} + \hat{I}_{i,j+2} \right)$$

$$= -\frac{1}{2} \left(m^2 \, \hat{I}_{i+j,j,0} - \hat{I}_{i+j+2,j,0} + \hat{I}_{i+j+2,j+2,0} \right) ,$$

(9.64)

similar as in Eq. (6.103).

Note that these quantities can also be used in (local) thermodynamic equilibrium; see Eq. (9.107) below. The conventional relativistic thermodynamic integrals (6.30) are then recovered as linear combinations of those defined in Eqs. (9.60), (9.61), as shown in Appendix 9.9.

It is instructive to explicitly write down the tensor decomposition of the generalized moments (9.58) of $\hat{f}_{0\mathbf{k}}$. The conserved quantities read

$$\hat{N}^\mu \equiv \hat{I}_{00}^\mu = \hat{I}_{100} u^\mu + \hat{I}_{110} l^\mu ,$$

(9.65)

$$\hat{T}^{\mu\nu} \equiv \hat{I}_{00}^{\mu\nu} = \hat{I}_{200} u^\mu u^\nu + 2 \hat{I}_{210} u^{(\mu} l^{\nu)} + \hat{I}_{220} l^\mu l^\nu - \hat{I}_{201} \Xi^{\mu\nu} .$$

(9.66)

The coefficients can be obtained by appropriate tensor projections of these quantities. According to Eqs. (9.34)–(9.38), we obtain

$$\hat{I}_{100} \equiv \hat{N}^\mu u_\mu = \hat{n} = \langle E_{\mathbf{k}u} \rangle_{\hat{0}} \equiv \hat{I}_{10} ,$$

(9.67)

$$\hat{I}_{110} \equiv -\hat{N}^\mu l_\mu = \hat{n}_l = \langle E_{\mathbf{k}l} \rangle_{\hat{0}} \equiv \hat{I}_{01} ,$$

(9.68)

$$\hat{I}_{200} \equiv \hat{T}^{\mu\nu} u_\mu u_\nu = \hat{\varepsilon} = \langle E_{\mathbf{k}u}^2 \rangle_{\hat{0}} \equiv \hat{I}_{20} ,$$

(9.69)

$$\hat{I}_{210} \equiv -\hat{T}^{\mu\nu} u_\mu l_\nu = \hat{M} = \langle E_{\mathbf{k}u} E_{\mathbf{k}l} \rangle_{\hat{0}} \equiv \hat{I}_{11} ,$$

(9.70)

$$\hat{I}_{220} \equiv \hat{T}^{\mu\nu} l_\mu l_\nu = \hat{P}_l = \langle E_{\mathbf{k}l}^2 \rangle_{\hat{0}} \equiv \hat{I}_{02} ,$$

(9.71)

$$\hat{I}_{201} \equiv -\frac{1}{2}\hat{T}^{\mu\nu}\Xi_{\mu\nu} = \hat{P}_\perp = -\frac{1}{2}\langle\Xi^{\alpha\beta}k_\alpha k_\beta\rangle_{\hat{0}}$$

$$\equiv -\frac{1}{2}\left(m^2\,\hat{I}_{00} - \hat{I}_{20} + \hat{I}_{02}\right) . \tag{9.72}$$

Note that $\hat{\varepsilon}$, $\hat{P}_l$, and $\hat{P}_\perp$ are related to each other, see Eq. (9.72), hence they are not independent variables.

In general, the conserved quantities (9.65), (9.66) contain eleven unknowns: the scalar quantities $\hat{n}$, $\hat{\varepsilon}$, $\hat{n}_l$, $\hat{M}$, $\hat{P}_l$, and $\hat{P}_\perp$, and the vectors u^μ (three independent components) and l^μ (two independent components). At the end of Sect. 9.1, we had already discussed our choice of l^μ, which is completely determined by u^μ, so there remain nine unknowns. The choice of a LR frame (Eckart or Landau) eliminates either $\hat{n}_l$ or $\hat{M}$, hence leaving eight unknowns. However, once $\hat{f}_{0k}\left(\hat{\alpha}, \hat{\beta}_u E_{ku}, \hat{\beta}_l E_{kl}\right)$ is specified, the remaining five scalar unknowns ($\hat{n}$, $\hat{\varepsilon}$, $\hat{P}_l$, $\hat{P}_\perp$, and—depending on the choice of the LR frame— either $\hat{n}_l$ or $\hat{M}$) are not independent variables anymore; they are functions of the three independent variables $\hat{\alpha}$, $\hat{\beta}_u$, $\hat{\beta}_l$. This reduces the number of independent variables to six. Five constraints are provided by the five equations of motion $\partial_\mu\hat{N}^\mu = 0$ and $\partial_\mu\hat{T}^{\mu\nu} = 0$. In the ideal-fluid limit, $\hat{\beta}_l \to 0$, and the system of equations of motion is closed. For arbitrary $\hat{\beta}_l$, however, we need an additional equation of motion to close the system of equations. This will effectively describe the decay of the momentum anisotropy of the distribution function and the approach of the system to local thermal equilibrium. This auxiliary equation can be provided, for example, from the higher moments of the Boltzmann equation as is usually done in kinetic theory.

It is instructive to repeat this discussion from a slightly different perspective. For very large times, any closed system described by the Boltzmann equation will reach global thermodynamical equilibrium. If, in this process, the system first reaches local thermodynamical equilibrium, the evolution toward global equilibrium is governed by ideal fluid dynamics. In this case, it is advantageous to explicitly exhibit the equilibrium equation of state $P_0(\varepsilon_0, n_0)$ in the fluid-dynamical equations of motion. Usually, this is done via the Landau matching conditions. These conditions require that particle density n and energy density ε in a general non-equilibrium state are equal to those of a fictitious (local) thermodynamical equilibrium state, $n = n_0(\alpha_0, \beta_0)$, $\varepsilon = \varepsilon_0(\alpha_0, \beta_0)$. These equations implicitly determine the intensive parameters α_0, β_0 in the distribution function (9.52) pertaining to the fictitious (local) equilibrium state.

Analogously, for the anisotropic state the Landau matching conditions read $\hat{n}\left(\hat{\alpha}, \hat{\beta}_u, \hat{\beta}_l\right) = n_0(\alpha_0, \beta_0)$ and $\hat{\varepsilon}\left(\hat{\alpha}, \hat{\beta}_u, \hat{\beta}_l\right) = \varepsilon_0(\alpha_0, \beta_0)$, which is equivalent to

$$\left(\hat{N}^\mu - N_{(0)}^\mu\right)u_\mu = 0 , \tag{9.73}$$

$$\left(\hat{T}^{\mu\nu} - T_{(0)}^{\mu\nu}\right)u_\mu u_\nu = 0 , \tag{9.74}$$

where $N_{(0)}^\mu$ and $T_{(0)}^{\mu\nu}$ were defined in Eq. (1.20). These dynamical matching conditions will determine the two parameters of a fictitious equilibrium state, $\alpha_0 = \alpha_0(\hat\alpha, \hat\beta_u, \hat\beta_l)$ and $\beta_0 = \beta_0(\hat\alpha, \hat\beta_u, \hat\beta_l)$, as function of the three scalar parameters $\hat\alpha, \hat\beta_u, \hat\beta_l$ pertaining to the anisotropic state.

In principle, there are infinitely many possibilities to extend an equilibrium distribution function by an additional free parameter, $\hat\beta_l$, each one resulting in a different set of thermodynamical relations, i.e., the latter are not universal. For example, choosing $\hat\beta_l$ as a free intensive variable that is related to some new conjugate extensive quantity, as in Ref. [21], we recover the conventional laws of thermodynamics only in the equilibrium limit, $\hat\beta_l \to 0$. Therefore, without specifying the exact form of the anisotropic distribution function, we cannot derive any thermodynamic relation, and in particular, the equation of state from kinetic theory.

Using the Landau matching conditions (9.73), (9.74) the tensor decomposition of the conserved quantities reads

$$\hat N^\mu = n_0\, u^\mu + \hat n_l\, l^\mu \,, \tag{9.75}$$

$$\hat T^{\mu\nu} = \varepsilon_0\, u^\mu u^\nu + 2\,\hat M\, u^{(\mu} l^{\nu)} + \hat P_l\, l^\mu l^\nu - \hat P_\perp\, \varXi^{\mu\nu} \,. \tag{9.76}$$

The equilibrium equation of state is now introduced by writing the isotropic pressure (9.44) in the form

$$\hat P \equiv \frac{1}{3}\left(\hat P_l + 2\hat P_\perp\right) \equiv P_0\,(\alpha_0, \beta_0) + \hat\Pi\left(\hat\alpha, \hat\beta_u, \hat\beta_l\right) \,, \tag{9.77}$$

which at the same time defines the bulk-viscous pressure $\hat\Pi$ with respect to the pressure in local equilibrium, and hence can be used to eliminate either $\hat P_l$ or $\hat P_\perp$ in Eq. (9.76).

Making the connection to the equilibrium equation of state is advantageous when the system is close to the ideal-fluid limit. However, for the anisotropic state, it does not solve the problem that the five conservation equations do not determine all independent variables. We are thus left with six independent variables, say n_0, ε_0, and $\hat P_l$ (or $\hat\Pi$) and the three components of u^μ, which means that we must supply one additional equation of motion.

The tensor decompositions (9.75), (9.76) were obtained previously in Refs. [9–11, 16–20] based on the distribution function (9.55). This spheroidal single-particle distribution function leads to $\hat n_l = \hat M = 0$, hence exclusively to a pressure anisotropy. Furthermore, in these references the case of vanishing $\hat\alpha$ was considered, which for a massless ideal-gas equation of state also leads to a vanishing bulk-viscous pressure. This also leaves the freedom to use the zeroth moment of the Boltzmann equation as additional input to close the equations of motion.

9.3 Expansion Around the Anisotropic Distribution Function

In principle, $f_{\mathbf k}$ is a solution of the Boltzmann equation. However, if we are only interested in the low-frequency, large-wavenumber limit of the latter, we may consider the (much simpler) fluid-dynamical equations of motion. In order to derive them from the Boltzmann equation, the method of moments, see Chap. 6, is particularly well suited [7,22,23], where $f_{\mathbf{k}}$ is expanded around the distribution function $f_{0\mathbf{k}}$ of a fictitious (local) equilibrium state. The corrections are written in terms of the irreducible moments of $\delta f_{\mathbf{k}} \equiv f_{\mathbf{k}} - f_{0\mathbf{k}}$. Then, the infinite set of moments of the Boltzmann equation provide an infinite set of equations of motion for these irreducible moments. Conventional dissipative fluid dynamics then emerges by truncating this set and expressing the irreducible moments in terms of the fluid-dynamical variables; for more details see Ref. [7].

Here we will follow the same strategy, except expanding $f_{\mathbf{k}}$ around $\hat{f}_{0\mathbf{k}}$ instead of $f_{0\mathbf{k}}$; see also Refs. [24,25]. Hence

$$f_{\mathbf{k}} \equiv \hat{f}_{0\mathbf{k}} + \delta\hat{f}_{\mathbf{k}} \,, \tag{9.78}$$

where it is implicitly assumed that the correction $\delta\hat{f}_{\mathbf{k}}$ fulfills $|\delta\hat{f}_{\mathbf{k}}| \ll \hat{f}_{0\mathbf{k}}$. The rationale behind the expansion around $\hat{f}_{0\mathbf{k}}$ instead of around $f_{0\mathbf{k}}$ is that, in the case of a pronounced anisotropy, $|\delta\hat{f}_{\mathbf{k}}| \ll |\delta f_{\mathbf{k}}|$, so that the convergence properties of the former series expansion are vastly improved over those of the latter. Without loss of generality we write these corrections as follows:

$$\delta\hat{f}_{\mathbf{k}} = \hat{f}_{0\mathbf{k}} \left(1 - a\,\hat{f}_{0\mathbf{k}}\right) \hat{\phi}_{\mathbf{k}} \,, \tag{9.79}$$

where $\hat{\phi}_{\mathbf{k}}$ is a measure of the deviation of $f_{\mathbf{k}}$ from $\hat{f}_{0\mathbf{k}}$.

We recall from Sect. 6.1 that, in the case of an expansion around a local-equilibrium distribution $f_{0\mathbf{k}}$, the deviation $\delta f_{\mathbf{k}} \equiv f_{\mathbf{k}} - f_{0\mathbf{k}}$ can be expanded in terms of a complete and orthogonal set of irreducible tensors 1, $k^{\langle \mu \rangle}$, $k^{\langle \mu} k^{\nu \rangle}$, $k^{\langle \mu} k^{\nu} k^{\lambda \rangle}$, ..., cf. Eq. (6.6). Similarly, the correction $\hat{\phi}_{\mathbf{k}}$ to the anisotropic state is expanded in terms of *another* complete, orthogonal set of irreducible tensors, 1, $k^{\{\mu\}}$, $k^{\{\mu} k^{\nu\}}$, $k^{\{\mu} k^{\nu} k^{\lambda\}}$, ..., where

$$k^{\{\mu_1} \cdots k^{\mu_\ell\}} = \Xi^{\mu_1 \cdots \mu_\ell}_{\nu_1 \cdots \nu_\ell} k^{\nu_1} \cdots k^{\nu_\ell} \,. \tag{9.80}$$

Here, the symmetric and traceless projection tensor $\Xi^{\mu_1 \cdots \mu_\ell}_{\nu_1 \cdots \nu_\ell}$ is defined in Eq. (9.110). By definition, this tensor and thus $k^{\{\mu_1} \cdots k^{\mu_\ell\}}$ are orthogonal to *both* u^{μ} and l^{μ}. For an arbitrary function of *both* $E_{\mathbf{k}u}$ and $E_{\mathbf{k}l}$, say $\hat{\mathrm{F}}(E_{\mathbf{k}u}, E_{\mathbf{k}l})$, the irreducible tensors satisfy the following orthogonality condition

$$\int dK \hat{\mathrm{F}}(E_{\mathbf{k}u}, E_{\mathbf{k}l}) k^{\{\mu_1} \cdots k^{\mu_\ell\}} k_{\{\nu_1} \cdots k_{\nu_n\}}$$
$$= \frac{\delta_{\ell n}}{2^\ell} \, \Xi^{\mu_1 \cdots \mu_\ell}_{\nu_1 \cdots \nu_\ell} \int dK \hat{\mathrm{F}}(E_{\mathbf{k}u}, E_{\mathbf{k}l}) \left(\Xi^{\alpha\beta} k_\alpha k_\beta\right)^\ell \,, \tag{9.81}$$

for the proof, see Appendix 9.10. This is similar to Eq. (6.7) for the irreducible tensors (6.6).

The expansion of $\hat{\phi}_{\mathbf{k}}$ now reads

$$\hat{\phi}_{\mathbf{k}} = \sum_{\ell=0}^{\infty} \sum_{n=0}^{N_\ell} \sum_{m=0}^{N_\ell-n} c_{nm}^{\{\mu_1 \cdots \mu_\ell\}} k_{\{\mu_1} \cdots k_{\mu_\ell\}} P_{\mathbf{k}nm}^{(\ell)} \, . \tag{9.82}$$

Here, similar to Eq. (6.10), N_ℓ truncates the (in principle) infinite sums over n and m at some natural number. Furthermore, $c_{nm}^{\{\mu_1 \cdots \mu_\ell\}}$ are some coefficients that will be determined later. Analogous to the orthogonal polynomials $P_{\mathbf{k}n}^{(\ell)}$, cf. Eq. (6.5), the polynomials

$$P_{\mathbf{k}nm}^{(\ell)} = \sum_{i=0}^{n} \sum_{j=0}^{m} a_{nimj}^{(\ell)} E_{\mathbf{k}u}^{i} E_{\mathbf{k}l}^{j} \tag{9.83}$$

form an orthogonal set, but now in *both* $E_{\mathbf{k}u}$ and $E_{\mathbf{k}l}$, where $a_{nimj}^{(\ell)}$ are coefficients that are independent of $E_{\mathbf{k}u}$ and $E_{\mathbf{k}l}$. These multivariate polynomials in $E_{\mathbf{k}u}$ and $E_{\mathbf{k}l}$ are constructed to satisfy the orthonormality relation (9.166). Note that for $m = 0$, the multivariate polynomials $P_{\mathbf{k}nm}^{(\ell)}$ defined in Eq. (9.83) naturally reduce to the polynomials $P_{\mathbf{k}n}^{(\ell)}$; for more details see Appendix 9.11.

Finally, in complete analogy to the expansion (6.12), the distribution function can be written as

$$f_{\mathbf{k}} = \hat{f}_{0\mathbf{k}} + \hat{f}_{0\mathbf{k}} \left(1 - a\,\hat{f}_{0\mathbf{k}}\right) \sum_{\ell=0}^{\infty} \sum_{n=0}^{N_\ell} \sum_{m=0}^{N_\ell-n} \hat{\rho}_{nm}^{\mu_1 \cdots \mu_\ell} k_{\{\mu_1} \cdots k_{\mu_\ell\}} \hat{\mathcal{H}}_{\mathbf{k}nm}^{(\ell)} \, . \tag{9.84}$$

Here, we introduced the irreducible moments of $\delta\hat{f}_{\mathbf{k}}$

$$\hat{\rho}_{ij}^{\mu_1 \cdots \mu_\ell} \equiv \left\langle E_{\mathbf{k}u}^{i} E_{\mathbf{k}l}^{j} k^{\{\mu_1} \cdots k^{\mu_\ell\}} \right\rangle_{\hat{\delta}} \, , \tag{9.85}$$

where

$$\langle \cdots \rangle_{\hat{\delta}} \equiv \langle \cdots \rangle - \langle \cdots \rangle_{\hat{0}} = \int dK \, (\cdots) \, \delta\hat{f}_{\mathbf{k}} \, . \tag{9.86}$$

Note the differences to the irreducible moments (6.14): apart from the weight factor $\delta\hat{f}_{\mathbf{k}}$ instead of $\delta f_{\mathbf{k}}$ they carry *two* indices i and j, to indicate the different powers of $E_{\mathbf{k}u}$ and $E_{\mathbf{k}l}$, and finally, the irreducible tensors $k^{\{\mu_1} \cdots k^{\mu_\ell\}}$ appear instead of $k^{\langle\mu_1} \cdots k^{\mu_\ell\rangle}$. Furthermore, the coefficients $\hat{\mathcal{H}}_{\mathbf{k}nm}^{(\ell)}$ are defined as

$$\hat{\mathcal{H}}_{\mathbf{k}nm}^{(\ell)} = \frac{\hat{W}^{(\ell)}}{\ell!} \sum_{i=n}^{N_\ell-m} \sum_{j=m}^{N_\ell-i} a_{injm}^{(\ell)} P_{\mathbf{k}ij}^{(\ell)} \, . \tag{9.87}$$

For the proof of Eq. (9.84), we first determine the coefficients $c_{nm}^{\{\mu_1 \cdots \mu_\ell\}}$ in Eq. (9.82). With the help of Eqs. (9.81), (9.82), and (9.166), one can prove that

$$c_{nm}^{\{\mu_1 \cdots \mu_\ell\}} \equiv \frac{\hat{W}^{(\ell)}}{\ell!} \left\langle P_{\mathbf{k}nm}^{(\ell)} k^{\{\mu_1} \cdots k^{\mu_\ell\}} \right\rangle_{\hat{\delta}} \, . \tag{9.88}$$

Inserting this into Eq. (9.82) and employing Eq. (9.83), we obtain

$$\hat{\phi}_{\mathbf{k}} = \sum_{\ell=0}^{\infty} \frac{\hat{W}^{(\ell)}}{\ell!} \sum_{n=0}^{N_\ell} \sum_{m=0}^{N_\ell-n} \sum_{i=0}^{n} \sum_{j=0}^{m} a_{nimj}^{(\ell)} \hat{\rho}_{ij}^{\mu_1\cdots\mu_\ell} k_{\{\mu_1} \cdots k_{\mu_\ell\}} P_{knm}^{(\ell)} , \tag{9.89}$$

where we have used the definition (9.85) of $\hat{\rho}_{ij}^{\mu_1\cdots\mu_\ell}$. Now, by renaming $n \leftrightarrow i$ and $m \leftrightarrow j$ and cleverly reordering the sums, we obtain Eq. (9.84) with $\hat{\mathcal{H}}_{knm}^{(\ell)}$ as defined in Eq. (9.87).

Note that, similar to Eq. (6.16), once we truncate the expansion at some finite N_ℓ, any irreducible moment of $\delta\hat{f}_{\mathbf{k}}$ with corresponding tensor rank is linearly related to the moments that appear in the truncated expansion

$$\hat{\rho}_{ij}^{\mu_1\cdots\mu_\ell} \equiv (-1)^\ell \ell! \sum_{n=0}^{N_\ell} \sum_{m=0}^{N_\ell-n} \hat{\rho}_{nm}^{\mu_1\cdots\mu_\ell} \gamma_{injm}^{(\ell)} , \tag{9.90}$$

where

$$\gamma_{injm}^{(\ell)} = \frac{\hat{W}^{(\ell)}}{\ell!} \sum_{n'=n}^{N_\ell-m} \sum_{m'=m}^{N_\ell-n'} \sum_{i'=0}^{n'} \sum_{j'=0}^{m'} a_{n'nm'm}^{(\ell)} a_{n'i'm'j'}^{(\ell)} \hat{J}_{i+i'+j+j'+2\ell,j+j',\ell} . \tag{9.91}$$

We also remark that the irreducible moments with negative powers i and j of $E_{\mathbf{k}u}$ and $E_{\mathbf{k}l}$, respectively, can be expressed in terms of the ones with positive indices.

9.4   Equations of Motion for the Irreducible Moments

The space-time evolution of the single-particle distribution function $f_{\mathbf{k}}$ of a single-component, weakly interacting, dilute gas is given by the relativistic Boltzmann equation (7.1). Considering only binary collisions, the collision term on the right-hand side is given by Eq. (5.2).

In order to derive equations of motion for the irreducible moments of $\delta\hat{f}_{\mathbf{k}}$, we proceed along the lines of Ref. [7], except that we express the derivatives using Eq. (9.14) instead of Eq. (5.29), and thus we rewrite the Boltzmann equation (7.1) as an evolution equation for the correction $\delta\hat{f}_{\mathbf{k}}$ instead of $\delta f_{\mathbf{k}}$

$$D\delta\hat{f}_{\mathbf{k}} = -D\hat{f}_{0\mathbf{k}} + E_{\mathbf{k}u}^{-1}\left(E_{\mathbf{k}l} D_l \hat{f}_{0\mathbf{k}} + E_{\mathbf{k}l} D_l \delta\hat{f}_{\mathbf{k}} - k^\mu \tilde{\nabla}_\mu \hat{f}_{0\mathbf{k}} - k^\mu \tilde{\nabla}_\mu \delta\hat{f}_{\mathbf{k}}\right)$$
$$+ E_{\mathbf{k}u}^{-1} C\left[\hat{f}_0 + \delta\hat{f}\right] . \tag{9.92}$$

Now, we form moments of Eq. (9.92), which leads to an infinite set of equations of motion for the irreducible moments (9.85). Defining

$$D\hat{\rho}_{ij}^{\{\mu_1\cdots\mu_\ell\}} \equiv \Xi_{\nu_1\cdots\nu_\ell}^{\mu_1\cdots\mu_\ell} D\hat{\rho}_{ij}^{\nu_1\cdots\nu_\ell} , \tag{9.93}$$

as well as

$$C_{ij}^{\{\mu_1\cdots\mu_\ell\}} = \Xi_{\nu_1\cdots\nu_\ell}^{\mu_1\cdots\mu_\ell} \int dK \, E_{ku}^i \, E_{kl}^j \, k^{\nu_1} \cdots k^{\nu_\ell} C\,[f] \,, \tag{9.94}$$

we obtain, after a long, but straightforward calculation, the equation of motion for the irreducible moments of tensor-rank zero

$$
\begin{aligned}
D\hat{\rho}_{ij} =\ & C_{i-1,j} - D\hat{\mathcal{I}}_{ij} + D_l \hat{\mathcal{I}}_{i-1,j+1} + \left(i\,\hat{\mathcal{I}}_{i-1,j+1} + j\,\hat{\mathcal{I}}_{i+1,j-1} \right) l_\alpha Du^\alpha \\
& - \left[(i-1)\,\hat{\mathcal{I}}_{i-2,j+2} + (j+1)\,\hat{\mathcal{I}}_{ij} \right] l_\alpha D_l u^\alpha \\
& + \frac{1}{2} \left[m^2 (i-1)\,\hat{\mathcal{I}}_{i-2,j} - (i+1)\,\hat{\mathcal{I}}_{ij} + (i-1)\,\hat{\mathcal{I}}_{i-2,j+2} \right] \tilde{\theta} \\
& - \frac{1}{2} \left[m^2 j\,\hat{\mathcal{I}}_{i-1,j-1} - j\,\hat{\mathcal{I}}_{i+1,j-1} + (j+2)\,\hat{\mathcal{I}}_{i-1,j+1} \right] \tilde{\theta}_l \\
& + D_l \hat{\rho}_{i-1,j+1} + \left(i\,\hat{\rho}_{i-1,j+1} + j\,\hat{\rho}_{i+1,j-1} \right) l_\alpha Du^\alpha \\
& - \left[(i-1)\,\hat{\rho}_{i-2,j+2} + (j+1)\,\hat{\rho}_{ij} \right] l_\alpha D_l u^\alpha \\
& + \frac{1}{2} \left[m^2 (i-1)\,\hat{\rho}_{i-2,j} - (i+1)\,\hat{\rho}_{ij} + (i-1)\,\hat{\rho}_{i-2,j+2} \right] \tilde{\theta} \\
& - \frac{1}{2} \left[m^2 j\,\hat{\rho}_{i-1,j-1} - j\,\hat{\rho}_{i+1,j-1} + (j+2)\,\hat{\rho}_{i-1,j+1} \right] \tilde{\theta}_l \\
& - \tilde{\nabla}_\mu \hat{\rho}_{i-1,j}^\mu + \left[(i-1)\,\hat{\rho}_{i-2,j+1}^\mu + j\,\hat{\rho}_{i,j-1}^\mu \right] l_\alpha \tilde{\nabla}_\mu u^\alpha \\
& + i\,\hat{\rho}_{i-1,j}^\mu Du_\mu - (i-1)\,\hat{\rho}_{i-2,j+1}^\mu D_l u_\mu \\
& - j\,\hat{\rho}_{i,j-1}^\mu D l_\mu + (j+1)\,\hat{\rho}_{i-1,j}^\mu D_l l_\mu + (i-1)\,\hat{\rho}_{i-2,j}^{\mu\nu} \tilde{\sigma}_{\mu\nu} \\
& - j\,\hat{\rho}_{i-1,j-1}^{\mu\nu} \tilde{\sigma}_{l,\mu\nu} \,.
\end{aligned}
\tag{9.95}
$$

Similarly, the time-evolution equation for the irreducible moments of tensor-rank one is

$$
\begin{aligned}
D\hat{\rho}_{ij}^{\{\mu\}} =\ & C_{i-1,j}^{\{\mu\}} - \frac{1}{2} \tilde{\nabla}^\mu \left(m^2\,\hat{\mathcal{I}}_{i-1,j} - \hat{\mathcal{I}}_{i+1,j} + \hat{\mathcal{I}}_{i-1,j+2} \right) \\
& + \frac{1}{2} \left[m^2 i\,\hat{\mathcal{I}}_{i-1,j} - (i+2)\,\hat{\mathcal{I}}_{i+1,j} + i\,\hat{\mathcal{I}}_{i-1,j+2} \right] \Xi_\alpha^\mu Du^\alpha \\
& - \frac{1}{2} \left[m^2 j\,\hat{\mathcal{I}}_{i,j-1} - j\,\hat{\mathcal{I}}_{i+2,j-1} + (j+2)\,\hat{\mathcal{I}}_{i,j+1} \right] \Xi_\alpha^\mu D l^\alpha \\
& - \frac{1}{2} \left[m^2 (i-1)\,\hat{\mathcal{I}}_{i-2,j+1} - (i+1)\,\hat{\mathcal{I}}_{i,j+1} + (i-1)\,\hat{\mathcal{I}}_{i-2,j+3} \right] \Xi_\alpha^\mu D_l u^\alpha \\
& + \frac{1}{2} \left[m^2 (j+1)\,\hat{\mathcal{I}}_{i-1,j} - (j+1)\,\hat{\mathcal{I}}_{i+1,j} + (j+3)\,\hat{\mathcal{I}}_{i-1,j+2} \right] \Xi_\alpha^\mu D_l l^\alpha \\
& + \frac{1}{2} \left[(i-1) \left(m^2\,\hat{\mathcal{I}}_{i-2,j+1} - \hat{\mathcal{I}}_{i,j+1} + \hat{\mathcal{I}}_{i-2,j+3} \right) \right. \\
& \qquad \left. + j \left(m^2\,\hat{\mathcal{I}}_{i,j-1} - \hat{\mathcal{I}}_{i+2,j-1} + \hat{\mathcal{I}}_{i,j+1} \right) \right] l_\alpha \tilde{\nabla}^\mu u^\alpha \\
& + \frac{1}{2} \left[m^2 i\,\hat{\rho}_{i-1,j} - (i+2)\,\hat{\rho}_{i+1,j} + i\,\hat{\rho}_{i-1,j+2} \right] \Xi_\alpha^\mu Du^\alpha
\end{aligned}
$$

$$
-\frac{1}{2}\left[m^2\, j\,\hat{\rho}_{i,j-1} - j\,\hat{\rho}_{i+2,j-1} + (j+2)\,\hat{\rho}_{i,j+1}\right]\Xi_\alpha^\mu\, Dl^\alpha
$$

$$
-\frac{1}{2}\left[m^2\,(i-1)\,\hat{\rho}_{i-2,j+1} - (i+1)\,\hat{\rho}_{i,j+1} + (i-1)\,\hat{\rho}_{i-2,j+3}\right]\Xi_\alpha^\mu\, Dlu^\alpha
$$

$$
+\frac{1}{2}\left[m^2\,(j+1)\,\hat{\rho}_{i-1,j} - (j+1)\,\hat{\rho}_{i+1,j} + (j+3)\,\hat{\rho}_{i-1,j+2}\right]\Xi_\alpha^\mu\, Dl l^\alpha
$$

$$
+\frac{1}{2}\Big[(i-1)\left(m^2\,\hat{\rho}_{i-2,j+1} - \hat{\rho}_{i,j+1} + \hat{\rho}_{i-2,j+3}\right)
$$
$$
+ j\left(m^2\,\hat{\rho}_{i,j-1} - \hat{\rho}_{i+2,j-1} + \hat{\rho}_{i,j+1}\right)\Big]l_\alpha\tilde{\nabla}^\mu u^\alpha
$$

$$
-\frac{1}{2}\tilde{\nabla}^\mu\left(m^2\,\hat{\rho}_{i-1,j} - \hat{\rho}_{i+1,j} + \hat{\rho}_{i-1,j+2}\right) + \Xi_\alpha^\mu\, Dl\,\hat{\rho}_{i-1,j+1}^\alpha
$$

$$
+\left(i\,\hat{\rho}_{i-1,j+1}^\mu + j\,\hat{\rho}_{i+1,j-1}^\mu\right)l_\alpha Du^\alpha + \hat{\rho}_{ij,v}\,\tilde{\omega}^{\mu v} + \hat{\rho}_{i-1,j+1,v}\,\tilde{\omega}_l^{\mu v}
$$

$$
+\frac{1}{2}\left[m^2\,(i-1)\,\hat{\rho}_{i-2,j}^\mu - (i+2)\,\hat{\rho}_{ij}^\mu + (i-1)\,\hat{\rho}_{i-2,j+2}^\mu\right]\tilde{\theta}
$$

$$
-\frac{1}{2}\left[m^2\, j\,\hat{\rho}_{i-1,j-1}^\mu - j\,\hat{\rho}_{i+1,j-1}^\mu + (j+3)\,\hat{\rho}_{i-1,j+1}^\mu\right]\tilde{\theta}_l
$$

$$
+\frac{1}{2}\left[m^2\,(i-1)\,\hat{\rho}_{i-2,j,v} - (i+1)\,\hat{\rho}_{ij,v} + (i-1)\,\hat{\rho}_{i-2,j+2,v}\right]\tilde{\sigma}^{\mu v}
$$

$$
-\frac{1}{2}\left[m^2\, j\,\hat{\rho}_{i-1,j-1,v} - j\,\hat{\rho}_{i+1,j-1,v} + (j+2)\,\hat{\rho}_{i-1,j+1,v}\right]\tilde{\sigma}_l^{\mu v}
$$

$$
-\left[(i-1)\,\hat{\rho}_{i-2,j+2}^\mu + (j+1)\,\hat{\rho}_{ij}^\mu\right]l_\alpha Dlu^\alpha + \left[(i-1)\,\hat{\rho}_{i-2,j+1}^{\mu v} + j\,\hat{\rho}_{i,j-1}^{\mu v}\right]l_\alpha\tilde{\nabla}_v u^\alpha
$$

$$
+ i\,\hat{\rho}_{i-1,j}^{\mu v}\, Du_v - j\,\hat{\rho}_{i,j-1}^{\mu v}\, Dl_v - (i-1)\,\hat{\rho}_{i-2,j+1}^{\mu v}\, Dlu_v + (j+1)\,\hat{\rho}_{i-1,j}^{\mu v}\, Dl l_v
$$

$$
- \Xi_\alpha^\mu\tilde{\nabla}_v\hat{\rho}_{i-1,j}^{\alpha v} + (i-1)\,\hat{\rho}_{i-2,j}^{\mu v\lambda}\tilde{\sigma}_{v\lambda} - j\,\hat{\rho}_{i-1,j-1}^{\mu v\lambda}\tilde{\sigma}_{l,v\lambda}\,. \tag{9.96}
$$

Finally, the equation of motion for the irreducible moments of tensor-rank two is

$$
D\hat{\rho}_{ij}^{\{\mu v\}} = C_{i-1,j}^{\{\mu v\}} + \frac{1}{4}\Big\{m^4\,(i-1)\,\hat{\mathcal{I}}_{i-2,j} - 2m^2\left[(i+1)\,\hat{\mathcal{I}}_{ij} - (i-1)\,\hat{\mathcal{I}}_{i-2,j+2}\right]
$$
$$
- 2(i+1)\,\hat{\mathcal{I}}_{i,j+2} + (i+3)\,\hat{\mathcal{I}}_{i+2,j} + (i-1)\,\hat{\mathcal{I}}_{i-2,j+4}\Big\}\tilde{\sigma}^{\mu v}
$$

$$
-\frac{1}{4}\Big\{m^4\, j\,\hat{\mathcal{I}}_{i-1,j-1} - 2m^2\left[j\,\hat{\mathcal{I}}_{i+1,j-1} - (j+2)\,\hat{\mathcal{I}}_{i-1,j+1}\right]
$$
$$
- 2(j+2)\,\hat{\mathcal{I}}_{i+1,j+1} + j\,\hat{\mathcal{I}}_{i+3,j-1} + (j+4)\,\hat{\mathcal{I}}_{i-1,j+3}\Big\}\tilde{\sigma}_l^{\mu v}
$$

$$
+\frac{1}{4}\Big\{m^4\,(i-1)\,\hat{\rho}_{i-2,j} - 2m^2\left[(i+1)\,\hat{\rho}_{ij} - (i-1)\,\hat{\rho}_{i-2,j+2}\right]
$$
$$
- 2(i+1)\,\hat{\rho}_{i,j+2} + (i+3)\,\hat{\rho}_{i+2,j} + (i-1)\,\hat{\rho}_{i-2,j+4}\Big\}\tilde{\sigma}^{\mu v}
$$

$$
-\frac{1}{4}\Big\{m^4\, j\,\hat{\rho}_{i-1,j-1} - 2m^2\left[j\,\hat{\rho}_{i+1,j-1} - (j+2)\,\hat{\rho}_{i-1,j+1}\right]
$$
$$
- 2(j+2)\,\hat{\rho}_{i+1,j+1} + j\,\hat{\rho}_{i+3,j-1} + (j+4)\,\hat{\rho}_{i-1,j+3}\Big\}\tilde{\sigma}_l^{\mu v}
$$

$$
+\frac{1}{2}\left[m^2\, i\,\hat{\rho}_{i-1,j}^{\{\mu} - (i+4)\,\hat{\rho}_{i+1,j}^{\{\mu} + i\,\hat{\rho}_{i-1,j+2}^{\{\mu}\right]Du^{v\}}
$$

$$
-\frac{1}{2}\left[m^2\, j\,\hat{\rho}_{i,j-1}^{\{\mu} - j\,\hat{\rho}_{i+2,j-1}^{\{\mu} + (j+4)\,\hat{\rho}_{i,j+1}^{\{\mu}\right]Dl^{v\}}
$$

$$
+ \frac{1}{2} \Xi_{\alpha\beta}^{\mu\nu} \Big[(i-1) \left(m^2 \, \hat{\rho}_{i-2,j+1}^\alpha - \hat{\rho}_{i,j+1}^\alpha + \hat{\rho}_{i-2,j+3}^\alpha \right)
$$

$$
+ j \left(m^2 \, \hat{\rho}_{i,j-1}^\alpha - \hat{\rho}_{i+2,j-1}^\alpha + \hat{\rho}_{i,j+1}^\alpha \right) \Big] l_\gamma \tilde{\nabla}^\beta u^\gamma
$$

$$
- \frac{1}{2} \Xi_{\alpha\beta}^{\mu\nu} \Big[m^2 (i-1) \, \hat{\rho}_{i-2,j+1}^\alpha - (i+3) \, \hat{\rho}_{i,j+1}^\alpha + (i-1) \, \hat{\rho}_{i-2,j+3}^\alpha \Big] D_l u^\beta
$$

$$
+ \frac{1}{2} \Xi_{\alpha\beta}^{\mu\nu} \Big[m^2 (j+1) \, \hat{\rho}_{i-1,j}^\alpha - (j+1) \, \hat{\rho}_{i+1,j}^\alpha + (j+5) \, \hat{\rho}_{i-1,j+2}^\alpha \Big] D_l l^\beta
$$

$$
- \frac{1}{2} \tilde{\nabla}^{\{\mu} \left(m^2 \, \hat{\rho}_{i-1,j}^{\nu\}} - \hat{\rho}_{i+1,j}^{\nu\}} + \hat{\rho}_{i-1,j+2}^{\nu\}} \right)
$$

$$
+ \Xi_{\alpha\beta}^{\mu\nu} D_l \hat{\rho}_{i-1,j+1}^{\alpha\beta} - 2 \tilde{\omega}_\lambda^{\{\mu} \hat{\rho}_{ij}^{\nu\}\lambda} - 2 \tilde{\omega}_{l,\lambda}^{\{\mu} \hat{\rho}_{i-1,j+1}^{\nu\}\lambda}
$$

$$
+ \left(i \, \hat{\rho}_{i-1,j+1}^{\mu\nu} + j \, \hat{\rho}_{i+1,j-1}^{\mu\nu} \right) l_\alpha D u^\alpha - \Big[(i-1) \, \hat{\rho}_{i-2,j+2}^{\mu\nu} + (j+1) \, \hat{\rho}_{ij}^{\mu\nu} \Big] l_\alpha D_l u^\alpha
$$

$$
+ \frac{1}{2} \Big[m^2 (i-1) \, \hat{\rho}_{i-2,j}^{\mu\nu} - (i+3) \, \hat{\rho}_{ij}^{\mu\nu} + (i-1) \, \hat{\rho}_{i-2,j+2}^{\mu\nu} \Big] \tilde{\theta}
$$

$$
+ \frac{2}{3} \Big[m^2 (i-1) \, \hat{\rho}_{i-2,j}^{\kappa\{\mu} - (i+2) \, \hat{\rho}_{ij}^{\kappa\{\mu} + (i-1) \, \hat{\rho}_{i-2,j+2}^{\kappa\{\mu} \Big] \tilde{\sigma}_\kappa^{\nu\}}
$$

$$
- \frac{1}{2} \Big[m^2 \, j \, \hat{\rho}_{i-1,j-1}^{\mu\nu} - j \, \hat{\rho}_{i+1,j-1}^{\mu\nu} + (j+4) \, \hat{\rho}_{i-1,j+1}^{\mu\nu} \Big] \tilde{\theta}_l
$$

$$
- \frac{2}{3} \Big[m^2 \, j \, \hat{\rho}_{i-1,j-1}^{\kappa\{\mu} - j \, \hat{\rho}_{i+1,j-1}^{\kappa\{\mu} + (j+3) \, \hat{\rho}_{i-1,j+1}^{\kappa\{\mu} \Big] \tilde{\sigma}_{l,\kappa}^{\nu\}} - \Xi_{\alpha\beta}^{\mu\nu} \tilde{\nabla}_\lambda \hat{\rho}_{i-1,j}^{\alpha\beta\lambda}
$$

$$
+ i \, \hat{\rho}_{i-1,j}^{\mu\nu\gamma} D u_\gamma - j \, \hat{\rho}_{i,j-1}^{\mu\nu\gamma} D l_\gamma - (i-1) \, \hat{\rho}_{i-2,j+1}^{\mu\nu\lambda} D_l u_\lambda + (j+1) \, \hat{\rho}_{i-1,j}^{\mu\nu\lambda} D_l l_\lambda
$$

$$
+ \Big[(i-1) \, \hat{\rho}_{i-2,j+1}^{\mu\nu\lambda} + j \, \hat{\rho}_{i,j-1}^{\mu\nu\lambda} \Big] l_\alpha \tilde{\nabla}_\lambda u^\alpha + (i-1) \, \hat{\rho}_{i-2,j}^{\mu\nu\lambda\kappa} \tilde{\sigma}_{\lambda\kappa} - j \, \hat{\rho}_{i-1,j-1}^{\mu\nu\lambda\kappa} \tilde{\sigma}_{l,\lambda\kappa} \; .
$$

$$
\tag{9.97}
$$

Since the fluid-dynamical equations of motion do not contain quantities of tensor rank higher than two, we do not explicitly quote the equations of motion for the irreducible moments $\hat{\rho}_{ij}^{\{\mu_1 \cdots \mu_\ell\}}$ with $\ell \geq 3$. The equations of motion of relativistic dissipative fluid dynamics for an anisotropic reference state can now be obtained from these general equations for different values of i and j. This is much more complicated than for the case of an isotropic reference state, and as of yet, has only been done in Ref. [8], in the lowest, 14-moment approximation, to which we refer the interested reader for further details. A resummation of higher moments in the framework of RTRFD, as explained in Chap. 6, is still missing.

9.5 Collision Integrals

In order to complete the derivation of the moment equations, we still need to consider the collision terms (9.94), which appear in Eqs. (9.95)–(9.97). Exchanging integration variables $(\mathbf{p}, \mathbf{p}') \leftrightarrow (\mathbf{k}, \mathbf{k}')$, we can rewrite Eq. (9.94) as

$$
C_{ij}^{\{\mu_1 \cdots \mu_\ell\}} = \frac{1}{2} \int dK \, dK' dP \, dP' \, f_\mathbf{k} f_{\mathbf{k}'} \left(1 - a f_\mathbf{p} \right) \left(1 - a f_{\mathbf{p}'} \right) W_{\mathbf{k}\mathbf{k}' \to \mathbf{p}\mathbf{p}'}
$$

$$
\times \left(E_{\mathbf{p}u}^i E_{\mathbf{p}l}^j p^{\{\mu_1} \cdots p^{\mu_\ell\}} - E_{\mathbf{k}u}^i E_{\mathbf{k}l}^j k^{\{\mu_1} \cdots k^{\mu_\ell\}} \right) .
$$

$$
\tag{9.98}
$$

As a consequence of the conservation of particle number, as well as energy and momentum in binary collisions, we have

$$C_{00} = C_{10} = C_{01} = C_{00}^{\{\mu\}} = 0 , \tag{9.99}$$

for any distribution function $f_\mathbf{k}$.

Now inserting the distribution function from Eq. (9.84) into Eq. (9.98) and neglecting terms proportional to $\delta \hat{f}_\mathbf{k} \delta \hat{f}_{\mathbf{k}'}$ we obtain

$$C_{ij}^{\{\mu_1 \cdots \mu_\ell\}} = \hat{C}_{ij}^{\{\mu_1 \cdots \mu_\ell\}} + \hat{\mathcal{L}}_{ij}^{\{\mu_1 \cdots \mu_\ell\}} . \tag{9.100}$$

Here,

$$\hat{C}_{ij}^{\{\mu_1 \cdots \mu_\ell\}} \equiv \frac{1}{2} \int dK dK' dP dP' W_{\mathbf{k}\mathbf{k}' \to \mathbf{p}\mathbf{p}'} \hat{f}_{0\mathbf{k}} \hat{f}_{0\mathbf{k}'} (1 - a\hat{f}_{0\mathbf{p}})(1 - a\hat{f}_{0\mathbf{p}'})$$
$$\times \left(E_{\mathbf{p}u}^i E_{\mathbf{p}l}^j p^{\{\mu_1} \cdots p^{\mu_\ell\}} - E_{\mathbf{k}u}^i E_{\mathbf{k}l}^j k^{\{\mu_1} \cdots k^{\mu_\ell\}} \right) . \tag{9.101}$$

In local thermodynamical equilibrium, i.e., replacing $\hat{f}_{0\mathbf{k}}$ by $f_{0\mathbf{k}}$, such a collision integral vanishes due to the symmetry of the collision rate, $W_{\mathbf{k}\mathbf{k}' \to \mathbf{p}\mathbf{p}'} = W_{\mathbf{p}\mathbf{p}' \to \mathbf{k}\mathbf{k}'}$, and energy conservation in binary elastic collisions, $E_{\mathbf{p}u} + E_{\mathbf{p}'u} = E_{\mathbf{k}u} + E_{\mathbf{k}'u}$. However, for the anisotropic state characterized by $\hat{f}_{0\mathbf{k}}$ this is a priori not the case. Thus, if we consider the fluid dynamics of such a system, without additional corrections from the irreducible moments of $\delta \hat{f}_\mathbf{k}$, as was done in Refs. [9–11, 16–20], the microscopic collision dynamics contained in the term (9.101) is solely responsible for the approach toward local thermodynamic equilibrium.

Let us now turn to the second term in Eq. (9.100) and, in order to simplify the discussion, consider the case of Boltzmann statistics, $a = 0$. Then

$$\hat{\mathcal{L}}_{ij}^{\{\mu_1 \cdots \mu_\ell\}} = \frac{1}{2} \sum_{r=0}^{\infty} \sum_{n=0}^{N_r} \sum_{m=0}^{N_r - n} \hat{\rho}_{nm}^{\nu_1 \cdots \nu_r} \int dK dK' dP dP' W_{\mathbf{k}\mathbf{k}' \to \mathbf{p}\mathbf{p}'} \hat{f}_{0\mathbf{k}} \hat{f}_{0\mathbf{k}'}$$
$$\times \left(k_{\{\nu_1} \cdots k_{\nu_r\}} \hat{\mathcal{H}}_{\mathbf{k}nm}^{(r)} + k'_{\{\nu_1} \cdots k'_{\nu_r\}} \hat{\mathcal{H}}_{\mathbf{k}'nm}^{(r)} \right)$$
$$\times \left(E_{\mathbf{p}u}^i E_{\mathbf{p}l}^j p^{\{\mu_1} \cdots p^{\mu_\ell\}} - E_{\mathbf{k}u}^i E_{\mathbf{k}l}^j k^{\{\mu_1} \cdots k^{\mu_\ell\}} \right) . \tag{9.102}$$

In analogy to Eq. (6.46), this expression can be rewritten as; see Appendix 9.10 for details,

$$\hat{\mathcal{L}}_{ij}^{\{\mu_1 \cdots \mu_\ell\}} \equiv \sum_{n=0}^{N_\ell} \sum_{m=0}^{N_\ell - n} \hat{\rho}_{nm}^{\mu_1 \cdots \mu_\ell} \mathcal{A}_{injm}^{(\ell)} , \tag{9.103}$$

where

$$\mathcal{A}_{injm}^{(\ell)} = \frac{1}{2} \int dK dK' dP dP' W_{\mathbf{kk'}\to\mathbf{pp'}} \hat{f}_{0\mathbf{k}} \hat{f}_{0\mathbf{k'}}$$

$$\times \left(k_{\{\mu_1} \cdots k_{\mu_\ell\}} \hat{\mathcal{H}}_{\mathbf{k}nm}^{(\ell)} + k'_{\{\mu_1} \cdots k'_{\mu_\ell\}} \hat{\mathcal{H}}_{\mathbf{k'}nm}^{(\ell)} \right)$$

$$\times \left(E_{\mathbf{p}u}^i E_{\mathbf{p}l}^j p^{\{\mu_1} \cdots p^{\mu_\ell\}} - E_{\mathbf{k}u}^i E_{\mathbf{k}l}^j k^{\{\mu_1} \cdots k^{\mu_\ell\}} \right) . \quad (9.104)$$

Similar to the tensor $\hat{C}_{ij}^{\{\mu_1\cdots\mu_\ell\}}$ in Eq. (9.101), the coefficients $\mathcal{A}_{injm}^{(\ell)}$ contain information about the microscopic interactions. The difference is that the tensor $\hat{\mathcal{L}}_{ij}^{\{\mu_1\cdots\mu_\ell\}}$ is linearly proportional to the irreducible moments $\hat{\rho}_{nm}^{\mu_1\cdots\mu_\ell}$. This means on the one hand that $\hat{\mathcal{L}}_{ij}^{\{\mu_1\cdots\mu_\ell\}} = 0$ if we consider a state characterized by $\hat{f}_{0\mathbf{k}}$ without corrections $\sim \delta\hat{f}_{\mathbf{k}}$, such as in Refs. [9–11,16–20]. On the other hand, the linear proportionality of $\hat{\mathcal{L}}_{ij}^{\{\mu_1\cdots\mu_\ell\}}$ to $\hat{\rho}_{nm}^{\mu_1\cdots\mu_\ell}$ means that the coefficients $\mathcal{A}_{injm}^{(\ell)}$ are inversely proportional to the relaxation time scales for the irreducible moments.

In the remainder of this section, we discuss a widely used simplification for the collision integral, the so-called relaxation-time approximation (RTA) [26]. In this case, the full collision integral is replaced by the following relativistically invariant expression [27], cf. Eq. (7.2),

$$C[f] \equiv -\frac{E_{\mathbf{k}u}}{\tau_R} (f_{\mathbf{k}} - f_{0\mathbf{k}}) = -\frac{E_{\mathbf{k}u}}{\tau_R} \left(\hat{f}_{0\mathbf{k}} + \delta\hat{f}_{\mathbf{k}} - f_{0\mathbf{k}} \right) , \quad (9.105)$$

where we used Eq. (9.78) and τ_R is the so-called relaxation time, which is assumed to be independent of the particle 4-momenta and determines the time scale over which $f_{\mathbf{k}}$ relaxes toward $f_{0\mathbf{k}}$. In our case, this approximation translates with the help of Eq. (9.94) and the definitions (9.56), (9.85) into

$$C_{i-1,j}^{\{\mu_1\cdots\mu_\ell\}} = -\frac{1}{\tau_R} \left(\hat{\mathcal{I}}_{ij}^{\{\mu_1\cdots\mu_\ell\}} + \hat{\rho}_{ij}^{\{\mu_1\cdots\mu_\ell\}} - \mathcal{I}_{ij}^{\{\mu_1\cdots\mu_\ell\}} \right) , \quad (9.106)$$

where we defined the generalized moments of the local-equilibrium distribution function $f_{0\mathbf{k}}$ of tensor-rank ℓ

$$\lim_{\hat{\beta}_l\to 0} \hat{\mathcal{I}}_{ij}^{\mu_1\cdots\mu_\ell} \equiv \mathcal{I}_{ij}^{\mu_1\cdots\mu_\ell} = \left\langle E_{\mathbf{k}u}^i E_{\mathbf{k}l}^j k^{\mu_1} \cdots k^{\mu_\ell} \right\rangle_0 , \quad (9.107)$$

for more details see Appendix 9.9. Note that for symmetry reasons $\mathcal{I}_{ij}^{\mu_1\cdots\mu_\ell} \equiv 0$ for all odd j. For the scalar collision integrals, this simplifies to

$$C_{i-1,j} = -\frac{1}{\tau_R} \left(\hat{\mathcal{I}}_{ij} + \hat{\rho}_{ij} - \mathcal{I}_{ij} \right) . \quad (9.108)$$

However, for any $\ell \geq 1$, $\hat{\mathcal{I}}_{ij}^{\{\mu_1\cdots\mu_\ell\}} = \mathcal{I}_{ij}^{\{\mu_1\cdots\mu_\ell\}} = 0$, and we have

$$C_{i-1,j}^{\{\mu_1\cdots\mu_\ell\}} = -\frac{1}{\tau_R}\hat{\rho}_{ij}^{\{\mu_1\cdots\mu_\ell\}} , \quad \ell \geq 1 . \tag{9.109}$$

This means that, in RTA, for any $\ell \geq 1$, we have $\hat{C}_{ij}^{\{\mu_1\cdots\mu_\ell\}} \equiv 0$ and $\mathcal{A}_{injm}^{(\ell)} = -\delta_{in}\delta_{jm}/\tau_R$.

9.6   Summary

In this chapter, we have laid out the foundations of the method of moments for an anisotropic reference state. What remains to be done is to truncate the infinite set of equations of motion (9.95)–(9.97) for the irreducible moments $\hat{\rho}_{ij}^{\mu_1\cdots\mu_\ell}$ of the deviation $\delta\hat{f}_{\mathbf{k}}$ of the single-particle distribution function $f_{\mathbf{k}}$ from the one for this reference state, $\hat{f}_{0\mathbf{k}}$. The most simple case is the 14-moment approximation, as shown in detail in Ref. [8].

Possible applications of this generalized fluid-dynamical theory include the highly anisotropic initial stage of a heavy-ion collision, but potentially also the dynamics of neutron-star matter in intense magnetic fields, where the latter singles out a direction in space and thus introduces an anisotropy. However, in order to treat the dynamical backreaction of the (electro-)magnetic field on the fluid requires a magneto-hydrodynamical framework. The derivation of a second-order dissipative relativistic magneto-hydrodynamical theory using the method of moments is beyond the scope of this monograph, but can be found, e.g., in Refs. [28,29].

9.7   Appendix 1: Irreducible Projection Operators

In this appendix, we present the irreducible projection operators necessary to derive the irreducible moments of $\delta\hat{f}_{\mathbf{k}}$. In analogy to Eq. (6.75), the irreducible projection operators read

$$\begin{aligned}
\Xi_{\nu_1\cdots\nu_n}^{\mu_1\cdots\mu_n} = \sum_{q=0}^{[n/2]} \hat{C}(n,q)\,\frac{1}{\mathcal{N}_{nq}} \\
\times \sum_{\mathcal{P}_\mu^n \mathcal{P}_\nu^n} \Xi^{\mu_1\mu_2} \cdots \Xi^{\mu_{2q-1}\mu_{2q}} \Xi_{\nu_1\nu_2} \cdots \Xi_{\nu_{2q-1}\nu_{2q}} \Xi_{\nu_{2q+1}}^{\mu_{2q+1}} \cdots \Xi_{\nu_n}^{\mu_n} .
\end{aligned} \tag{9.110}$$

The coefficient in front of the sum over distinct permutations is the same as in Eq. (6.75). The coefficient $\hat{C}(n,q)$ will be determined below. Just as the projectors (6.75), the projectors (9.110) are also symmetric under the interchange of μ- and ν-type indices

$$\Xi_{\nu_1\cdots\nu_n}^{\mu_1\cdots\mu_n} = \Xi_{(\nu_1\cdots\nu_n)}^{(\mu_1\cdots\mu_n)} , \tag{9.111}$$

and traceless with respect to contraction of either μ- or ν-type indices

$$\varXi^{\mu_1 \cdots \mu_n}_{\nu_1 \cdots \nu_n} g_{\mu_i \mu_j} = \varXi^{\mu_1 \cdots \mu_n}_{\nu_1 \cdots \nu_n} g^{\nu_i \nu_j} = 0 \quad \text{for any } i, j \,. \tag{9.112}$$

This relation forms the basis for the determination of the coefficients $\hat{C}(n, q)$, as we shall show now. (The calculation closely follows that given in Chap. 6 for the coefficients $C(n, q)$, cf. Eq. (6.76), cf. also Ref. [6].)

Without loss of generality, let us consider the contraction (9.112) of the projector (9.110) with respect to the two indices μ_1, μ_2. For the following arguments, it is advantageous to replace the sum over distinct permutations $\mathcal{P}^n_\mu \mathcal{P}^n_\nu$ of the n μ- and ν-type indices by the sum over *all* permutations $\bar{\mathcal{P}}^n_\mu \bar{\mathcal{P}}^n_\nu$ (with in total $(n!)^2$ different terms). In the various terms of this sum, the indices μ_1, μ_2 can appear in four different ways (for the sake of notational simplicity, we omit the other $\varXi$ projectors in these terms):

		$\xrightarrow{\times g_{\mu_1 \mu_2}}$		
(i)	$\varXi^{\mu_1 \mu_i} \varXi^{\mu_j \mu_2}$	$\longrightarrow$	$\varXi^{\mu_i \mu_j}$	$2q(2q-2)$ terms ,
(ii)	$\varXi^{\mu_1 \mu_2}$	$\longrightarrow$	2	$2q$ terms ,
(iii)	$\varXi^{\mu_1 \mu_i} \varXi^{\mu_2}_{\nu_j}$	$\longrightarrow$	$\varXi^{\mu_i}_{\nu_j}$	$2 \cdot 2q(n-2q)$ terms ,
(iv)	$\varXi^{\mu_1}_{\nu_i} \varXi^{\mu_2}_{\nu_j}$	$\longrightarrow$	$\varXi_{\nu_i \nu_j}$	$(n-2q)(n-2q-1)$ terms .

The arrow symbolizes contraction with $g_{\mu_1 \mu_2}$, with the corresponding result shown to the right of the arrow. On the far right, we also denoted the number of times that such terms occur in the sum over all permutations. Computing this number is a simple combinatorial exercise: for case (i), the index μ_1 can appear in $2q$ different positions and the index μ_2 in the remaining $2q - 2$ positions. For case (ii), there are $2q$ different positions for the index μ_1, but then the position of μ_2 is fixed. For case (iii), there are $2q$ positions for μ_1 and $n - 2q$ positions for μ_2. Interchanging $\mu_1 \leftrightarrow \mu_2$ gives another factor of 2. Finally, for case (iv) there are $n - 2q$ positions for μ_1 and $n - 2q - 1$ remaining positions for μ_2.

The cases (i)–(iii) generate terms of the form

$$\varXi^{\mu_3 \mu_4} \cdots \varXi^{\mu_{2q-1} \mu_{2q}} \varXi_{\nu_1 \nu_2} \cdots \varXi_{\nu_{2q-1} \nu_{2q}} \varXi^{\mu_{2q+1}}_{\nu_{2q+1}} \cdots \varXi^{\mu_n}_{\nu_n} \,, \tag{9.113}$$

while case (iv) generates terms of the form

$$\varXi^{\mu_3 \mu_4} \cdots \varXi^{\mu_{2q+1} \mu_{2q+2}} \varXi_{\nu_1 \nu_2} \cdots \varXi_{\nu_{2q+1} \nu_{2q+2}} \varXi^{\mu_{2q+3}}_{\nu_{2q+3}} \cdots \varXi^{\mu_n}_{\nu_n} \,. \tag{9.114}$$

Note that, because of the contraction of μ_1, μ_2, only $n - 2$ indices of the μ-type are to be permutated. Collecting all terms (i)–(iv) with the correct prefactors, Eq. (9.112) reads

$$0 = \sum_{q=0}^{[n/2]} \hat{C}(n,q) \frac{1}{(n!)^2} \sum_{\bar{\mathcal{P}}_\mu^{n-2}\bar{\mathcal{P}}_\nu^n} \{2q\,[2q-2+2+2(n-2q)]$$

$$\times \; \Xi^{\mu_1\mu_2} \cdots \Xi^{\mu_{2q-3}\mu_{2q-2}} \Xi_{\nu_1\nu_2} \cdots \Xi_{\nu_{2q-1}\nu_{2q}} \Xi^{\mu_{2q-1}}_{\nu_{2q+1}} \cdots \Xi^{\mu_{n-2}}_{\nu_n}$$

$$+ \; (n-2q)(n-2q-1)\Xi^{\mu_1\mu_2}\cdots\Xi^{\mu_{2q-1}\mu_{2q}}\Xi_{\nu_1\nu_2}\cdots\Xi_{\nu_{2q+1}\nu_{2q+2}}\Xi^{\mu_{2q+1}}_{\nu_{2q+3}}\cdots\Xi^{\mu_{n-2}}_{\nu_n}\},$$

$$(9.115)$$

where we relabeled the μ-type indices $\mu_i \to \mu_{i-2}$. We now observe that the $q=0$-term does not contribute to the first term in curly brackets, while the $q=[n/2]$-term does not contribute to the last term (no matter whether n is even or odd). Thus, we may write

$$0 = \frac{1}{(n!)^2} \sum_{\bar{\mathcal{P}}_\mu^n\bar{\mathcal{P}}_\nu^n} \left\{ \sum_{q=1}^{[n/2]} \hat{C}(n,q)\,2q\,(2n-2q) \right.$$

$$\times \; \Xi^{\mu_1\mu_2} \cdots \Xi^{\mu_{2q-3}\mu_{2q-2}} \Xi_{\nu_1\nu_2} \cdots \Xi_{\nu_{2q-1}\nu_{2q}} \Xi^{\mu_{2q-1}}_{\nu_{2q+1}} \cdots \Xi^{\mu_{n-2}}_{\nu_n}$$

$$+ \; \sum_{q=0}^{[n/2]-1} \hat{C}(n,q)\,(n-2q)(n-2q-1)$$

$$\left. \times \; \Xi^{\mu_1\mu_2} \cdots \Xi^{\mu_{2q-1}\mu_{2q}} \Xi_{\nu_1\nu_2} \cdots \Xi_{\nu_{2q+1}\nu_{2q+2}} \Xi^{\mu_{2q+1}}_{\nu_{2q+3}} \cdots \Xi^{\mu_{n-2}}_{\nu_n} \right\}. \quad (9.116)$$

Substituting $q \to q-1$ in the second sum, we observe that the product of the Ξ-projectors is actually identical in both sums, so that we obtain

$$0 = \sum_{q=1}^{[n/2]} \left[\hat{C}(n,q)\,2q\,(2n-2q) + \hat{C}(n,q-1)\,(n-2q+2)(n-2q+1) \right]$$

$$\times \frac{1}{(n!)^2} \sum_{\bar{\mathcal{P}}_\mu^{n-2}\bar{\mathcal{P}}_\nu^n} \Xi^{\mu_1\mu_2} \cdots \Xi^{\mu_{2q-3}\mu_{2q-2}} \Xi_{\nu_1\nu_2} \cdots \Xi_{\nu_{2q-1}\nu_{2q}} \Xi^{\mu_{2q-1}}_{\nu_{2q+1}} \cdots \Xi^{\mu_{n-2}}_{\nu_n}. \quad (9.117)$$

In order to fulfill this relation, we have to demand that the term in brackets vanishes, which leads to the recursion relation

$$\hat{C}(n,q) = -\frac{(n-2q+2)(n-2q+1)}{2q\,(2n-2q)} \hat{C}(n,q-1). \quad (9.118)$$

If we set

$$\hat{C}(n,0) \equiv 1 \quad \forall\, n, \quad (9.119)$$

the solution is

$$\hat{C}(n, q) = (-1)^q \frac{1}{2^q q!} \frac{n!}{(n-2q)!} \frac{(2n-2q-2)!!}{(2n-2)!!}$$

$$= (-1)^q \frac{1}{4^q} \frac{(n-q)!}{q!(n-2q)!} \frac{n}{n-q} , \qquad (9.120)$$

where we used the definition of the double factorial for even numbers.

Upon complete contraction,

$$\mathcal{E}^{\mu_1 \cdots \mu_n}_{\mu_1 \cdots \mu_n} \equiv \mathcal{E}^{\mu_1 \cdots \mu_n}_{\nu_1 \cdots \nu_n} g^{\nu_1}_{\mu_1} \cdots g^{\nu_n}_{\mu_n} = 2 . \qquad (9.121)$$

In order to prove this relation, it is advantageous to first prove the recursion relation

$$\mathcal{E}^{\mu_1 \cdots \mu_n}_{\nu_1 \cdots \nu_n} g^{\nu_j}_{\mu_i} = \mathcal{E}^{\mu_1 \cdots \mu_{n-1}}_{\nu_1 \cdots \nu_{n-1}} , \qquad (9.122)$$

valid for any i, j, and $n > 1$. Equation (9.121) then immediately follows on account of $\mathcal{E}^{\mu_1 \cdots \mu_n}_{\mu_1 \cdots \mu_n} = \mathcal{E}^{\mu_1}_{\mu_1} = 2$.

Equation (9.122) is proved as follows. First note that, since $\mathcal{E}^{\mu_1 \cdots \mu_n}_{\nu_1 \cdots \nu_n}$ is symmetric in all indices, we may without loss of generality choose $i = j = n$ in Eq. (9.122). Then, as already done in the derivation of Eq. (9.120), it is advantageous to replace in Eq. (9.110) the sum over distinct permutations $\mathcal{P}^n_\mu \mathcal{P}^n_\nu$ by the sum over all permutations $\bar{\mathcal{P}}^n_\mu \bar{\mathcal{P}}^n_\nu$ (with in total $(n!)^2$ different terms). Inserting this into the left-hand side of Eq. (9.122), we see that contraction of the indices μ_n, ν_n generates five different types of terms:

		$\times g^{\nu_n}_{\mu_n}$		
(i)	$\mathcal{E}^{\mu_i \mu_n} \mathcal{E}_{\nu_j \nu_n}$	$\longrightarrow$	$\mathcal{E}^{\mu_i}_{\nu_j}$	$2q \cdot 2q$ terms ,
(ii)	$\mathcal{E}^{\mu_i \mu_n} \mathcal{E}^{\mu j}_{\nu_n}$	$\longrightarrow$	$\mathcal{E}^{\mu_i \mu_j}$	$2q(n-2q)$ terms ,
(iii)	$\mathcal{E}_{\nu_i \nu_n} \mathcal{E}^{\mu_n}_{\nu_j}$	$\longrightarrow$	$\mathcal{E}_{\nu_i \nu_j}$	$2q(n-2q)$ terms ,
(iv)	$\mathcal{E}^{\mu_n}_{\nu_j} \mathcal{E}^{\mu_i}_{\nu_n}$	$\longrightarrow$	$\mathcal{E}^{\mu_i}_{\nu_j}$	$(n-2q)(n-2q-1)$ terms ,
(v)	$\mathcal{E}^{\mu_n}_{\nu_n}$	$\longrightarrow$	2	$(n-2q)$ terms .

The number of terms is easily explained as follows: in case (i), there are $2q$ possible positions for the index μ_n and $2q$ possible positions for the index ν_n. In cases (ii) and (iii), there are $2q$ resp. $n - 2q$ positions for the index μ_n and $n - 2q$ resp. $2q$ positions for the index ν_n. In case (iv), there are $n - 2q$ positions for the index μ_n and a remaining $n - 2q - 1$ positions for the index ν_n. Finally, in case (v), there are $n - 2q$ possibilities (equal to the number of projectors with mixed indices) to have the indices μ_n and ν_n occurring at the same projector. One observes that upon contraction, in case (i) (and after suitably relabelling indices) one generates terms of the form

$$\mathcal{E}^{\mu_1 \mu_2} \cdots \mathcal{E}^{\mu_{2q-3} \mu_{2q-2}} \mathcal{E}_{\nu_1 \nu_2} \cdots \mathcal{E}_{\nu_{2q-3} \nu_{2q-2}} \mathcal{E}^{\mu_{2q-1}}_{\nu_{2q-1}} \cdots \mathcal{E}^{\mu_{n-1}}_{\nu_{n-1}} ,$$

i.e., terms with $q - 1$ projectors $\Xi^{\mu_i \mu_j}$, $q - 1$ projectors $\Xi_{\nu_i \nu_j}$, and $n - 2q + 1$ projectors $\Xi^{\mu_i}_{\nu_j}$. On the other hand, in all other cases (ii)–(v) one generates terms of the form

$$\Xi^{\mu_1 \mu_2} \cdots \Xi^{\mu_{2q-1} \mu_{2q}} \Xi_{\nu_1 \nu_2} \cdots \Xi_{\nu_{2q-1} \nu_{2q}} \Xi^{\mu_{2q+1}}_{\nu_{2q+1}} \cdots \Xi^{\mu_{n-1}}_{\nu_{n-1}} \, ,$$

i.e., terms with q projectors $\Xi^{\mu_i \mu_j}$, q projectors $\Xi_{\nu_i \nu_j}$, and $n - 2q - 1$ projectors $\Xi^{\mu_i}_{\nu_j}$.

To proceed, it is advantageous to consider the case of n even and n odd separately. Let us first focus on the (somewhat simpler) case of n even, where $[n/2] \equiv n/2$. Collecting the results obtained so far, the left-hand side of Eq. (9.122) can be written as

$$\Xi^{\mu_1 \cdots \mu_n}_{\nu_1 \cdots \nu_n} g^{\nu_n}_{\mu_n} = \sum_{q=0}^{n/2} \hat{C}(n, q) \frac{1}{(n!)^2} \sum_{\bar{\mathcal{P}}^{n-1}_\mu \bar{\mathcal{P}}^{n-1}_\nu} [(n - 2q)(n + 2q + 1)$$
$$\times \, \Xi^{\mu_1 \mu_2} \cdots \Xi^{\mu_{2q-1} \mu_{2q}} \Xi_{\nu_1 \nu_2} \cdots \Xi_{\nu_{2q-1} \nu_{2q}} \Xi^{\mu_{2q+1}}_{\nu_{2q+1}} \cdots \Xi^{\mu_{n-1}}_{\nu_{n-1}}$$
$$+ \, 4q^2 \, \Xi^{\mu_1 \mu_2} \cdots \Xi^{\mu_{2q-3} \mu_{2q-2}} \Xi_{\nu_1 \nu_2} \cdots \Xi_{\nu_{2q-3} \nu_{2q-2}} \Xi^{\mu_{2q-1}}_{\nu_{2q-1}} \cdots \Xi^{\mu_{n-1}}_{\nu_{n-1}} \Big] \, .$$

$$(9.123)$$

We now note that the first term in brackets does not contribute for $q = n/2$, while the second term does not contribute for $q = 0$

$$\Xi^{\mu_1 \cdots \mu_n}_{\nu_1 \cdots \nu_n} g^{\nu_n}_{\mu_n} = \sum_{q=0}^{n/2-1} \hat{C}(n, q) \, (n - 2q)(n + 2q + 1) \frac{1}{(n!)^2}$$
$$\times \sum_{\bar{\mathcal{P}}^{n-1}_\mu \bar{\mathcal{P}}^{n-1}_\nu} \Xi^{\mu_1 \mu_2} \cdots \Xi^{\mu_{2q-1} \mu_{2q}} \Xi_{\nu_1 \nu_2} \cdots \Xi_{\nu_{2q-1} \nu_{2q}} \Xi^{\mu_{2q+1}}_{\nu_{2q+1}} \cdots \Xi^{\mu_{n-1}}_{\nu_{n-1}}$$
$$+ \sum_{q=1}^{n/2} \hat{C}(n, q) \, 4q^2 \frac{1}{(n!)^2}$$
$$\times \sum_{\bar{\mathcal{P}}^{n-1}_\mu \bar{\mathcal{P}}^{n-1}_\nu} \Xi^{\mu_1 \mu_2} \cdots \Xi^{\mu_{2q-3} \mu_{2q-2}} \Xi_{\nu_1 \nu_2} \cdots \Xi_{\nu_{2q-3} \nu_{2q-2}} \Xi^{\mu_{2q-1}}_{\nu_{2q-1}} \cdots \Xi^{\mu_{n-1}}_{\nu_{n-1}} \, .$$

$$(9.124)$$

Substituting the summation index $q \to q + 1$ in the second sum, we observe that the product of projectors becomes identical to the one in the first sum, so that we can write

$$\Xi^{\mu_1 \cdots \mu_n}_{\nu_1 \cdots \nu_n} g^{\nu_n}_{\mu_n} = \sum_{q=0}^{n/2-1} \Big[\hat{C}(n, q) \, (n - 2q)(n + 2q + 1) + \hat{C}(n, q + 1) \, 4(q + 1)^2 \Big]$$
$$\times \frac{1}{(n!)^2} \sum_{\bar{\mathcal{P}}^{n-1}_\mu \bar{\mathcal{P}}^{n-1}_\nu} \Xi^{\mu_1 \mu_2} \cdots \Xi^{\mu_{2q-1} \mu_{2q}} \Xi_{\nu_1 \nu_2} \cdots \Xi_{\nu_{2q-1} \nu_{2q}} \Xi^{\mu_{2q+1}}_{\nu_{2q+1}} \cdots \Xi^{\mu_{n-1}}_{\nu_{n-1}} \, .$$

$$(9.125)$$

With the definition (9.120), one now convinces oneself that

$$\frac{1}{n^2}\left[\hat{C}(n,q)(n-2q)(n+2q+1)+\hat{C}(n,q+1)\,4(q+1)^2\right]=\hat{C}(n-1,q).$$

(9.126)

Reverting the sum over all permutations of $n-1$ indices of μ- and of ν-type to the one over distinct permutations and using the definition (9.110) and the fact that for even n one has $n/2-1=[(n-1)/2]$, one then arrives at Eq. (9.122).

Let us now consider the case of n odd, where $[n/2]=[(n-1)/2]=(n-1)/2$. For $q<(n-1)/2$, the arguments of the previous case of n even can be taken over unchanged. However, when $q=(n-1)/2$ things become more subtle: in the cases (ii), (iii), and (iv), one observes that, after contraction of the indices, no further projectors of the type $\varXi_{\nu_j}^{\mu_i}$ occur. We thus treat the case $q=(n-1)/2$ separately. Collecting the results obtained so far, the left-hand side of Eq. (9.122) can be written as

$$\varXi_{\nu_1\cdots\nu_n}^{\mu_1\cdots\mu_n}\,g_{\mu_n}^{\nu_n}=\sum_{q=0}^{(n-1)/2-1}\hat{C}(n,q)\frac{1}{(n!)^2}\sum_{\bar{\mathscr{P}}_\mu^{n-1}\bar{\mathscr{P}}_\nu^{n-1}}[(n-2q)(n+2q+1)$$

$$\times\ \varXi^{\mu_1\mu_2}\cdots\varXi^{\mu_{2q-1}\mu_{2q}}\varXi_{\nu_1\nu_2}\cdots\varXi_{\nu_{2q-1}\nu_{2q}}\varXi_{\nu_{2q+1}}^{\mu_{2q+1}}\cdots\varXi_{\nu_{n-1}}^{\mu_{n-1}}$$

$$+\,4q^2\,\varXi^{\mu_1\mu_2}\cdots\varXi^{\mu_{2q-3}\mu_{2q-2}}\varXi_{\nu_1\nu_2}\cdots\varXi_{\nu_{2q-3}\nu_{2q-2}}\varXi_{\nu_{2q-1}}^{\mu_{2q-1}}\cdots\varXi_{\nu_{n-1}}^{\mu_{n-1}}\Big]$$

$$+\hat{C}\left(n,\frac{n-1}{2}\right)\frac{1}{(n!)^2}\sum_{\bar{\mathscr{P}}_\mu^{n-1}\bar{\mathscr{P}}_\nu^{n-1}}\Big[2n\,\varXi^{\mu_1\mu_2}\cdots\varXi^{\mu_{n-2}\mu_{n-1}}\varXi_{\nu_1\nu_2}\cdots\varXi_{\nu_{n-2}\nu_{n-1}}$$

$$+\,(n-1)^2\,\varXi^{\mu_1\mu_2}\cdots\varXi^{\mu_{n-4}\mu_{n-3}}\varXi_{\nu_1\nu_2}\cdots\varXi_{\nu_{n-4}\nu_{n-3}}\varXi_{\nu_{n-2}}^{\mu_{n-2}}\cdots\varXi_{\nu_{n-1}}^{\mu_{n-1}}\Big].$$

(9.127)

Noting that the term in the third line does not contribute for $q=0$, we may combine the terms in the third and fifth line to a sum that runs from $q=1$ to $(n-1)/2$. Substituting $q\to q+1$ in that sum, we observe that this sum can be combined with the one in the first line, similar to Eq. (9.125). With Eq. (9.126) this yields

$$\varXi_{\nu_1\cdots\nu_n}^{\mu_1\cdots\mu_n}\,g_{\mu_n}^{\nu_n}=\sum_{q=0}^{(n-1)/2-1}\hat{C}(n-1,q)\frac{1}{[(n-1)!]^2}$$

$$\times\sum_{\bar{\mathscr{P}}_\mu^{n-1}\bar{\mathscr{P}}_\nu^{n-1}}\varXi^{\mu_1\mu_2}\cdots\varXi^{\mu_{2q-1}\mu_{2q}}\varXi_{\nu_1\nu_2}\cdots\varXi_{\nu_{2q-1}\nu_{2q}}\varXi_{\nu_{2q+1}}^{\mu_{2q+1}}\cdots\varXi_{\nu_{n-1}}^{\mu_{n-1}}$$

$$+\hat{C}\left(n,\frac{n-1}{2}\right)\frac{2}{n}\frac{1}{[(n-1)!]^2}\sum_{\bar{\mathscr{P}}_\mu^{n-1}\bar{\mathscr{P}}_\nu^{n-1}}\varXi^{\mu_1\mu_2}\cdots\varXi^{\mu_{n-2}\mu_{n-1}}\varXi_{\nu_1\nu_2}\cdots\varXi_{\nu_{n-2}\nu_{n-1}}.$$

With Eq. (9.120), one proves that

(9.128)

$$\frac{2}{n}\hat{C}\left(n,\frac{n-1}{2}\right)=\hat{C}\left(n-1,\frac{n-1}{2}\right).$$

(9.129)

Then, the last term in Eq. (9.128) just represents the missing $q = (n - 1)/2$—term of the sum and we again obtain Eq. (9.122), q.e.d.

Note that the relation (9.121) means that the projection of an arbitrary tensor of rank n with the projector (9.110), i.e., $A^{\nu_1 \cdots \nu_n} \Xi^{\mu_1 \cdots \mu_n}_{\nu_1 \cdots \nu_n}$, has 2 independent tensor components. A proof of this was given in Appendix 6.6.

9.8   Appendix 2: Generalized Thermodynamic Integrals

In this appendix, we compute the generalized thermodynamical integrals $\hat{I}_{i+j+n,j+r,q}$ in Eq. (9.58). In that equation, we introduced the symmetrized tensor products

$$\Xi^{(\mu_1\mu_2} \cdots \Xi^{\mu_{2q-1}\mu_{2q}} l^{\mu_{2q+1}} \cdots l^{\mu_{2q+r}} u^{\mu_{2q+r+1}} \cdots u^{\mu_n)}$$

$$\equiv \frac{1}{b_{nrq}} \sum_{\mathcal{P}^n_\mu} \Xi^{\mu_1\mu_2} \cdots \Xi^{\mu_{2q-1}\mu_{2q}} l^{\mu_{2q+1}} \cdots l^{\mu_{2q+r}} u^{\mu_{2q+r+1}} \cdots u^{\mu_n} , \quad (9.130)$$

where b_{nrq} is the number of terms in the sum over distinct permutations of the n indices $\mu_1, \ldots, \mu_n$. Again, there are in total $n!$ different permutations of the indices. There are q projection operators $\Xi^{\mu_i\mu_j}$, r factors of l^{μ_k}, and $n - r - 2q$ factors of u^{μ_m}. Permutations of the order of the $\Xi^{\mu_i\mu_j}$, the l^{μ_k}, and of the u^{μ_m} among themselves to not lead to distinct terms. Likewise, permutations of the two indices of the symmetric projection operator $\Xi^{\mu_i\mu_j}$ do not lead to distinct terms. Thus, the total number of distinct terms in the symmetrized tensor product is

$$b_{nrq} \equiv \frac{n!}{2^q \, q! \, r! \, (n - r - 2q)!} = \frac{n! \, (2q - 1)!!}{(2q)! \, r! \, (n - r - 2q)!} . \quad (9.131)$$

A suitable projection of the tensor $\hat{I}^{\mu_1 \cdots \mu_n}_{ij}$ is now found by employing the orthogonality relation

$$\Xi^{(\mu_1\mu_2} \cdots \Xi^{\mu_{2q-1}\mu_{2q}} l^{\mu_{2q+1}} \cdots l^{\mu_{2q+r}} u^{\mu_{2q+r+1}} \cdots u^{\mu_n)}$$

$$\times \Xi_{(\mu_1\mu_2} \cdots \Xi_{\mu_{2q'-1}\mu_{2q'}} l_{\mu_{2q'+1}} \cdots l_{\mu_{2q'+r'}} u_{\mu_{2q'+r'+1}} \cdots u_{\mu_n)}$$

$$= (-1)^r \frac{(2q)!!}{b_{nrq}} \delta_{qq'} \delta_{rr'} . \quad (9.132)$$

In order to prove this relation, we first note that the Kronecker deltas are easily explained by the fact that if $q \neq q'$ or $r \neq r'$, there are terms where a u^{μ_i} or an l^{μ_j} are either contracted with each other or with projection operators $\Xi_{\mu_i\mu_k}$, $\Xi_{\mu_j\mu_m}$, which gives zero. We thus need to prove Eq. (9.132) only for $q = q', r = r'$. Again, since both sets of indices are symmetrized, we may keep one set fixed, i.e.

$$\Xi^{(\mu_1\mu_2} \cdots \Xi^{\mu_{2q-1}\mu_{2q}} l^{\mu_{2q+1}} \cdots l^{\mu_{2q+r}} u^{\mu_{2q+r+1}} \cdots u^{\mu_n)}$$

$$\times \Xi_{(\mu_1\mu_2} \cdots \Xi_{\mu_{2q-1}\mu_{2q}} l_{\mu_{2q+1}} \cdots l_{\mu_{2q+r}} u_{\mu_{2q+r+1}} \cdots u_{\mu_n)}$$

$$= \frac{1}{b_{nrq}}\, \varXi^{\mu_1\mu_2}\cdots \varXi^{\mu_{2q-1}\mu_{2q}} l^{\mu_{2q+1}}\cdots l^{\mu_{2q+r}} u^{\mu_{2q+r+1}}\cdots u^{\mu_n}$$

$$\times \sum_{\mathscr{P}^n_\mu} \varXi_{\mu_1\mu_2}\cdots \varXi_{\mu_{2q-1}\mu_{2q}} l_{\mu_{2q+1}}\cdots l_{\mu_{2q+r}} u_{\mu_{2q+r+1}}\cdots u_{\mu_n}$$

$$= \frac{1}{b_{nrq}}\,(-1)^r\, \varXi^{\mu_1\mu_2}\cdots \varXi^{\mu_{2q-1}\mu_{2q}} \sum_{\mathscr{P}^{2q}_\mu} \varXi_{\mu_1\mu_2}\cdots \varXi_{\mu_{2q-1}\mu_{2q}}$$

$$= \frac{1}{b_{nrq}}\,(-1)^r\,\frac{(2q)!}{2^q q!}\, \varXi^{\mu_1\mu_2}\cdots \varXi^{\mu_{2q-1}\mu_{2q}} \varXi_{(\mu_1\mu_2}\cdots \varXi_{\mu_{2q-1}\mu_{2q})}\,, \quad (9.133)$$

where in the next-to-last step, we used the fact that only those permutations in the sum are non-vanishing where the indices $\mu_{2q+1},\ldots,\mu_{2q+r}$ are on l's and $\mu_{2q+r+1},\ldots,\mu_{2n}$ are on u's. Then, we exploited $u^\mu u_\mu = 1$ and $l^\mu l_\mu = -1$. We now prove by complete induction that

$$\varXi^{\mu_1\mu_2}\cdots \varXi^{\mu_{2q-1}\mu_{2q}} \varXi_{(\mu_1\mu_2}\cdots \varXi_{\mu_{2q-1}\mu_{2q})} = \frac{(2^q q!)^2}{(2q)!}\,. \qquad (9.134)$$

This holds obviously for $q = 1$, since $\varXi^{\mu_1\mu_2} \varXi_{\mu_1\mu_2} = \varXi^{\mu_1}_{\mu_1} \equiv 2 \equiv 2^2/2$. We now assume that Eq. (9.134) holds for q and prove it for $q + 1$

$$\varXi^{\mu_1\mu_2}\cdots \varXi^{\mu_{2q+1}\mu_{2q+2}} \varXi_{(\mu_1\mu_2}\cdots \varXi_{\mu_{2q+1}\mu_{2q+2})}$$
$$= \frac{2^{q+1}(q+1)!}{(2q+2)!}\, \varXi^{\mu_1\mu_2}\cdots \varXi^{\mu_{2q+1}\mu_{2q+2}} \sum_{\mathscr{P}^{2q+2}_\mu} \varXi_{\mu_1\mu_2}\cdots \varXi_{\mu_{2q+1}\mu_{2q+2}}\,. \qquad (9.135)$$

Consider the contraction of $\varXi^{\mu_{2q+1}\mu_{2q+2}}$ with the sum over distinct permutations of $2q + 2$ indices $\mu_1,\ldots,\mu_{2q+2}$. There is one term in the sum where both indices are on the same $\varXi$ projector. This term is $\sim \varXi^{\mu_{2q+1}\mu_{2q+2}} \varXi_{\mu_{2q+1}\mu_{2q+2}} \equiv 2$. Then, there are $2q$ terms where the indices μ_{2q+1} and μ_{2q+2} are on different projectors, say $\varXi_{\mu_{2q+1}\mu_j} \varXi_{\mu_i\mu_{2q+2}}$. Contracting with $\varXi^{\mu_{2q+1}\mu_{2q+2}}$ gives a term $\sim \varXi_{\mu_i\mu_j}$, where both indices are from the set $\mu_1,\ldots,\mu_{2q}$. Putting this together and using Eq. (9.134) gives

$$\varXi^{\mu_1\mu_2}\cdots \varXi^{\mu_{2q+1}\mu_{2q+2}} \varXi_{(\mu_1\mu_2}\cdots \varXi_{\mu_{2q+1}\mu_{2q+2})}$$
$$= \frac{2(q+1)}{(2q+2)(2q+1)}\frac{2^q q!}{(2q)!}\, \varXi^{\mu_1\mu_2}\cdots \varXi^{\mu_{2q-1}\mu_{2q}} (2q+2)$$
$$\times \sum_{\mathscr{P}^{2q}_\mu} \varXi_{\mu_1\mu_2}\cdots \varXi_{\mu_{2q-1}\mu_{2q}}$$
$$= \frac{2q+2}{2q+1}\, \varXi^{\mu_1\mu_2}\cdots \varXi^{\mu_{2q-1}\mu_{2q}} \varXi_{(\mu_1\mu_2}\cdots \varXi_{\mu_{2q-1}\mu_{2q})}$$
$$= \frac{2(q+1)}{2q+1}\frac{(2^q q!)^2}{(2q)!} = \frac{[2(q+1)]^2}{(2q+2)(2q+1)}\frac{(2^q q!)^2}{(2q)!} = \frac{[2^{q+1}(q+1)!]^2}{(2q+2)!}\,,$$
$$\qquad (9.136)$$

which is Eq. (9.134) for $q + 1$. Now we insert Eq. (9.134) into Eq. (9.133) and obtain

$$
\begin{aligned}
&\Xi^{(\mu_1\mu_2} \cdots \Xi^{\mu_{2q-1}\mu_{2q}} l^{\mu_{2q+1}} \cdots l^{\mu_{2q+r}} u^{\mu_{2q+r+1}} \cdots u^{\mu_n)} \\
&\times \Xi_{(\mu_1\mu_2} \cdots \Xi_{\mu_{2q-1}\mu_{2q}} l_{\mu_{2q+1}} \cdots l_{\mu_{2q+r}} u_{\mu_{2q+r+1}} \cdots u_{\mu_n)} \\
&= \frac{1}{b_{nrq}} (-1)^r \frac{(2q)!}{2^q q!} \frac{(2^q q!)^2}{(2q)!} = \frac{1}{b_{nrq}} (-1)^r 2^q q! \, .
\end{aligned}
\tag{9.137}
$$

With $(2q)!! = 2^q q!$, we obtain Eq. (9.132), q.e.d..

The thermodynamic integrals $\hat{I}_{i+j+n,j+r,q}$ are now obtained by a projection of Eq. (9.58). The orthogonality relation (9.132) leads to the result

$$
\begin{aligned}
\hat{I}_{i+j+n,j+r,q} &\equiv \frac{(-1)^{q+r}}{(2q)!!} \hat{I}_{ij}^{\mu_1\cdots\mu_n} \Xi_{(\mu_1\mu_2} \cdots \Xi_{\mu_{2q-1}\mu_{2q}} l_{\mu_{2q+1}} \cdots l_{\mu_{2q+r}} u_{\mu_{2q+r+1}} \cdots u_{\mu_n)} \\
&= \frac{(-1)^q}{(2q)!!} \int dK \, E_{\mathbf{k}u}^{i+n-r-2q} E_{\mathbf{k}l}^{j+r} \left(\Xi^{\mu\nu} k_\mu k_\nu \right)^q \hat{f}_{0\mathbf{k}} \, ,
\end{aligned}
\tag{9.138}
$$

where we used the definition (9.56) of the tensor $\hat{I}_{ij}^{\mu_1\cdots\mu_n}$. With the definition of the average $\langle\cdots\rangle_{\hat{0}}$, the second line yields Eq. (9.60).

Note that since $E_{\mathbf{k}u}^i = \left(k^\mu u_\mu\right)^i$ and $E_{\mathbf{k}l}^j = \left(-k^\mu l_\mu\right)^j$ the tensors (6.89), (9.56) immediately follow from the projection of higher-rank tensors on tensors built from u's and l's

$$
I_i^{\mu_1\cdots\mu_n} = u^{(\mu_1} \cdots u^{\mu_i)} I_0^{(\mu_1\cdots\mu_{n+i})} \, ,
\tag{9.139}
$$

$$
\hat{I}_{ij}^{\mu_1\cdots\mu_n} = (-1)^j u^{(\alpha_1} \cdots u^{\alpha_i} l^{\beta_1} \cdots l^{\beta_j)} \hat{I}_{00}^{(\alpha_1\cdots\alpha_i\beta_1\cdots\beta_j\mu_1\cdots\mu_n)} \, .
\tag{9.140}
$$

Other useful relations are obtained from contracting two indices of the tensors (6.89), (9.56) with a Δ-resp. a Ξ-projector

$$
I_i^{\mu_1\cdots\mu_n} \Delta_{\mu_{n-1}\mu_n} = m^2 I_i^{\mu_1\cdots\mu_{n-2}} - I_{i+2}^{\mu_1\cdots\mu_{n-2}} \, ,
\tag{9.141}
$$

$$
\hat{I}_{ij}^{\mu_1\cdots\mu_n} \Xi_{\mu_{n-1}\mu_n} = m^2 \hat{I}_{ij}^{\mu_1\cdots\mu_{n-2}} - \hat{I}_{i+2,j}^{\mu_1\cdots\mu_{n-2}} + \hat{I}_{i,j+2}^{\mu_1\cdots\mu_{n-2}} \, .
\tag{9.142}
$$

The generalized thermodynamic integrals (9.60) and (9.61) obey useful recursion relations, which are given here. Using $\Xi^{\mu\nu} k_\mu k_\nu = m^2 - E_{\mathbf{k}u}^2 + E_{\mathbf{k}l}^2$ in Eq. (9.60), we get

$$
\hat{I}_{n+2,r,q} - \hat{I}_{n+2,r+2,q} = m^2 \hat{I}_{nrq} + (2q + 2) \hat{I}_{n+2,r,q+1} \, ,
\tag{9.143}
$$

$$
\hat{J}_{n+2,r,q} - \hat{J}_{n+2,r+2,q} = m^2 \hat{J}_{nrq} + (2q + 2) \hat{J}_{n+2,r,q+1} \, .
\tag{9.144}
$$

For $n = r = q = 0$, Eq. (9.143) reads

$$
\hat{I}_{200} = m^2 \hat{I}_{000} + \hat{I}_{220} + 2\hat{I}_{201} \, .
\tag{9.145}
$$

In the massless limit, this leads to the familiar relation $\hat{\varepsilon} = \hat{P}_l + 2\hat{P}_\perp$.

9.9 Appendix 3: Generalized Thermodynamic Integrals in the Equilibrium Limit

In this appendix, we derive some properties of the generalized moments (9.56) and the corresponding generalized thermodynamic integrals (9.60) in the limit of local thermodynamic equilibrium; see Eq. (9.107).

The generalized moments (9.107) can also be expanded in terms of the 4-vectors u^μ, l^μ, and $\Xi^{\mu\nu}$, just as in Eq. (9.58), where the corresponding thermodynamic integrals are defined similar to Eq. (9.60)

$$I_{nrq} = \frac{(-1)^q}{(2q)!!} \int dK \, E_{ku}^{n-r-2q} E_{kl}^r \left(\Xi^{\mu\nu} k_\mu k_\nu \right)^q f_{0k} . \tag{9.146}$$

Making use of Eq. (9.4), of the binomial theorem, and of the definition of the double factorial for even arguments, it is straightforward to obtain a relation between the thermodynamical integrals I_{nq} and I_{nrq}

$$I_{nq} = \frac{(-1)^q}{(2q+1)!!} \int dK \, E_{ku}^{n-2q} \left(\Xi^{\mu\nu} k_\mu k_\nu - E_{kl}^2 \right)^q f_{0k}$$

$$= \frac{1}{(2q+1)!!} \sum_{r=0}^{q} \frac{2^{q-r} q!}{r!} I_{n,2r,q-r} . \tag{9.147}$$

E.g. for $q = 0, 1, 2$, we have

$$I_{n0} = I_{n00} , \tag{9.148}$$

$$I_{n1} = \frac{1}{3} \left(2I_{n01} + I_{n20} \right) , \tag{9.149}$$

$$I_{n2} = \frac{1}{15} \left(8I_{n02} + 4I_{n21} + I_{n40} \right) . \tag{9.150}$$

Note that the corresponding auxiliary thermodynamical integrals J_{nrq} may be defined similarly to Eq. (9.61), and will obviously will lead to analogous relations.

Furthermore, using $df_{0k}/dE_{ku} = -\beta_0 f_{0k}(1 - a f_{0k})$, after an integration by parts, Eq. (6.30) can be rewritten as a relation between the conventional thermodynamic and auxiliary integrals

$$\beta_0 J_{nq} = I_{n-1,q-1} + (n - 2q) I_{n-1,q} , \tag{9.151}$$

see Eq. (5.52). Similarly, for the auxiliary thermodynamical integrals J_{nrq}, we obtain (as long as $r \geq 2$, $q \geq 1$)

$$\beta_0 J_{nrq} \equiv I_{n-1,r,q-1} + (n - r - 2q) I_{n-1,r,q} \tag{9.152}$$

$$= (r - 1) I_{n-1,r-2,q} + (n - r - 2q) I_{n-1,r,q} .$$

Comparing the right-hand sides, we obtain the identity

$$I_{n-1,r,q-1} = (r-1)\, I_{n-1,r-2,q} \; . \tag{9.153}$$

E.g., for $n = 3$, $r = 2$, and $q = 1$, we obtain the equivalence of the longitudinal and transverse pressures in thermodynamical equilibrium

$$I_{220} = I_{201} \; . \tag{9.154}$$

The main thermodynamic relations are also obtained from integration by parts, namely

$$d I_{nq}\,(\alpha_0, \beta_0) \equiv \left(\frac{\partial I_{nq}}{\partial \alpha_0}\right)_{\beta_0} d\alpha_0 + \left(\frac{\partial I_{nq}}{\partial \beta_0}\right)_{\alpha_0} d\beta_0 = J_{nq}\,d\alpha_0 - J_{n+1,q}\,d\beta_0 \; , \tag{9.155}$$

and similarly

$$d I_{nrq}\,(\alpha_0, \beta_0) \equiv \left(\frac{\partial I_{nrq}}{\partial \alpha_0}\right)_{\beta_0} d\alpha_0 + \left(\frac{\partial I_{nrq}}{\partial \beta_0}\right)_{\alpha_0} d\beta_0 = J_{nrq}\,d\alpha_0 - J_{n+1,r,q}\,d\beta_0 \; . \tag{9.156}$$

9.10   Appendix 4: Orthogonality of the Irreducible Tensors

In this appendix, we derive the orthogonality condition (9.81). The derivation utilizes the relation

$$k^{\{\mu_1} \cdots k^{\mu_\ell\}} k_{\{\mu_1} \cdots k_{\mu_\ell\}} = \frac{1}{2^{\ell-1}} \left(\Xi^{\alpha\beta} k_\alpha k_\beta\right)^\ell \; . \tag{9.157}$$

We now prove Eq. (9.157). Analogously to Eq. (6.108), we obtain

$$k^{\{\mu_1} \cdots k^{\mu_\ell\}} k_{\{\mu_1} \cdots k_{\mu_\ell\}} = \sum_{q=0}^{[\ell/2]} \hat{C}(\ell, q) \left(\Xi^{\alpha\beta} k_\alpha k_\beta\right)^\ell \; . \tag{9.158}$$

The Chebyshev polynomial of the first kind $T_\ell(z)$ has the representation [30]

$$\begin{aligned} T_\ell(z) &= \frac{\ell}{2} \sum_{q=0}^{[\ell/2]} (-1)^q \frac{(\ell - q - 1)!}{q!(\ell - 2q)!} (2z)^{\ell - 2q} \\ &= 2^{\ell-1} \sum_{q=0}^{[\ell/2]} (-1)^q \frac{1}{4^q} \frac{(\ell - q)!}{q!(\ell - 2q)!} \frac{\ell}{\ell - q} z^{\ell - 2q} \; . \end{aligned} \tag{9.159}$$

Since $T_\ell(1) = 1$ for all ℓ [30], we obtain with Eq. (9.120)

$$\sum_{q=0}^{[\ell/2]} \hat{C}(\ell, q) = \frac{1}{2^{\ell-1}} , \quad \text{q.e.d.} . \tag{9.160}$$

In order to prove Eq. (9.81), we define a rank-$(\ell + n)$ tensor

$$\hat{M}_{\{\nu_1\cdots\nu_n\}}^{\{\mu_1\cdots\mu_\ell\}} = \int dK \hat{F}(E_{\mathbf{k}u}, E_{\mathbf{k}l}) k^{\{\mu_1} \cdots k^{\mu_\ell\}} k_{\{\nu_1} \cdots k_{\nu_n\}} , \tag{9.161}$$

which is (separately) symmetric under permutations of the μ- and ν-type indices and depends solely on the fluid 4-velocity u^μ and the 4-vector l^μ. Therefore, $\hat{M}_{\{\nu_1\cdots\nu_n\}}^{\{\mu_1\cdots\mu_\ell\}}$ must be constructed from tensor structures made of u^μ, l^μ, and $\Xi^{\mu\nu}$. Furthermore, $\hat{M}_{\{\nu_1\cdots\nu_n\}}^{\{\mu_1\cdots\mu_\ell\}}$ must be orthogonal to u^μ as well as to l^μ, which implies that it can only be constructed from combinations of the projection operators $\Xi^{\mu\nu}$, and henceforth the rank of the tensor, $\ell + n$, must be an even number. Now, following the arguments presented in Appendix 6.7, one can prove that

$$\hat{M}_{\{\nu_1\cdots\nu_n\}}^{\{\mu_1\cdots\mu_\ell\}} = \delta_{\ell n} \hat{M} \Xi_{\nu_1\cdots\nu_n}^{\mu_1\cdots\mu_\ell} , \tag{9.162}$$

where $\hat{M}$ is a scalar. This is the analogue to Eq. (6.113). Using Eqs. (9.121) and (9.157), we finally obtain

$$\hat{M} \equiv \frac{1}{\Xi_{\mu_1\cdots\mu_\ell}^{\mu_1\cdots\mu_\ell}} \int dK \, \hat{F}(E_{\mathbf{k}u}, E_{\mathbf{k}l}) \, k^{\{\mu_1} \cdots k^{\mu_\ell\}} k_{\{\mu_1} \cdots k_{\mu_\ell\}}$$

$$= \frac{1}{2^\ell} \int dK \, \hat{F}(E_{\mathbf{k}u}, E_{\mathbf{k}l}) \left(\Xi^{\alpha\beta} k_\alpha k_\beta \right)^\ell . \tag{9.163}$$

Finally, we also prove Eq. (9.103). We first rewrite the collision integral (9.102) as

$$\hat{\mathcal{L}}_{ij}^{\{\mu_1\cdots\mu_\ell\}} \equiv \sum_{r=0}^{\infty} \sum_{n=0}^{N_r} \sum_{m=0}^{N_r-n} \hat{\rho}_{nm}^{\nu_1\cdots\nu_r} \left(\mathcal{A}_{injm} \right)_{\nu_1\cdots\nu_r}^{\mu_1\cdots\mu_\ell} , \tag{9.164}$$

where, following similar arguments as above, the tensor $\left(\mathcal{A}_{injm} \right)_{\nu_1\cdots\nu_r}^{\mu_1\cdots\mu_\ell}$ can be shown to possess the property

$$\left(\mathcal{A}_{injm} \right)_{\nu_1\cdots\nu_r}^{\mu_1\cdots\mu_\ell} = \delta_{\ell r} \mathcal{A}_{injm}^{(\ell)} \Xi_{\nu_1\cdots\nu_r}^{\mu_1\cdots\mu_\ell} . \tag{9.165}$$

Substituting this into Eq. (9.164) leads to Eq. (9.103).

9.11   Appendix 5: Orthogonal Polynomials

The orthonormal polynomials are defined by the following orthonormality condition

$$\int dK \,\hat{\omega}^{(\ell)} P^{(\ell)}_{\mathbf{k}nm} P^{(\ell)}_{\mathbf{k}n'm'} = \delta_{nn'}\delta_{mm'} \,, \tag{9.166}$$

where the weight is given as

$$\hat{\omega}^{(\ell)} = \frac{\hat{W}^{(\ell)}}{(2\ell)!!} \left(\varXi^{\alpha\beta}k_\alpha k_\beta\right)^\ell \hat{f}_{0\mathbf{k}} \left(1 - a\,\hat{f}_{0\mathbf{k}}\right) \,. \tag{9.167}$$

We now construct the complete set of orthonormal polynomials in both $E_{\mathbf{k}u}$ and $E_{\mathbf{k}l}$ using the Gram-Schmidt orthogonalization procedure in the 14-moment approximation, i.e., where $N_0 = 2$, $N_1 = 1$, and $N_2 = 0$, cf. Refs. [7,31]. With Eq. (9.82), one observes that we only need to determine the polynomials $P^{(0)}_{\mathbf{k}00}$, $P^{(0)}_{\mathbf{k}01}$, $P^{(0)}_{\mathbf{k}02}$, $P^{(0)}_{\mathbf{k}10}$, $P^{(0)}_{\mathbf{k}11}$, $P^{(0)}_{\mathbf{k}20}$, $P^{(1)}_{\mathbf{k}00}$, $P^{(1)}_{\mathbf{k}01}$, $P^{(1)}_{\mathbf{k}10}$, and $P^{(2)}_{\mathbf{k}00}$.

Using the orthonormality condition (9.166) for $n = m = n' = m' = 0$ and Eq. (9.61), we first obtain the value of the normalization constant in Eq. (9.167)

$$\hat{W}^{(\ell)} = \frac{(-1)^\ell}{\hat{J}_{2\ell,0,\ell}} \,. \tag{9.168}$$

Then, the orthonormality condition (9.166) can be written with the help of Eqs. (9.61) and (9.83) as

$$\hat{J}_{2\ell,0,\ell}\,\delta_{nn'}\delta_{mm'} = \sum_{i=0}^{n}\sum_{j=0}^{m}\sum_{r=0}^{n'}\sum_{s=0}^{m'} a^{(\ell)}_{nimj}\, a^{(\ell)}_{n'rm's}\, \hat{J}_{i+r+j+s+2\ell,\,j+s,\ell} \,. \tag{9.169}$$

From this equation, we will successively construct the polynomials

(i) $P^{(0)}_{\mathbf{k}00}$, $P^{(1)}_{\mathbf{k}00}$, $P^{(2)}_{\mathbf{k}00}$: For any ℓ and $n = n' = m = m' = 0$, we obtain from Eq. (9.169)

$$1 = a^{(\ell)}_{0000} \equiv P^{(\ell)}_{\mathbf{k}00} \,. \tag{9.170}$$

This determines the polynomials $P^{(0)}_{\mathbf{k}00}$, $P^{(1)}_{\mathbf{k}00}$, and $P^{(2)}_{\mathbf{k}00}$.

(ii) $P^{(0)}_{\mathbf{k}10}$: Consider Eq. (9.169) for $\ell = 0$, $n = n' = 1$, $m = m' = 0$, as well as for $\ell = 0$, $n = 1$, $n' = 0$, $m = m' = 0$. Solving these two equations for the coefficients $a^{(0)}_{1000}$ and $a^{(0)}_{1100}$ leads to the result

$$\frac{a^{(0)}_{1000}}{a^{(0)}_{1100}} = -\frac{\hat{J}_{100}}{\hat{J}_{000}} \,, \quad \left(a^{(0)}_{1100}\right)^2 = \frac{\hat{J}^2_{000}}{\hat{D}_{10}} \,, \tag{9.171}$$

where

$$\hat{D}_{nq} = \hat{J}_{n-1,0,q} \hat{J}_{n+1,0,q} - \hat{J}_{n0q}^2 .$$ (9.172)

This uniquely determines the polynomial $P_{k10}^{(0)} = a_{1000}^{(0)} + a_{1100}^{(0)} E_{ku}$.

(iii) $P_{k20}^{(0)}$: Consider Eq. (9.169) for $\ell = 0, n = n' = 2, m = m' = 0$, for $\ell = 0$, $n = 2, n' = 1, m = m' = 0$, and for $\ell = 0, n = 2, n' = 0, m = m' = 0$. From these three equations, one obtains

$$\frac{a_{2000}^{(0)}}{a_{2200}^{(0)}} = \frac{\hat{D}_{20}}{\hat{D}_{10}}, \quad \frac{a_{2100}^{(0)}}{a_{2200}^{(0)}} = \frac{\hat{G}_{12}}{\hat{D}_{10}}, \quad \left(a_{2200}^{(0)}\right)^2 = \frac{\hat{J}_{000} \hat{D}_{10}}{\hat{J}_{200} \hat{D}_{20} + \hat{J}_{300} \hat{G}_{12} + \hat{J}_{400} \hat{D}_{10}},$$ (9.173)

where

$$\hat{G}_{nm} = \hat{J}_{n00} \hat{J}_{m00} - \hat{J}_{n-1,0,0} \hat{J}_{m+1,0,0} .$$ (9.174)

This uniquely determines the polynomial $P_{k20}^{(0)} = a_{2000}^{(0)} + a_{2100}^{(0)} E_{ku} + a_{2200}^{(0)} E_{ku}^2$.

(iv) $P_{k10}^{(1)}$: Consider Eq. (9.169) for $\ell = 1, n = n' = 1, m = m' = 0$, as well as for $\ell = 1, n = 1, n' = 0, m = m' = 0$. From these two equations, one obtains

$$\frac{a_{1000}^{(1)}}{a_{1100}^{(1)}} = -\frac{\hat{J}_{301}}{\hat{J}_{201}}, \quad \left(a_{1100}^{(1)}\right)^2 = \frac{\hat{J}_{201}^2}{\hat{D}_{31}},$$ (9.175)

which uniquely determines the polynomial $P_{k10}^{(1)} = a_{1000}^{(1)} + a_{1100}^{(1)} E_{ku}$.

(v) $P_{k01}^{(0)}$: Consider Eq. (9.169) for $\ell = 0, n = n' = 0, m = m' = 1$, as well as for $\ell = 0, n = n' = 0, m = 1, m' = 0$. From these two equations, one obtains

$$\frac{a_{0010}^{(0)}}{a_{0011}^{(0)}} = -\frac{\hat{J}_{110}}{\hat{J}_{000}}, \quad \left(a_{0011}^{(0)}\right)^2 = \frac{\hat{J}_{000}^2}{\hat{D}_{110}},$$ (9.176)

where we defined

$$\hat{D}_{nrq} = \hat{J}_{n-1,r-1,q} \hat{J}_{n+1,r+1,q} - \hat{J}_{nrq}^2 .$$ (9.177)

This uniquely determines the polynomial $P_{k01}^{(0)} = a_{0010}^{(0)} + a_{0011}^{(0)} E_{kl}$.

(vi) $P_{k02}^{(0)}$: Consider Eq. (9.169) for $\ell = 0, n = n' = 0, m = m' = 2$, for $\ell = 0$, $n = n' = 0, m = 2, m' = 1$, and for $\ell = 0, n = n' = 0, m = 2, m' = 0$. From these three equations, one obtains

$$\frac{a_{0020}^{(0)}}{a_{0022}^{(0)}} = \frac{\hat{D}_{220}}{\hat{D}_{110}}, \quad \frac{a_{0021}^{(0)}}{a_{0022}^{(0)}} = \frac{\hat{G}_{1122}}{\hat{D}_{110}},$$

$$\left(a_{0022}^{(0)}\right)^2 = \frac{\hat{J}_{000} \hat{D}_{110}}{\hat{J}_{220} \hat{D}_{220} + \hat{J}_{330} \hat{G}_{1122} + \hat{J}_{440} \hat{D}_{110}},$$ (9.178)

where we defined

$$\hat{G}_{nrmp} = \hat{J}_{nr0}\hat{J}_{mp0} - \hat{J}_{n-1,r-1,0}\hat{J}_{m+1,p+1,0} \ . \tag{9.179}$$

This uniquely determines the polynomial $P_{\mathbf{k}02}^{(0)} = a_{0020}^{(0)} + a_{0021}^{(0)}E_{\mathbf{k}l} + a_{0022}^{(0)}E_{\mathbf{k}l}^2$.

(vii) $P_{\mathbf{k}01}^{(1)}$: Consider Eq. (9.169) for $\ell = 1, n = n' = 0, m = m' = 1$ and for $\ell = 1, n = n' = 0, m = 1, m' = 0$. From these two equations, one obtains

$$\frac{a_{0010}^{(1)}}{a_{0011}^{(1)}} = -\frac{\hat{J}_{311}}{\hat{J}_{201}} \ , \quad \left(a_{0011}^{(1)}\right)^2 = \frac{\hat{J}_{201}^2}{\hat{D}_{311}} \ , \tag{9.180}$$

which uniquely determines the polynomial $P_{\mathbf{k}01}^{(1)} = a_{0010}^{(1)} + a_{0011}^{(1)}E_{\mathbf{k}l}$.

(viii) $P_{\mathbf{k}11}^{(0)}$: Consider Eq. (9.169) for $\ell = 0, n = n' = 1, m = m' = 1$, for $\ell = 0, n = n' = 1, m = 1, m' = 0$, for $\ell = 0, n = 1, n' = 0, m = m' = 1$, and for $\ell = 0, n = 1, n' = 0, m = 1, m' = 0$. From these four equations, one obtains

$$\frac{a_{1010}^{(0)}}{a_{1111}^{(0)}} = -\frac{\hat{J}_{310}\hat{G}_{2210} - \hat{J}_{210}\hat{D}_{210} - \hat{J}_{200}\hat{G}_{2221}}{\hat{J}_{210}\hat{G}_{2100} - \hat{J}_{200}\hat{D}_{110} - \hat{J}_{100}\hat{G}_{2111}} \ , \tag{9.181}$$

$$\frac{a_{1011}^{(0)}}{a_{1111}^{(0)}} = -\frac{\hat{J}_{310}\hat{G}_{2100} + \hat{J}_{200}\hat{G}_{1121} + \hat{J}_{100}\hat{D}_{210}}{\hat{J}_{210}\hat{G}_{2100} - \hat{J}_{200}\hat{D}_{110} - \hat{J}_{100}\hat{G}_{2111}} \ , \tag{9.182}$$

$$\frac{a_{1110}^{(0)}}{a_{1111}^{(0)}} = -\frac{\hat{J}_{320}\hat{G}_{2100} - \hat{J}_{310}\hat{D}_{110} - \hat{J}_{210}\hat{G}_{2111}}{\hat{J}_{210}\hat{G}_{2100} - \hat{J}_{200}\hat{D}_{110} - \hat{J}_{100}\hat{G}_{2111}} \ , \tag{9.183}$$

and

$$\frac{1}{\left(a_{1111}^{(0)}\right)^2} = \frac{\hat{J}_{420}}{\hat{J}_{000}} + \left(\frac{a_{1010}^{(0)}}{a_{1111}^{(0)}}\right)^2 + \frac{\hat{J}_{220}}{\hat{J}_{000}}\left(\frac{a_{1011}^{(0)}}{a_{1111}^{(0)}}\right)^2 + \frac{\hat{J}_{200}}{\hat{J}_{000}}\left(\frac{a_{1110}^{(0)}}{a_{1111}^{(0)}}\right)^2$$

$$+ 2\frac{\hat{J}_{210}}{\hat{J}_{000}}\frac{a_{1010}^{(0)}}{a_{1111}^{(0)}} + 2\frac{\hat{J}_{320}}{\hat{J}_{000}}\frac{a_{1011}^{(0)}}{a_{1111}^{(0)}} + 2\frac{\hat{J}_{310}}{\hat{J}_{000}}\frac{a_{1110}^{(0)}}{a_{1111}^{(0)}}$$

$$+ 2\frac{\hat{J}_{100}}{\hat{J}_{000}}\frac{a_{1010}^{(0)}a_{1110}^{(0)}}{\left(a_{1111}^{(0)}\right)^2} + 2\frac{\hat{J}_{110}}{\hat{J}_{000}}\frac{a_{1010}^{(0)}a_{1011}^{(0)}}{\left(a_{1111}^{(0)}\right)^2} + 2\frac{\hat{J}_{210}}{\hat{J}_{000}}\frac{a_{1011}^{(0)}a_{1110}^{(0)}}{\left(a_{1111}^{(0)}\right)^2} \ . \tag{9.184}$$

In this equation, we refrained from explicitly inserting the coefficients (9.181)–(9.183), because the resulting expression becomes too unwieldy. With Eqs. (9.181)–(9.184), the polynomial $P_{\mathbf{k}11}^{(0)} = a_{1010}^{(0)} + a_{1011}^{(0)}E_{\mathbf{k}l} + a_{1110}^{(0)}E_{\mathbf{k}u} + a_{1111}^{(0)}E_{\mathbf{k}u}E_{\mathbf{k}l}$ is uniquely determined.

We remark that the results for $m = 0$ are formally similar to the polyomials $P_{\mathbf{k}n}^{(\ell)}$ with coefficients $a_{ni}^{(\ell)}$ from Eq. (6.5); see for example Eqs. (91) – (99) of Ref. [31]. The reason is that the polynomials $P_{\mathbf{k}n}^{(\ell)}$ are the $m = 0$ case of the more general multivariate polynomials $P_{\mathbf{k}n0}^{(\ell)}$.

Finally, with these results we can explicitly compute the $\hat{\mathcal{H}}_{\mathbf{k}nm}^{(\ell)}$ coefficients from Eq. (9.87) for the cases relevant for the 14-moment approximation.

References

1. Bjorken, J.D.: Phys. Rev. **D 27**, 140–151 (1983). https://doi.org/10.1103/PhysRevD.27.140
2. Gedalin, M.: Phys. Fluids B **3**, 1871 (1991)
3. Gedalin, M., Oiberman, I.: Phys. Rev. E **51**, 4901 (1995)
4. Huang, X.G., Huang, M., Rischke, D.H., Sedrakian, A.: Phys. Rev. D **81**, 045015 (2010). https://doi.org/10.1103/PhysRevD.81.045015, arXiv:0910.3633 [astro-ph.HE]
5. Huang, X.G., Sedrakian, A., Rischke, D.H.: Annals Phys. **326**, 3075–3094 (2011). https://doi.org/10.1016/j.aop.2011.08.001, arXiv:1108.0602 [astro-ph.HE]
6. de Groot, S.R., van Leeuwen, W.A., van Weert, Ch.G.: Relativistic Kinetic Theory - Principles and Applications. North Holland (1980)
7. Denicol, G.S., Niemi, H., Molnar, E., Rischke, D.H.: Phys. Rev. D **85**, 114047 (2012) [erratum: Phys. Rev. D **91**(3), 039902 (2015)]. https://doi.org/10.1103/PhysRevD.85.114047, arXiv:1202.4551 [nucl-th]
8. Molnar, E., Niemi, H., Rischke, D.H.: Phys. Rev. D **93**(11), 114025 (2016). https://doi.org/10.1103/PhysRevD.93.114025, arXiv:1602.00573 [nucl-th]
9. Florkowski, W., Ryblewski, R.: Phys. Rev. C **83**, 034907 (2011). https://doi.org/10.1103/PhysRevC.83.034907, arXiv:1007.0130 [nucl-th]
10. Ryblewski, R., Florkowski, W.: Eur. Phys. J. C **71**, 1761 (2011). https://doi.org/10.1140/epjc/s10052-011-1761-8, arXiv:1103.1260 [nucl-th]
11. Ryblewski, R., Florkowski, W.: Phys. Rev. C **85**, 064901 (2012). https://doi.org/10.1103/PhysRevC.85.064901, arXiv:1204.2624 [nucl-th]
12. Eckart, C.: Phys. Rev. **58**, 919–924 (1940). https://doi.org/10.1103/PhysRev.58.919
13. Landau, L.D., Lifshitz, E.M.: Fluid Dynamics. Pergamon, New York (1959)
14. Baumjohann, W., Treumann, R.A.: Basic Space Plasma Physics. Imperial College Press (1997)
15. Romatschke, P., Strickland, M.: Phys. Rev. D **68**, 036004 (2003). https://doi.org/10.1103/PhysRevD.68.036004, arXiv:hep-ph/0304092 [hep-ph]
16. Martinez, M., Strickland, M.: Phys. Rev. C **81**, 024906 (2010). https://doi.org/10.1103/PhysRevC.81.024906, arXiv:0909.0264 [hep-ph]
17. Martinez, M., Strickland, M.: Nucl. Phys. A **848**, 183-197 (2010). https://doi.org/10.1016/j.nuclphysa.2010.08.011, arXiv:1007.0889 [nucl-th]
18. Martinez, M., Strickland, M.: Nucl. Phys. A **856**, 68–87 (2011). https://doi.org/10.1016/j.nuclphysa.2011.02.003, arXiv:1011.3056 [nucl-th]
19. Martinez, M., Ryblewski, R., Strickland, M.: Phys. Rev. C **85**, 064913 (2012). https://doi.org/10.1103/PhysRevC.85.064913, arXiv:1204.1473 [nucl-th]
20. Ryblewski, R., Florkowski, W.: J. Phys. G **38**, 015104 (2011). https://doi.org/10.1088/0954-3899/38/1/015104, arXiv:1007.4662 [nucl-th]
21. Barz, H.W., Kampfer, B., Lukacs, B., Martinas, K., Wolf, G.: Phys. Lett. B **194**, 15–19 (1987). https://doi.org/10.1016/0370-2693(87)90761-1
22. Grad, H.: Commun. Pure Appl. Math. **2**, 331 (1949); ibid. 325 (1949)
23. Israel, W., Stewart, J.M.: Annals Phys. **118**, 341–372 (1979). https://doi.org/10.1016/0003-4916(79)90130-1

24. Bazow, D., Heinz, U.W., Strickland, M.: Phys. Rev. C **90**(5), 054910 (2014). https://doi.org/
    10.1103/PhysRevC.90.054910, arXiv:1311.6720 [nucl-th]
25. Heinz, U.W., Bazow, D., Strickland, M.: Nucl. Phys. A **931**, 920–925 (2014). https://doi.org/
    10.1016/j.nuclphysa.2014.08.082, arXiv:1408.0756 [nucl-th]
26. Bhatnagar, P.L., Gross, E.P., Krook, M.: Phys. Rev. **94**, 511–525 (1954). https://doi.org/10.
    1103/PhysRev.94.511
27. Anderson, J.L., Witting, H.R.: Physica **74**, 466 (1974)
28. Denicol, G.S., Huang, X.G., Molnár, E., Monteiro, G.M., Niemi, H., Noronha, J., Rischke,
    D.H., Wang, Q.: Phys. Rev. D **98**(7), 076009 (2018). https://doi.org/10.1103/PhysRevD.98.
    076009, arXiv:1804.05210 [nucl-th]
29. Denicol, G.S., Molnár, E., Niemi, H., Rischke, D.H.: Phys. Rev. D **99**(5), 056017 (2019). https://
    doi.org/10.1103/PhysRevD.99.056017, arXiv:1902.01699 [nucl-th]
30. Abramowitz, M., Stegun, I.A.: Handbook of Mathematical Functions, vol. Eqs. (22.2.4) and
    (22.3.6), 9th edn. Dover (1972)
31. Denicol, G.S., Molnár, E., Niemi, H., Rischke, D.H.: Eur. Phys. J. A **48**, 170 (2012). https://
    doi.org/10.1140/epja/i2012-12170-x, arXiv:1206.1554 [nucl-th]